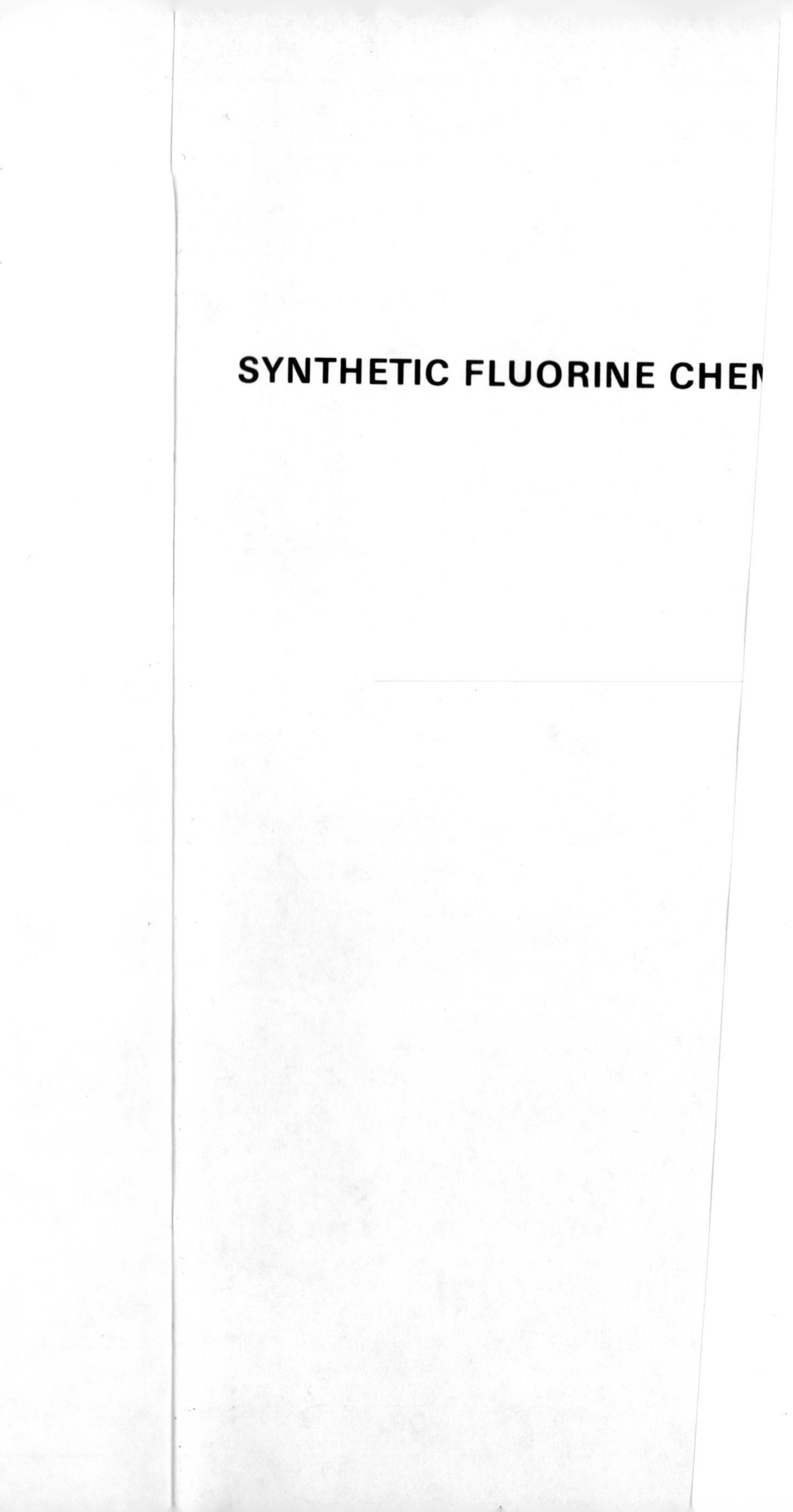
SYNTHETIC FLUORINE CHEM

SYNTHETIC FLUORINE CHEMISTRY

Edited by

George A. Olah
University of Southern California
Los Angeles, California

Richard D. Chambers
University of Durham
Durham, England

and

G. K. Surya Prakash
University of Southern California
Los Angeles, California

A WILEY-INTERSCIENCE PUBLICATION

JOHN WILEY & SONS, INC.

New York · Chichester · Brisbane · Toronto · Singapore

In recognition of the importance of preserving what has been written, it is a policy of John Wiley & Sons, Inc., to have books of enduring value published in the United States printed on acid-free paper, and we exert our best efforts to that end.

Library of Congress Cataloging in Publication Data:

Synthetic fluorine chemistry / G. A. Olah, R. D. Chambers, and G. K. S. Prakash (eds.).
p. cm.
"A Wiley-Interscience publication."
Includes bibliographical references.
ISBN 0-471-54370-5 (alk. paper) :
1. Fluorine compounds—Synthesis. 2. Organofluorine compounds--Synthesis. I. Olah, George A. (George Andrew), 1927-
II. Chambers, R. D. (Richard D.) III. Prakash, G. K. Surya.
QD181.F1S96 1992
546′.731—dc20 91-25155
CIP

Printed in the United States of America

10 9 8 7 6 5 4 3 2 1

In memory of Donald P. Loker, Benefactor and Friend

CONTRIBUTORS

James L. Adcock, Department of Chemistry, University of Tennessee, Knoxville, Tennessee

F. Aubke, Department of Chemistry, The University of British Columbia, Vancouver, B. C., Canada

Kurt Baum, Fluorochem Corporation, Azusa, California

Thomas R. Bierschenk, Exfluor Research Corporation, Austin, Texas

Donald J. Burton, University of Iowa

M. S. R. Cader, Department of Chemistry, The University of British Columbia, Vancouver, B. C., Canada

R. D. Chambers, Department of Chemistry, University of Durham, South Road, Durham, England

Karl O. Christe, Rocketdyne Division of Rockwell International Corporation, Canoga Park, California

David A. Dixon, Central Research and Development, Du Pont Experimental Station, P.O. Box 80328, Wilmington, Delaware

William B. Farnham, Central Research and Development, Du Pont Experimental Station, P.O. Box 80328, Wilmington, Delaware

Raymond H. Gimi, Department of Chemistry, State University of New York, Albany

Timothy J. Juhlke, Exfluor Research Corporation, Austin, Texas

Hajimu Kawa, Exfluor Research Corporation, Austin, Texas

Carl G. Krespan, Central Research and Development, Du Pont Experimental Station, P.O. Box 80328, Wilmington, Delaware

Richard J. Lagow, Department of Chemistry, University of Texas at Austin, Austin, Texas

David M. Lemal, Department of Chemistry, Dartmouth College, Hanover, New Hampshire

Xing-Ya Li, Loker Hydrocarbon Research Institute and Department of Chemistry, University of Southern California, University Park, Los Angeles, California

F. Mistry, Department of Chemistry, The University of British Columbia, Vancouver, B.C., Canada

George A. Olah, Loker Hydrocarbon Research Institute and Department of Chemistry, University of Southern California, University Park, Los Angeles, California

G. K. Surya Prakash, Loker Hydrocarbon Research Institute and Department of Chemistry, University of Southern California, University Park, Los Angeles, California.

Shlomo Rozen, School of Chemistry, Raymond and Beverly Sackler Faculty of Exact Sciences, Tel-Aviv University, Tel-Aviv, Israel

Carl J. Schack, Rocketdyne Division of Rockwell International Corporation, Canoga Park, California

G. J. Schrobilgen, Department of Chemistry, McMaster University, Hamilton, Ontario, Canada

Konrad Seppelt, Institute for Inorganic and Analytical Chemistry, Free University, Berlin, Germany

V. D. Shteingarts, Institute of Organic Chemistry, Siberian Division of the USSR Academy of Sciences, Novosibirsk, USSR

Natalie E. Takenaka, Chemistry Department, Dartmouth College, Hanover, New Hampshire

John T. Welch, Department of Chemistry, State University of New York, Albany

William W. Wilson, Rocketdyne Division of Rockwell International Corporation, Canoga Park, California

Takashi Yamazaki, Department of Chemistry, State University of New York, Albany

PREFACE

Since the isolation of elemental fluorine by Henri Moissan in 1886, the chemistry of fluorine containing compounds has seen an enormous growth. Fluoro compounds are of great interest not only in the inorganic but also in the polymer, organic, and pharmaceutical industry. Fluorine being the most electronegative element forms very strong bonds with almost all other elements. The most prevalent fluorine containing compounds are those possessing a C—F bond. Of the 10 million compounds registered in the Chemical Abstracts (published by the American Chemical Society) 6.2% of them contain compounds with a C—F bond. In 1989 roughly 4900 research papers were published in the area of organofluorine chemistry alone. Germane to the growth and application of fluorine containing compounds is the development of new synthetic methods in fluorine chemistry.

Recognizing the importance of fluorine chemistry, the Loker Hydrocarbon Research Institute of the University of Southern California organized one of its biannual research symposia in February 1990 on the topic "Synthetic Fluorine Chemistry." The symposium was dedicated to the memory of the late Donald P. Loker, a great friend and benefactor of the Institute.

The success of this symposium led to the suggestion to publish the contributions in an enlarged form as a collected volume. We were pleased to accept the invitation to edit the book, which we feel will be a valuable contribution to the fluorine chemistry literature.

The first four chapters deal with various aspects of inorganic fluorine chemistry.

In Chapter 1, Schrobilgen describes the preparation and fascinating Lewis acid behavior and reactivity of fluorinated noble gas fluorides. Christe, Wilson, and Schack in Chapter 2 discuss the systematic replacement of fluorine by oxygen in a series of fluorides and oxyfluorides using nitrate and sulfate anions. In Chapter 3 Aubke, Cader, and Mistry report in detail on fluorine containing transition metal derivatives of strong protonic acids and superacids. This chapter encompasses transition metal derivatives of monoprotonic fluoro superacids of the general type $HF—MF_5$ (M = As, Sb, Nb, or Ta) and fluoroxy, fluorophosphonic, and sulfonic acids. Synthesis of unusual carbon–sulfur multiply bonded compounds stabilized by fluorine ligands is discussed by Seppelt in Chapter 4.

The next four chapters discuss fluorination methods.

Lagow, Bierschenk, Juhlke, and Kawa in Chapter 5 cover the synthesis of perfluoropolyethers, an extraordinary class of new fluorinated compounds, which have promise as high-performance lubricants. Their synthesis is achieved by direct fluorination of polyethers, a technology developed previously by Lagow. Adcock in Chapter 6 describes an improved universal synthesis of perfluorinated organic compounds by an aerosol direct fluorination method. Electrophilic fluorination of organic compounds using fluorine and some of its derived reagents is discussed by Rozen in Chapter 7. The F_2 in $CHCl_3$ serves as a good source for electrophilic fluorine in a variety of electrophilic substitutions. Olah and Li in Chapter 8 discuss fluorination using liquid and solid onium poly(hydrogen fluoride) complexes. These reagents with their great convenience substitute for volatile hydrogen fluoride in a wide variety of organic fluorinations.

The next three chapters deal with organometallic fluorine chemistry and its use in synthesis.

In Chapter 9 Burton describes the use of perfluoroalkyl containing organometallic reagents in organofluorine chemistry. The discussion includes preparation and use of trifluoromethyl cadmium, zinc, copper reagents, and related perfluoroalkenyl organometallics. Nucleophilic perfluoroalkylation of organic functional groups using perfluoroalkyltrialkylsilanes is covered by Prakash in Chapter 10. Development of silicon mediated synthons and reactions in organofluorine chemistry is discussed by Farnham in Chapter 11.

Synthetic aspects of electrophilic ipso-reactions of polyfluorinated aromatic compounds is described by Shteingarts in Chapter 12. Takenaka and Lemal, in Chapter 13, discuss the synthetic and mechanistic aspects of the perfluorobenzene oxide–perfluorooxepin system. The chemistry of perhalodioxins and perhalodihydrodioxins are covered by Krespan and Dixon in Chapter 14. In Chapter 15, Welch, Yamazaki, and Gimi describe the use of fluoroacetamide acetal Claisen rearrangement as a tool for asymmetric synthesis of a number of biologically important intermediates. Chambers in Chapter 16 describes the preparation of unusual fluorinated alkenes and dienes by fluoride ion induced reaction process. Finally, in Chapter 17, Baum reports the synthesis of fluorinated condensation monomers.

Obviously, a single volume cannot cover all aspects of synthetic fluorine chemistry. However, the noted investigators of the field based on the lectures given at the symposium give a broad perspective of current major research trends in synthetic fluorine chemistry.

George A. Olah
Richard D. Chambers
G. K. Surya Prakash

Los Angeles, California
Durham, England
Los Angeles, California
June, 1991

CONTENTS

SYNTHETIC FLUORINE CHEMISTRY

CHAPTER 1

Lewis Acid Properties of Fluorinated Nobel Gas Cations

GARY J. SCHROBILGEN

1.1. INTRODUCTION

While many examples of xenon bonded to oxygen or fluorine and of xenon bonded to other highly electronegative inorganic ligands through oxygen were synthesized immediately following the discovery of noble gas reactivity,[1] over a decade had elapsed before an example with a ligating atom other than oxygen and fluorine, namely, nitrogen, was synthesized.[2] This occurred two decades before the Xe—N bond in $FXeN(SO_2F)_2$ was definitively characterized in the solid state by X-ray crystallography and in solution by multinuclear magnetic resonance (multi-NMR) spectroscopy.[3] Other imidodisulfurylfluoride xenon–nitrogen bonded species have since been characterized using primarily NMR spectroscopy, namely, $Xe[N(SO_2F)_2]_2$,[4,5] $F[XeN(SO_2F)_2]_2^+$,[4,5] $XeN(SO_2F)_2^+AsF_6^-$,[6] and $XeN(SO_2F)_2^+Sb_3F_{16}^-$.[6] The latter salt was also charac-

Synthetic Fluorine Chemistry,
Edited by George A. Olah, Richard D. Chambers, and G. K. Surya Prakash.
ISBN 0-471-54370-5

terized by single-crystal X-ray diffraction. In addition, the compound $Xe[N(SO_2CF_3)_2]_2$ was prepared and characterized,[7] and is the most thermally stable of the imido derivatives of xenon.

Recently, a significant extension of noble gas chemistry, and in particular, noble gas nitrogen bonds, has been achieved by taking advantage of the Lewis acid properties of noble gas cations. This has given rise to numerous new examples of xenon–nitrogen and krypton–nitrogen bonds. The adduct salts, which have stabilities ranging from explosive at −60°C for $HC{\equiv}N{-}KrF^+AsF_6^-$,[8] to stable at room temperature for *s*-$C_3F_3N_2N{-}XeF^+AsF_6^-$,[9] have donor–acceptor bonds which are among the weakest bonds that still deserve to be called bonds. The present chapter outlines the syntheses, structural characterization, and bonding of noble gas adduct cations for a variety of organic and inorganic nitrogen base centers.

1.2. LEWIS ACID AND OXIDANT PROPERTIES OF NOBLE GAS CATIONS

In view of the propensity of the XeF^+ cation to form strong fluorine bridges to counteranions in the solid state,[10] the XeF^+ cation may be regarded as having a significant Lewis acid strength. Based on considerations of the high electron affinities of the cations (ArF^+, 13.7 eV, [11]; KrF^+, 13.2 eV, [12]; XeF^+, 10.9 eV, [13]) and first adiabatic ionization potentials of selected bases, where the first adiabatic ionization potential (IP_1), determined from photoelectron spectroscopy, are equal to or greater than the estimated electron affinity (EA) of the noble gas cation. It has been possible to single out specific nitrogen bases and classes of nitrogen bases that offer reasonable promise for preparing noble gas adduct cations in which the strongly oxidizing noble gas cations are bound to organic and perfluoroorganic fragments through the nitrogen of the base. A list of some nitrogen bases that have potential or proven compatibility with the estimated electron affinities of the XeF^+ and/or KrF^+ cations is given in Table 1.1 along with their IP_1 values.

In many ways $HC{\equiv}N$, with $IP_1 = 13.80$ eV,[14] is archetypical of the other oxidatively resistant organic and perfluoroorganic nitrogen bases that form adducts with noble gas Lewis acid centers. For that reason, its interactions with the XeF^+ and KrF^+ cations will be discussed in considerable detail.

1.3. SYNTHETIC APPROACHES AND CHARACTERIZATION

Multinuclear magnetic resonance (NMR) spectroscopy has played an important role in the characterization of noble gas species.[31–33] A substantial portion of this chapter is concerned with structural characterization and inferences

TABLE 1.1. First Adiabatic Ionization Potentials of Some Organic and Inorganic Nitrogen Bases.

Compound	First Ionization Potential	Reference
$CF_3C{\equiv}N$	13.9	15
$N{\equiv}C{-}C{\equiv}N$	13.57 ± 0.02	16
$HC{\equiv}N$	13.80	14
trans-N_2F_2	13.10 ± 0.1	17
$CH_2FC{\equiv}N$	13.00 ± 0.1	18
$CH_2ClC{\equiv}N$	12.2 ± 0.1	18
$N{\equiv}SF_3$	12.50	19
$ClC{\equiv}N$	12.49 ± 0.04	20
$CHF_2C{\equiv}N$	12.40	18
$CD_3C{\equiv}N$	12.235 ± 0.005	21
$CHCl_2C{\equiv}N$	12.9 ± 0.3	18
$CH_3C{\equiv}N$	12.194 ± 0.005	21
N_2F_4	12.04 ± 0.1	22
$BrC{\equiv}N$	11.95 ± 0.08	20
$C_2H_5C{\equiv}N$	11.85	16
$N{\equiv}SF$	11.82	19
n-$C_3H_7C{\equiv}N$	11.67	23
ND_3	11.52	24
s-$C_3F_3N_3$	11.50	25
$(CH_3)_2CHC{\equiv}N$	11.49	23
ND_2H	11.47 ± 0.02	24
$N{\equiv}C{-}C{\equiv}C{-}C{\equiv}N$	11.4 ± 0.2	26
$N{\equiv}C{-}C{\equiv}C{-}C{\equiv}C{-}C{\equiv}N$	11.40 ± 0.2	26
$S(C{\equiv}N)_2$	11.32	27
$(CH_3)_3CC{\equiv}N$	11.11	23
$IC{\equiv}N$	10.98 ± 0.05	20
$H_2NC{\equiv}N$	10.71	28
B-$B_3F_3N_3$	10.46	25
NH_3	10.34 ± 0.07	29
C_5F_5N	10.08 ± 0.05	25
s-$C_3H_3N_3$	10.07 ± 0.05	30

regarding bonding derived from NMR studies. The role of NMR in noble gas chemistry is exemplified by xenon, although the observation of the ^{19}F, $^{14,15}N$, ^{13}C, and ^{1}H NMR spectra is equally crucial. Xenon is the most favorable noble gas from an NMR standpoint, since it has a spin $\frac{1}{2}$ nuclide, ^{129}Xe (26.44% natural abundance), with a receptivity of 31.8 relative to that of natural abundance ^{13}C. The high receptivity, nonquadrupolar nature, and short spin–lattice relaxation times of ^{129}Xe allow spectral data to be readily acquired using modern Fourier transform nuclear magnetic resonance (FT NMR) spectrometers.

1.3.1. The Hydrocyano Cations, $HC{\equiv}N{-}NgF^+$ (Ng = Kr or Xe)

Hydrogen cyanide is oxidatively among the most resistant ligands investigated thus far (Table 1.1), having a first adiabatic ionization potential of 13.80 eV. The estimated electron affinity of XeF^+ (10.9 eV)[13] suggested that $HC{\equiv}N$ would be resistant to oxidative attack by the XeF^+ cation and that $HC{\equiv}N{-}XeF^+$ might have sufficient thermal stability to permit its spectroscopic characterization in solution and in the solid state.

The reaction of XeF^+ with $HC{\equiv}N$ and the subsequent isolation of $HC{\equiv}N{-}XeF^+AsF_6^-$ and its characterization have indeed been realized. The reactions of $XeF^+AsF_6^-$ and $Xe_2F_3^+AsF_6^-$ with $HC{\equiv}N$ have been reported[34,35] and were carried out according to Eqs. 1.1 and 1.2 by combining stoichiometric amounts of the reactants in anhydrous HF(−20 to −10°C).

$$XeF^+AsF_6^- + HC{\equiv}N \longrightarrow HC{\equiv}N{-}XeF^+AsF_6^- \qquad (1.1)$$

$$Xe_2F_3^+AsF_6^- + HC{\equiv}N \longrightarrow HC{\equiv}N{-}XeF^+AsF_6^- + XeF_2 \qquad (1.2)$$

The compound, $HC{\equiv}N{-}XeF^+AsF_6^-$, has been isolated as a white microcrystalline solid upon removal of the solvent at −30°C and was stable for up to 4–6 h at 0°C. Solutions of $HC{\equiv}N{-}XeF^+AsF_6^-$ in HF solution at ambient temperature slowly decompose over a period of 13 h.

Every element in the $HC{\equiv}N{-}XeF^+$ cation possesses at least one nuclide that is suitable for observation by NMR spectroscopy, namely, the spin $\frac{1}{2}$ nuclides ^1H, ^{13}C, ^{15}N, ^{129}Xe, and ^{19}F, and the spin 1 nucleus ^{14}N (Fig. 1.1). Multinuclear magnetic resonance spectra were recorded for $HC{\equiv}N{-}XeF^+AsF_6^-$ in HF and BrF_5 solvents for all six spin $\frac{1}{2}$ nuclei of the cation using natural abundance and ^{13}C and ^{15}N enriched compounds. All possible nuclear spin–spin couplings have been observed, establishing the solution structure of the $HC{\equiv}N{-}XeF^+$ cation (Table 1.2). Included among these scalar couplings are $^1J(^{129}Xe{-}^{14}N)$, $^2J(^{129}Xe{-}^{13}C)$, and $^3J(^{139}Xe{-}^1H)$, representing the first time scalar couplings have been observed between these nuclides.

An interesting feature of the NMR spectroscopy of the $HC{\equiv}N{-}XeF^+$ cation is the ready observation of the directly bonded $^{139}Xe{-}^{14}N$ and $^{14}N{-}^{13}C$ scalar couplings. The observation of both couplings and the relative ease of observing $^1J(^{129}Xe{-}^{14}N)$ in the alkyl nitrile, trifluorotriazine and perfluoropyridine adducts of XeF^+ (see Section 1.4.1 and Table 1.3) is attributed to several factors that minimize quadrupole relaxation of the $^{129}Xe{-}^{14}N$ and $^{14}N{-}^{13}C$ couplings: the low electric field gradient at the ^{14}N nucleus of the adduct cations; low viscosity of the HF solvent leading to a short molecular correlation time, and the small line width factor for ^{14}N.[36] However, in the higher viscosity solvent BrF_5 (−58°C), the $^{129}Xe{-}^{14}N$ and $^{14}N{-}^{13}C$ couplings are quadrupole collapsed into single lines. Because they are generally obscured owing to quadrupolar relaxation caused by the ^{14}N nucleus, ^{15}N enrichment was required for the observation of scalar couplings between

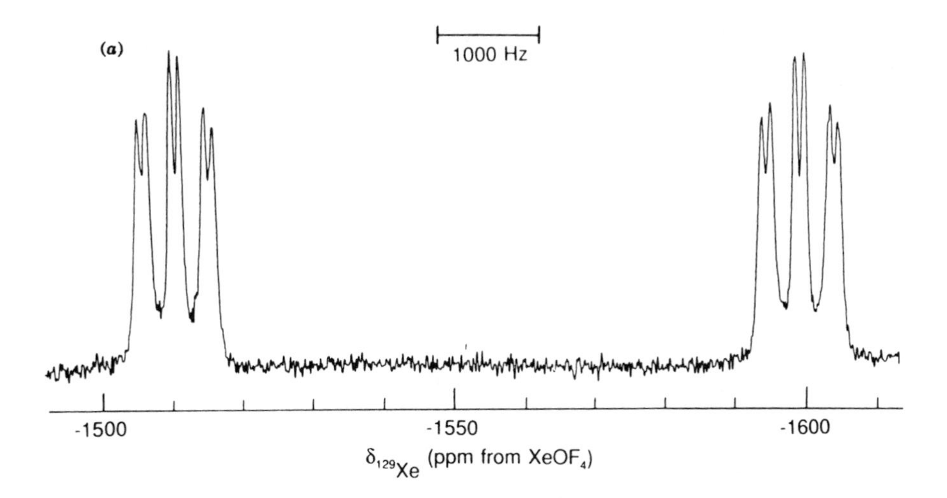

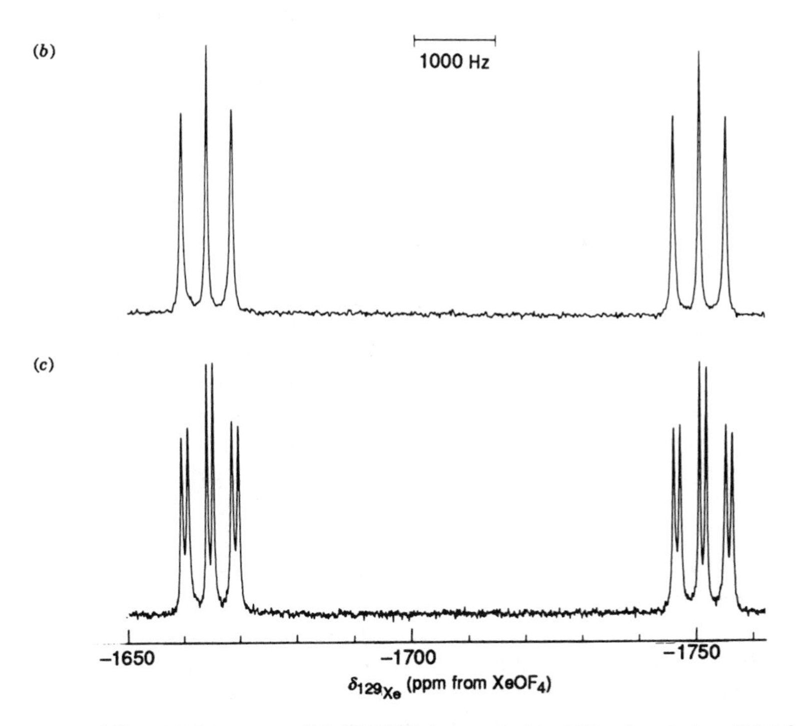

Figure 1.1. ^{129}Xe NMR spectra (69.563 MHz) recorded in HF solvent at −10°C for (*a*) $HC{\equiv}N{-}XeF^+AsF_6^-$ for a 99.2% ^{13}C enriched sample; and $CH_3C{\equiv}N{-}XeF^+AsF_6^-$ where (*b*) is the natural spectrum and (*c*) is 99.7% ^{13}C enriched at the C-2. [Reprinted with permission from A. A. A. Emara and G. J. Schrobilgen, *J. Chem. Soc. Chem. Commun.*, 1649 (1987). Copyright ©, 1987 from the Chemical Society, London.

TABLE 1.2. NMR Chemical Shifts and Spin-Spin Coupling Constants for the HC≡N—XeF$^+$ Cation.[a]

Chemical Shifts (ppm)[b]				
$\delta(^{129}Xe)$	$\delta(^{19}F)$	$\delta(^{14}N/^{15}N)$	$\delta(^{13}C)$	$\delta(^{1}H)$
−1555 (−1570)	−199.0 (−193.1)	−234.5 (−230.2)	104.1	4.70 (6.01)

Coupling Constants (Hz)			
$^1J(^{129}Xe—^{19}F)$	6171 (6165)	$^1J(^{129}Xe—^{15}N)$	471 (483)
$^1J(^{129}Xe—^{14}N)$	334	$^1J(^{13}C—^{1}H)$	308
$^1J(^{14}N—^{13}C)$	22	$^2J(^{15}N—^{19}F)$	23.9 (23.9)
$^2J(^{129}Xe—^{13}C)$	84	$^3J(^{129}Xe—^{1}H)$	24.7 (26.8)
$^2J(^{15}N—^{1}H)$	(13.0)	$^4J(^{19}F—^{1}H)$	2.6 (2.7)
$^3J(^{19}F—^{13}C)$	18		

[a]References 34 and 35. Spectra were recorded in anhydrous HF at −10°C or at −58°C on BrF_5 solvent (values in parentheses).

[b]Samples were referenced externally at 24°C with respect to the neat liquid references standards $XeOF_4$ (^{129}Xe), $CFCl_3$ (^{19}F), CH_3NO_2 (^{14}N and ^{15}N), $(CH_3)_4Si$ (^{13}C and ^{1}H) where a positive chemical shift denotes a resonance occurring to high frequency of the reference compound.

nitrogen and nondirectly bonded nuclei where the magnitudes of the couplings are small (Table 1.2).

Prior to the synthesis of the HC≡N—NgF$^+$ cation, no examples of krypton compounds had been reported in which krypton was bonded to an element other than fluorine. In view of the previous success in forming the HC≡N—XeF$^+$ cation,[34,35] the synthesis of the krypton(II) analog was undertaken.[8] The estimated electron affinity for KrF^+ of 13.2 eV suggested that HC≡N might have at least a marginal resistance to oxidative attack by the KrF^+ cation and that HC≡N—KrF$^+$ might have sufficient thermal stability to permit its spectroscopic characterization.

Unlike the xenon(II) analog, the direct interaction of $KrF^+AsF_6^-$ with HC≡N solutions in HF and BrF_5 solvents was not attempted owing to the strongly oxidizing character of the KrF^+ cation towards HC≡N and BrF_5 as well as its tendency to undergo autoredox reactions in both solvent media. Rather, the interaction of the less reactive KrF_2 with HC≡NH$^+AsF_6^-$ in HF was the preferred synthetic route. Reaction of sparingly soluble HC≡NH$^+AsF_6^-$ with KrF_2 in HF at −60°C led to instantaneous deposition of a white solid which, upon warming above −50°C, rapidly evolved Kr, NF_3, and CF_4 gases and was usually followed by a violent detonation and accompanying emission of white light. When these reactions were allowed to

TABLE 1.3. A Comparison of Xenon-Nitrogen Reduced Coupling Constants and ^{129}Xe Chemical Shifts in F-Xe-L and Xe-L$^+$ Type Systems.[a]

Species	Hybridization at Nitrogen	^{1}K(Xe—N) (10^{22} NA^{-2} m^{-3})	$\delta(^{129}Xe)$ (ppm)	T(°C)	References
$HC{\equiv}N—XeF^+$	*sp*	1.381[b]	−1555 (−1570)	−10 (−58)	34, 35
$RC{\equiv}N—XeF^+$	*sp*	1.297–1.393	−1541 to −1721	−10 to −50	34, 37
$F_3S{\equiv}N—XeF^+$	*sp*	1.435	(−1661)	(−60)	38
s-$C_3F_3N_2N—XeF^+$	sp^2	1.013	−1808 (−1863)	−5 (−50)	9
$C_5F_5N—XeF^+$	sp^2	0.983	−1872 (−1922)	−30 (−30)	39
4-$CF_3C_5F_4N—XeF^+$	sp^2	0.991	−1803 (−1853)	−15 (−50)	39
$(FO_2S)_2N—XeF$[c]	sp^2	0.913[c]	−2009	−40	6
$(FO_2S)_2N—Xe^+$[d]	sp^2	0,272[b]	−1943	25	6
$F_4S{=}N—Xe^+$	sp^2		−2672	−20	38
$F_5S—N(H)—Xe^+$	sp^3		−2886	−20	38
$F_5Te—N(H)—Xe^+$	sp^3	0.398[b]	−2841 (−2903)	−45 (−50)	40

[a]Values, unless otherwise indicated, were determined in HF and in BrF_5 (in parentheses) solvent; $1J(^{129}Xe—^{19}F)]/[h\gamma(^{14,15}N)\gamma(^{129}Xe)]$.
[b]Recorded for the ^{15}N enriched cation.
[c]Measured in SO_2ClF solvent.
[d]Measured in SbF_5 solvent.

proceed at approximately −60°C, the mixtures were periodically quenched to −196°C in order to study the development of the Raman spectrum of the product (see Section 1.4). The interaction of $HC{\equiv}NH^+AsF_6^-$ and KrF_2 in BrF_5 led to a soluble product, which was stable to at least −55°C in BrF_5 with only slight decomposition. The ^{19}F NMR spectra of these solutions at −58°C and in HF at −60°C are consistent with Eq. 1.3

$$KrF_2 + HC{\equiv}NH^+AsF_6^- \longrightarrow HC{\equiv}N{-}KrF^+AsF_6^- + HF \qquad (1.3)$$

The structure of the $HC{\equiv}N{-}KrF^+$ cation in solution has been confirmed by reaction with 99.5% ^{15}N enriched $HC{\equiv}NH^+AsF_6^-$ in BrF_5 solvent. The ^{19}F and ^1H resonances exhibit new doublet splittings attributed to ^{15}N coupling (Fig. 1.2*a*). The new splitting (26 Hz) on the ^{19}F resonance was assigned to the two-bond spin–spin coupling $^2J(^{19}F{-}^{15}N)$ and compares favorably in magnitude with previously reported values for $F{-}Xe{-}N(SO_2F)_2$ $[^2J(^{19}F{-}^{15}N) = 39.2\ Hz]$ and $CH_3C{\equiv}N{-}XeF^+$ $[^2J(^{19}F{-}^{15}N) = 36.1\ Hz$[37]. Krypton isotopic shifts arising from ^{82}Kr (11.56%), ^{84}Kr (56.90%), and ^{86}Kr (17.37%) were resolved on the ^{19}F resonance (0.0138 ppm amu^{-1}) and served as an added confirmation that the fluorine resonance arise from fluorine directly bonded to krypton. The doublet fine structure (12.2 Hz) on the ^1H resonance of the ^{15}N enriched cation (Fig. 1.2*b*) was assigned to $^2J(^{15}N{-}^1H)$ [cf. $^2J(^{15}N{-}^1H) = 19.0$ Hz for $HC{\equiv}NH^+$ in HF solvent]. The ^{15}N NMR spectrum was comprised of a well-resolved doublet of doublets (Fig. 1.2*c*) arising from $^2J(^{19}F{-}^{15}N)$ and $^2J(^{15}N{-}^1H)$, which simplified to a doublet (26 Hz) upon broad band ^1H decoupling, confirming the aforementioned coupling constant assignments.

1.3.2. Perfluoroalkyl- and Alkylnitrile Adducts of XeF^+

The measured value of the first adiabatic IP of $CF_3C{\equiv}N$ (13.9 eV)[15] suggests that this base should be even more resistant to oxidative attack by KrF^+ and XeF^+ than $HC{\equiv}N$. Moreover, $CF_3C{\equiv}N$ would be expected to possess a more weakly basic nitrogen that would be conducive to the formation of a correspondingly more ionic Ng—N bond (see Section 1.4.2). The perfluoroalkyl nitrile adduct cations of XeF^+ have been prepared by the interaction of equimolar amounts of $XeF^+AsF_6^-$ or $Xe_2F_3^+AsF_6^-$ and $R_FC{\equiv}N$ ($R_F = CF_3$, C_2F_5, and n-C_3F_7) in BrF_5 solvent[9] according to Eq. 1.4.

$$R_FC{\equiv}N + XeF^+AsF_6^-(Xe_2F_3^+AsF_6^-) \longrightarrow R_FC{\equiv}N{-}XeF^+AsF_6^-(+XeF_2) \qquad (1.4)$$

The syntheses of the krypton(II) analogs have also been reported and were undertaken at low temperatures in BrF_5 solvent using the general synthetic approach given in Eq. 1.5.

$$R_FC{\equiv}N{-}AsF_5 + KrF_2 \longrightarrow R_FC{\equiv}N{-}KrF^+AsF_6^- \qquad (1.5)$$

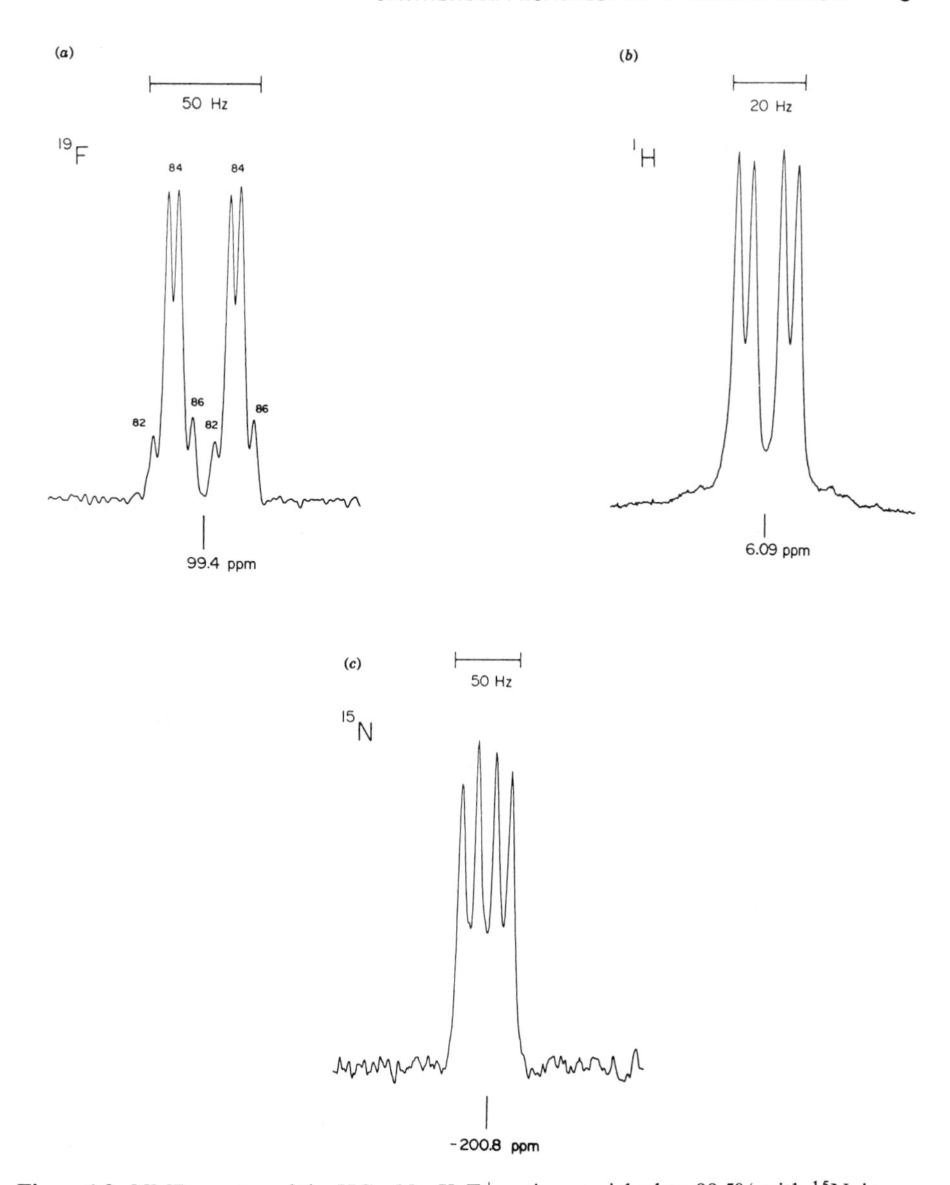

Figure 1.2. NMR spectra of the $HC{\equiv}N{-}KrF^+$ cation, enriched to 99.5% with ^{15}N, in BrF_5 as solvent at $-57°C$. (*a*) The ^{19}F spectrum (235.36 MHz) depicting $^2J(^{19}F{-}^{15}N)$ and $^4J(^{19}F{-}^1H)$ and krypton isotope shifts. Lines assigned to fluorine bonded to ^{82}Kr (11.56%), ^{84}Kr (56.90%), and ^{86}Kr (17.37%) are denoted by the krypton mass number. The innermost lines of the ^{82}Kr and ^{86}Kr doublets overlap their corresponding ^{84}Kr doublets. The isotopic shift arising from ^{83}Kr (11.53%) is not resolved; those of ^{78}Kr (0.35%) and ^{80}Kr (2.27%) are too weak to be observed. (*b*) The ^1H Spectrum (80.02 MHz) depicting $^2J(^{15}N{-}^1H)$ and $^4J(^{19}F{-}^1H)$. (*c*) The ^{15}N Spectrum (50.70 MHz) depicting $^2J(^{19}F{-}^{15}N)$ and $^2J(^{15}N{-}^1H)$. [Reprinted with permission from G. J. Schrobilgen, *J. Chem. Soc. Chem. Commun.*, 863, (1988). Copyright ©, 1988 from the Chemical Society, London.

TABLE 1.4. Correlation of Physical Properties for Representative Ng—F Bonds.

Species[b]	r(Ng—F)[c] (Å)	ν(Ng—F) (cm^{-1})	NMR Parameters[a] $^1J(^{129}Xe—^{19}F)$[d] (Hz)	$\delta(^{129}Xe)$[d,e] (ppm)	$\delta(^{19}F)$[d,e] (ppm)	T(°C)	References
KrF^+	(1.740)						
$KrF^+Sb_2F_{11}^-$		624					42
$KrF^+AsF_6^-$		609					42
$(FKr)_2F^+$[f]		605			73.6	−65	42
HC≡N—KrF^+	(1.748)	560			99.4	−58	8
CF_3C≡N—KrF^+					93.1	−59	9
C_2F_5C≡N—KrF^+					91.1	−59	9
C_3F_7C≡N—KrF^+					91.9	−59	9
KrF_2	1.875	462			68.0	−56	43, 44
	(1.843)						
XeF^+	(1.886)						
$XeF^+Sb_2F_{11}^-$	1.82 (3)	619	7230	−574	−290.2	23[g]	31, 42, 45, 46
$XeF^+AsF_6^-$	1.873 (6)	610	6892	−869		−47	9, 42, 47, 48
$(FXe)_2F^+$[f]	1.90 (3)	593	6740	−1051	−252.0	−62	31, 42, 46, 49
HC≡N—XeF^+	(1.904)	564	6181	−1569	−198.4[h]	−58	34
F_3S≡N—XeF^+		554	6248	−1661	−180.5	−60	38
CF_3C≡N—XeF^+			6397	−1337	−210.4	−63	9
C_2F_5C≡N—XeF^+			6437	−1294	−212.9	−63	9
C_3F_7C≡N—XeF^+			6430	−1294	−213.2	−63	9
CH_3C≡N—XeF^+[h]		560	6020	−1708	−185.5	−10	34
s-$C_3F_3N_2N$—XeF^+		548	5932	−1862	−145.6	−50	9
			5909	−1808	−154.9	−5	
FO_2SO—XeF	1.940 (8)	528	5830	−1666		−40	31, 46, 50, 51
cis/*trans*-F_4OIO—XeF		527	5803/	−1824/	−161.7[i]	0	52
			5910	−1720	−170.1[i]	0	

$C_5F_5N—XeF^+$		528	5926	−1922	−139.6	−30	39
$4\text{-}CF_3C_5F_4N—XeF^+$		524	5963	−1853	−144.6	−50	39
$F_5TeO—XeF^j$		520		−2051	-151.0^k	26	53, 54
$(FO_2S)_2N—XeF$	1.967 (3)	506	5586	−1977	-126.1^j	−58	3, 4
			5664^k	-2009^k	-126.0^k	−40	
$(FO_2S)_2N—Xe^{+g}$				−1943		25	6
$F_4S{=}N—Xe^+$				-2672^l		−20	38
$F_5S—N(H)—XeF^+$				-2886^l		−20	38
$F_5Te—N(H)—XeF^+$				-2903^l		−50	40
				-2841^l		−45	
XeF_2	1.977 (1.984)	496	5621	−1685	−184.3	−52	9, 55, 56

[a]Spectra were obtained in BrF_5 solvent unless otherwise indicated.
[b]Unless otherwise indicated, all cations have AsF_6^- as the counterion.
[c]Values in parentheses are calculated values determined in Ref. 41.
[d]The NMR parameters of KrF and XeF groups are very sensitive to solvent and temperature conditions; it is therefore important to make comparisons in the same solvent medium at the same or nearly the same temperatures.
[e]References with respect to the neat liquids $XeOF_4$ (^{129}Xe) and $CFCl_3$ (^{19}F) at 24°C; a positive sign denotes the chemical shift of the resonance in question occurs to higher frequency of (is more deshielded than) the resonance of the reference substance.
[f]Table entries refer to the terminal fluorine on the noble gas atom.
[g]Recorded in SbF_5 solvent.
[h]$\delta(^{19}F)$ measured in anhydrous HF solvent at −10°C.
[i]$\delta(^{19}F)$ measured in SO_2ClF solvent at −40°C.
[j]NMR parameters measured in SO_2ClF solvent.
[k]NMR parameters measured in SO_2ClF solvent at −50°C.
[l]$\delta(^{129}Xe)$ measured in HF solvent.

The $R_FC{\equiv}N{-}NgF^+$ cations have been characterized in BrF_5 by low-temperature (−57 to −61°C) ^{19}F and ^{129}Xe NMR spectroscopy and consisted of two sets of new signals: a singlet in the F-on-Kr^{II} and in the F-on-Xe^{II} regions, and resonances in the F-on-C region with characteristic $^3J(^{19}F{-}^{19}F)$ and $^1J(^{19}F{-}^{13}C)$ couplings having chemical shifts to high frequency of the parent base molecules (Table 1.4). In each case, the singlet assigned to F-on-Xe^{II} was flanked by natural abundance (26.44%) ^{129}Xe satellites arising from $^1J(^{129}Xe{-}^{19}F)$. The integrated relative intensities of the fluorine-on-noble gas environment and perfluoroalkyl group are consistent with the proposed formulations. Furthermore, the F-on-Kr^{II} resonance of $CF_3C{\equiv}N{-}KrF^+$ could be resolved to show the ^{82}Kr, ^{84}Kr, and ^{86}Kr isotopic shifts (0.0105 ppm amu^{-1}), which compare favorably with previously measured values for $HC{\equiv}N{-}KrF^+$ (0.0138 ppm amu^{-1})[8] and KrF_2 (0.0104 ppm amu^{-1}).[54] In addition, the F-on-Kr^{II} resonances occur to high frequency of KrF_2 while the F-on-Xe^{II} resonances occur to low frequency of XeF_2 (Table 1.4). Similar, but slightly more positive ^{19}F chemical shifts have been observed for $HC{\equiv}N{-}KrF^+$ [$\delta(^{19}F)$ 99.4 ppm; −57°C; BrF_5 solvent] with respect to KrF_2 [$\delta(^{19}F)$ 68.0 ppm; −56°C; BrF_5 solvent].[8] This is in contrast to the $R_FC{\equiv}N{-}XeF^+$ series of cations that display significantly more positive ^{19}F [F-on-Xe^{II}] and ^{129}Xe chemical shifts when compared with $HC{\equiv}N{-}XeF^+$ [$\delta(^{19}F)$ −198.4 ppm; $\delta(^{129}Xe)$ −1552 ppm; $^1J(^{129}Xe{-}^{19}F)$ 6150 Hz; −10°C; HF solvent][34] and XeF_2 [$\delta(^{19}F)$ −184.3 ppm; $\delta(^{129}Xe)$ −1685.2 ppm; $^1J(^{129}Xe{-}^{19}F)$ 5621 Hz; −52°C; BrF_5]. The ^{129}Xe and ^{19}F complexation shifts indicate the Xe—N bonds of the $R_FC{\equiv}N{-}XeF^+$ cations are significantly more ionic than in $HC{\equiv}N{-}XeF^+$ or $RC{\equiv}N{-}XeF^+$[9] and is further supported by significantly larger $^1J(^{129}Xe{-}^{19}F)$ values measured for the $R_FC{\equiv}N{-}XeF^+$ cations, which are known to increase with ionic character of the Xe—L bond in F—Xe—L type compounds (see Table 1.4).[31–33] The $R_FC{\equiv}N{-}XeF^+$ cations represent the most ionic Xe—N bonded species presently known. In contrast, the analogous comparison of ^{19}F chemical shifts for $R_FC{\equiv}N{-}KrF^+$ cations suggests that the Kr—N bonds of these cations may be slightly more covalent than in $HC{\equiv}N{-}KrF^+$.

All three fluoro(perfluoroalkylnitrile)krypton(II) cations are thermally less stable with respect to redox decomposition than $HC{\equiv}N{-}KrF^+$ or their xenon(II) analogs, preventing their isolation and characterization in the solid state. Decompositions monitored by ^{19}F NMR spectroscopy occurred over periods of approximately 1–2 h at −57 to −61°C. The major decomposition products consisted of Kr and the fluorinated products (^{19}F NMR parameters listed in parentheses): CF_4 (−63.1 ppm), C_2F_6 (−88.6 ppm), and NF_4^+ [219.4 ppm, $^1J(^{19}F{-}^{14}N)$ 229 Hz] for all three $R_FC{\equiv}N{-}KrF^+$ cations studied, and $n\text{-}C_3F_8$ (−83.8 ppm, F_3C—; −132.8 ppm, —CF_2—) for $C_2F_5C{\equiv}N{-}KrF^+$ and $n\text{-}C_3F_8$, $n\text{-}C_4F_{10}$ (−82.8 ppm, F_3C—; −129.2 ppm, —CF_2—) for $n{-}C_3F_7C{\equiv}N{-}KrF^+$.[9]

Reactions of $XeF^+AsF_6^-$ with alkyl nitriles, $RC{\equiv}N$, and $C_6F_5C{\equiv}N$ have also been carried out by combining stoichiometric amounts of the reactants in anhydrous HF and warming to −50 to −10°C to effect reaction and dissolution in the solvent.[34,37] The reactions proceed by analogy with Eq. 1.1. In the case of the alkyl nitriles and HC≡N, Equilibrium 1.6 is significant so that equilibrium amounts of XeF_2, $RC{\equiv}NH^+$, and $HC{\equiv}NH^+$ are observed in the ^{129}Xe, ^{19}F, $^{14,15}N$, ^{13}C, and ^{1}H NMR spectra but XeF_2 frequently is not observed in the ^{19}F and ^{129}Xe NMR spectra. The apparent absence of XeF_2 in the NMR spectra is attributed to chemical exchange involving free XeF^+ arising from Equilibrium 1.7 and $Xe_2F_3^+$ as an exchange intermediate (Equilibrium 1.8).

$$RC{\equiv}N{-}XeF^+ + HF \rightleftharpoons RC{\equiv}NH^+ + XeF_2 \tag{1.6}$$

$$RC{\equiv}N{-}XeF^+ \rightleftharpoons RC{\equiv}N + XeF^+ \tag{1.7}$$

$$XeF^+ + XeF_2 \rightleftharpoons Xe_2F_3^+ \tag{1.8}$$

Multinuclear magnetic resonance spectroscopy (^{129}Xe, ^{19}F, ^{14}N, ^{15}N, ^{13}C, and ^{1}H) has provided unambiguous proof for the structures of the $RC{\equiv}N{-}XeF^+$ cations in HF solution (Table 1.5). Several alkyl nitrile adducts of the XeF^+ cation have also been isolated and characterized in the solid state by low-temperature Raman spectroscopy. The Xe—F stretching frequencies are consistent with weak covalent bonding between xenon and nitrogen (see Section 1.4): CH_3 (560), CH_2F (565), CH_2Cl (564), C_2H_5 (541), $(CH_3)_2CH$ (556), $(CH_3)_3C$ (560); values in parentheses are the Xe—F stretching frequencies (cm^{-1}).[37]

The decompositions of the nitrile adduct cations $CH_3(CH_2)_nC{\equiv}N{-}XeF^+$ ($n = 0$–3) have been monitored in HF solution by multi-NMR spectroscopy.[37] The rate of fluorination of the alkyl chain increases with increasing chain length, with the degree of fluorination increasing at the alkyl carbons in the order $\beta < \gamma < \delta$, where no fluorination is observed at the carbon α to the C≡N group.

1.3.3. Perfluoropyridine and *s*-Trifluorotriazine Adducts of XeF^+

The fluoro(perfluoropyridine)xenon(II) cations, $4\text{-}RC_5F_4N{-}XeF^+$ (R = F or CF_3), have been observed in HF and BrF_5 solutions and are stable in both media up to −30°C.[39] The salts, $4\text{-}RC_5F_4N{-}XeF^+AsF_6^-$, have been isolated at −30°C from BrF_5 solutions initially containing equimolar amounts of $4\text{-}RC_5F_4NH^+AsF_6^-$ and XeF_2. The resulting white solids were stable to −25°C. Low-temperature Raman and ^{129}Xe, ^{19}F, and ^{14}N NMR spectroscopic results are consistent with planar cations (Structure **I**) in which the xenon atom is

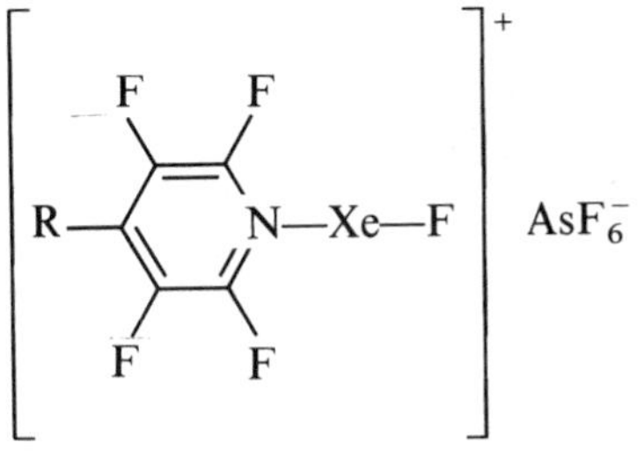

$$\left[\text{R}{-}C_5F_4N{-}Xe{-}F \right]^+ AsF_6^-$$

I

TABLE 1.5. Selected NMR Chemical Shifts and Coupling Constants for $RC{\equiv}N{-}XeF^+$ Cations.[a]

Cation	Chemical Shifts (ppm)[b]			Spin–Spin Couplings (Hz)	
	$\delta(^{129}Xe)$	$\delta(^{19}F)$	$\delta(^{14}N)$	$J(^{129}Xe{-}^{19}F)$	$J(^{129}Xe{-}^{14}N)$
$CH_3C{\equiv}N{-}XeF^+$	−1707	−185.5	−251.1	6020	313
$CH_2FC{\equiv}N{-}XeF^+$	−1541	−198.2 (XeF) −241.7 (CF)	−229.2	6164	333
$CH_2ClC{\equiv}N{-}XeF^+$	−1583	−195.5	−236.6	6147	331
$CH_3CH_2C{\equiv}N{-}XeF^+$	−1717	−184.6	−251.9	6016	312
$CH_2FCH_2C{\equiv}N{-}XeF^+$	−1662	−182.8 (XeF) −218.8 (CF)		6063	322
$CH_3CH_2CH_2C{\equiv}N{-}XeF^+$	−1718	−189.1	−249.7	6020	309
$CH_2FCH_2CH_2C{\equiv}N{-}XeF^+$	−1663	−187.7 (XeF) −222.7 (CF)		6065	321
$CH_3CHFCH_2C{\equiv}N{-}XeF^+$	−1700	−186.1 (XeF) −172.1 (CF)	−257.8	6038	315
$CHF_2CH_2CH_2C{\equiv}N{-}XeF^+$		−120.9 (CF)			
$CH_3CH_2CH_2CH_2C{\equiv}N{-}XeF^+$	−1720	−183.2	−247.1	6022	309
$CH_2FCH_2CH_2CH_2C{\equiv}N{-}XeF^+$	−1703	−184.6 (XeF)		6027	311
$CH_3CHFCH_2CH_2C{\equiv}N{-}XeF^+$	~ −1705[c]	−185.1 (XeF) −175.9 (CF)		6015	*c*
$(CH_3)_2CHC{\equiv}N{-}XeF^+$	−1721	−184.5	−251.4	6016	309
$(CH_3)_3CC{\equiv}N{-}XeF^+$	−1721	−184.3	−251.4	6024	309
$CH_2ClC(CH_3)HC{\equiv}N{-}XeF^+$	−1703	−198.7		6027	314
$CH_2FC(CH_3)HC{\equiv}N{-}XeF^+$	−1669	−187.9 (XeF) −235.3 (CF)	−243.8	6027	301
$C_6F_5C{\equiv}N{-}XeF^+$	−1424	−201.8		6610	

[a] References 34 and 37.

[b] Recorded at −10 to −30°C and referenced externally at 24°C with respect to neat liquid references: $XeOF_4$ (^{129}Xe), $CFCl_3$(^{19}F) and CH_3NO_2(^{14}N).

[c] Resonance overlaps with that of $CH_2FCH_2CH_2CH_2C{\equiv}N{-}XeF^+$.

coordinated to the aromatic ring through the lone pair of electrons on the nitrogen (Ref. 39 and Tables 1.3 and 1.4).

Equimolar amounts of $XeF^+AsF_6^-$ and the perfluoropyridine, 4-RC_5F_4N (R = F or CF_3), react in anhydrous HF at −30 to −20°C according to Eq. 1.9 and Equilibria 1.10 and 1.11 to give the novel Xe—N bonded cations, 4-RC_5F_4N—XeF^+, as the AsF_6^- salts in solution.

$$\text{4-RC}_5\text{F}_4\text{N} + (1 + x)\text{HF} \rightarrow \text{4-RC}_5\text{F}_4\text{NH}^+(\text{HF})_x\text{F}^- \quad (1.9)$$

$$\text{4-RC}_5\text{F}_4\text{NH}^+(\text{HF})_x\text{F}^- + \text{XeF}^+\text{AsF}_6^- \rightleftharpoons$$

$$\text{4-RC}_5\text{F}_4\text{NH}^+\text{AsF}_6^- + \text{XeF}_2 + x\text{HF} \quad (1.10)$$

$$\text{XeF}^+\text{AsF}_6^- + \text{4-RC}_5\text{F}_4\text{NH}^+\text{AsF}_6^- \rightleftharpoons$$

$$\text{4-RC}_5\text{F}_4\text{N—XeF}^+\text{AsF}_6^- + \text{“HAsF}_6\text{”} \quad (1.11)$$

At −30°C these solutions consisted of equilibrium mixtures of XeF_2, 4-$RC_5F_4NH^+AsF_6^-$, and 4-RC_5F_4N—$XeF^+AsF_6^-$ as determined by NMR spectroscopy (Table 1.3). Removal of HF solvent by pumping at −50°C resulted in white solids, which Raman spectroscopy at −196°C also showed to be mixtures of 4-RC_5F_4N—$XeF^+AsF_6^-$, XeF_2, and 4-$RC_5F_4NH^+AsF_6^-$.

An alternative approach, which led to isolation of the Xe—N bonded cations, allowed stoichiometric amounts of XeF_2 and the perfluoropyridinium cations, as their AsF_6^- salts, to react in HF and BrF_5 solvents at −30°C according to Equilibrium 1.12.[39]

$$\text{XeF}_2 + \text{4-RC}_5\text{F}_4\text{NH}^+\text{AsF}_6^- \rightleftharpoons \text{4-RC}_5\text{F}_4\text{N—XeF}^+\text{AsF}_6^- + \text{HF} \quad (1.12)$$

The equilibria in both solvents were monitored by ^{129}Xe, ^{19}F, and ^{14}N NMR spectroscopy. In the case of BrF_5, formation of 4-RC_5F_4N—$XeF^+AsF_6^-$ was strongly favored over that in HF solvent; the equilibrium ratio [4-RC_5F_4N—XeF^+]/[4-$RC_5F_4NH^+$] being 0.25 and 2.1 in HF and BrF_5 solvents, respectively, at −30°C for R = F and 3.7 for R = CF_3 in BrF_5 at −30°C ($K_F = 4.5$ at −30°C and $K_{CF_3} = 13.6$ at −50°C in BrF_5 for Equilibrium 1.12). Consequently, removal of BrF_5 solvent under vacuum at −30°C yielded white solids corresponding to the salts 4-RC_5F_4N—$XeF^+AsF_6^-$. The CF_3 derivatives substituted at the 2- and 3- positions have also been synthesized from their perfluoropyridinium salts in BrF_5 solvent and characterized by NMR spectroscopy [$\delta(^{129}Xe)$, −1899 and −1877, respectively].[37]

The interaction of liquid *s*-trifluorotriazine, *s*-$C_3N_3F_3$, with $XeF^+AsF_6^-$ at room temperature for 3 h followed by removal of excess *s*-trifluorotriazine under vacuum resulted in a white powder that is stable indefinitely at room temperature.[9] The combining ratio $XeF^+AsF_6^-$: *s*-$C_3N_3F_3$ = 1.00:1.00 is consistent with Equation 1.13.

$$\text{XeF}^+\text{AsF}_6^- + s\text{-C}_3\text{F}_3\text{N}_3 \rightarrow s\text{-C}_3\text{F}_3\text{N}_2\text{N—XeF}^+ \quad (1.13)$$

Both the ^{19}F and ^{129}Xe NMR findings (Tables 1.3 and 1.4) for the salt dissolved in BrF_5 and HF solvents are consistent with the cation formulation given by Structure **II**. The ^{129}Xe NMR spectrum recorded in BrF_5 at $-50°C$ consists of a doublet arising from $^{1}J(^{129}Xe—^{19}F) = 5932$ Hz. The $^{129}Xe—^{14}N$ coupling is quadrupole collapsed, as has been observed previously for $4\text{-}CF_3\text{-}C_5F_4N—XeF^+AsF_6^-$ and $C_5F_5N—XeF^+AsF_6^-$ in BrF_5 at low temperatures.[39]

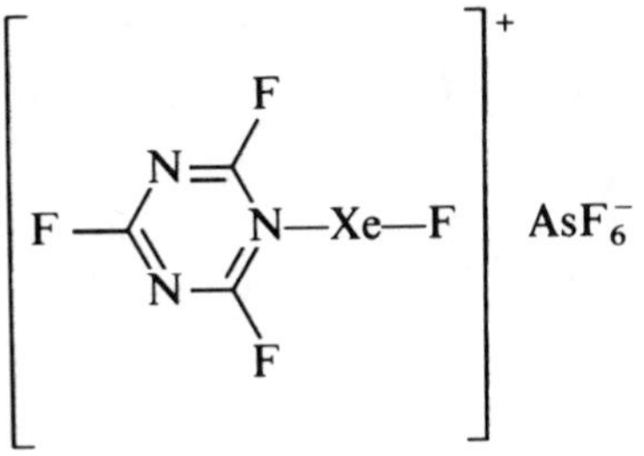

$$\left[F—C_3N_3F_2—N—Xe—F \right]^+ AsF_6^-$$

II

In HF solvent, however, $^{1}J(^{129}Xe—^{14}N) = 245$ Hz was observed at $-5°C$ and compares favorably in magnitude with those reported previously for the related perfluoropyridine cations (235–238 Hz). The ^{19}F NMR spectrum shows two F-on-C environments in the ratio of 1:2 and a F-on-Xe^{II} environment with accompanying ^{129}Xe natural abundance (26.44%) satellites arising from $^{1}J(^{129}Xe—^{19}F)$ and a 1:2:1 triplet arising from $^{4}J(F_1—F_2) = 10.9$ Hz. The latter coupling has also been observed for the perfluoropyridine cations $4\text{-}CF_3\text{-}C_5F_4N—XeF^+$ (25.8 Hz) and $C_5F_5N—XeF^+$ (25.0 Hz).[39]

The reaction of $XeF_3^+SbF_6^-$ with $s\text{-}C_3F_3N_3$ in BrF_5 solvent at 20°C fails to give $C_3F_3N_2N—XeF_3^+$.[57] Rather, reduction of Xe^{IV} to Xe^{II} occurs according to Equation 1.14.

$$2XeF_3^+SbF_6^- + 3s\text{-}C_3F_3N_3 \rightarrow s\text{-}C_3F_3N_2N—XeF^+SbF_6^- + \quad (F_2C_3N_3F_2)N—N(C_3N_3F_2F_2) \tag{1.14}$$

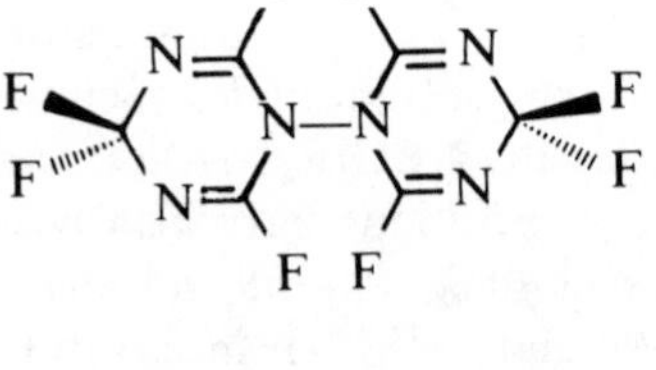

1.3.4. The Lewis Acid Properties of the $XeOTeF_5^+$ and $XeOSeF_5^+$ Cations

More recently, this work has been extended to the Lewis acid properties of the noble gas cations $XeOTeF_5^+$ and $XeOSeF_5^+$.[38,57] The $XeOSeF_5^+$ cation, which was previously unknown, was prepared according to Eqs. 1.15–1.17.

TABLE 1.6. Selected NMR Parameters for $XeOMF_5^+$ (M = Se or Te) Adduct Cations with Nitrogen Bases.

Cation	Chemical Shifts (ppm)		Coupling Constants (Hz)	
	$\delta(^{129}Xe)$	$\delta(^{19}F)$[c]	$^2J(F_A—F_B)$	Solvent (T = °C)
$CH_3C{\equiv}N—Xe—OTeF_5^{+}$[a]	−2061	−45.8 F_A −44.0 F_B	181	SO_2ClF (−50)
$C_5F_5N—Xe—OTeF_5^{+}$[a]	−2246	−46.3 F_A −42.8 F_B	~180	SO_2ClF (−70)
s-$C_3F_3N_2N—Xe—OTeF_5^{+}$[a]	−2192	−47.1 F_A ~ −42.3 F_B	~187	SO_2ClF (−50)
$F_3S{\equiv}N—Xe—OSeF_5^{+}$[b]	−1979	67.9 F_A 70.4 F_B	219	BrF_5 (−60)

[a]Prepared as the $Sb(OTeF_5)_6^-$ salt; Ref. 57.
[b]Prepared as the AsF_6^- salt; Ref. 38.
[c]Subscripts refer to the AB_4 spin systems of the F_5MO-groups.

$$2SeOF_2 + 3XeF_2 \rightarrow Xe(OSeF_5)_2 + 2Xe \quad (1.15)$$

$$Xe(OSeF_5)_2 + XeF_2 \rightleftharpoons 2FXeOSeF_5 \quad (1.16)$$

$$AsF_5 + FXeOSeF_5 \rightarrow XeOSeF_5^+AsF_6^- \quad (1.17)$$

While the $XeOMF_5^+$ cations (M = Se and Te) are expected to be weaker Lewis acids, they are expected to be less strongly oxidizing than NgF^+ cations, and have been shown to form stable adducts at low temperature with several organic and inorganic nitrogen bases leading to the first examples of O—Xe—N linkages (Table 1.6).

Although the s-$C_3F_3N_2N$—$XeOMF_5^+AsF_6^-$ salts were successfully prepared near room temperature by reaction of the neat compounds (Eq. 1.18), other potential organic ligands such as nitriles and pyridines are vigorously oxidized by $XeOMF_5^+$ cations when these reactions are attempted in the absence of a solvent under similar conditions.

$$XeOMF_5^+AsF_6^- + s\text{-}C_3F_3N_3 \rightarrow F_5MO\text{—}Xe\text{—}NN_2C_3F_3^+AsF_6^- \quad (1.18)$$

To partially address this problem, $XeOTeF_5^+Sb(OTeF_5)_6^-$ was synthesized and is the first fully substituted $OTeF_5$ salt prepared to date.[57] The salt was prepared in SO_2ClF using the redox synthesis described by Eq. 1.19.

$$2Xe(OTeF_5)_2 + Sb(OTeF_5)_3 \rightarrow XeOTeF_5^+Sb(OTeF_5)_6^- + Xe \quad (1.19)$$

It was shown that the decreased polarity of the $Sb(OTeF_5)_6^-$ anion relative to the AsF_6^- anion facilitated solubility in SO_2ClF at low-temperatures and was found to be soluble in high concentrations down to the freezing point of the solvent (−124°C). This synthetic approach allowed the formation of adduct species between $XeOTeF_5^+$ and SO_2ClF and representative members of the series of nitrogen bases, that is, acetonitrile and pentafluoropyridine, at low temperatures and under nonsolvolytic conditions in a low-polarity solvent (Table 1.6 and Ref. 57). The only other known examples of adduct formation with the weakly basic SO_2ClF molecule have been reported for SbF_5 and AsF_5 at low temperatures,[58] suggesting that the ability of the $XeOTeF_5^+$ cation to coordinate with the weak electron pair donor, SO_2ClF, is a consequence of the weak coordinating ability of the $Sb(OTeF_5)_6^-$ anion. Removal of excess SO_2ClF from these solutions under vacuum at 25°C resulted in the 1:1 adduct salt, F_5TeOXe—$OSOClF^+Sb(OTeF_5)_6^-$, which slowly dissociated upon further pumping at room temperature.[57] In addition, $Bi(OTeF_5)_5$ and $Et_4N^+M'(OTeF_5)_6^-$ (M′ = Sb or Bi) were also synthesized and characterized by ^{121}Sb, ^{209}Bi, ^{19}F and ^{125}Te NMR.[57] Prior to these studies only $As(OTeF_5)_5$ and $As(OTeF_5)_6^-$ had been reported.[59]

1.3.5. Noble Gas Cation Adducts with Inorganic Bases

The synthesis of $N{\equiv}SF_3$ Lewis acid–base adducts with the noble gas cations XeF^+, $XeOSeF_5^+$, and $XeOTeF_5^+$ have also been investigated.[38] Several synthetic approaches have been used (Eq. 1.20).

$$XeL^+AsF_6^- + N{\equiv}SF_3 \rightarrow L{-}Xe{-}N{\equiv}SF_3^+AsF_6^- \qquad (1.20)$$

where L = XeF^+, $OSeF_5^+$ and the solvent was BrF_5 at −50°C. In addition, solid $F{-}Xe{-}N{\equiv}SF_3^+AsF_6^-$ has been synthesized by the direct interaction of liquid $N{\equiv}SF_3$ with solid $XeF^+AsF_6^-$ at −20°C. Structures have been characterized in solution using primarily ^{19}F and ^{129}Xe NMR spectroscopy (Tables 1.3, 1.4, and 1.6) and in the solid state by low-temperature Raman spectroscopy (Table 1.4).

The solvolytic behavior of the $F_3S{\equiv}N{-}XeF^+$ cation has been monitored in anhydrous HF.[38] Two successive addition/eliminations of HF occur across the sulfur–nitrogen bond to give the $F_4S{=}N{-}Xe^+$ (Structure **III**) and $F_5S{-}N(H){-}Xe^+$ (Structure **IV**) cations. The ^{129}Xe NMR spectrum of the SF_5 derivative shows that this xenon is the most shielded ^{129}Xe environment (most covalent Xe—N bond) observed for a xenon–nitrogen bonded species (Tables 1.3 and 1.4). Moreover, the $F_5S{-}N(H){-}Xe^+$ cation represents the first example of a bond between xenon and an sp^3-hybridized nitrogen. More recently, the tellurium analog has been prepared in HF and BrF_5 solvent at −45 and −50°C, respectively,[40] according to Eq. 1.21.

$$F_5TeNH_3^+AsF_6^- + XeF_2 \rightarrow F_5Te{-}N(H){-}Xe^+AsF_6^- + 2HF \qquad (1.21)$$

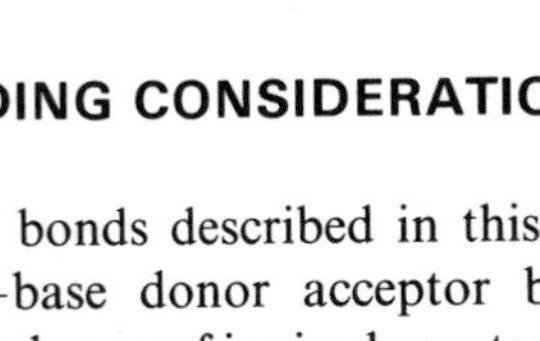

III 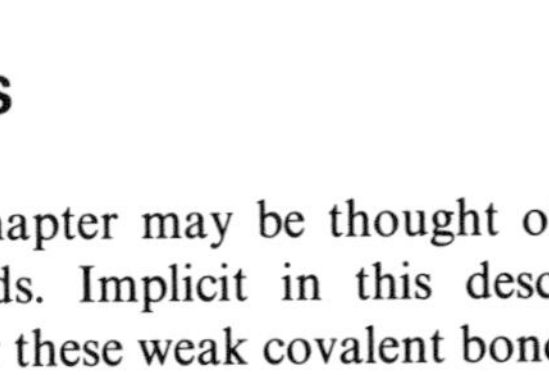 **IV**

1.4. BONDING CONSIDERATIONS

The Ng—N bonds described in this Chapter may be thought of as classical Lewis acid–base donor acceptor bonds. Implicit in this description is a considerable degree of ionic character for these weak covalent bonds, which is a dominant feature of their stability. This premise has been supported and further illuminated by several theoretical calculations on $HC{\equiv}N{-}NgF^+$ cations (Ng = Ne, Ar, Kr, and Xe) and spectroscopic measurements on a variety of XeF^+, KrF^+, and $XeOMF_5^+$ adducts of nitrogen bases in which the formal hybridization of nitrogen ranges from *sp*, sp^2 to sp^3.

1.4.1. Spectroscopic Findings

The reaction of the gas-phase NgF^+ ions with F^- to yield the difluorides results in increases in the calculated Ng—F bond lengths of 0.1 Å, while their reaction with HC≡N causes the same bond lengths (calculated) to increase on average by only 0.016 Å.[41] There is a correlation of Ng—F bond length and the Ng—F vibrational frequency with the base strength of the ligand attached to NgF^+. This is illustrated by the examples shown in Table 1.4. The NgF^+ species are only weakly coordinated by a fluorine bridge to the $Sb_2F_{11}^-$ anion, providing the highest Ng—F stretching frequencies (Xe, 619; Kr, 624 cm^{-1}) while the interaction of XeF^+ with $N(SO_2F)_2^-$ is representative of a much stronger covalent interaction and provides the lowest Xe—F stretching frequency observed to date for a covalent derivative of XeF_2 (506 cm^{-1}).[3] The C—N bond of HC≡N is calculated to shorten by ~0.05 Å on forming the adduct, while the C—H bond is calculated to lengthen by 0.008 Å.[41] These predicted changes in bond length also correlate with the observed shifts in their corresponding stretching frequencies, ν(CN), increasing by 60 cm^{-1} for HC≡N—NgF^+ (Ng = Kr and Xe) compounds and ν(C—H) decreasing by 171 cm^{-1} in the Xe adduct.[34,41]

A simple valence bond description satisfactorily accounts for qualitative trends in bond lengths and associated spectroscopic parameters. The bonding in NgF_2 and $LNgF^+$ may be presented by valence bond Structures **V**–**X**, where Structures **VII** and **X** are the least important contributing structures. The NgF_2 molecule has a formal bond order (b.o.) of $\frac{1}{2}$ and $LNgF^+$ has a formal Ng—F bond order of $\frac{1}{2} \leqslant$ b.o. < 1, approaching 1 in the most weakly coordinated case of $FNg^+ \cdots F{-}Sb_2F_{10}^-$.

$$\underset{\mathbf{V}}{F{-}Ng^+\ F^-} \leftrightarrow \underset{\mathbf{VI}}{F^-\ {}^+Ng{-}F} \leftrightarrow \underset{\mathbf{VII}}{F^-\ Ng^{2+}\ F^-}$$

$$\underset{\mathbf{VIII}}{L{-}Ng^{2+}\ F^-} \leftrightarrow \underset{\mathbf{IX}}{L\ {}^+Ng{-}F} \leftrightarrow \underset{\mathbf{X}}{L\ Ng^{2+}\ F^-}$$

Thus, as the strength of the base L increases and the Ng—L bond becomes progressively less ionic, the Ng—F bond is lengthened and weakened. However, Structure **VIII** is never dominant by virtue of the large build up of formal positive charge on the Ng atom, resulting in an Ng—F bond that is dominated by Structure **IX** and which retains the bulk of its covalent character while the Ng—L bond is only weakly covalent. The observed values of the Ng—F stretching frequencies clearly place the adducts with HC≡N and perfluoroalkyl nitriles toward the most ionic end of the scale. The same trends are reflected in the ^{19}F and ^{129}Xe chemical shifts, which are listed in Table 1.4, if one accepts the correlation of increased shielding (more negative chemical shift) with a transfer of negative charge to Xe and, hence, with increased covalent character of the Ng—L interaction (increased ionic character of the Ng—F bond).[31–33] The

$F_4S{=}N{-}Xe^+$, $F_5S{-}N(H){-}Xe^+$ and $F_5Te{-}N(H){-}Xe^+$ cations are characterized by the high shieldings of their ^{129}Xe nuclei and the absence of $^1J(^{129}Xe{-}^{19}F)$ (Tables 1.3 and 1.4). The NMR findings are consistent with a strong covalent bonding interaction between xenon and nitrogen along with a commensurate increase in Xe—F bond ionic character (essentially 100%) which may be conveniently represented by the dominant valence bond Structure **VIII** (formal charge of +1 on xenon).

Quadrupolar nuclei in noncubic environments generally yield poorly resolved one-bond coupling patterns in their NMR spectra due to quadrupolar relaxation effects.[36] In spite of the low symmetries about the nitrogen atoms in the cations investigated to date, the ^{14}N NMR spectra frequently show well resolved and only partially quadrupole collapsed scalar couplings, $^1J(^{129}Xe{-}^{14}N)$ [and $^1J(^{14}N{-}^{13}C)$ in the case of $HC{\equiv}N{-}XeF^+$; Tables 1.2, 1.3, and 1.5]. In prior studies of the imidodisulfurylfluoride derivatives of xenon(II),[3,4,6] the lack of cubic symmetry and the large electric field gradient at ^{14}N in the trigonal planar $-N(SO_2F)_2$ group necessitated ^{15}N enrichment in order to observe xenon–nitrogen scalar couplings and nitrogen chemical shifts in $FXeN(SO_2F)_2$,[3] $XeN(SO_2F)_2^+$,[6] $Xe[N(SO_2F)_2]_2$,[4] and $F[XeN(SO_2F)_2]_2^+$[4] cations in SbF_5, BrF_5, and SO_2ClF solvents. The extent of minimal quadrupolar relaxation via a reduced electric field gradient has been examined theoretically for $HC{\equiv}N{-}NgF^+$ (Ng = Kr and Xe) by calculating the field gradient at the nitrogen nucleus and comparing it to that in isolated $HC{\equiv}N$.[41] Relative to free $HC{\equiv}N$, the principal component of the electric field gradient tensor in the $H \rightarrow N$ direction, is halved upon formation of both $HC{\equiv}N{-}XeF^+$ and $HC{\equiv}N{-}KrF^+$ and is in agreement with the fact that $^1J(^{129}Xe{-}^{14}N)$ can be readily observed under conditions favoring low molecular correlation times. The electric field gradients at the Xe nucleus in XeF^+ and $HC{\equiv}N{-}XeF^+$ have also been calculated,[41] and differ by less than 7%, a result in agreement with the experimental observation that the quadrupole splitting observed in the ^{129}Xe Mössbauer spectrum of $HC{\equiv}N{-}XeF^+$ (40.2 ± 0.3 mm s^{-1}) is, within experimental error, the same as that obtained for the salt $XeF^+AsF_6^-$ (40.5 ± 0.1 mm s^{-1}).[60]

The difference in the magnitudes of $^1J(^{129}Xe{-}^{15}N)$ in $HC{\equiv}N{-}XeF^+$ (471 Hz)[34] and $FXeN(SO_2F)_2$ (307 Hz)[3] may be discussed using a previous assessment of the nature of bonding to xenon in solution for $FXeN(SO_2F)_2$ where the Xe—N bond of $FXeN(SO_2F)_2$ is regarded as a σ bond having sp^2 hybrid character.[3,4,6] In high-resolution NMR spectroscopy spin–spin coupling involving heavy nuclides is, with few exceptions, dominated by the Fermi contact mechanism.[61] For Xe—N spin–spin couplings dominated by the Fermi contact mechanism, one-bond coupling constants can be discussed on the basis of the formalism developed by Pople and Santry.[62] In discussions of xenon–nitrogen scalar couplings in xenon(II) imidodisulfurylfluoride compounds,[6] the *s* electron density at the xenon nucleus was assumed constant and a change in the hybridization at nitrogen accounted for changes in xenon–nitrogen spin–spin coupling. Even though the Xe—N bond is markedly less covalent in the hydrocyano cation than in $FXeN(SO_2F)_2$, the $^1J(^{129}Xe—^{15}N)$ values observed for $HC{\equiv}N—XeF^+$ are substantially larger than for xenon bonded to the trigonal planar nitrogen in $FXeN(SO_2F)_2$ (Table 1.3). The dependence of xenon–nitrogen spin–spin coupling on nitrogen *s* character in the xenon–nitrogen bonds may also be invoked to account for the relative magnitudes of the scalar couplings. A comparison of $^1J(^{129}Xe—^{15}N)$ for $HC{\equiv}N—XeF^+$ with that of the trigonal planar sp^2 hybridized nitrogen atom in $FXeN(SO_2F)_2$ illustrates this point and allows assessment of the relative degrees of hybridization of the nitrogen orbitals used in σ bonding to xenon. The ratio, $[^1J(^{129}Xe—^{15}N)_{sp}]/[^1J(^{129}Xe—^{15}N)_{sp^2}] = 1.53$, for the $HC{\equiv}N—XeF^+$ cation and $FXeN(SO_2F)_2$ is in excellent agreement with the theoretical ratio, 1.50, calculated from the predicted fractional *s* characters of the nitrogen orbitals used in bonding to xenon in both compounds.

The formal *s* character in the Xe—N bond of other adduct cations, for which either $^1J(^{129}Xe—^{14}N)$ or $^1J(^{129}Xe—^{15}N)$ are known, can be related to the reduced coupling constant, $^1K(Xe—N)$, thus eliminating the effect of the different gyromagnetic ratios of ^{14}N and ^{15}N on the magnitude of the scalar coupling.[6,62] A comparison of $^1K(Xe—N)$ of XeF^+ adduct cations with nitrogen bases and of xenon(II) derivatives containing the $N(SO_2F)_2$ group is given in Table 1.3. In general, $^1K(Xe—N)$ values occur in two discrete ranges corresponding to the formal hybridization of nitrogen, *sp* or sp^2. There are presently no measured values for XeF^+ coordinated to an sp^3 example, even although the $F_5S—N(H)—Xe^+$ and $F_5Te—N(H)—Xe^+$ cations are known (see Structure **IV** and Table 1.3).

1.4.2. Theoretical Calculations

In order to better characterize the nature of the Ng—N bonds in $HC{\equiv}N—KrF^+$ and $HC{\equiv}N—XeF^+$, their syntheses and spectroscopic measurements have been complemented by several theoretical calculations on the $HC{\equiv}N—NgF^+$ (Ng = Ne, Ar, Kr, or Xe) cations.[41,63–66]

Theoretical investigation of $HC{\equiv}N—KrF^+$ and $HC{\equiv}N—XeF^+$ at the SCF

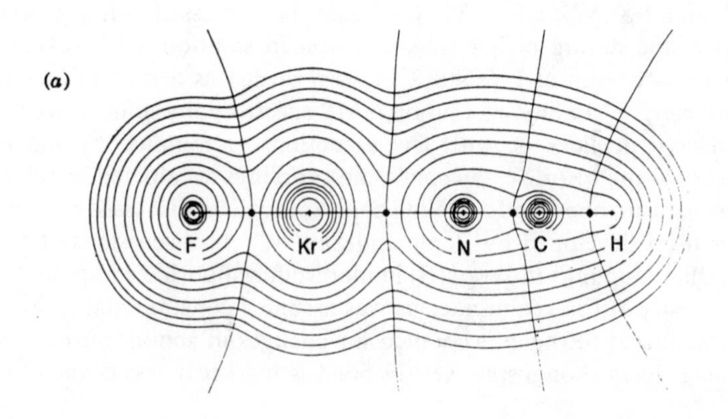

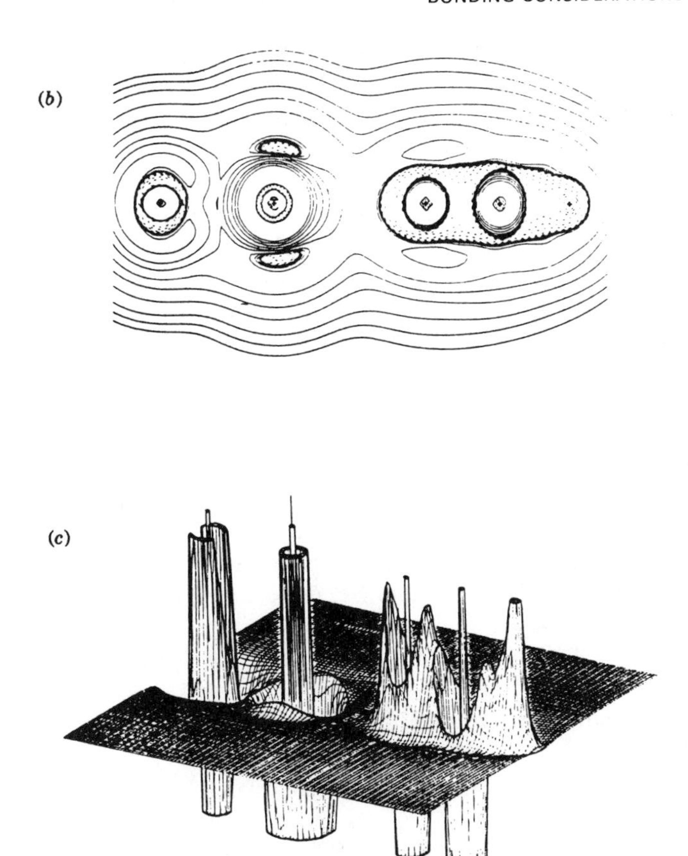

Figure 1.3. (*a*) Contour map of the charge density in the adduct HC≡N—KrF$^+$ showing the bond paths and the intersection of the interatomic surfaces. Bond critical points are denoted by black circles. Note the near planarity of the Kr—N interatomic surface, which is also a characteristic of a hydrogen bond. The outer contour value is 0.001 au. The remaining contours increase in value in the order 2×10^n, 4×10^n, 8×10^n with n starting at -3 and increasing in steps of 1 to give a maximum contour value of 20. (*b*) Contour map of the Laplacian distribution for HC≡N—KrF$^+$. The positions of the nuclei are the same as in part *a*. Solid contours denote positive and dashed lines denote negative values of $\nabla^2\rho$. The magnitudes of the contour values are as in part *a*, without the initial value of 0.001 au. (*c*) A relief map of $-\nabla^2\rho$. A maximum in the relief map is a maximum in charge concentration. If the inner spikelike feature at its nucleus is counted, the Kr atom exhibits four alternating regions of charge concentration and charge depletion corresponding to the presence of four quantum shells. Note the absence of a lip on the N side of the Kr VSCC demonstrating the presence of a hole in its outer sphere of charge concentration. Contrast the localized, atomiclike nature of the Laplacian distribution for the F and Kr atoms with the continuous valence shell of charge concentration enveloping all three of the nuclei in HC≡N. [Reprinted with permission from P. J. MacDougall, G. J. Schrobilgen, and R. F. W. Bader, *Inorg. Chem.*, **28**, 763 (1989). Copyright © 1989 from the American Chemical Society.

level using the theory of atoms in molecules[41] has shown that the ability of KrF^+ and XeF^+ cations to act as Lewis acids is related to the presence of holes in the valence shell charge concentrations of the noble gas atoms that expose their cores. The mechanism of formation of the Ng—N bonds in the adducts of KrF^+ and XeF^+ with HC≡N is similar to the formation of a hydrogen bond, that is, the mutual penetration of the outer diffuse nonbonded densities of the Ng and N atoms is facilitated by their dipolar and quadrupolar polarizations, which remove density along their axis of approach, to yield a final density in the interatomic surface that is only slightly greater than the sum of the unperturbed densities (Fig. 1.3 and 1.4). Thus, not surprisingly, the KrF^+ and XeF^+ cations are best described as hard acids. The energies of formation of these adducts are dominated by the large stabilizations of the Ng atoms that result from the increase in the concentration of charge in their inner quantum shells. The Ng—N bonds that result from the interaction of the closed-shell reactants KrF^+/XeF^+ and HC≡N lie closer to the closed shell limit than do bonds formed in the reaction of KrF^+/XeF^+ with F^-. The calculated gas phase energies of the reaction between the closed-shell species are −136.0 and −144.3 kJ mol^{-1} for Ng = Kr and Xe, respectively, for

$$NgF^+ + HC{\equiv}N \rightarrow HC{\equiv}N{-}NgF^+ \tag{1.22}$$

and −874.5 and −886.6 kJ mol^{-1} for

$$NgF^+ + F^- \rightarrow F{-}Ng{-}F \tag{1.23}$$

The molecular structure and force field of HC≡N—KrF^+ have also been

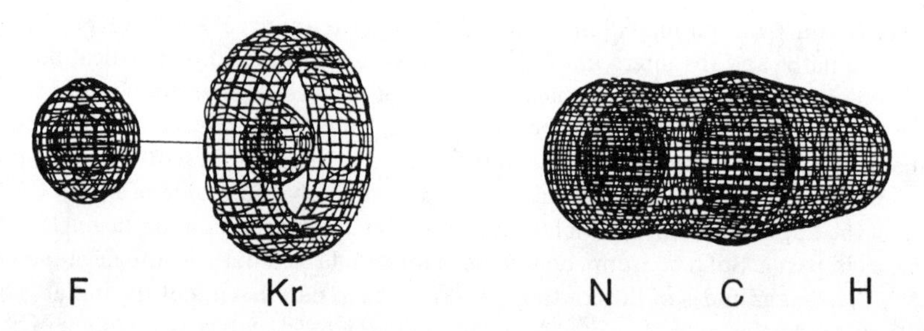

Figure 1.4. Zero envelopes of the Laplacian distributions ($\nabla^2\rho = 0$ for all points on the surface) for isolated KrF^+ and HC≡N shown aligned for adduct formation and shown to the same scale. The separation between the Kr and N nuclei is 5.0 Å. All that remains of the VSCC of the Kr atom is a belt of charge concentration, and the diagram clearly illustrates the exposure of its penultimate spherical shell of charge concentration—the core of the krypton atom. The diagram also contrasts the interatomic nature of the charge concentration in HCN with its pronounced intraatomic form in KrF^+. [Reprinted with permission from P. J. MacDougall, G. J. Schrobilgen, and R. F. W. Bader, *Inorg. Chem.*, **28**, 763 (1989). Copyright ©, 1989 from the American Chemical Society.

calculated at higher levels of theory (Tables 1.7 and 1.8 and Refs. 63–66). The incorporation of electron correlation is necessary to describe satisfactorily the structures, stabilities, and vibrational frequencies. The force field has been calculated, and the cation is predicted to be linear. Reasonable agreement with the two observed vibrational Raman bands in the solid has been found in all four sets of calculations. The stretching mode leading to dissociation into KrF^+ and HC≡N was calculated to be near 200 cm^{-1}.[66] The KrF and HC≡N fragments in the complex have geometries similar to those for the isolated molecules (Table 1.7). The Kr—N bond distance is predicted to be relatively long, 2.18–2.32 Å (cf. sum of Kr and N van der Waals radii; 3.4 Å) and the Kr—F distance in the adduct is predicted to increase by only 0.01–0.13 Å relative to that of KrF^+. The HC≡N—KrF^+ cation is predicted to be bound by 130.1–176.1 kJ mol^{-1} with respect to KrF^+ and HC≡N at higher levels of theory (cf. 126.4–136.0 kJ mol^{-1} at SCF level) with zero-point energy corrections.[63–66]

All four of the calculated results are consistent with an ionic and a covalent component in the bonding of KrF^+ with HC≡N, that is, the fragments are σ bonded with a high degree of ionic character. The Kr—N bond is the result of a donor–acceptor interaction between the *sp*-lone pair on N and the empty σ^* orbital on KrF^+. In the study by Dixon and Arduengo,[66] the molecular orbitals and bonding are analyzed in terms of hypervalent bonds. Their results indicate that the bonding between Kr and N is not a simple covalent σ bond but is best described in terms of three-center and four-center hypervalent bonds, where both types of hypervalent bonds are needed to describe the covalent σ bonding in HC≡N—KrF^+. The charge distribution for HC≡N—KrF^+ shows some transfer of electronic charge from the carbon to the KrF^+ region, consistent with a contribution from the resonance structure H—$^+$C=N—Kr—F.

High-level *ab initio* calculations[65] also predict that the HC≡N—ArF^+ cation is a stable argon–nitrogen bonded species with a heat of association, −160 kJ mol^{-1}, corresponding to Eq. 1.22, which is comparable to that of the

TABLE 1.7. Bond Distances (Å) for Optimized Geometries for KrF^+, HC≡N, and HC≡N—KrF^+.

Species	*r*(C—H)	*r*(C≡N)	*r*(Kr—F)	*r*(Kr—N)	Reference
KrF^+			1.697		66
			1.725		64
			1.752		68
HC≡N	1.057	1.126			66
	1.064	1.156			69
HC≡N—KrF^+	1.065	1.122	1.709	2.320	66
	1.068	1.128	1.748	2.307	41
	1.067	1.168	1.707	2.313	64
	1.073	1.129	1.772	2.183	64
	1.076	1.175	1.83	2.281	63

TABLE 1.8. Calculated Vibrational Frequencies (cm^{-1}) and IR Intensities ($km\ mol^{-1}$) for $HC{\equiv}N{-}KrF^+$.

Vibrational Assignment	ν(calc) HF[a]	ν(scaled) HF[a]	ν(calc) HF[b]	ν(calc) MP—2d[b]	ν(calc) MP—2[c]	ν(expt)[d]	I(calc)[a]
C—H stretch	3556	3253	3588	3467	3446		133
C≡N stretch	2446	2128	2461	2130	2100	2158	221
H—C≡N bend	935	760	940	762	752		43
Kr—F stretch	782	620	782	644	562	560	66
F—Kr—N≡C bend	253	228	256	271	244		63
Kr—N stretch	217	195	222	287	267		72
F—Kr—N≡C bend	115	104	120	116	117		28

[a]Reference 66.
[b]Reference 64.
[c]Reference 63.
[d]Reference 8.

observed krypton analog (-157 kJ mol^{-1}). These results, together with another recent high-level theoretical calculation, which estimates the electron affinity for ArF^+ to be 13.66 eV,[11] suggest that HC≡N, with its high first ionization potential (13.80 eV),[14] may be oxidatively resistant enough to withstand the formidable electron affinity of the ArF^+ cation. The synthesis of the HC≡N—ArF^+ cation is likely to require the ArF^+ cation, which is expected to be an oxidizer of unprecedented strength, if and when it is synthesized. Reactions analogous to those given in Eq. 1.3 and 1.5 involving ArF_2 are not feasible, as ArF_2 is predicted to be unbound.[11] The synthesis of ArF^+ is in itself, formidable and will entail the use of a counteranion that is capable of withstanding its high electron affinity.[11]

At the correlated level, the HC≡N—NeF^+ cation[65] is found to be an unstable species, dissociating spontaneously to HC≡N—Ne^+ + F, where the HC≡N—Ne^+ fragment is itself a weakly bound species having a binding energy of only 6 kJ mol^{-1}.

1.5. EPILOGUE

The gap resulting from the syntheses of the first examples of Kr—N bonds[8,9] and the previous existence of several examples of Kr—F bonds (KrF_2, KrF^+, and $F(KrF)_2^+$ (Ref. 42)) has prompted the reinvestigation of the possibility of forming the first Kr—O bonded species.[67] Using ^{19}F and ^{17}O NMR spectroscopy of ^{17}O enriched samples, it has been possible to document the formation of the first Kr—O bonds by the synthesis and decomposition of thermally unstable $Kr(OTeF_5)_2$ according to Eq. 1.24 and 1.25.

$$3KrF_2 + 2B(OTeF_5)_3 \rightarrow 3Kr(OTeF_5)_2 + 2BF_3 \qquad (1.24)$$

$$Kr(OTeF_5)_2 \rightarrow Kr + F_5TeOOTeF_5 \qquad (1.25)$$

The syntheses of the first examples of xenon bonded to aromatic rings; *n*-$CF_3C_5F_4N$—XeF^+ ($n = 2$, 3, and 4),[37,39] C_5F_5N—XeF^+,[39] C_5F_5N—Xe—$OTeF_5^+$,[57] *s*-$C_3F_3N_2N$—XeF^+,[9] and *s*-$C_3F_3N_2N$—Xe—$OTeF_5^+$[57] have recently been complemented by the first documented case of xenon bonded to an aromatic ring through carbon, namely, the $C_6F_5Xe^+$ cation.[70–72] While the aryl group of the $C_6F_5Xe^+$ cation serves to reduce the electron density at Xe^{II}, as evidenced from the extreme low-frequency position of the ^{129}Xe chemical shift (-3769 ppm relative to neat $XeOF_4$) in CH_3CN solvent, the ^{129}Xe shieldings in the *s*-trifluorotriazine and perfluoropyridine derivatives are less than in the carbon analogue and may in large part be associated with the electron-withdrawing character of the fluorine on xenon. The electron-donating character of the nitrogen heterocycles relative to HC≡N and RC≡N in the $XeOTeF_5^+$ and XeF^+ adducts is, however, apparent from the higher ^{129}Xe shieldings in the former (Tables 1.3 and 1.4). The crystal structure of the $C_6F_5Xe^+(C_6F_5)_2BF_2^-$

shows the Xe—C distance [2.092(8) Å][72] is comparable to the I—C distance in $C_6F_5I(O_2CC_6F_5)_2$ [2.072(4) Å].[73] Coordination of $CH_3C{\equiv}N$ with the $C_5F_5Xe^+$ cation serves to lower the effective positive charge at xenon by coordination to the nitrogen of $CH_3C{\equiv}N$ to give an Xe ··· N contact of 2.681 Å that is significantly shorter than the sum of the van der Waals radii for Xe and N (3.6 Å) and substantially longer than in the $Xe{-}N(SO_2F)_2^+$ cation [2.02(1) Å],[6] and $FXe{-}N(SO_2F)_2$ [2.200(3) Å].[3] The Xe—N bond in $C_5F_5Xe{-}N{\equiv}CCH_3^+$ is clearly weakly covalent and can be described in terms very similar to those used to describe the XeL^+ adducts considered in this Chapter.

ACKNOWLEDGMENTS

The United States Air Force Astronautics Laboratory, Edwards Air Force Base, California (Contract F49620-87-C-0049), and the Natural Sciences and Engineering Research Council of Canada are gratefully acknowledged for their financial support.

REFERENCES

1. Bartlett, N. and Sladky, F. O., in *Comprehensive Inorganic Chemistry*, Pergamon, New York, 1973, Vol. 1, Chapt. 6, pp. 213–330.
2. LeBlond, R. D. and DesMarteau, D. D., *J. Chem. Soc. Chem. Commun.*, 555 (1974).
3. Sawyer, J. F., Schrobilgen, G. J., and Sutherland, S. J., *Inorg. Chem.*, **21**, 4064 (1982).
4. Schumacher, G. A. and Schrobilgen, G. J., *Inorg. Chem.*, **22**, 2178 (1983).
5. DesMarteau, D. D., LeBlond, R. D., Hossain, S. F., and Nöthe, D., *J. Am. Chem. Soc.*, **103**, 7734 (1981).
6. Faggiani,R., Kennepohl,D. K., Lock, C. J. L., and Schrobilgen, G. J., *Inorg. Chem.*, **25**, 563 (1986).
7. Foropoulos, Jr. J. and DesMarteau, D. D., *J. Am. Chem. Soc.*, **104**, 4260 (1982).
8. Schrobilgen, G. J., *J. Am. Chem. Soc. Chem. Commun.*, 863 (1988).
9. Schrobilgen, G. J., *J. Am. Chem. Soc. Chem. Commun.*, 1506 (1988).
10. Selig, H. and Holloway, J. H., *Top. Curr. Chem.*, **124**, 33 (1984).
11. Frenking, G., Koch, W., Deakyne, C. A., Liebman, J. F., and Bartlett, N., *J. Am. Chem. Soc.*, **111**, 31 (1989).
12. $EA(KrF^+) = IP(Kr) + BE(KrF^\cdot) - BE(KrF^+) = 14.0 + 0.8 - 1.6 = 13.2\,eV$.
13. $EA(XeF^+) = IP(Xe) + BE(XeF^\cdot) - BE(XeF^+) = 12.1 + 0.86 - 2.1 = 10.9\,eV$.
14. V. H. Dibeler and S. K. Liston, *J. Chem. Phys.*, **48**, 4765 (1968).
15. Bock, H., Dammel, R., and Lentz, D., *Inorg. Chem.*, **23**, 1535 (1984).
16. Field, F. H. and Franklin, J. L., in *Electron Impact Phenomena and the Properties of Gaseous Ions*, Academic: New York, 1957, Chapt. 4, p. 113.
17. Herron, J. T. and Dibeler, V. H., *J. Res. Natl. Bur. Standards*, **65** A, 405 (1961).
18. van der Kelen, G. P. and DeBièvre, P. J., *Bull. Soc. Chim. Belg.*, **69**, 379 (1960).

19. Beach, D. B., Jolly, W. L., Mews, R., and Waterfeld, A., *Inorg. Chem.*, **23**, 4080 (1984).
20. Herron, J. T. and Dibeler, V. H., *J. Am. Chem. Soc.*, **82**, 1555 (1960).
21. Rider, D. M., Ray, G. W., Darland, E. J., and Leroi, G. E., *J. Chem. Phys.*, **74**, 1652 (1981).
22. Herron, J. T. and Dibeler, V. H., *J. Chem. Phys.*, **33**, 1595 (1960).
23. Levitt, B. W., Widing, H. F., and Levitt, L. S., *Chem. Ind.*, 793 (1973).
24. Neuert, H., *Z. Naturforsch.*, **7** A, 293 (1952).
25. Brundle, C. R., Robin, M. B., and Kuebler, N. A., *J. Am. Chem. Soc.*, **94**, 1466 (1972).
26. Dibeler, V. H., Reese, R. M., and Franklin, J. L., *J. Am. Chem. Soc.*, **83**, 1813 (1961).
27. Rosmus, P., Stafast, H., and Bock, H., *Chem. Phys. Lett.*, **34**, 275 (1975).
28. Prasad, S. R. and Singh, A. N., *Indian J. Phys.*, **59** B, 1 (1985).
29. Collin, J., *Can. J. Chem.*, **37**, 1053 (1959).
30. Omura, I., Baba, H., Higasi, K., and Kanaoka, Y., *Bull. Chem. Soc. Jpn.*, **30**, 633 (1957).
31. Schrobilgen, G. J., Holloway, J. H., Granger, P. and Brevard, C., *Inorg. Chem.*, **17**, 980 (1978).
32. Schrobilgen, G. J., in *NMR and the Periodic Table*, Harris, R. K. and Mann, B. E. (Eds.); Academic: London, 1978, Chapt. 14, pp. 439–454.
33. Jameson, C. J., in *Multinuclear NMR*, Mason, J. (Ed.); Plenum: New York, 1987, Chapt. 18, pp. 463–475.
34. Emara, A. A. A. and Schrobilgen, G. J., *J. Chem. Soc. Chem. Commun.*, 1646 (1987).
35. Emara, A. A. A. and Schrobilgen, G. J., *Inorg. Chem.*, in press.
36. Sanders, J. C. P. and Schrobilgen, G. J., in *A Methodological Approach to Multinuclear in Liquids and Solids—Chemical Applications*; NATO Advanced Study Institute, Magnetic Resonance; Granger, P. and Harris, R. K. (Eds.), Kluwer Academic Publishers, Dordrecht, pp. 157–186.
37. Emara, A. A. A. and Schrobilgen, G. J. unpublished work.
38. Arner, N. T., Sanders, J. C. P., Schrobilgen, G. J., and Thrasher, J. S., Proceedings of the Fourth United States Air Force High-Energy Density Matter (HEDM) Conference, Long Beach, California, February 25–28, 1990.
39. Emara, A. A. A. and Schrobilgen, G. J., *J. Chem. Soc. Chem. Commun.*, 257 (1988).
40. Sanders, J. C. P., Schrobilgen, G. J., and Whalen, J. M., Proceedings of the Fifth United States Air Force High-Energy Density Matter (HEDM) Conference, Albuquerque, New Mexico, February 24–27, 1991.
41. MacDougall, P. J., Schrobilgen, G. J., and Bader, R. F. W., *Inorg. Chem.*, **28**, 763 (1989).
42. Gillespie, R. J. and Schrobilgen, G. J., *Inorg. Chem.*, **15**, 22 (1976).
43. Murchison, C., Reichman, S., Anderson, D., Overend, J., and Schreiner, F., *J. Am. Chem. Soc.*, **90**, 5680 (1968).
44. Claassen, H. H., Goodman, G. L., Malm, J. G., and Schreiner, F., *J. Chem. Phys.*, **42**, 1229 (1965).
45. Burgess, J., Fraser, C. J. W., McRae, V. M., Peacock, R. D., and Russell, D. R., *J. Inorg. Nucl. Chem., Suppl.*, 183 (1976).
46. Gillespie, R. J., Netzer, A., and Schrobilgen, G. J., *Inorg. Chem.*, **13**, 1455 (1974).

47. Schrobilgen, G. J., unpublished work.
48. Zalkin, A., Ward, D. L., Biagioni, R. N., and Templeton, D. H., *Inorg. Chem.*, **17**, 1318 (1978).
49. Bartlett, N., DeBoer, B. G., Hollander, F. J., Sladky, F. O., Templeton, D. H. and Zalkin, A., *Inorg. Chem.*, **12**, 780 (1974).
50. Bartlett, N., Wechsberg, M., Jones, G. R., and Burbank, R. D., *Inorg. Chem.*, **11**, 1124 (1972).
51. Landa, B. and Gillespie, R. J., *Inorg. Chem.*, **12**, 1383 (1973).
52. Syvret, R. G. and Schrobilgen, G. J., *Inorg. Chem.*, **28**, 1564 (1989).
53. Birchall, T., Myers, R. D., deWaard, H., and Schrobilgen, G. J., *Inorg. Chem.*, **21**, 1068 (1982).
54. Sanders, J. C. P. and Schrobilgen, G. J., unpublished work.
55. Reichman, S. and Schreiner, F., *J. Chem. Phys.*, **51**, 2355 (1969).
56. Agron, P. A., Begun, G. M., Levy, H. A., Mason, A. A., Jones, G., and Smith, D. F., *Science*, **139**, 842 (1963).
57. Emara, A. A. A., Hutchinson, D., Paprica, A., Sanders, J. C. P., and Schrobilgen, G. J., Proceedings of the Third United States Air Force High-Energy Density Materials (HEDM) Conference, New Orleans, Louisiana, March 12–15, 1989.
58. (a) Dean, P. A. W. and Gillespie, R. J., *J. Am. Chem. Soc.*, **91**, 7260 (1969). (b) Brownstein, M. and Gillespie, R. J., *J. Am. Chem. Soc.*, **92**, 2718 (1970).
59. Collins, M. J. and Schrobilgen, G. J., *Inorg. Chem.*, **24**, 2608 (1985).
60. Valsdóttir, J., M.Sc. Thesis, McMaster University, 1990.
61. (a) Jameson, C. J., in *Multinuclear NMR*; Mason, J. J., (Ed.); Plenum: New York, 1987; Chap. 4, pp. 116–118. (b) Jameson, C. J. and Gutowsky, H. S., *J. Chem. Phys.*, **51**, 2790 (1969). (c) Kunz, R. W., *Helv. Chim. Acta*, **63**, 2054 (1980). (d) Mason, J., *J. Polyhedron*, **8**, 1657 (1989). (e) Wrackmeyer, B. and Horchler, K., in *Annual Reports on NMR Spectroscopy*; Webb, G. A. (Ed.); Academic: London, 1989; Vol. 22, p. 261.
62. Pople, J. A. and Santry, D. P., *Mol. Phys.*, **8**, 1 (1964).
63. Hillier, I. H. and Vincent, M. A., *J. Chem. Soc. Chem. Commun.*, 30 (1989).
64. Koch, W., *J. Chem. Soc. Chem. Commun.*, 215 (1989).
65. Wong, M. W. and Radom, L., *J. Chem. Soc. Chem. Commun.*, 719 (1989).
66. Dixon, D. A. and Arduengo, A. J., *Inorg. Chem.*, **29**, 970 (1990).
67. Sanders, J. C. P. and Schrobilgen, G. J., *J. Chem. Soc. Chem. Commun.*, 1576 (1989).
68. Liu, B. and Schaefer, H. F., III, *J. Chem. Phys.*, **55**, 2369 (1971).
69. Harmony, M. D., Laurie, V. W., Kuczkowski, R. L., Schwendeman, R. H., Ramsay, D. A., Lovas, F. J., Lafferty, W. J., and Maki, A. J., *J. Phys. Chem. Ref. Data*, **8**, 619 (1979).
70. Naumann, D. and Tyrra, W., *J. Chem. Soc. Chem. Commun.*, 47 (1989).
71. Frohn, H. J. and Jakobs, S., *J. Chem. Soc. Chem. Commun.*, 625 (1989).
72. Frohn, H. J., Jakobs, S., and Henkel, G., *Angew. Chem. Int. Ed. Engl.*, **28**, 1506 (1989).
73. Frohn, H. J., Helber, J., and Richter, A., *Chem. Ztg.*, **107**, 169 (1983).

CHAPTER 2

Controlled Replacement of Fluorine by Oxygen in Fluorides and Oxyfluorides

KARL O. CHRISTE, WILLIAM W. WILSON, and CARL J. SCHACK

2.1. INTRODUCTION

Fluorine–oxygen exchange reactions play an important role in synthetic chemistry. Although numerous methods and reagents have been described for these exchange reactions, the emphasis of these studies has been almost exclusively on the selective replacement of oxygen by fluorine. For example, SF_4[1] and its derivatives, such as $SF_3N(CH_3)_2$,[2] have been widely used to convert carbonyl groups to CF_2 groups, and inorganic oxides can be transformed into fluorides by reagents such as HF, F_2, or halogen fluorides.[3] However, much less attention has been paid to the opposite reaction, that is, the conversion of fluorides to oxyfluorides. This is not surprising because generally

Synthetic Fluorine Chemistry,
Edited by George A. Olah, Richard D. Chambers, and G. K. Surya Prakash.
ISBN 0-471-54370-5

oxides are more readily preparable than fluorides, and many fluorides undergo facile hydrolysis to the corresponding oxyfluorides and oxides.

For the replacement of fluorine by oxygen, hydrolysis is the most frequently used method. For highly fluorinated compounds of the more electronegative elements, however, these hydrolysis reactions often present significant experimental difficulties, particularly when a controlled and stepwise replacement of fluorine by oxygen is desired. The hydrolysis reactions of these compounds are often violent, as found for XeF_6[4,5] or ClF_3,[6] and require careful moderation. Thus, SiO_2 combined with a trace of HF can be used for the slow formation of water (Eq. 2.1), followed by a continuous regeneration of the HF during the hydrolysis of the fluoride starting material (Eq. 2.2). This approach has been demonstrated previously for compounds such as IF_7.[7-10]

$$SiO_2 + 4HF \rightarrow SiF_4 + 2H_2O \tag{2.1}$$

$$IF_7 + H_2O \rightarrow IOF_5 + 2HF \tag{2.2}$$

Another approach to moderate otherwise violent or uncontrollable hydrolysis reactions involves the use of suitable solvents, such as HF, and of stoichiometric amounts of water, as reported for XeF_6 [5](Eq. 2.3).

$$XeF_6 + H_2O \xrightarrow{HF} XeOF_4 + 2HF \tag{2.3}$$

In spite of the above improvements in the techniques of hydrolyzing highly reactive fluorides, these reactions remain experimentally challenging and often are dangerous[5] and difficult to scale up. Consequently, alternate reagents that allow the safe, easily controllable, and stepwise replacement of fluorine by oxygen, are highly desirable.

Previously investigated examples for such alternate reagents include SiF_3OSiF_3, SeO_2F_2, POF_3, and several oxides. Most of these alternate reagents exhibit drawbacks. Thus, SiF_3OSiF_3 reacted with XeF_6 (Eq. 2.4),

$$XeF_6 + SiF_3OSiF_3 \rightarrow XeOF_4 + 2SiF_4 \tag{2.4}$$

but did not work for IF_7, BrF_5, and so on.[11] Similarly, the highly toxic SeO_2F_2 was demonstrated only for XeF_6 (Eq. 2.5).[12]

$$XeF_6 + SeO_2F_2 \rightarrow XeOF_4 + SeOF_4 \tag{2.5}$$

The most versatile of these alternate reagents appears to be POF_3, which reacted with XeF_6 (Eq. 2.6),[13] UF_6 (Eq. 2.7),[14] ClF_5,[14] and IF_7 (Eq. 2.8).[15]

$$XeF_6 + POF_3 \rightarrow XeOF_4 + PF_5 \tag{2.6}$$

$$UF_6 + POF_3 \rightarrow UOF_4 + PF_5 \tag{2.7}$$

$$IF_7 + POF_3 \rightarrow IOF_5 + PF_5 \tag{2.8}$$

Most of the previously reported, oxide based fluorine–oxygen exchange reactions, such as (Eqs. 2.9,[10] 2.10,[16] or 2.11[17]

$$IF_7 + I_2O_5 \rightarrow IOF_5 + 2IO_2F \quad (2.9)$$

$$4ClF_3 + 6NaClO_3 \rightarrow 6ClO_2F + 6NaF + 2Cl_2 + 3O_2 \quad (2.10)$$

$$2BrF_5 + CsIO_4 \rightarrow 2BrOF_3 + CsIO_2F_4 \quad (2.11)$$

represent useful syntheses for specific compounds, but the exchange reagents have not been studied systematically.

Several years ago, while studying the compatibility of the nitrate and sulfate anions with various halogen fluorides,[18] we surprisingly found that these anions are excellent, general reagents for fluorine–oxygen exchange. Since then, we have systematically investigated the scope of these reactions and we present a summary of our results in this chapter.

2.2. REACTIONS OF THE NITRATE ANION

2.2.1. Xenon(VI) Fluoride and Oxyfluorides

Xenon hexafluoride is an ideal test case for the general usefulness of a fluorine–oxygen exchange reagent since it can undergo stepwise fluorine replacement (Eq. 2.12). When preparing $XeOF_4$ and XeO_2F_2,

$$XeF_6 \rightarrow XeOF_4 \rightarrow XeO_2F_2 \rightarrow XeO_3 \quad (2.12)$$

precise control of the stepwise exchange is of utmost importance because of the shock sensitivity of the potential by-product XeO_3. Other important aspects, besides high yields and ready availability of the exchange reagent, are the ease of product separation and mild reaction conditions to avoid product decomposition.

In our studies[19] it was found that $NaNO_3$ is best suited for the conversion of XeF_6 to $XeOF_4$ (Eq. 2.13).

$$XeF_6 + NaNO_3 \xrightarrow[10h]{70^\circ C} XeOF_4 + NaF + NO_2F \quad (2.13)$$

The formation of XeO_2F_2 can be suppressed by the use of a moderate excess of XeF_6. The excess of XeF_6 is readily separable from the desired $XeOF_4$ because at the reaction temperature it forms stable $NaXeF_7$ and Na_2XeF_8 salts with the NaF by-product. The only other volatile by-product is NO_2F, which is much more volatile than $XeOF_4$ and can be separated easily from the $XeOF_4$ by fractional condensation through traps kept at -78 and $-196^\circ C$. The yields of $XeOF_4$ are about 80% based on the limiting reagent $NaNO_3$. The use of other alkali metal nitrates is less desirable. In the case of $CsNO_3$ the resulting CsF complexes $XeOF_4$ with formation of $CsXeOF_5$[20-22] and for $LiNO_3$ the resulting LiF does not complex any unreacted XeF_6 starting material.

For the conversion of $XeOF_4$ to XeO_2F_2, the use of alkali metal nitrates is possible but, due to the relative involatility of XeO_2F_2 and its ease of forming stable $XeO_2F_3^-$ salts, N_2O_5 is the preferred reagent.[23] In the solid state, N_2O_5 has the ionic structure $NO_2^+NO_3^-$[24,25] and reacts with $XeOF_4$ according to (Eq. 2.14).

$$XeOF_4 + NO_2^+NO_3^- \xrightarrow[1.5h]{25°C} XeO_2F_2 + 2NO_2F \tag{2.14}$$

In this manner and by the use of an excess of $XeOF_4$, the only product of low volatility is XeO_2F_2, thus allowing for an efficient product separation. Again, the formation of XeO_3 was suppressed by the use of a slight excess of $XeOF_4$ starting material, and the yield of XeO_2F_2 was essentially quantitative. The only minor complication in this XeO_2F_2 synthesis is the formation of an unstable $NO_2^+[XeO_2F_3 \cdot nXeO_2F_2]^-$ type adduct between NO_2F and XeO_2F_2, which requires prolonged pumping on the product at room temperature to ensure complete NO_2F removal from the XeO_2F_2.[23]

Conversion of either XeF_6, $XeOF_4$, or XeO_2F_2 to the highly explosive XeO_3 can be achieved by their reactions with excess N_2O_5.[23] However, no detailed studies were carried out on these systems due to the sensitivity of XeO_3.

2.2.2. Chlorine Fluorides and Oxyfluorides

Excess $NaNO_3$ readily reacts with ClF at subambient temperatures to give NaF and $ClONO_2$ (Eq. 2.15).[26]

$$ClF + NaNO_3 \rightarrow NaF + ClONO_2 \tag{2.15}$$

The yield of $ClONO_2$, however, was only about 75% because of the competing reaction (Eq. 2.16) which is favored by an excess of ClF.

$$ClF + ClONO_2 \rightarrow Cl_2O + NO_2F \tag{2.16}$$

With a sufficiently large excess of ClF, the overall reaction then becomes (Eq. 2.17).

$$2ClF + NaNO_3 \rightarrow Cl_2O + NaF + NO_2F \tag{2.17}$$

In addition to the NaF, $ClONO_2$, and NO_2F products, smaller amounts of Cl_2 and ClO_2F were also observed as by-products due to the side reactions (Eqs. 2.18 and 2.19), which result in the following net reaction (Eq. 2.20).

$$2ClF + 2Cl_2O \rightarrow 2[FClO] + 2Cl_2 \tag{2.18}$$

$$2[FClO] \rightarrow ClF + ClO_2F \quad (2.19)$$

$$ClF + 2Cl_2O \rightarrow ClO_2F + 2Cl_2 \quad (2.20)$$

Thus, ClF readily undergoes fluorine–oxygen exchange with the NO_3^- anion with the ratio of the major products, $ClONO_2$ and Cl_2O, depending on the stoichiometry of the reactants. Smaller amounts of Cl_2 and ClO_2F formed in this system are due to side reactions.

In the case of ClF_3, the main reaction with the NO_3^- anion is again a facile fluorine–oxygen exchange (C1. 2.21)[26]

$$ClF_3 + NaNO_3 \xrightarrow{25^\circ C} [FClO] + NaF + NO_2F \quad (2.21)$$

with the thermally unstable FClO either undergoing disproportionation (Eq. 2.22) or decomposition (Eq. 2.23).

$$2[FClO] \rightarrow ClF + ClO_2F \quad (2.22)$$

$$2[FClO] \rightarrow 2ClF + O_2 \quad (2.23)$$

The formation of ClF, ClO_2F, O_2, and NO_2F is favored by the use of an excess of ClF_3. If, however, a large excess of $NaNO_3$ is used, side reactions (Eqs. 2.15, 2.24, 2.25) are also observed.

$$NO_2F + NaNO_3 \rightarrow N_2O_5 + NaF \quad (2.24)$$

$$ClO_2F + NaNO_3 \rightarrow ClONO_2 + O_2 + NaF \quad (2.25)$$

The compound ClF_5 also reacts readily at room temperature with nitrates.[26] Even in the presence of a large excess of ClF_5, the fluorine–oxygen exchange cannot be stopped at the $ClOF_3$ stage but proceeds all the way to ClO_2F (Eq. 2.26).

$$ClF_5 + 2MNO_3 \xrightarrow{25^\circ C} ClO_2F + 2NO_2F + 2MF \quad (2.26)$$

(M = Li, Na, K, Rb, and Cs)

This is in marked contrast to BrF_5 and IF_5 (see below) and is due to the extraordinary reactivity of $ClOF_3$, which is much more reactive than ClF_5.[27] Attempts to trap the intermediately formed $ClOF_3$ as $M^+ClOF_4^-$ salts were also unsuccessful indicating that the complexation of $ClOF_3$ with MF is slower than its fluorine–oxygen exchange with the nitrate anion.

When a large excess of nitrate is used in the reaction of ClF_5 with MNO_3 (Eq. 2.26), the ClO_2F product can undergo further fluorine–oxygen exchange with NO_3^- (Eq. 2.25). This was confirmed by separate experiments between ClO_2F and either $LiNO_3$ or $NO_2^+NO_3^-$.[26]

Thus, all the chlorine fluorides and oxyfluorides, except for the highly unreactive ClO_3F,[27] undergo rapid fluorine–oxygen exchange with the nitrate anion. Due to the high reactivity of $ClOF_3$ and in contrast to BrF_5 and IF_5, a controlled single step fluorine–oxygen exchange in ClF_5 could not be realized (see note added in proof).

2.2.3. Bromine Pentafluoride

The reactions of BrF_5 with $M^+NO_3^-$ serve as excellent examples of how the nature of the products can be influenced by the appropriate choices of the M^+ cation and the reaction stoichiometries.[18,28] With an excess of BrF_5 and M being either Na, K, Rb, or Cs, the corresponding $M^+BrOF_4^-$ salts can be prepared in 70–100% yield under very mild (−30 to 25 °C) conditions (Eq. 2.27).

$$BrF_5 + MNO_3 \xrightarrow[\text{excess } BrF_5]{-30 \text{ to } 25\,^\circ C} MBrOF_4 + NO_2F \qquad (2.27)$$

(M = Na, K, Rb, and Cs)

Since lithium does not form a stable $LiBrOF_4$ salt, the reaction of $LiNO_3$ with excess BrF_5 (Eq. 2.28) can be used for a convenient synthesis of free $BrOF_3$.

$$BrF_5 + LiNO_3 \xrightarrow[\text{excess } BrF_5]{0\,^\circ C} BrOF_3 + LiF + NO_2F \qquad (2.28)$$

Since $BrOF_3$ is considerably less volatile than BrF_5, the two can be separated readily by fractional condensation or distillation.

While the use of an excess of BrF_5 results in the single-step replacement of two fluorines by one oxygen, the application of a 1 : 3 mol ratio of $BrF_5 : LiNO_3$ causes complete fluorine–oxygen exchange with $BrONO_2$ formation (Eq. 2.29).[28]

$$BrF_5 + 3LiNO_3 \xrightarrow[\text{excess } LiNO_3]{0\,^\circ C} BrONO_2 + 3LiF + O_2 + 2NO_2F \qquad (2.29)$$

If the $BrF_5 : LiNO_3$ mole ratio is further changed to 1 : 5 or greater, N_2O_5 is produced (Eq. 2.30)

$$2NO_2F + 2LiNO_3 \rightarrow NO_2^+NO_3^- + 2LiF \qquad (2.30)$$

which can react with $BrONO_2$ (Eq. 2.31).[28]

$$NO_2^+NO_3^- + BrONO_2 \geqq NO_2^+[Br(ONO_2)_2]^- \qquad (2.31)$$

2.2.4. Iodine Fluorides and Oxyfluorides

The reactions of IF and IF_3 with nitrates were not studied since IF and IF_3 are relatively unstable and easily disproportionate to I_2 and IF_5. With excess IF_5, the alkali metal nitrates undergo a controlled, single-step fluorine–oxygen exchange to form the corresponding $MIOF_4$ salts (Eq. 2.32).[29]

$$IF_5 + MNO_3 \rightarrow MIOF_4 + NO_2F \quad (2.32)$$

(M = Li, K, Cs)

However, these reactions are more sluggish than the corresponding BrF_5 reactions. As a consequence, the side reaction (Eq. 2.33)

$$NO_2F + MNO_3 \rightarrow N_2O_5 + MF \quad (2.33)$$

becomes faster than the previous reaction (Eq. 2.32) resulting in an equal consumption of MNO_3 by its reactions with IF_5 and NO_2F (Eqs. 2.32 and 2.33). Furthermore, the following reactions (Eqs. 2.34 and 2.35) also become competitive,

$$KF + IF_5 \rightarrow KIF_6 \quad (2.34)$$

$$CsF + 3IF_5 \rightarrow CsI_3F_{16} \quad (2.35)$$

and some of the N_2O_5 decomposes to $N_2O_4 + O_2$ under these conditions. Consequently, the $MIOF_4$ salts prepared in this manner, usually contain substantial amounts of IF_6^- and $I_3F_{16}^-$ salts as by-products.

From the reaction of excess IF_7 with either $LiNO_3$ or $NaNO_3$, no IOF_5 is isolated. Instead, IF_5 and O_2 are obtained (Eq. 2.36),

$$IF_7 + MNO_3 \xrightarrow{60°C} IF_5 + \tfrac{1}{2}O_2 + MF + NO_2F \quad (2.36)$$

indicative of a competing deoxygenation reaction.[29] In the case of $CsNO_3$, the same deoxygenation occurs but the CsF product reacts with IF_5 and IF_7 to give CsI_3F_{16} and $CsIF_8$, respectively. If an excess of MNO_3 is used, the IF_5 product can react further with MNO_3 and form $MIOF_4$ salts (Eq. 2.32). It was further experimentally confirmed that IOF_5 does not undergo fluorine–oxygen exchange giving IO_2F_3, but loses oxygen giving IF_5, which then undergoes fluorine–oxygen exchange with formation of IOF_4^- salts (Eq. 2.32).[29]

The fact that IF_7 readily undergoes fluorine–oxygen exchange either during controlled hydrolysis[7-10] or with POF_3,[15] but not with the NO_3^- anion remains somewhat a puzzle. It has previously been speculated[29] that this lack of fluorine–oxygen exchange in the IF_7–nitrate system might be due to either the instability of an intermediate IOF_6^- anion or the lack of a free valence electron pair on the iodine central atom of IF_7. Since then, however, we have synthesized and characterized stable IOF_6^- salts[30]; thus, the first explanation can be ruled out.

2.2.5. Carbonyl Fluoride

The nitrate anion can also exchange carbon bonded fluorine for oxygen. This was demonstrated[31] for carbonyl fluoride, COF_2 (Eq. 2.37).

$$COF_2 + MNO_3 \xrightarrow[CsF]{85°C} CO_2 + MF + NO_2F \tag{2.37}$$

(M = Li and Na)

The reactions were carried out in a steel cylinder and, in this manner, essentially quantitative yields of NO_2F are obtainable. This reaction is remarkable because it is a very rare example for the formation of a nitrogen–fluorine bond using a fluorinating agent as mild as COF_2. Furthermore, it is interesting that the heavier alkali metal nitrates, such as $CsNO_3$, do not react under these conditions with COF_2. This was explained[31] by thermochemical calculations, which show that for Li and Na the ΔH values are still favorable but become increasingly more positive for the heavier alkali metals. It should be noted that all the ΔH values given in Ref. 31 are slightly in error[32] by -11 kcal mol^{-1}, but that the general trend remains the same for the different alkali metals.

2.2.6. Mechanism of the Fluorine-Oxygen Exchange Involving Nitrates

Of the nitrate based fluorine–oxygen exchange reactions studied so far, the simplest case is that of MNO_3 and ClF, which yields MF and $ClONO_2$ (Eq. 2.15).[26] Assuming for the more highly fluorinated starting materials an analogous first reaction step (Eq. 2.38),

$$XF_n + MNO_3 \rightarrow F_{(n-1)}XONO_2 + MF \tag{2.38}$$

the formation of an intermediate $F_{(n-1)}XONO_2$ is expected. This intermediate could easily undergo an internal nucleophilic substitution (S_Ni) reaction,[18] accompanied by NO_2F elimination (Eq. 2.39).

$$\begin{array}{c} \diagdown | \\ -X-\overline{O} \\ \diagup | \quad \diagdown \\ |\underline{F}| \longrightarrow N \\ \quad O \; O \end{array} \longrightarrow \;\; \diagdown X{=}\overline{O} + F{-}NO_2 \tag{2.39}$$

Such a mechanism could account for the generally observed reaction products, MF, NO_2F, and the corresponding oxyfluoride. If the oxyfluoride end product is amphoteric and can form a stable salt with the cogenerated alkali metal fluoride, then this salt is observed as the final product (Eq. 2.40).

$$XOF_{(n-2)} + MF \rightarrow M^+[XOF_{(n-1)}]^- \quad (2.40)$$

If, as for the ClF + NO_3^- reaction (Eq. 2.15), the resulting nitrate intermediate no longer contains a fluoride ligand, NO_2F elimination becomes impossible, and the halogen nitrate becomes the final product.

All the nitrate reactions studied so far seem to follow this pattern, except for the IF_7 case where deoxygenation of the expected IOF_5 product occurred (Eq. 2.36). Since the IF_7 + MNO_3 reactions require elevated temperatures and conditions under which IOF_5 can undergo deoxygenation,[29] the latter might be a secondary reaction, and the IF_7 + MNO_3 reaction might involve the same primary steps as all the other nitrate reactions.

2.3. REACTIONS OF THE SULFATE ANION

The reactions of the sulfate anion with highly fluorinated compounds of the more electronegative elements resemble those of the nitrate anion. Again, fluorine–oxygen exchange occurs but this exchange generally stops at the SO_3F^- level (Eq. 2.41),

$$XF_n + M_2SO_4 \rightarrow XOF_{(n-2)} + MSO_3F + MF \quad (2.41)$$

and does not proceed further to the SO_2F_2 stage (Eq. 2.42).[33]

$$XF_n + MSO_3F \rightarrow XOF_{(n-2)} + SO_2F_2 + MF \quad (2.42)$$

Since MSO_3F is a nonvolatile solid, whereas NO_2F is a volatile gas, the use of M_2SO_4 may be more convenient than that of MNO_3 if the desired product is volatile but either complexes with NO_2F or is difficult to separate from it. Compared to the nitrate anion, the sulfate anion is less reactive and requires longer reaction times and/or higher temperatures. Consequently, the reactions of the sulfate anion were not studied as extensively as those of the nitrate anion and were limited to the following examples.

2.3.1. Bromine Pentafluoride

At room temperature Cs_2SO_4 readily undergoes fluorine–oxygen exchange with BrF_5 (Eq. 2.43).[33]

$$BrF_5 + Cs_2SO_4 \xrightarrow[\text{1 day}]{25^\circ C} CsBrOF_4 + CsSO_3F \quad (2.43)$$

Even with an 80-fold excess of BrF_5, the fluorine–oxygen exchange did not proceed past the $CsSO_3F$ stage. Attempts were made to use this reaction for the synthesis of free $BrOF_3$ by the replacement of Cs_2SO_4 with Li_2SO_4 since lithium does not form a stable $BrOF_4^-$ salt. Under conditions (0 °C, 1 day), which worked well for $LiNO_3$[28], no reaction was observed for Li_2SO_4. This shows that the SO_4^{2-} anion is less reactive than NO_3^-.

2.3.2. Iodine Fluorides

Reaction temperatures in excess of 70 °C were required to initiate a slow reaction between IF_7 and Li_2SO_4. Even at this temperature, the conversion of the Li_2SO_4 was only about 6–7%. As in the case of NO_3^-, deoxygenation of the IOF_5 occurred and IF_5 and O_2 were the observed products (Eq. 2.44).[34]

$$IF_7 + Li_2SO_4 \xrightarrow{70\,^{\circ}C} IF_5 + 0.5O_2 + LiSO_3F + LiF \qquad (2.44)$$

Attempts to convert IOF_5 with Li_2SO_4 to IO_2F_3 at 75 °C were, as in the case of NO_3^-, also unsuccessful.[34]

2.3.3. Xenon Fluorides

At room temperature, excess XeF_6 reacts with Li_2SO_4 to give the expected $XeOF_4$ in high yield (Eq. 2.45).[34]

$$XeF_6 + Li_2SO_4 \xrightarrow{25\,^{\circ}C} XeOF_4 + LiSO_3F + LiF \qquad (2.45)$$

The $XeOF_4$ can be reacted further with Li_2SO_4 to give XeO_2F_2 in modest yield (Eq. 2.46).[34]

$$XeOF_4 + Li_2SO_4 \xrightarrow{25\,^{\circ}C} XeO_2F_2 + LiSO_3F + LiF \qquad (2.46)$$

As in the case of NO_3^-, care must be taken to use excess $XeOF_4$ to avoid the formation of explosive XeO_3.

2.4. SUMMARY

Oxoanions, such as NO_3^- or SO_4^{2-}, are effective, readily available, nontoxic, and low cost reagents for controlled, stepwise fluorine–oxygen exchange in highly fluorinated compounds of the more electronegative elements. Product separations can be facilitated greatly by appropriate choices of the anion, the countercations and the mole ratios of the reagents. The reactions appear to be quite general, controllable, safe, and scalable.

ACKNOWLEDGMENTS

The authors are indebted to Mr. R. D. Wilson for help and to the Army Research Office, the Air Force Astronautics Laboratory, and the Office of Naval Research for financial support of this work.

REFERENCES

1. Burmakov, A. I., Kunshenko, B. V., Alekseeva, L. A., and Yagupolskii, L. M., in *New Fluorinating Agents in Organic Synthesis* (L. German and S. Zemskov, Eds.), Springer-Verlag, Berlin, 1989, pp. 197–253.
2. Hudlicky, M., in *Organic Reactions*, Wiley, New York, 1988, pp. 513–637.
3. Many examples for these types of reactions can be found in Vols. 3 and 4 of *Inorganic Reactions and Methods* (J. J. Zuckerman and A. P. Hagen, Eds.), VCH Publishers, New York, 1989.
4. Chernik, C. L., Claassen, H. H., Malm, J. G., and Plurien, P. L., in *Noble-Gas Compounds* (H. H. Hyman, Ed.), University of Chicago Press, 1963, p. 106.
5. Schumacher, G. A. and Schrobilgen, G. J., *Inorg. Chem.*, **23**, 2923 (1984).
6. Bougon, R., Carles, M., and Aubert, J., *C. R. Acad. Sci. Ser. C*, **265**, 179 (1967).
7. Gillespie, R. J. and Quail, J. W., *Proc. Chem. Soc.*, 278, (1963).
8. Schack, C. J., Pilipovich, D., Cohz, S. N., and Sheehan, D. F., *J. Phys. Chem.*, **72**, 4697 (1968).
9. Alexakos, L. G., Cornwell, C. D., and Pierce, S. B., *Proc. Chem. Soc.*, 341 (1963).
10. Bartlett, N. and Levchuck, L. E. *Proc. Chem. Soc.*, 342 (1963).
11. Jacob, E., *Z. Naturforsch.*, **35b**, 1095 (1980).
12. Seppelt, K. and Rupp, H. H., *Z. Anorg. Allg. Chem.*, **409**, 331 (1974).
13. Nielsen, J. B., Kinkead, S. A., and Eller, P. G., *Inorg. Chem.*, **29**, 3621 (1990).
14. Kinkead, S. A. and Nielsen, J. B., paper 47 presented at the ACS Ninth Winter Fluorine Conference, St. Petersburg, FL (February, 1989).
15. Schack, C. J. and Christe, K. O., *J. Fluorine Chem.*, **49**, 167 (1990).
16. Christe, K. O., Wilson, R. D., and Schack, C. J., *Inorg. Nucl. Chem. Lett.*, **11**, 161 (1975).
17. Christe, K. O., Wilson, R. D., and Schack, C. J., *Inorg. Nucl. Chem. Lett.*, **20**, 2104 (1981).
18. Wilson, W. W. and Christe, K. O., *Inorg. Chem.*, **26**, 916 (1987).
19. Christe, K. O. and Wilson, W. W., *Inorg. Chem.*, **27**, 1296 (1988).
20. Selig, H., *Inorg. Chem.*, **5**, 183 (1966).
21. Waldman, M. C. and Selig, H., *J. Inorg. Nucl. Chem.*, **35**, 2173 (1973).
22. Schrobilgen, G. J., Martin-Rovet, D., Charpin, P., and Lance, M., *J. Chem. Soc. Chem. Commun.*, 894 (1980).
23. Christe, K. O. and Wilson, W. W., *Inorg. Chem.*, **27**, 3763 (1988).
24. Grison, E., Eriks, K., and De Vries, J. L., *Acta Crystallogr.*, **3**, 290 (1950).
25. Wilson, W. W. and Christe, K. O., *Inorg. Chem.*, **26**, 1631 (1987).

26. Christe, K. O., Wilson, W. W., and Wilson, R. D., *Inorg. Chem.*, **28**, 675 (1989).
27. Christe, K. O. and Schack, C. J., *Adv. Inorg. Chem. Radiochem.*, **18**, 319 (1976).
28. Wilson, W. W. and Christe, K. O., *Inorg. Chem.*, **26**, 1573 (1987).
29. Christe, K. O., Wilson, W. W., and Wilson, R. D., *Inorg. Chem.*, **28**, 904 (1989).
30. Christe, K. O., Sanders, J. C. P., Schrobilgen, G. J., and Wilson, W. W., *J. Chem. Soc. Chem. Commun.*, 837 (1991).
31. Schack, C. J. and Christe, K. O., *Inorg. Chem.*, **27**, 4771 (1988).
32. The authors are grateful to Dr. W. Jolly for bringing the error to their attention.
33. Christe, K. O., Wilson, W. W., and Schack, C. J., *J. Fluorine Chem.*, **43**, 125 (1989).
34. Wilson, W. W. and Christe, K. O., unpublished results.

Note added in proof: The controlled replacement of two fluorines by one oxygen atom in ClF_5 has recently been achieved by B. B. Chaivanov, Y. B. Sokolov and S. N. Spirin, "Researchers in the Field of Inorganic Fluorine Chemistry," Preprint IAE-4936/13, Moscow, 1989, by the reaction of ClF_5 with $H_3O^+BF_4^-$ according to

$$ClF_5 + H_3O^+BF_4^- \rightarrow ClOF_2^+BF_4^- + 3HF.$$

CHAPTER 3

Transition Metal Derivatives of Strong Protonic Acids and Superacids

F. AUBKE, M. S. R. CADER, and F. MISTRY

3.1. INTRODUCTION

3.1.1. Scope of This Chapter

The use of strong protonic fluoroacids and superacids as reaction media, solvents, and synthetic reagents in both organic and inorganic chemistry is well documented.[1] In this chapter we shall focus attention on the role of these acids and their anions in the synthesis and stabilization of transition metal derivatives.

Synthetic Fluorine Chemistry,
Edited by George A. Olah, Richard D. Chambers, and G. K. Surya Prakash.
ISBN 0-471-54370-5

The protonic acids and superacids chosen fall into two general classes:

1. Monoprotonic fluoro superacids of the general type HF—MF_5 (M = As, Sb, Nb, or Ta), some of the strongest Lewis acids,[2] all derived from hydrogen fluoride as solvent system.[3]
2. Fluoroxyacids of the following types: $HOEF_5$, (E = Se and Te), the fluorophosphoric acids (H_2PO_3F and HPO_2F_2), and the sulfonic acids (HSO_3F and HSO_3CF_3).

The self-ionization anions of these superacids and fluoroacids are poorly coordinating, weakly nucleophilic ligands, well capable of stabilizing electrophilic metallic centers. They are all nonoxidizable and reasonably resistant to reduction; however, their coordination chemistry is expected to differ. While anions of the type OEF_5^- will be predominantly monodentate, the others often act as bidentate or tridentate ligands as well, usually with bridging, rather than chelating, configurations.

To illustrate the coordination behavior of these fluoro ligands, their transition metal derivatives (Groups 4–12) have been chosen for the following reasons:

1. The chosen group of compounds allows a good illustration of the various synthetic routes available.
2. With the oxidation states of the metals in these derivatives varying from +1 to +6, a wide range of derivatives with different properties is expected, ranging from molecular compounds with the metal in a high oxidation state to coordination polymers where low to intermediate oxidation states of the metal are encountered.
3. The spectroscopic and magnetic properties of some of the derivatives allow their facile characterization and provide useful clues regarding their molecular structures and the electronic state of the central metal cation.

The synthetic reactions described below are not generally restricted to the preparation of transition metal derivatives. They may also be applied to the synthesis of pretransition and posttransition metal compounds. Two restrictions have been imposed in order to focus the topic: (a) the discussion will be limited to binary transition metal derivatives of the selected oxyacids; and (b) only when their intended synthesis results in the formation of ternary fluoro or oxo derivatives, either due to incomplete substitution or due to subsequent decomposition of an initially formed binary derivative, will these compounds be mentioned.

Likewise, a number of binary compounds with the metal in an intermediate to high oxidation state have not been isolated, but have only been obtained as alkali metal derivatives {e.g., $K_2[Pd(SO_3F)_6]$}, and will be briefly mentioned in the summary. In addition, emphasis will be placed on synthetic aspects.

Molecular structures, available to a limited extent, will be discussed in some detail. Otherwise, only the physical and spectroscopic methods used in the characterization of the reaction products will be listed for each compound in the tables.

There are some obvious omissions among the fluoroacids and superacids chosen. Pentafluorosulfuric acid ($HOSF_5$) is thermally unstable and so far only a very limited number of transition metal derivatives have been reported for $HSeO_3F$ and the superacids HF—PF_5 and HF—BiF_5. Likewise, the limited number of derivatives of the metals in Group 3 (Sc–La) does not warrant their inclusion.

Derivatives of metals in Groups 11 and 12 are included, however, because some of them, in particular the salts of Ag^+ and Hg^{2+}, are very useful reagents in metathetical reactions often leading to other transition metal or organometallic derivatives. Their inclusion illustrates the fact that the synthetic methods discussed here may be readily extended to derivatives of pretransition, post-transition, or *f*-block metals as well, and many examples to illustrate this point are known, but will not be discussed here.

Only limited aspects of the material covered in this chapter have been reviewed previously. The coordinating ability of the fluorosulfate and the trifluoromethylsulfate groups has been discussed relatively recently by Lawrance.[4] Older reviews of fluorosulfates by Jache[5] and Woolf[6] are more general in scope, as is a summary dealing with halogen derivatives of Group 16 oxyacids and their use in synthesis by Aubke and DesMarteau.[7] A review on trifluoromethylsulfate derivatives by Howells and McCown[8] is available, but very few transition metal derivatives were known at the time.

Derivatives of the HF—AsF_5 and HF—SbF_5 are generally considered ternary fluorides and a book on inorganic solid state fluorides edited by Hagenmüller[9] provides a good introduction to more applied aspects of these compounds and may serve as a useful guide to original references. Transition metal fluorophosphates, together with other halophosphoric acid derivatives, have been surveyed by Dehnicke and Shihada,[10] while Seppelt,[11] and Engelbrecht and Sladky[12] summarized the coordination behavior of OEF_5^- ligands (E = Se or Te), and synthetic approaches to $OTeF_5^-$ and $OSeF_5^-$ compounds. A recent comprehensive review on coordination chemistry[13] has accorded the transition metal derivatives of these somewhat nonclassical ligands a rather uneven treatment, with coverage, depending on the metal, ranging from good to nonexistent. This is not totally unexpected, because most binary transition metal derivatives with bidentate and tridentate ligands are better viewed as coordination polymers, often with extended layer lattices, rather than as coordination complexes.

It seems therefore appropriate to summarize pertinent aspects of these compounds at this time, since many novel and interesting examples have been reported within the past 10 years, and a comparative review of this kind has not been attempted to our knowledge.

3.1.2. Order of Presentation and General Comments

The transition metal compounds discussed are arranged in groups depending on the type of anion present. The hexafluoroarsenates and hexafluoroantimonates are discussed first, followed by a brief summary of fluoroanion derivatives with NbF_6^- and TaF_6^- as counteranions, including a brief mention of $AgPF_6$, which has found extensive use as a synthetic reagent in organometallic chemistry. The fluorooxy compounds starting with the pentafluorochalcogenate anions $OSeF_5^-$ and $OTeF_5^-$ are summarized next, followed by the fluorophosphates (PO_3F^{2-} and $PO_2F_2^-$), and finally the fluorosulfates and the trifluoromethylsulfates.

The principal general synthetic routes are discussed within each group of compounds, and the reported derivatives are presented in tabular form in order of increasing group number (from 4 to 12) together with any relevant information on the methods used in characterizing the various compounds.

With the presentation closely oriented at synthetic aspects, some of the unusual features displayed by selected compounds within this large, diverse group, and some interesting applications in synthesis are discussed in the final section. These examples are largely taken from our own work in this area.

It is generally agreed that the anions discussed here are all very weakly basic and have high group electronegativities. However, no consistent method of ranking is available for all 10 anions, due to their rather diverse nature. For the fluoroanions (EF_6^-) an order of anion basicity may be inferred from the order of Lewis acidity of the parent EF_5 compounds. As discussed in detail by Olah et al.[1] various orders of Lewis acidity have been proposed, based on a number of experimental methods; however, for the strongest Lewis acids the generally accepted order appears to be $SbF_5 > AsF_5 > TaF_5 > BF_3 > NbF_5 > PF_5$. For their anions, the basicity is expected to increase in roughly the same order from SbF_6^- to PF_6^-.

For a number of dimethyltin(IV) compounds of the type $(CH_3)_2SnX_2$ (X = a fluoro or fluoroxy anion) with linear or near linear C—Sn—C groupings and bidentate bridging anions, the ^{119}Sn Mössbauer parameters suggest the following order of anion basicity: $F^- > SO_3CH_3^- > PO_2F_2^- > TaF_6^- > HbF_6^- > SO_3CF_3^- \approx SO_3F^- > SbF_6^- \approx Sb_2F_{11}^-$ (Ref. 14) with AsF_6^- ranked between SO_3F^- and SbF_6^-. The anions $OSeF_5^-$ and $OTeF_5^-$ do not form dimethyltin(IV) compounds with the required structural features and are hence not included in this ranking.

For some of these anions, group electronegativities in excess of the value 4, commonly quoted for fluorine, have been suggested based on structural and spectroscopic features displayed by a number of main group derivatives.[11] It seems, however, that the experimental criteria used do not permit a clear differentiation between electronegativity, as defined initially by Pauling,[15] and the overall electron-withdrawing ability of an atom or group, best expressed in terms of Tafts inductive constant, σ^*.[16] Such differentiation is possible for the fluorosulfate group based on ^{119}Sn Mössbauer spectra of selected compounds. From isomer shifts in the series $[SnX_6]^{2-}$ (X = Br, Cl, F, or SO_3F) a slightly lower electronegativity is suggested[17] for SO_3F^- than for F^-. On the other hand, quadrupole splittings for the isostructural series $X_2Sn(SO_3F)_2$ ($X = CH_3$, Br, Cl, F, or SO_3F) suggest higher σ^* values for SO_3F than for F.[18] This is consistent with the view that molecular groups composed of highly electronegative atoms, like OEF_5^- (E = Se or Te) or SO_3X^- (X = F or CF_3), are more capable of withdrawing and redistributing electron density than are monoatomic ligands like F or Cl. This conclusion is also evident from the order of anion basicities, suggested by Mallela et al.,[14] discussed above.

3.2. DISCUSSION OF SYNTHETIC ROUTES AND EXAMPLES

3.2.1. Fluoro Derivatives

3.2.1.A. Hexafluoroarsenates and Hexafluoroantimonates. Well-defined molecular monoprotonic acids of the type HEF_6 (E = As, Sb or Nb, Ta, and P) do not exist as such. The parent Lewis acids EF_5, however, behave as acidic solutes in HF (Ref. 3) and anions of the type EF_6^- are well characterized.

The Lewis acids AsF_5 and SbF_5 are widely used in the synthesis of metallic as well as nonmetallic derivatives. As discussed above, the anions of the acids, AsF_6^- and SbF_6^-, are extremely weak nucleophiles, well capable of stabilizing a wide range of electrophilic cations. Transition metal hexafluoroarsenates and hexafluoroantimonates are, in most reported instances, formulated as metal difluoride adducts of the parent acids, and are of the type $x\mathrm{MF_2} \cdot y\mathrm{EF_5}$ (E = As or Sb; x or $y = 1, 2, 3, \ldots$).[19–23] This formulation, preferred by the authors, is derived from a common synthetic route where a metal difluoride is reacted with AsF_5 or SbF_5, in the presence of anhydrous HF or SO_2, to yield the corresponding products. The formulation $x\mathrm{MF_2} \cdot y\mathrm{EF_5}$ is chosen, and retained in this chapter, because there is frequently a lack of definitive structural information. For the product of the reaction between MF_2 and $2SbF_5$, alternative structural forms ($M[SbF_6]_2$ and $MF[Sb_2F_{11}]$) are possible.

When other preparative methods, such as the solvolysis of compounds like metal(II) fluorosulfates[24] or the oxidation of metals by the Lewis acids in SO_2 are employed,[25,26] the resulting products are conveniently formulated by the authors as $M_x[AsF_6]_y$ or $M_x[SbF_6]_y$.

3.2.1.A.1. Hexafluoroarsenates. Most metal(II) hexafluoroarsenates are prepared at room temperature by the reaction of the corresponding metal difluoride with an excess of AsF_5 in anhydrous HF.[19,21–22] The resulting products are variously formulated as $MF_2 \cdot AsF_5$, $MF_2 \cdot 2AsF_5$, or $2MF_2 \cdot 3AsF_5$, depending on the nature of the metal difluoride used in the reaction.

$$MF_2 + AsF_5 \xrightarrow[RT]{HF} MF_2 \cdot AsF_5 \quad (3.1)$$

where M = Cr or Ag and RT = room temperature,

$$MF_2 + 2AsF_5 \xrightarrow[RT]{HF} MF_2 \cdot 2AsF_5 \quad (3.2)$$

where M = Mn, Co, Ni, Cd, or Hg, and

$$2MF_2 + 3AsF_5 \xrightarrow[RT]{HF} 2MF_2 \cdot 3AsF_5 \quad (3.3)$$

where M = Fe, Cu, or Zn.

Two different compositions for products of the reactions of the same metal difluoride are reported in two cases. When FeF_2 and AsF_5 are reacted at 55 °C, $FeF_2 \cdot AsF_5$ is obtained,[20,21] whereas at room temperature, $2FeF_2 \cdot 3AsF_5$ is formed.[22] Cobalt difluoride and arsenic pentafluoride also reportedly yielded $CoF_2 \cdot AsF_5$ at room temperature.

The oxidation of selected metals by AsF_5 in liquid SO_2 at room temperature leads to the corresponding metal hexafluoroarsenates as well:

$$M + 2AsF_5 \xrightarrow{SO_2} MF_2 \cdot AsF_5 + AsF_3 \quad (3.4)$$

where M = Fe or Ni, and

$$Mn + 3AsF_5 \xrightarrow{SO_2} MnF_2 \cdot 2AsF_5 + AsF_3 \quad (3.5)$$

The reaction with nickel is also reported[28] to yield $Ni[AsF_6]_2 \cdot 2SO_2$.

Low-valent metal derivatives have been formed by the oxidation of copper and mercury. A salt of univalent copper, formulated as $Cu^+[AsF_6^-]$, is obtained when copper is solvolyzed in a mixture of AsF_5 and SO_2.[28] When mercury is reacted in varying mole ratios, two distinct, well-defined stoichiometric compounds, $Hg_3[AsF_6]_2$ and $Hg_4[AsF_6]_2$, are formed[26] in addition to a material of the unusual composition $Hg_{2.86}AsF_6$.[29]

The $MF_2:AsF_5$ ratio can be altered by partial thermal decomposition for some of the materials[21,22] according to:

$$MF_2 \cdot 2AsF_5 \xrightarrow{65-173\,°C} MF_2 \cdot AsF_5 + AsF_5 \quad (3.6)$$

where M = Ni, Cd, or Hg, and

$$2MF_2 \cdot 3AsF_5 \xrightarrow{40-125\,°C} 2MF_2 \cdot AsF_5 + 2AsF_5 \quad (3.7)$$

where M = Fe, Cu, or Zn.

Between 60–115 °C, $NiF_2 \cdot AsF_5$ is converted to $2NiF_2 \cdot 3AsF_5$, and $2AgF_2 \cdot AsF_5$ is formed from $AgF_2 \cdot AsF_5$ at temperatures up to 352 °C.

The reported hexafluoroarsenates are listed in Table 3.1. It is noted that the majority of the compounds is derived from the *3d* block metals with the exception of Ag and Cd (*4d*), and Hg (*5d*). In addition, the oxidation states of the metals do not go beyond +2. Mercury, as an exception, also displays noninteger oxidation states. Finally, materials of the type $MF_2 \cdot AsF_5$ must be regarded as ternary, rather than binary, hexafluoroarsenates. Their formation reflects the relatively weak Lewis acidity of AsF_5. For the stronger F^- acceptor SbF_5, $CoF[SbF_6]_2$ is the only product of this composition.[25]

The X-ray powder data of the $MF_2 \cdot AsF_5$ compounds (M = Cr, Fe, Cu, Ag, or Zn) show that none of them are isostructural, but they are postulated[22] to be loosely related to $SnF_2 \cdot AsF_5$, where the structure is seen to have cyclic $[(Sn—F)_3]^{3+}$ cations.[30] Additional structural information is obtained from vibrational spectra, and for the Cr, Cu, and Zn species, structures containing oligomeric, cyclic cations of the type $[(M—F)_n]^{n+}$ and distorted $[AsF_6]^-$ anions are postulated.[22]

The powder patterns from the $MF_2 \cdot 2AsF_5$ compounds are indexed on the basis of tetragonal unit cells with similar cell parameters. The structures are discussed either in terms of ionic compounds, $M^{2+}[AsF_6^-]_2$, or covalent fluorine bridged polymers.[20] The observed IR stretching frequencies do not support the formulation $MF^+[As_2F_{11}]^-$. Materials of the composition $2MF_2 \cdot 3AsF_5$ may be viewed as $2[MF]^+[AsF_6]^-[As_2F_{11}]^-$ or $[M_2F_3]^+[As_3F_{16}]^-$. The compound $2AgF_2 \cdot AsF_5$ is believed to be structurally related to $2XeF_2 \cdot AsF_5$,[22] where a fluorine bridged $Xe_2F_3^+$ cation is present.[31]

Crystal structures have been reported for $AgF_2 \cdot AsF_5$, $Hg_x[AsF_6]_2$ ($x = 3$ or 4),[26] and $Hg_{2.86}[AsF_6]$.[29] The structure of the silver compound consists of infinite, fluorine bridged $[Ag—F]_n^{n+}$ chains lying along the *a* axis, and $[AsF_6]^-$ octahedra cross-linked to these chains via additional fluorine bridges. The Ag center is located in an approximately pentagonal bipyramidal environment.[32]

In $Hg_3[AsF_6]_2$, discrete linear and symmetrical Hg_3^{2+} ions and octahedral $[AsF_6]^-$ ions are observed. The Hg—Hg bond length is 2.55 Å, and the Hg—F distance of the terminal Hg atoms is 2.38 Å, which indicates some covalent cation–anion interactions.[26] The compound $Hg_4[AsF_6]_2$ has a similar chain structure, but relatively short interactions of 2.985 Å between individual Hg_4^{2+} ions result in the formation of nonlinear chains. The Hg—Hg distances of 2.620 and 2.588 Å are found within the Hg_4^{2+} cation.[33] In contrast, $Hg_{2.86}[AsF_6]$ features noninteracting, mutually perpendicular infinite chains of Hg atoms with a Hg—Hg distance of 2.64 Å.[29] In addition, unpublished data for $Hg_2[AsF_6]_2 \cdot SO_2$ quoted by Cutforth et al.[33] suggest a rather short Hg—Hg distance of 2.45 Å. It appears that the Hg—Hg bond length increases with increasing chain length of the cation.

3.2.1.A.2. Hexafluoroantimonates. The binary transition metal hexafluoroantimonates are prepared by the following general methods: (a) reaction of a metal(II) difluoride with SbF_5 in HF or SO_2; (b) reaction of a metal(II)

TABLE 3.1. Transition Metal Hexafluoroarsenates.

Compound	Synthesis[a]	Analysis[b]	References
$CrF_2 \cdot AsF_5$	$CrF_2 + AsF_5$ in HF	M/A, IR, Raman, X-ray powder, μ_{eff}	22
$MnF_2 \cdot 2AsF_5$	$MnF_2 + AsF_5$ in HF	M/A, IR, Raman, X-ray powder	20
	$Mn + AsF_5$ in SO_2	M/A, μ_{eff}	25
$FeF_2 \cdot AsF_5$ yellow powder	$2FeF_2 \cdot 3AsF_5$ 40–60°C	M/A, IR, X-ray powder, DTA, DTG	21, 22
	$FeF_2 + AsF_5$ in HF, 55°C	M/A, IR, X-ray powder, DTA, DTG	21, 22
	$Fe + AsF_5$ in SO_2	M/A, μ_{eff}, Mössbauer	25
$2FeF_2 \cdot 3AsF_5$	$FeF_2 + AsF_5$ in HF	M/A, IR, Raman, X-ray powder, μ_{eff}	22
$CoF_2 \cdot 2AsF_5$	$CoF_2 + AsF_5$ in HF	M/A, IR, Raman, X-ray powder	20, 21
$CoF_2 \cdot AsF_5$	$CoF_2 + AsF_5$ in HF	M/A, UV–vis	27
$NiF_2 \cdot AsF_5$ pale yellow powder	$Ni + AsF_5$ in SO_2	M/A, μ_{eff}	25
$NiF_2 \cdot AsF_5$	$NiF_2 \cdot 2AsF_5$ 150°C	DTA, DTG	21
$NiF_2 \cdot 2AsF_5$	$NiF_2 + AsF_5$ in HF	M/A, IR, Raman, X-ray powder	20, 21
$2NiF_2 \cdot 3AsF_5$	$NiF_2 \cdot 2AsF_5$ 60–115°C	DTA, DTG	21
$Ni[AsF_6]_2 \cdot 2SO_2$	$Ni + AsF_5$ in SO_2	M/A, IR, μ_{eff}	28
$Cu[AsF_6]$ white powder	$Cu + AsF_5$ in SO_2, 2d	M/A, IR, Raman, X-ray powder, μ_{eff}	28
$CuF_2 \cdot AsF_5$	$2CuF_2 \cdot 3AsF_5$ 69–118°C	M/A, IR, Raman, X-ray powder, DTA, DTG	21, 22
$2CuF_2 \cdot 3AsF_5$	$Cu + AsF_5$ in HF	M/A, IR, Raman, X-ray powder	21, 22
$AgF_2 \cdot AsF_5$	$AgF_2 + AsF_5$ in HF	M/A, IR, Raman, X-ray powder, X'tal struc. μ_{eff}	22, 32
$2AgF_2 \cdot AsF_5$	$AgF_2 \cdot AsF_5$ 65–352°C	DTA, DTG	21, 22
$ZnF_2 \cdot AsF_5$	$2ZnF_2 \cdot 3AsF_5$ 58–125°C	M/A, IR, Raman, X-ray powder, DTA, DTG	21, 22
$2ZnF_2 \cdot 3AsF_5$	$ZnF_2 + AsF_5$ in HF	M/A, IR, Raman	22
$CdF_2 \cdot AsF_5$	$CdF_2 \cdot 2AsF_5$ 105–173°C	DTA, DTG	21
$CdF_2 \cdot 2AsF_5$	$CdF_2 + AsF_5$ in HF	M/A, IR, Raman, X-ray powder	20, 21
$HgF_2 \cdot AsF_5$	$HgF_2 \cdot 2AsF_5$ 65–95°C	DTA, DTG	21
$HgF_2 \cdot 2AsF_5$	$HgF_2 + AsF_5$ in HF	M/A, IR, Raman, X-ray powder	20, 21
$Hg_2[AsF_6]_2$ white powder	$Hg + AsF_5$ in SO_2	Raman, elec. spec.	26

TABLE 3.1. Continued

Compound	Synthesis[a]	Analysis[b]	References
$Hg_{2.86}[AsF]_6$ gold–yellow crystals	$Hg + AsF_5$ in SO_2	M/A, X'tal struc.	29
$Hg_3[AsF_6]_2$ yellow crystals	$Hg + Hg_2[AsF_6]_2$ in SO_2	M/A, Raman, elec. spec., X'tal struc.	26
	$Hg + AsF_5$ in SO_2	M/A, Raman, elec. spec., X'tal struc.	26
$Hg_4[AsF_6]_2$	$Hg + AsF_5$ in SO_2, 7d	X'tal struc.	33

[a]Reaction times are listed in minutes (m), hours (h), or days (d).
[b]The following abbreviations are used in the tables: M/A = microanalysis, IR = Infrared, X-ray powder = X-ray powder diffraction pattern, UV–vis = UV–visible, DTA = Differential Thermal Analysis, DTG = Differential Thermal Gravimetry, TDC = Thermal Differential Calorimetry, X'tal struc. = Crystal Structure, elec. spec. = electronic spectrum, and MS = Mass Spectrometry. μ_{eff}, when listed without a temperature, is at room temperature.

fluorosulfate in excess SbF_5; (c) oxidation of the metal with SbF_5 in SO_2 as a solvent; or (d) metal fluorination with F_2 in the presence of SbF_5.

When a series of metal(II) difluorides are reacted at room temperature with SbF_5 in HF or SO_2 as the solvent,[19,23] binary compounds formulated as $MF_2 \cdot 2SbF_5$ are obtained in high yield according to:

$$MF_2 + 2SbF_5 \xrightarrow{\text{HF or } SO_2} MF_2 \cdot 2SbF_5 \tag{3.8}$$

where M = Cr, Fe, CO, Ni, Cu, Ag, Zn, or Cd.

The solvolysis[24] of metal(II)fluorosulfates in an excess of SbF_5 cleanly yields the corresponding hexafluoroantimonates according to:

$$M(SO_3F)_2 + 6SbF_5 \xrightarrow{25-60^\circ C} M[SbF_6]_2 + 2Sb_2F_9SO_3F \tag{3.9}$$

where M = Ni, Pd, Cu, or Ag.

Other preparative methods involve direct fluorination, restricted to nickel, leading to the synthesis of a material formulated as $Ni[SbF_6]_2$,[34] and the oxidation of a metal by SbF_5 according to:[25]

$$M + 4SbF_5 \xrightarrow{SO_2} M[SbF_6]_2 + SbF_3 \cdot SbF_5 \tag{3.10}$$

where M = Mn, Fe, or Ni.
In the case of cobalt, the reaction leads to $CoF[SbF_6]$,[25] whereas mercury is converted to $Hg_3[Sb_2F_{11}]$.[26]

Meuwsen and Mögling[35] were able to synthesize a group of hydrated metal(II) hexafluoroantimonates, from the corresponding metal salts (e.g., $CuSO_4$) and $K[Sb(OH)_6]$ in aqueous solution followed by the subsequent treatment of the intermediate, having the composition $M[Sb(OH)_6]_2$, with an excess of 40% aqueous HF at elevated temperatures. The compounds were formulated as $M[SbF_6]_2 \cdot 6H_2O$ (M = Mn, Co, Ni, or Cu), but apparently no attempts to dehydrate these compounds have been reported. Baillie et al.[36] mentioned briefly that $Co[SbF_6]_2$ and $Cu[SbF_6]_2$ form stable solutions in nonprotonic solvents such as Et_2O, $MeNO_2$, and C_6H_6 but very little detail on the products is available.

A summary of the synthetic routes to transition metal hexafluoroantimonates is given in Table 3.2. Some information on the structures of the metal(II) hexafluoroantimonates has been reported. For the Mn, Fe, and Ni compounds, obtained from SbF_5 in SO_2,[25] μ_{eff} values at room temperature are consistent with octahedral coordination for the metal centers. Based on vibrational and X-ray powder data, $Ni[SbF_6]_2$, obtained by the high temperature fluorination of nickel,[34] is shown to be related to the $LiSbF_6$ structure by the occupation of only every second octahedral Li^+ site with Ni^{2+}, leading to a layer structure.

Contrasting magnetic properties are reported for $Ag[SbF_6]_2$ depending on the method of preparation. When $Ag(SO_3F)_2$ is solvolyzed in an excess of SbF_5, a nearly white diamagnetic compound, insoluble in HF, is formed.[24] In contrast, blue paramagnetic $Ag[SbF_6]_2$ crystallizes from a solution of AgF_2 and SbF_5 in HF.[19, 32] Formulation of the white species as $Ag^IAg^{III}[SbF_6]_4$ is also consistent with vibrational spectra and the chemical behavior of the compound. A precedent for the suggested mixed-valency compound may be seen in the recently reported white diamagnetic species formulated as $Ag[Ag(CF_3)_4]$.[37]

A crystal structure is reported for the blue form of $Ag[SbF_6]_2$.[32] The Ag^{2+} ion is located in a distorted elongated octahedral environment, which in turn implies a layer structure and tridentate bridging coordination of the $[SbF_6]^-$ moiety. Except for the observed distortion due to the Jahn–Teller effect, the reported structure is consistent with the proposal by Christe et al.[34] for $Ni[SbF_6]_2$. Thermal conversion of blue to white $Ag[SbF_6]_2$ occurs at 100 °C as shown by differential scanning calorimetry.

In addition to the compounds discussed here, $AgSbF_6$ and $AgAsF_6$ are known and both are commercially available. Reaction of AgF with SbF_5 or AsF_5 in HF is the simplest synthetic route. Except for these two salts, $Cu[ASF_6]$,[28] and the chain cations formed by mercury,[26,29,33] the principal oxidation state displayed by the metal is +2. It appears from the examples reported so far that both anions are most suited to stabilize cations in low oxidation states and, in case of the polymercury cations, have the ability to allow retention of metal–metal bonds. On the other hand, even when an excess of SbF_5 is used formation of polyanions like $Sb_2F_{11}^-$ is observed only for $Hg_3[Sb_2F_{11}]$.[26]

Even though little is known about the solution chemistry of the compounds

TABLE 3.2. Transition Metal Hexafluoroantimonates.[a]

Compound	Synthesis	Analysis	References
$Cr[SbF_6]_2$ white powder	$CrF_2 + SbF_5$ in HF	M/A, IR, Raman, X-ray powder	19, 23
$Mn[SbF_6]_2$	$Mn + SbF_5$ in SO_2	M/A, μ_{eff}	25
$Mn[SbF_6]_2 \cdot 6H_2O$	$Mn[Sb(OH)_6]_2+$ (40% HF)	M/A	35
$Fe[SbF_6]_2$ white powder	$FeF_2 + SbF_5$ in HF	M/A, IR, Raman, X-ray powder	19, 23
	$Fe + SbF_5$ in SO_2	M/A, μ_{eff}, Mössbauer	25
$Co[SbF_6]_2$ violet powder	$CoF_2 + SbF_5$ in HF or SO_2	M/A, IR, Raman, X-ray powder	19, 23
$CoF[SbF_6]_2$ lilac powder	$Co + SbF_5$ in SO_2	M/A, μ_{eff}	25
$Co[SbF_6]_2 \cdot 6H_2O$	$Co[Sb(OH)_6]_2$ + (40% HF)	M/A	35
$Ni[SbF_6]_2$ yellow powder	$NiF_2 + SbF_5$ in HF	M/A, IR, Raman, X-ray powder	19
	$Ni + SbF_5$ in SO_2	M/A, μ_{eff}	25
	$Ni(SO_3F)_2 + SbF_5$ 60°C, 14 d	M/A, IR, $\mu_{eff}^{2-82\,K}$	24
	$Ni + SbF_5 + F_2$ 270°C, 250 atm	M/A, IR, X-ray powder	34
$Ni[SbF_6]_2 \cdot 6H_2O$	$Ni[Sb(OH)_6]_2$ + (40% HF)	M/A	35
$Pd[SbF_6]_2$ gray powder	$Pd(SO_3F)_2 + SbF_5$ 50°C, 14d	M/A, IR, UV–vis, $\mu_{eff}^{2-82\,K}$	24
$Cu[SbF_6]_2$ white powder	$Cu(SO_3F)_2 + SbF_5$ 50°C, 10d	M/A, IR, Raman, $\mu_{eff}^{2-82\,K}$	24
$Cu[SbF_6]_2 \cdot 6H_2O$	$Cu[Sb(OH)_6]_2+$ (40% HF)	M/A	35
$Ag[SbF_6]_2$ blue crystals	$AgF_2 + SbF_5$ in HF	M/A, IR, Raman, X'tal struc.	19
$Ag[SbF_6]_2$ white powder mp ≈ 180°C	$Ag(SO_3F)_2 + SbF_5$ 10d	M/A, IR, Raman	24
$Zn[SbF_6]_2$ white powder	$ZnF_2 + SbF_5$ in HF	M/A, IR, Raman, X-ray powder	19, 23
$Cd[SbF_6]_2$ white powder	$CdF_2 + SbF_5$ in HF	M/A, IR, Raman, X-ray powder	19, 23
$Hg_3[Sb_2F_{11}]_2$ white powder	$Hg + SbF_5$ in SO_2, 24h	M/A, Raman, UV–vis	26

[a]See footnotes in Table 3.1 for definition of abbreviations.

discussed here, their synthesis from MF_2 and SbF_5 (Ref. 32) suggests that anhydrous HF, and possibly SO_2, would be suitable media for such studies.

3.2.1.B. Miscellaneous Fluoro Derivatives. A limited number of fluoro derivatives of the transition metals, used in a variety of synthetic reactions, will be considered here. The silver(I)hexafluoroniobates, silver(I)hexafluorotantalates, and silver(I)hexafluorophosphates were first prepared in the 1950s and use in metathesis as a convenient source of hexafluorometallate anions has increased ever since. Both $AgNbF_6$ and $AgTaF_6$ are prepared by reacting silver (or silver chloride) together with niobium and tantalum, respectively, in bromine trifluoride.[38, 39] The hexafluorophosphate derivative, $AgPF_6$, is synthesized in a similar manner, by reacting BrF_3 with (a) Ag and PBr_5,[40] (b) Ag and P_4O_{10};[39] or (c) AgCl and P_4O_{10}.[41]

Brown et al.[42] reported the preparation of mercury compounds of the type Hg_3NbF_6 and Hg_3TaF_6 from Hg and $Hg[MF_6]_2$ (M = Nb or Ta) in liquid SO_2. The parent compounds $Hg[MF_6]_2$ are made by reacting MF_5 (M = Nb or Ta) with HgF_2. The trigonal crystals of the mercury derivatives consist of hexagonal sheets of Hg atoms that are separated by sheets of MF_6^- anions. Although Hg—Hg bonds are found in many compounds, formation of layered sheets in these compounds is unusual. In addition, in the reaction involving tantalum(V) fluoride, $Hg_4[Ta_2F_{11}]_2$ is isolated, where $[Ta_2F_{11}]^-$ anions and almost linear Hg_4^{2+} cations are found.[43]

Two other layered compounds, $Ag[NbF_6]_2$ and $Ag[TaF_6]_2$, have been made by high temperature (250–400 °C) fluorination of Ag_2O and M_2O_5 (M = Nb or Ta).[44] The crystal structure of $Ag[TaF_6]_2$ indicates elongated $[AgF_6]$ octahedra with all structural features as reported for $AgF_2 \cdot 2SbF_5$.[19,32] The compound is magnetically dilute down to 13 K, with $\mu_{eff} \approx 1.95\,\mu_B$.[44] Pyrolysis of $Ag[TaF_6]_2$ at 150°C provides an alternate route to $AgTaF_6$, with the formation of F_2 and TaF_5.[44]

It appears that the limited solubility of the tetrameric fluorides of Nb and Ta in HF has not only affected their use as superacids, but has also impeded their extensive use in the synthesis of transition metal derivatives, with none of the routes employed involving HF as a solvent.

3.2.2. Fluorooxy Derivatives

3.2.2.A. Ortho Pentafluorochalcogenates. The parent acids $HOSF_5$,[45] $HOSeF_5$,[46] and $HOTeF_5$ (Refs. 47, 48) are difficult to prepare, but rank among the strongest simple protonic acids known.[49] In contrast to the other fluorooxyacids discussed in this chapter, the acids are solids at room temperature with small liquid ranges ($HOSeF_5$, mp 38°C, bp 44°C; $HOTeF_5$ mp 40 ± 1°C, bp 60 ± 1 °C). This indicates that solvolysis reactions of metal compounds in the parent acid may not play an important role in the synthesis of transition metal derivatives and other synthetic pathways must be sought. In addition, $HOSF_5$ exists only at low temperature and thermal degradation occurs at about −65 °C,[10] according to

$$M(OSF_5)_n \longrightarrow MF_n + nSOF_4 \qquad (3.11)$$

This decomposition mode is also observed for many of the $OSeF_5^-$ compounds to yield, for example, $TiF_x(OSeF_5)_{4-x}$ and $WF_x(OSeF_5)_{6-x}$.[50] These compounds have not been isolated and have only been observed in mass spectrometric studies of these systems. Consequently, only a single binary $OSeF_5$–transition metal derivative, $Hg(OSeF_5)_2$, has been reported. The thermal stability of the $OTeF_5$ group is greater, and a fair number $M(OTeF_5)_n$ derivatives have been reported.

An alternate decomposition mode involves the elimination of the anhydride, $F_5TeOTeF_5$, and the formation of the oxide according to

$$M(OTeF_5)_{2n} \longrightarrow MO_n + nF_5TeOTeF_5 \tag{3.12}$$

As a result of these decomposition reactions, attempts to synthesize the binary transition metal salts of these acids often lead to the formation of fluoro- and oxo-derivatives instead. Where both are known, the $OTeF_5$ derivatives are generally more stable than the $OSeF_5$ analogues.

Nevertheless, the high acidities and relative thermal stabilities of $HOSeF_5$ and $HOTeF_5$ allow for the solvolysis of a small number of halides leading to interesting and useful compounds according to the following equations:

$$MCl_n + nHOTeF_5 \longrightarrow nHCl + M(OTeF_5)_n \tag{3.13}$$

where M = B or Ti, $n = 3$ or 4, and

$$HgF_2 + 2HOEF_5 \longrightarrow 2HF + Hg(OEF_5)_n \tag{3.14}$$

where E = Se or Te.

Of the products, $B(OTeF_5)_3$, and to a lesser extent $Hg(OEF_5)_2$, play an important role as OEF_5^- transfer reagents in metathesis. The general reaction proceeds as shown below:

$$MF_n + n/3\,B(OTeF_5)_3 \longrightarrow n/3\,BF_3 + M(OTeF_3)_n \tag{3.15}$$

In particular, $B(OTeF_5)_3$, first synthesized by Sladky et al.,[51] is an extremely versatile reagent. The advantages of using $B(OTeF_5)_3$ as a synthetic reagent are (a) its availability and high Lewis acidity;[51] (b) the by-product BF_3 is volatile and easily removed; and (c) the potentially wide application range on account of the large number of binary transition metal fluorides known. Both $Hg(OTeF_5)_2$ and $Hg(OSeF_5)_2$ have found similar, but more limited, use because product separation from the by-product HgF_2 is not always easy.

The OEF_5 group (E = Se or Te) is often regarded as a pseudohalogen, or more precisely as a "pseudofluorine."[11] This is apparent in the case of the $OTeF_5$ group, which has a very extensive chemistry, much like fluorine. Like fluorine, it can stabilize metals in high oxidation states resulting in compounds such as $Xe(OTeF_5)_6$,[52] $Te(TeOF_5)_6$,[53,54] $W(OTeF_5)_6$,[55] and $U(OTeF_5)_6$.[56,57] This similarity to fluorine has been attributed to the high group electronegativity of the OEF_5 groups.[12] However, the limited thermal stability of the OEF_5 group and its molecular nature are often responsible for the formation of ternary oxo- or fluoro- derivatives in partial decomposition reactions.

Unlike fluorine, however, the OEF_5 groups are strictly monodentate, terminal ligands, with only one notable exception discussed later. As a consequence of this preferred monodentate bonding mode and the repulsion caused by the five fluorine atoms, binary compounds of OEF_5^- exist as discrete monomers giving rise to low-melting and low-boiling materials, in spite of their sometimes high molecular weight, which are soluble in nonpolar solvents. This makes these compounds very suitable for mass spectrometric and NMR studies, as can be seen in Table 3.3.

The compound $Au(OTeF_5)_3$ is the only binary, oligomeric transition metal derivative of OEF_5^-. A crystal structure has been reported,[58] which reveals a centrosymmetric dimer with four terminal $OTeF_5$ groups and two μ-oxo bridging bidentate $OTeF_5$ groups with a tri-coordinated oxygen atom. The gold atoms are seemingly disordered, and the reported Au—O bond lengths for terminal $OTeF_5$, which are extremely short in comparison to other reported values,[59-61] must be viewed with some caution. While the data are seemingly overinterpreted, there is no doubt as to the unusual bidentate bridging nature of $OTeF_5^-$ in this compound.

On the other hand, the crystal structure of $U(OTeF_5)_6$ (Ref. 56) reveals discrete monomers in which the six $OTeF_5$ groups are octahedrally coordinated to the central uranium atom. The typical properties of these compounds are reflected by $U(OTeF_5)_6$. For $W(OTeF_5)_6$, obtained interestingly not from WF_6 or WCl_6, but from WF_5, a triclinic unit cell is suggested in a preliminary crystal structure study.[55] All other transition metal derivatives of $OTeF_5^-$ are, in essence, monomeric.

In $Ti(OTeF_5)_4$ a rare case of tetrahedral oxygen coordination to titanium is observed.[50,62] The coordination number 4 is also exhibited by $VO(OTeF_5)_3$ and $CrO_2(OTeF_5)_2$.[55] The binary derivatives of Nb and Ta, $MoO(OTeF_5)_4$, where a crystal structure is reported, and $ReO_2(OTeF_5)_3$ all have pentacoordinated central metal atoms.[55] However, $M(OTeF_5)_5$ (M = Nb or Ta), which have been isolated in pure form, are thermally unstable.[55] There are several other fluoro- and oxo- OEF_5 derivatives of the early transition metals, often with several geometric isomers, the ones included here are only those that resulted from a synthesis aimed towards the binary OEF_5 derivatives.

With the OEF_5^- groups (E = Se or Te) generally monodentate, it becomes apparent that these ligands will tend to form molecular transition metal derivatives with the central metal in a high oxidation state, rather than form layer materials as discussed in the previous section, or coordination polymers, as discussed later, where the metal has a low to intermediate oxidation state. Consequently, magnetic properties are less useful in product characterization while mass spectrometry and, even more so, NMR become very useful and important tools as is demonstrated widely.[11] It is hence not surprising that the syntheses of many binary $OTeF_5^-$ derivatives of the early transition metal are reported or attempted while $Au(OTeF_5)_3$[58] is the only compound formed by a

TABLE 3.3. Transition Metal Derivatives of OXF_5 (X = Se and Te).[a]

Compound	Synthesis	Analysis	Reference
$Ti(OTeF_5)_4$ highly volatile white crystals, mp 84°C	$TiCl_4 + Hg(OTeF_5)_2$ −30°C	M/A, Raman, ^{19}F NMR, MS	50
$Ti(OTeF_5)_4$ $\rho_{20} = 3.212\ g\ cm^{-1}$	$TiCl_4 + HOTeF_5$	M/A, IR, ^{19}F NMR, TDC, X-Ray powder	62
$O{=}V(OTeF_5)_3$ mp ≈27°C bp 41°C$_{(16\ torr)}$	$VCl_5 + B(OTeF_5)_3$ −30°C	M/A, Raman, ^{19}F NMR, MS	55
$Nb(OTeF_5)_5$ white crystals mp ≈88–91°C	$NbF_5 + B(OTeF_5)_3$ in $C_2F_3Cl_3$	M/A, Raman, ^{19}F NMR, MS	55
$Ta(OTeF_5)_5$ off-white solid dec pt 77–82°C	$TaF_5 + B(OTeF_5)_3$ in $C_2F_3Cl_3$	M/A, Raman, ^{19}F NMR, MS	55
$CrO_2(OTeF_5)_2$	$CrF_5 + B(OTeF_5)_3$ in $C_4F_9SO_2F$	M/A, ^{19}F NMR,	55
$O{=}Mo(OTeF_5)_4$	$MoF_6 + B(OTeF_5)_3$ in $C_2Cl_2F_4$, 50°C, 12 h	M/A, Raman, ^{19}F NMR, MS, X'tal struc.	55
$MoF_n(OTeF_5)_{6-n}$	$MoF_6 + B(OTeF_5)_3$ 80°C, 1 h	^{19}F NMR, MS	152
$O{=}MoF_3(OTeF_5)$	$MoF_6 + B(OTeF_5)_3$ 80°C, 1 h	M/A, IR, ^{19}F NMR, MS	152
$W(OTeF_5)_6$ white solid mp = 93°C	$WF_5 + B(OTeF_5)_3$ in $C_2Cl_3F_3$, 20°C	M/A, Raman, ^{19}F NMR	55
$WCl(OTeF_5)_5$	$WCl_6 + Hg(OTeF_5)_2$ in $CFCl_3$, −20°C	M/A, Raman, ^{19}F NMR, MS	50
$WF_2(OTeF_5)_4$	$WF_6 + B(OTeF_5)_3$ 60°C, 30 d	M/A, Raman, ^{19}F NMR, MS	55
$WF_4(OTeF_5)_2$	$WF_6 + B(OTeF_5)_3$ in $CFCl_3$, 65°C, 3 d	M/A, Raman, ^{19}F NMR, MS	55
$WF_5(OTeF_5)$	$WF_5 + B(OTeF_5)_3$ 120°C, 1.5 h	M/A, ^{19}F NMR, MS	153
$WF_4(OTeF_5)_2$	$WF_5 + B(OTeF_5)_3$ 120°C, 1.5 h	M/A, ^{19}F NMR, MS	153
$ReO_2(OTeF_5)_3$	$ReF_7 + B(OTeF_5)_3$	M/A, Raman, ^{19}F NMR	55
$O{=}Os(OTeF_5)_4$	$OsF_6 + B(OTeF_5)_3$ 30–35°C	M/A, ^{19}F NMR, MS	55
$AgOTeF_5$ white solid mp = 92°C	$AgCN + HOTeF_5$ in CH_3CN, 25°C, 2 h	IR, Raman ^{19}F NMR	154 155
$Au(OTeF_5)_3$ mp = 128°C	$AuF_3 + B(OTeF_5)_3$ 60°C, 5 d	M/A, Raman, ^{19}F NMR, MS, X'tal struc.	58
$Hg(OSeF_5)_2$ white crystals	$HgF_2 + HOSeF_5$	M/A, IR, ^{19}F NMR, MS	156
$Hg(OTeF_5)_2$ white solid subl. 200°C (at 10^{-3} torr)	$HgF_2 + HOTeF_5$ $HgF_2 + B(OTeF_5)_3$ 100°C	IR, ^{19}F NMR, MS IR, ^{19}F NMR, MS	155 155

TABLE 3.3. Continued

Compound	Synthesis	Analysis	Reference
$Hg(OTeF_5)_2$ white crystals dec pt = 200°C	$HgF_2 + HOTeF_5$	M/A, IR, ^{19}F NMR, MS	64

[a]See footnotes in Table 3.1 for definition of abbreviations.

late, electron-rich metal and the resulting dimeric compound is reported to be molecular in nature as well.

Finally, it should be mentioned that the peroxides $(F_5E)_2O_2$ (E = S, Se, or Te) are known[11,63,64] and are best obtained by UV photolysis of either $ClOEF_5$ (E = S, Se, or Te) or of $Xe(OEF_5)_2$ (E = Se or Te), but they have not found any use in the synthesis of transition metal derivatives.

3.2.2.B. Fluorophosphates. The binary transition metal compounds discussed in this section include transition metal monofluorophosphates, with PO_3F^{2-} as the anion, and the difluorophosphates, with $PO_2F_2^-$ as the anion. The fluorophosphoric acids H_2PO_3F and HPO_2F_2 have been reviewed by Lange.[65] They are related to phosphoryl fluoride (POF_3) in one way, and the ortho- or metaphosphoric acids H_3PO_4 or HPO_3 by hydrolysis or HF solvolysis equilibria of the type suggested by Lange and Livingston:[66]

$$POF_3 \xrightarrow[-HF]{H_2O} H[PO_2F_2] \underset{HF}{\overset{H_2O}{\rightleftharpoons}} H_2PO_3F \underset{HF}{\overset{H_2O}{\rightleftharpoons}} H_3PO_4 \quad (3.16)$$

which suggest methods of preparation. For example, monofluoro phosphoric acid (H_2PO_3F) is obtained from HPO_3 and HF or more conveniently via the reaction of 67% aqueous HF with P_2O_5

$$P_2O_5 + 2HF + H_2O \rightleftharpoons 2H_2PO_3F \quad (3.17)$$

while phosphorus(V) oxide and anhydrous HF produce a mixture of di- and monofluorophosphoric acid:

$$P_2O_5 + 3HF \rightleftharpoons H_2PO_3F + HPO_2F_2 \quad (3.18)$$

and separation is possible by distillation. The compound H_2PO_3F is less volatile and cannot be distilled without decomposition. As a consequence, only HPO_2F_2, which may be obtained in pure form by distillation, is used as a synthetic reagent in solvolysis reactions.[67] Similar interconversions, as shown in Eq. 3.16 for the parent acids, are also applicable to their salts, in particular to the transition metal derivatives. A direct route to $H_2PO_2F_2$ involving the fluorination of $POCl_3$ and $P_2O_3Cl_4$ by aqueous HF (40%) has been reported recently.[68]

3.2.2.B.1. Monofluorophosphates. The monofluorophosphate ion, PO_3F^{2-}, is isoelectronic to the sulfate ion (SO_4^{2-}) with identical ionic charges of -2, and some structural similarity between the two sets of salts is anticipated. It is, however, generally accepted that in metal monofluorophosphates the fluorine atom does not participate in bonding to the central metal atom,[10] as seen in the case of the alkali metal monofluorophosphates. Hence, the maximum denticity of the PO_3F group is expected to be three.

A series of water soluble metal monofluorophosphates of the types MPO_3F and $M_2(PO_3F)_3$ can be prepared by metathesis from Ag_2PO_3F and the corresponding metal chlorides in water or any other suitable solvent as shown by Singh and Sinha:[69,70]

$$Ag_2PO_3F_{(aq)} + MCl_{2(aq)} \longrightarrow MPO_3F(aq) + 2AgCl(s) \quad (3.19)$$

$$3Ag_2PO_3F(aq) + 2MCl_3(aq) \longrightarrow M_2(PO_3F)_3 + 6AgCl(s) \quad (3.20)$$

where M = Cr^{III}, Mn^{II}, Fe^{III}, Co^{II}, Ni^{II}, Cu^{II}, or Cd^{II}.
All are obtained as hydrates. It is nuclear, considering the ready pH-dependent hydrolysis of the PO_3F^{2-} anion,[65] how well the P—F linkage is retained when water is used as a solvent.

The parent compound Ag_2PO_3F is synthesized by the fusion of Na_2PO_3F or K_2PO_3F and $AgNO_3$ at 800 °C.[71] Anhydrous Zn, Cd, and Hg salts are prepared by thermal decomposition[72a] of the corresponding metal(II) difluorophosphates at appropriate temperatures, yielding volatile phosphoryl fluoride as a by-product, according to:

$$M(PO_2F_2)_2 \xrightarrow{120-190\,^\circ C} MPO_3F + POF_3 \quad (3.21)$$

where M = Zn, Cd, or Hg.

The compound Hg_2PO_3F can be prepared from the reaction of POF_3 and HgO. In this procedure, PF_3 is initially reacted with HgO where, at 350°C, HgO is completely reduced by PF_3 to the metal and POF_3, but at 200–300 °C, a solid phase of the composition Hg_2PO_3F is observed. This compound is reportedly formed via a secondary reaction.[72b] Evidence for the suggested formation of fluorine according to Eq. 3.22 is, however, lacking:

$$POF_3 + 2HgO \rightarrow Hg_2PO_3F + F_2 \quad (3.22)$$

The various preparative methods and characterization techniques for the monofluorophosphates are summarized in Table 3.4. The hydrated monofluorophosphates of Cr, Mn, Fe, Ni, and Cd are isostructural with the corresponding sulfates and their transition metal salts are reported to be brightly colored, paramagnetic solids.

An X-ray diffraction study on single crystals of $CoPO_3F \cdot 3H_2O$ reveals a layered material, where CoO_6 octahedra are formed by O-bridging tridentate

TABLE 3.4. Transition Metal Monofluorophosphates.[a]

Compound	Synthesis	Analysis	References
$Cr_2(PO_3F)_3 \cdot 18H_2O$		M/A, IR, UV	69, 70
$MnPO_3F \cdot 4H_2O$		M/A, IR, UV	69, 70
$Fe_2(PO_3F)_3 \cdot 12H_2O$	$MCl_x + Ag_2PO_3F$ in H_2O, 4°C	M/A, IR, UV	69, 70
$CoPO_3F \cdot 3H_2O$		X'tal struc.	73a
$CoPO_3F \cdot 6H_2O$		M/A	157
$NiPO_3F \cdot 7H_2O$		M/A, IR, UV	69, 70, 157
$CuPO_3F \cdot 5H_2O$		M/A	157
Ag_2PO_3F	$AgNO_3 + Na_2PO_3F$ in H_2O	M/A, IR	71, 158, 159
	$AgNO_3 + K_2PO_3F$ in H_2O	^{31}P NMR	160
$ZnPO_3F$	$Zn(PO_2F_2)_2$ 170–190°C	M/A, IR	72a
$CdPO_3F$	$Cd(PO_2F_2)_2$ 170–190°C	M/A, IR	72a
$CdPO_3F \cdot 8/3H_2O$	$Ag_2PO_3F + CdCl_2$ in H_2O, 4°C	M/A, IR, UV	69, 70
$HgPO_3F$	$Hg(PO_2F_2)_2$ 120°C	M/A, IR	72a
Hg_2PO_3F	$Ag_2PO_3F + Hg_2Cl_2$ in H_2O	M/A	158
	$POF_3 + HgO$ 200–300°C	M/A, X-ray powder	72b

[a]See footnotes in Table 3.1 for definition of abbreviations.

PO_3F^{2-} anions and coordinated H_2O. The layers are held together by hydrogen bonds.[73] The fluorine atom of the PO_3F group is not involved in bonding.

3.2.2.B.2. Difluorophosphates. Transition metal difluorophosphates are synthesized by four methods: (a) solvolysis of the appropriate metal chloride in difluorophosphoric acid, HPO_2F_2; (b) the reaction of metals with HPO_2F_2; (c) various reactions of the anhydride $P_2O_3F_4$ with a number of metal halides, oxides, or nitrates; and (d) fluorination of dichlorophosphates.

The solvolysis of chlorides in HPO_2F_2, used widely to prepare alkali difluorophosphates, can be applied efficiently to the synthesis of transition metal(II) difluorophosphates,[74] according to:

$$MCl_2 + 2HPO_2F_2 \longrightarrow M(PO_2F_2)_2 + 2HCl \qquad (3.23)$$

where M = Mn, Fe, Co, Ni, Cu, Zn, or Cd.
Iron(III) difluorophosphate is made, in a similar manner, from $FeCl_3$ and HPO_2F_2,[74,75] but when $CrBr_3$ is solvolyzed in HPO_2F_2 the solvate $Cr(PO_2F_2)_3 \cdot HPO_2F_2$ is formed.[75] The binary salt $Cu(PO_2F_2)_2$ (Ref. 76) is obtained when copper metal reacts with HPO_2F_2, but incomplete dissolution of

copper in HPO_2F_2 and prolonged reaction times make this method unsatisfactory for the large scale preparation of $Cu(PO_2F_2)_2$.

Some metals dissolve in an excess of anhydrous HPO_2F_2 to yield the acid solvates[75] according to:

$$M + 3HPO_2F_2 \longrightarrow M(PO_2F_2)_2 \cdot HPO_2F_2 + H_2 \tag{3.24}$$

where M = Mn, Fe, Co, or Ni.
In the reaction involving iron, exclusion of air is necessary in order to obtain the iron(II) difluorophosphate. In the presence of dry air, the acid insoluble species $Fe(PO_2F_2)_3$ is isolated.

A versatile synthetic route to many binary metal difluorophosphates employs the anhydride $P_2O_3F_4$ as reported by Vast et al.[77] The appropriate metal fluorides, oxides, or nitrates when reacted with $P_2O_3F_4$, yield the corresponding difluorophosphates as shown below:

$$CoF_3 + 3P_2O_3F_4 \longrightarrow Co(PO_2F_2)_3 + 3POF_3 \tag{3.25}$$

$$Cu(NO_3)_2 + 2P_2O_3F_4 \longrightarrow Cu(PO_2F_2)_2 + 2NO_2PO_2F_2 \tag{3.26}$$

$$ZnO + P_2O_3F_4 \longrightarrow Zn(PO_2F_2)_2 \tag{3.27}$$

In addition to the three compounds in Eqs. 3.25–3.27, difluorophosphates of Ag^{I}, Hg^{II}, Cd^{II}, and Hg_2^{II} have been obtained in this manner.

An unusual reaction resulting in the formation of $Cr(PO_2F_2)_3$ is reported by Brown et al.[78] When CrO_3 is reacted with $P_2O_3F_4$, O_2 is evolved according to:

$$4CrO_3 + 6P_2O_3F_4 \longrightarrow 4Cr(PO_2F_2)_3 + 3O_2 \tag{3.28}$$

A high temperature fluorination process, using fluorine[79] or arsenic trifluoride[80] transforms $Fe(PO_2Cl_2)_3$ to the corresponding difluorophosphate:

$$Fe(PO_2Cl_2)_3 + 3F_2 \xrightarrow{175^{\circ}C} Fe(PO_2F_2)_3 + 3Cl_2 \tag{3.29}$$

Although many binary transition metal difluorophosphates can be made by solvolysis, this method may not be a suitable technique in the case of metal chlorides of the early transition metals. The compound $TiCl_2(PO_2F_2)_2$ is obtained when $TiCl_4$ is solvolyzed in excess HPO_2F_2.[81] A detailed list of the difluorophosphate derivatives discussed above together with the physical methods used in their characterization is given in Table 3.5.

For $Fe(PO_2F_2)_3$, made either by (a) fluorination of $Fe(PO_2Cl_2)_3$;[80] (b) metal oxidation with HPO_2F_2 when air is present;[75] or (c) the solvolysis of $FeCl_3$ in HPO_2F_2 (Ref. 75) the ^{57}Fe-Mössbauer spectra are different. The fluorination product gives no X-ray powder pattern and, based on Raman spectra, coordination to the iron center is postulated to involve both oxygen and fluorine. This is also consistent with the observation of a quadrupole splitting in

TABLE 3.5. Transition Metal Difluorophosphates.[a]

Compound	Synthesis	Analysis	References
$Cr(PO_2F_2)_3$ green solid	$CrO_3 + P_2O_3F_4$	M/A, IR	78
$Cr(PO_2F_2)_3 \cdot HPO_2F_2$	$CrBr_3 + HPO_2F_2$	M/A, IR	75
$Mn(PO_2F_2)_2$ white powder mp 184°C	$MnCl_2 + HPO_2F_2$	M/A, IR	74
$Mn(PO_2F_2)_2 \cdot HPO_2F_2$	$Mn + HPO_2F_2$	M/A, μ_{eff}, X-ray powder	75
$Fe(PO_2F_2)_2$ pale blue powder dec pt 180°C	$FeCl_2 + HPO_2F_2$	M/A, IR	74
$Fe(PO_2F_2)_3$ white powder	$FeCl_3 + HPO_2F_2$ in Et_2O	M/A, IR, X-ray powder, Mössbauer	74, 75
	$Fe(PO_2Cl_2)_3 + F_2$ 175°C	M/A, IR	79
	$Fe(PO_2Cl_2)_3 + AsF_3$ 250°C	IR, Raman, Mössbauer	80
$Fe(PO_2F_2)_2 \cdot HPO_2F_2$	$Fe + HPO_2F_2$	M/A, μ_{eff}, X-ray powder, Mössbauer	75
$Co(PO_2F_2)_2$ pink powder, mp 173°C	$CoCl_2 + HPO_2F_2$	M/A, IR	74
$Co(PO_2F_2)_3$	$CoF_3 + P_2O_3F_4$	M/A	77
$Co(PO_2F_2)_2 \cdot HPO_2F_2$	$Co + HPO_2F_2$	M/A, IR, μ_{eff}, X-ray powder	75
$Ni(PO_2F_2)_2$ yellow powder dec pt 255°C	$NiCl_2 + HPO_2F_2$	M/A, IR	74
$Ni(PO_2F_2)_2 \cdot HPO_2F_2$	$Ni + HPO_2F_2$	M/A, μ_{eff} X-ray powder	75
$Cu(PO_2F_2)_2$	$Cu + HPO_2F_2$	X'tal struc.	76
pale blue powder	$CuCl_2 + HPO_2F_2$	M/A, IR	74
dec pt 265°C	$Cu(NO_3)_2 + P_2O_3F_4$	M/A	77
$AgPO_2F_2$	$AgF + P_2O_3F_4$	M/A	77
$Zn(PO_2F_2)_2$ glassy solid	$ZnO + P_2O_3F_4$ 150°C	M/A, IR	72a, 77
	$Zn(NO_3)_2 + P_2O_3F_4$	M/A	80
	$ZnCl_2 + HPO_2F_2$	M/A, IR	79
$Zn(PO_2F_2)_2 \cdot 2HPO_2F_2$	$ZnCl_2 + HPO_2F_2$	M/A, IR	161
$Cd(PO_2F_2)_2$	$CdCl_2 + HPO_2F_2$	M/A, IR	74
white powder dec pt 245°C	$CdO + P_2O_3F_4$ 150°C	M/A, IR	72a, 77
	$Cd(NO_3)_2 + P_2O_3F_4$	M/A	77
$Hg(PO_2F_2)_2$	$HgO + P_2O_3F_4$, 150°C	M/A, IR	72a, 77
$Hg_2(PO_2F_2)_2$	$HgF_2 + P_2O_3F_4$	M/A, Raman	77

[a]See footnotes in Table 3.1 for definition of abbreviations.

the Mössbauer spectrum for this form of $Fe(PO_2F_2)_3$. Based on the vibrational spectra and the single line Mössbauer spectra of the $Fe(PO_2F_2)_3$ obtained via oxidation and solvolysis, only oxygen coordination to the central metal atom is suggested.

All the difluorophosphates of divalent and trivalent metals discussed so far have polymeric structures where the metal center is located in a distorted octahedral coordination site. The magnetic moments reported for the monosolvated difluorophosphates of Fe^{II}, Co^{II}, and Ni^{II} fall within the expected range for octahedral complexes.[75] This implies involvement of the oxygens of the solvate in coordination.

For the binary difluorophosphates of divalent metals, involvement of fluorine in metal coordination has been postulated by Dove et al.[75] to reconcile evidence of octahedral metal coordination with the presence of only two oxygens per PO_2F_2 group. Interestingly, the crystal structure of $Cu(PO_2F_2)_2$ may point to an alternate coordination mode because only coordination to oxygen, again in a polymeric structure, is evident. It is seen that one oxygen atom of the PO_2F_2 group coordinates to two different copper atoms via one short (mean Cu—O distance 1.96 Å) and one longer bond (mean Cu—O distance 2.57 Å), while coordination through fluorine is not observed. In addition to the μ-oxo bridging configuration in $Cu(PO_2F_2)_2$, O-bidentate symmetrically bridging PO_2F_2 groups are present in $Co(PO_2F_2)_2 \cdot 2CH_3CN$, with a CoO_4N_2 coordination environment observed, as a result of coordinated acetonitrile. Once again, fluorine is not involved in coordination.[76]

In summary, with the principal reagents HPO_2F_2 and its anhydride $P_2O_3F_4$, only compounds with the metal center, generally from the 3*d* series, in either a +2 or +3 oxidation state are obtained, and suitable reactive precursors are required. Hence, the number of derivatives is limited and restricted to coordination polymers. With the possible exception of $Fe(PO_2F_2)_3$ produced by fluorination[80] no definite evidence for the involvement of fluorine in metal coordination has been reported.

3.2.2.C. Fluorosulfates. Fluorosulfuric acid (HSO_3F) has been known since 1892,[82] and ranks among the strongest known Brønsted acids. It is available commercially, prepared by the reaction of HF with SO_3, and can be further purified by double distillation at atmospheric pressure.[83] The clear, colorless liquid can be stored in dry glass containers. The wide liquid range and the high thermal stability of HSO_3F makes it a suitable medium for a number of reactions, some leading to the binary fluorosulfates. Detailed summaries on physical and chemical properties of this acid have been published by Thompson,[84] Gillespie,[85] and Jache.[5]

In addition to HSO_3F, bis(fluorosulfuryl) peroxide ($S_2O_6F_2$) first reported by Dudley and Cady,[86] is found to be another useful synthetic reagent. The initial method of preparation, the catalytic (AgF_2) fluorination of sulfur trioxide by fluorine according to:

$$F_2 + 2SO_3 \xrightarrow{AgF_2} S_2O_6F_2 \qquad (3.30)$$

has remained extremely useful, because the synthesis of large quantities (1–2 kg) is possible. The mechanism of the reaction has been discussed in detail by Leung and Aubke.[87] Unlike the thermally unstable peroxide $S_2O_6(CF_3)_2$,[88] bis(fluorosulfuryl) peroxide may be stored in glass for prolonged periods of time. The chemistry of $S_2O_6F_2$ has been reviewed by De Marco and Shreeve.[63]

Its solution in fluorosulfuric acid, where it behaves as a weak base,[89,90] is of particular synthetic use. The solution combines the oxidizing power of $S_2O_6F_2$ and the ionizing ability of HSO_3F, and allows extension of the working range to +160°C, with only a slight attack on glass at the higher end. There are additional advantage to the use of the $HSO_3F—S_2O_6F_2$ reagent combination: (a) with metal oxidation the principal route, precursors are available in high purity, in the form of fine powders, for all transition metals. Only the lack of reactivity noted for some metals, and the low thermal stability of some of the reaction products, present limitations; (b) binary fluorosulfates with the metal in relatively high oxidation states (up to +5) are obtainable; and (c) the method allows the *in situ* preparation of metal fluorosulfates in HSO_3F, suitable either for solution studies or for the preparation of ternary fluorosulfates by addition of alkali metal fluorosulfates according to:

$$2CsSO_3F + Pd + 2S_2O_6F_2 \longrightarrow Cs_2[Pd(SO_3F)_6] \quad (3.31)$$

where the binary fluorosulfate, in this case $Pd(SO_3F)_4$ cannot be isolated; and (d) metal oxidation in HSO_3F is generally fast and complete, whereas $S_2O_6F_2$ by itself reacts far more slowly and often produces product mixtures.

Fluorosulfate chemistry displays many parallels to halogen chemistry, and the fluorosulfate group may be viewed as a pseudohalogen. Hence, the synthetic methods used in the preparation of the fluorosulfates have striking parallels to those used in the synthesis of halides, with the necessary modifications.

Two synthetic methods in general have been used to prepare a large numer of transition metal fluorosulfate derivatives: (a) solvolysis of a corresponding metal salt such as MCl_2, MSO_4, or $M(RCOO)_2$ in excess fluorosulfuric acid, HSO_3F; and (b) the oxidation of a metal with the strongly oxidizing and fluorosulfonating reagent bis(fluorosulfuryl) peroxide, $S_2O_6F_2$, in the presence or absence of HSO_3F.

Solvolysis is almost exclusively the route of choice to the synthesis of 3*d*-block metal(II) fluorosulfates (with the exception of Cd), whereas metal oxidation by $S_2O_6F_2$ yields a variety of electron rich 4*d*- and 5*d*-metal fluorosulfate compounds.

Solvolysis involving metal(II) salts and fluorosulfuric acid is illustrated below for metal chlorides and metal benzoates:

$$MCl_2 + 2HSO_3F \longrightarrow M(SO_3F)_2 + 2HCl \quad (3.32)$$

$$M(C_6H_5COO)_2 + 4HSO_3F \longrightarrow M(SO_3F)_2 + 2C_6H_5COOH^+_{2(solv)} + 2SO_3F^- \quad (3.33)$$

Either the high volatility of HCl or the excellent solubility of the protonated carboxylic acid in HSO_3F permit facile product separation. Similar displacement reactions with HSO_3F are also reported for sulfates, fluorides, and acetates;[40,91] all leading to the corresponding binary metal(II) fluorosulfates.

In addition, the solvolysis of $FeCl_3$,[92] $Zr(CO_2CF_3)_4$,[93] and $Ag(CO_2CF_3)$,[87] in HSO_3F produce $Fe(SO_3F)_3$, $Zr(SO_3F)_4$, and $AgSO_3F$, respectively. It is interesting to note that according to Goubeau and Milne,[92] the solvolysis of metal chlorides in HSO_3F is facilitated by the addition of KSO_3F to the reaction mixture. Complete solvolysis of $ZrCl_4$ in HSO_3F is reported as well, however, the isolated solid is claimed to be ZrF_3SO_3F.[94] The interaction of boiling HSO_3F with a large number of transition metals has been studied by Brazier and Woolf[95] but appears to be useful only in the synthesis of $Cu(SO_3F)_2$ or possibly $AgSO_3F$.

Metal oxidation by $S_2O_6F_2$, either on its own or in fluorosulfuric acid as reaction medium, has proven to be a simple and quantitative route to a substantial number of binary fluorosulfates:

$$M + n/2\ S_2O_6F_2 \xrightarrow{HSO_3F} M(SO_3F)_n \tag{3.34}$$

where $n = 2$, 3, or 4.
Often, however, prolonged reaction times (several weeks) and elevated temperatures (140–150°C) are required.

Osmium, silver, and mercury can also be oxidized by $S_2O_6F_2$ in the absence of HSO_3F to give the $Os(SO_3F)_3$,[96] $Ag(SO_3F)_2$,[97] and $Hg(SO_3F)_2$.[98,99] Interestingly, depending on the method of preparation, two forms of $Os(SO_3F)_3$ are obtained which differ slightly in color, vibrational spectra, and magnetic properties. The compound $Pd(SO_3F)_2$, obtained from $BrSO_3F$ and Pd, is further oxidized by $S_2O_6F_2$ to yield $Pd(SO_3F)_3$.[97] The formulation of $Pd(SO_3F)_3$ as $Pd^{II}[Pd^{IV}(SO_3F)_6]$ is based on magnetic and vibrational data, and will be discussed in some detail later in this section.

Reactions with bis(fluorosulfuryl)peroxide ($S_2O_6F_2$) may involve not only metal powder, but also metal carbonyls or metal chlorides. There is, however, no real advantage because in reactions with metal carbonyls, CO is not only substituted but is also, in part, oxidized to CO_2, while in chloride substitution reactions, excessive reagent is consumed due to the formation of chlorine(I) fluorosulfate, and eventually even ClO_2SO_3F. In addition, incomplete substitution is often encountered leading to compounds like $TiCl_2(SO_3F)_2$.[81] Compounds such as $MO(SO_3F)_3$ (M = Nb or Ta) or $WO(SO_3F)_4$,[100] are often encountered as a result of partial decomposition.

Only three binary fluorosulfates have been obtained from carbonyls. Of these, $Mn(SO_3F)_4$, reportedly obtained from $Mn_2(CO)_{10}$ (Ref. 101) was misanalyzed and has been shown subsequently to have the composition $Mn(SO_3F)_3$.[102] It is more conveniently made by the reaction of Mn with $S_2O_6F_2$ in HSO_3F. The other binary fluorosulfates obtained from carbonyls are $Co(SO_3F)_2$ and $Cr(SO_3F)_3$.[101] Metal carbonates and oxides are not suited for the synthesis of binary fluorosulfates since their reaction with $S_2O_6F_2$ results in the formation of metal oxyfluorosulfates.[103]

Some fluorosulfate derivatives prepared by the above route can be converted to lower oxidation state derivatives by thermal decomposition. Either $S_2O_5F_2 + O_2$ or $S_2O_6F_2$ are released as volatile products in a reductive elimination, a reversal of the oxidation process cited above. The compounds $Ir(SO_3F)_3$,[104] $Pd(SO_3F)_2$,[105,106] and $AgSO_3F$ (Ref. 97) are obtained from $Ir(SO_3F)_4$, $Pd(SO_3F)_3$, and $Ag(SO_3F)_2$, respectively, by this method according to:

$$Ir(SO_3F)_4 \xrightarrow{120°C} Ir(SO_3F)_3 + \tfrac{1}{2}S_2O_5F_2 + \tfrac{1}{2}O_2 \qquad (3.35)$$

$$2Pd(SO_3F)_3 \xrightarrow{160°C} 2Pd(SO_3F)_2 + S_2O_6F_2 \qquad (3.36)$$

$$2Ag(SO_3F)_2 \xrightarrow{210°C} 2AgSO_3F + S_2O_6F_2 \qquad (3.37)$$

A more convenient reduction of $Pd(SO_3F)_3$ is affected with bromine in SO_2 as a solvent:[106]

$$2Pd(SO_3F)_3 + Br_2 \xrightarrow[SO_2]{25°C} 2Pd(SO_3F)_2 + 2BrSO_3F \qquad (3.38)$$

A few noble metal fluorosulfates such as $Pd(SO_3F)_2$,[105] $Pt(SO_3F)_4$,[107] $Ag_3(SO_3F)_4$,[87] and $Au(SO_3F)_3$ (Ref. 107) are also prepared by the oxidation of the respective metals by bromine monofluorosulfate, $BrSO_3F$.

The use of $BrSO_3F$ instead of $HSO_3F—S_2O_6F_2$ in the synthesis of binary metal fluorosulfates generally offers no real advantages for the following reasons:

1. $S_2O_6F_2$ is initially required to synthesize $BrSO_3F$.
2. Longer reaction times are encountered.
3. Intermediates such as $Br_3[Au(SO_3F)_4]$ (Ref. 108) or $(Br_3)_2[Pt(SO_3F)_6]$ (Ref. 109) are often formed, which need to be pyrolyzed to yield $Au(SO_3F)_3$ and $Pt(SO_3F)_4$ respectively.[107]
4. $Au(SO_3F)_3$ and $Pt(SO_3F)_4$ made in this manner are unexpectedly weakly paramagnetic.[110]

The only advantage is seen where relatively low-valent compounds like $Pd(SO_3F)_2$ or $Ag_3(SO_3F)_4$ are obtained directly, while reaction with $S_2O_6F_2$ generates $Pd(SO_3F)_3$ or $Ag(SO_3F)_2$, respectively.

Another synthetic method deserves mention: $Cr(SO_3F)_3$ is obtained when chromium(V) fluoride is reacted with an excess of SO_3 according to the general equation:

$$CrF_5 + 5SO_3 \longrightarrow Cr(SO_3F)_3 + S_2O_6F_2 \qquad (3.39)$$

with bis(fluorosulfuryl)peroxide produced as a volatile by-product.[111] A potentially useful route involves the reaction of transition metal methoxides with the anhydride of fluorosulfuric acid ($S_2O_5F_2$), but only disubstitution is observed when $Ti(OCH_3)_4$ is chosen as the reactant.[112] However, this method holds some promise because $Bi(SO_3F)_3$ is obtained in this manner. A list of the binary transition metal fluorosulfates, together with synthetic routes and analytical methods is presented in Table 3.6.

The examples in Table 3.6 suggest that the SO_3F^- group is reasonably resistant to thermal decomposition. Where decomposition does occur, three general modes are observed:

1. Elimination of SO_3 to yield fluoride fluorosulfates as observed for Nb and Ta.
2. Elimination of disulfurylfluoride ($S_2O_5F_2$) leading to oxyfluorosulfate as observed for Nb, Ta, V, Mo, and Re.
3. Reductive elimination of $SO_3F^{\cdot}$ radicals, which may recombine to give $S_2O_6F_2$, seen in the case Pd, Ag, and Au.

A number of transition metal fluorosulfates have been obtained only in the form of anionic complexes, usually with potassium or cesium as countercations. For example, $Ru(SO_3F)_4$ and $Pd(SO_3F)_4$ have remained elusive, but complexes such as $Cs_2[Ru(SO_3F)_6]$, $Cs[Ru(SO_3F)_5]$,[113] and $[Pd(SO_3F)_6]^{2-}$ stabilized by Cs^+, NO^+, ClO_2^+, Ba^{2+}, and Pd^{2+} (Ref. 114) have all been isolated and fully characterized. Another set of ternary fluorosulfates of the general type $M[Sn(SO_3F)_6]$ (M = Mn, Fe, Co, Ni, and Cu) is reported along with the corresponding hexakis(trifluoro methylsulfato) complexes.[115]

Only a single molecular structure, that of gold(III) fluorosulfate, which was obtained by single-crystal X-ray diffraction, has been reported so far.[116] The compound is a dimer with both monodentate and symmetrically bridging bidentate fluorosulfate groups, in analogy to the structures reported for $[AuCl_3]_2$ (Ref. 117) and $[AuBr_3]_2$ (Ref. 118) but unlike the polymeric structure found for AuF_3.[119] The nearly square planar coordination environment for gold is a common feature of all four structures. Weak intermolecular $Au \cdots O$ contacts result in a very distorted octahedral coordination polyhedron for gold in $[Au(SO_3F)_3]_2$.

The polymeric nature of most transition metal fluorosulfates, and their resulting lack of volatility and solubility in HSO_3F, or other suitable solvents, has prevented the formation of suitable single crystals, and structural evidence rests largely on vibrational spectra and magnetic properties. The intense colors frequently observed in many compounds often make a Raman study difficult.

Silver(I) fluorosulfate, the only example in this group with the metal in a +1 oxidation state appears to be an ionic compound. However, the six fundamental frequencies expected for a SO_3F^- ion with C_{3v} symmetry are doubled.[87]

Fluorosulfates of the type $M(SO_3F)_2$ appear to belong to a single structural type, derived from the $CdCl_2$ layer structure, with one exception—the mixed-

TABLE 3.6. Transition Metal Fluorosulfates.[a]

Compound	Synthesis	Analysis	References
$Ti(OCH_3)_2(SO_3F)_2$	$Ti(OCH_3)_4 + S_2O_5F_2$ 60°C, 2 d	M/A, IR	112
$Zr(SO_3F)_4$ white solid dec pt 90°C	$Zr(CH_3CO_2)_4 + HSO_3F$	M/A, IR, DTG	93
$V(SO_3F)_3$ dec pt 140°C	$VCl_3 + HSO_3F + KCl$	M/A, IR, μ_{eff}, reflectance spectra, thermal analysis	162
$VO(SO_3F)_3$	$V(CO)_6 + S_2O_6F_2$	M/A, IR, X-ray powder	101
$NbO(SO_3F)_3$	$NbCl_5 + S_2O_6F_2$	M/A	163
$NbF_2(SO_3F)_3$ white powder mp = 120°C	$Nb + S_2O_6F_2$ in HSO_3F	M/A, IR, Raman	164
$TaF_4(SO_3F)$	$Ta + TaF_5 + S_2O_6F_2$ in HSO_3F	M/A, IR, Raman	164
$TaO(SO_3F)_3$	$TaCl_5 + S_2O_6F_2$	M/A	163
$Cr(SO_3F)_3$ pale green solid	$Cr(CO)_6 + S_2O_6F_2$ 60°C, 20 d	IR, elec spec	101
	$CrF_5 + SO_3$	M/A, elec spec, X-ray powder, μ_{eff}	111
$MoO_2(SO_3F)_2$	$Mo(CO)_6 + S_2O_6F_2$	M/A	165
$WF_2(SO_3F)_4$	$WF_6 + SO_3$	M/A $WF_6 \cdot 4.5(SO_3)$	166
$Mn(SO_3F)_2$	$Mn(CH_3CO_2)_2 + HSO_3F$	M/A,	98, 167
$\rho_{23C} = 2.64$		IR, μ_{eff}	91
$Mn(SO_3F)_3$	$Mn + S_2O_6F_2$	M/A, IR, UV, μ_{eff}	168
	$Mn(SO_3F)_2 + S_2O_6F_2$	M/A, IR, UV, μ_{eff}	168
$Mn(SO_3F)_4$ dec pt = 105°C	$Mn_2(CO)_{10} + S_2O_6F_2$	M/A, IR, UV, X-ray powder	101
$MnO(SO_3F)$	$MnCO_3 + S_2O_6F_2$	M/A, IR	103
$ReO_2(SO_3F)_3$	$Re + S_2O_6F_2$	M/A	163
$ReO_3(SO_3F)$	$Re + S_2O_6F_2$	M/A	163
$Fe(SO_3F)_2$	$FeSO_4 + HSO_3F$, 5 h	M/A,	167
		IR, μ_{eff}	91
	$FeCl_2 + HSO_3F$, 1 d	M/A, IR, elec spec, $\mu_{eff}^{201-300K}$, Mössbauer	98, 169
$Fe(SO_3F)_3$ green–gray powder	$FeCl_3 + HSO_3F$	M/A, IR, X-ray powder, $\mu_{eff}^{90-298K}$	92
$Ru(SO_3F)_3$, red brown powder, dec pt 140°C	$Ru + S_2O_6F_2$ in HSO_3F, 60°C, 24 h	M/A, IR, $\mu_{eff}^{77-298K}$ ESR	113
$Os(SO_3F)_3$ green powder dec pt 130°C	$Os + S_2O_6F_2$ in HSO_3F, 60°C, 3 d	M/A, Raman, elec spec, $\mu_{eff}^{77-273K}$	96
$WO(SO_3F)_4$	$W(CO)_6 + S_2O_6F_2$	M/A, Raman	100
	$WCl_6 + S_2O_6F_2$	M/A, Raman	100
$Co(SO_3F)_2$	$CoSO_4 + HSO_3F$	M/A	167
pink powder	$Co[m\text{-}BrC_6H_4CO_2]_2 + HSO_3F$	M/A, IR, elec spec,	91, 98
		$\mu_{eff}^{120-312K}$	170
	$Co(CO)_8 + S_2O_6F_2$, 6 h	M/A, UV	101

TABLE 3.6. Continued

Compound	Synthesis	Analysis	References
$Rh(SO_3F)_3$ orange powder dec pt = 190°C	$Rh + S_2O_6F_2$ in HSO_3F, 130°C, 21 d	M/A, IR	96
$Ir(SO_3F)_3$, purple powder, dec pt = 200°C	$Ir(SO_3F)_4$ 120°C	M/A, IR	104
$Ir(SO_3F)_4$ brown powder dec pt = 100°C	$Ir + S_2O_6F_2$ in HSO_3F, 60°C, 12 h	M/A, IR, $\mu_{eff}^{78-284\,K}$	104
$Ni(SO_3F)_2$ yellow powder	$NiSO_4 + HSO_3F$	M/A	167
		IR, elec spec, μ_{eff}	91
	$Ni(CH_3CO_2)_2 + HSO_3F$	M/A, elec spec, μ_{eff}	91, 167
	$NiCl_2 + HSO_3F$ 14 d	M/A, IR	98
$NiO(SO_3F)$	$NiCO_3 + S_2O_6F_2$	M/A, IR	103
$Pd(SO_3F)_2$ purple powder dec pt 250°C	$Pd + BrSO_3F$, 110°C, 14 d	M/A, IR, Raman, elec spec, $\mu_{eff}^{100-300\,K}$	105, 106
	$Pd[Pd(IV)(SO_3F)_6]$ 160°C	$\mu_{eff}^{2-82\,K}$	134
$Pd[Pd(SO_3F)_6]$ brown powder dec pt 180°C	$Pd + S_2O_6F_2$ in HSO_3F, 80°C, 24 h	M/A, Raman, $\mu_{eff}^{100-300\,K}$	105
		$\mu_{eff}^{5-82\,K}$	134
$Pt(SO_3F)_4$ dark yellow powder mp 220°C	$Pt + BrSO_3F$, 95°C, 21 d	M/A	107
	$Pt + S_2O_6F_2$ in HSO_3F, 120°C, 2 d	IR, Raman, ^{19}F NMR, conductivity, elec spec	109
$Cu(SO_3F)_2$ White powder	$CuSO_4$, $Cu(CH_3COO)_2$,	M/A	167
	$CuCl_2$ or $CuF_2 + HSO_3F$	IR, μ_{eff}	91
	$Cu(C_6H_6CO_2)_2 + HSO_3F$	M/A, IR, elec spec,	98
		$\mu_{eff}^{100-312\,K}$	133
	$CuCl_2 + HSO_3F$	M/A, IR, X-ray powder, $\mu_{eff}^{90-298\,K}$	98
	$Cu + HSO_3F$ (partial reaction)		95
$AgSO_3F$	$Ag + BrF_3 + (NO)_2S_2O_7$	M/A	171
	$Ag(SO_3F)_2$, 210°C	IR	172
	$Ag(SO_3F)_2$, 210°C		97
	$Ag(CO_2CF_3) + HSO_3F$		87
$Ag_3(SO_3F)_4$ black powder dec pt 170°C	Ag, Ag_2O or $AgSO_3F + BrSO_3F$ 25–70°C, 2 d	M/A, IR, $\mu_{eff}^{80-309\,K}$	87
$Ag(SO_3F)_2$ brown powder dec pt 210°C	$Ag + S_2O_6F_2$ 70°C, 7 d	M/A, IR, ESR, $\mu_{eff}^{80-300\,K}$	97
	$Ag + S_2O_6F_2$ in HSO_3F, 25°C, 30 m	M/A, IR, elec spec, ESR, $\mu_{eff}^{80-301\,K}$	87
	Ag_2O, Ag_2CO_3, $AgSO_3F$, $AgSO_3CF_3$ $+ S_2O_6F_2$	M/A, IR, elec spec, ESR, $\mu_{eff}^{80-301\,K}$	87
		$\mu_{eff}^{4-82\,K}$	134
	$AgF_2 + SO_3$	M/A, IR, elec spec, ESR, $\mu_{eff}^{80-301\,K}$	87

TABLE 3.6. Continued

Compound	Synthesis	Analysis	References
	Ag, Ag_2O, or $AgSO_3F$ + $BrSO_3F$	M/A, IR, elec spec, ESR, $\mu_{eff}^{80-301\,K}$	87
	Ag + BrF_3 + SO_3		173
	Ag + HSO_3F		95
$Ag_2O(SO_3F)_2$	Ag_2CO_3 or Ag_2O + $S_2O_6F_2$	M/A	103
$Au(SO_3F)_2$ orange powder	$Au(SO_3F)_3$ + Au in HSO_3F, 22 d	M/A, IR, Raman, ESR	120
$Au(SO_3F)_3$ orange–yellow	Au + $BrSO_3F$, 65°C, 1 d	M/A, Raman, ^{19}F NMR	107
powder dec pt 140°C	Au + $S_2O_6F_2$ in HSO_3F, 3 h	IR, Raman, conductivity, X'tal struc.	108, 114 120
$Zn(SO_3F)_2$	$Zn(CH_3CO_2)_2$ + HSO_3F	M/A	167
white powder	$Zn(C_6H_6CO_2)_2$ + HSO_3F	M/A, IR	98
	$ZnCl_2$ + HSO_3F	M/A, IR, X-ray powder	92
	Zn + $S_2O_6F_2$ in HSO_3F, 90°C, 21 d	IR, Raman	174
$Cd(SO_3F)_2$ white powder	$Cd(C_6H_6CO_2)_2$ + HSO_3F	M/A, IR	98
	Cd + $S_2O_6F_2$ in HSO_3F, 90°C, 28 d	IR, Raman	174
$Hg(SO_3F)_2$	Hg + $S_2O_6F_2$	M/A, IR	98, 99
white powder	Hg + $S_2O_6F_2$ in HSO_3F	IR, Raman	174

[a]See footnotes in Table 3.1 for definition of abbreviations.

valency $Au^{I}[Au^{III}(SO_3F)_4]$.[120] The O-tridentate bridging fluorosulfate group results in MO_6-coordination polyhedra within the layer structure. Regular octahedra are found for M = Fe, Co, Ni, Pd, Zn, Cd, and Hg, while the symmetry of the SO_3F group appears to be reduced for M = Mn, Cu, and Ag with Jahn–Teller distortions expected for Cu^{2+} and Ag^{2+}. This structure type is also postulated for other sulfonates of divalent metals (e.g., $SO_3CF_3^-$ and $SO_3CH_3^-$ derivatives), and also extends to divalent pretransition and posttransition metal salts. The only molecular structure reported so far for this type is that of $Ca(SO_3CH_3)_2$.[121] For the $M(SO_3F)_2$ compounds discussed here electronic spectra and magnetic properties, where applicable and reported, confirm the structural conclusions reached.

A greater structural diversity is encountered for binary fluorosulfates of the general composition $M(SO_3F)_3$. The dimeric structure of $[Au(SO_3F)_3]$ (Ref. 116) and the mixed-valency compound $Pd^{II}[Pd^{IV}(SO_3F)_6]$ are clear exceptions. The mixed-valency formulation follows the precedent of PdF_3 (Ref. 122) and is supported by the magnetic behavior and the synthesis and structural characterization of bimetallic compounds of the types $Pd^{II}[M^{IV}(SO_3F)_6]$ (M = Pt or Sn) and $Ba[Pd^{IV}(SO_3F_6)]$.[114] The fluorosulfate group appears to bond strongly to the M^{IV} metal center and coordinates weakly to M^{II}, in an anisobidentate bonding mode.

Layer structures of the $CrCl_3$ prototype appear to be present in $V(SO_3F)_3$ ($3d^2$), $Cr(So_3F)_3$ ($3d^3$), $Mn(SO_3F)_3$ ($3d^4$), and $Fe(SO_3F)_3$ ($3d^5$). In all compounds, *O*-bidentate bridging fluorosulfate groups are present, but band profilerations found for $Mn(SO_3F)_3$ and, to a lesser extent, for $V(SO_3F)_3$ suggest distorted MO_6^- octahedra. This is expected for $Mn(SO_3F)_3$ on account of Jahn–Teller distortion.

The remaining examples, all involving 4*d* and 5*d* metals, are low-spin complexes. The compounds $Ir(SO_3F)_3$ and $Rh(SO_3F)_3$ are diamagnetic, and $Ru(SO_3F)_3$ as well as both forms of $Os(SO_3F)_3$ are weakly paramagnetic. With Raman spectra generally difficult to obtain on account of the intense colors, very limited structural conclusions may be drawn based on IR spectra, where the presence of both monodentate and bidentate fluorosulfate groups is suggested. For $Ru(SO_3F)_3$,[113] the presence of magnetic moments $\approx 1.0\,\mu_B$ and Curie–Weiss law behavior and the ESR spectra indicate a low-symmetry site for the metal and possibly the presence of two or more different phases in analogy to $RuCl_3$.[123]

Of the tetrakis(fluorosulfates) known, $Ir(SO_3F)_4$ and $Pt(SO_3F)_4$ are similar as seen from their vibrational spectra, which show the presence of both bridging and terminal SO_3F^- groups. Both compounds are soluble in HSO_3F, and $Pt(SO_3F)_4$ enhances the acidity of HSO_3F.[109] The tetrakis(fluorosulfate) $Ir(SO_3F)_4$ displays limited thermal stability and is paramagnetic as would be expected of a ${}^2T_{2g}$ ground state, however, antiferromagnetic coupling is present in $Ir(SO_3F)_4$ and its derivative $Cs_2[Ir(SO_3F)_6]$.[104] The reported IR spectrum for $Zr(SO_3F)_4$ is apparently poorly resolved and unambiguous conclusions cannot be drawn based on this alone.[93]

In summary, it appears from this discussion and the listings in Table 3.6 that of all the anions discussed, the fluorosulfate group is capable of coordinating to the largest number of transition metal ions. When recent work from our group (see Section 3.3) is included, it appears that at least one binary fluorosulfate of all the transition metals is known, with oxidation states commonly ranging from +2 to +4. Exceptions are tungsten and rhenium, where only oxyfluorosulfates have been isolated. No synthesis appears to have been attempted for technetium. This large number of known compounds may be attributed to two major causes:

1. The important and unique role fluorosulfuric acid (HSO_3F) and bis(fluorosulfuryl) peroxide ($S_2O_6F_2$) play as synthetic reagents, either individually or in combination.

2. The versatile coordinating ability of the fluorosulfate group, which may function as a mono-, bi-, or tridentate ligand. This results in the stabilization of metal ions in high, intermediate, and low oxidation states. Where the SO_3F group functions as a polydentate ligand, a bridging configuration is evident, resulting in stable solids, often viewed as coordination polymers.

In addition, as the examples cited in this chapter indicate, transition metal fluorosulfates play a useful role as starting materials in the conversion of fluorosulfates to hexafluoro antimonates or trifluoromethylsulfates by solvolysis in either SbF_5 or HSO_3CF_3.

3.2.2.D. Trifluoromethylsulfates. In contrast to the large number of binary transition metal fluorosulfate compounds reported, only a small number of trifluoromethylsulfate derivatives are known. Almost all the $M(SO_3CF_3)_n$ compounds are made by the solvolysis of suitable metal salts in an excess of trifluoromethylsulfuric acid, HSO_3CF_3.

The two compounds that are not prepared by solvolyses are $CuSO_3CF_3$, made from the reduction of $Cu(SO_3CF_3)_2$ by copper metal,[124] and $AgSO_3CF_3$, made via metathesis from Ag_2CO_3 and $Ba(SO_3CF_3)_2$ in H_2O.[125,126] The latter method is feasible, because the SO_3CF_3 group is not hydrolyzed by water.

In the solvolysis of $M(SO_3F)_x$ in HSO_3CF_3,[102,127] the reaction initially proceeds according to:

$$M(SO_3F)_x + xHSO_3CF_3 \xrightarrow{HSO_3CF_3} M(SO_3CF_3)_x + xHSO_3F \qquad (3.40)$$

where $x = 2$ or 3, M = Mn, Pd, Ag, or Au.
However, the by-product HSO_3F and the reactant HSO_3CF_3 undergo a degradation reaction and produce a series of products according to:[128,129]

$$2HSO_3CF_3 + HSO_3F \xrightarrow{25°C} COF_2, CO_2, SO_2, CF_3SO_3F, CF_3SO_3CF_3, SiF_4, \ldots \qquad (3.41)$$

The largely volatile products do not appear to interfere in the isolation of the trifluoromethylsulfates.[102]

Two potential routes to transition metal trifluoromethylsulfates may be mentioned briefly. Bromine(I) trifluoromethylsulfate ($BrSO_3CF_3$) may be useful in Cl versus SO_3CF_3 exchange reactions. This approach has been used successfully in the synthesis of $VO(SO_3CF_3)_3$ and $CrO_2(SO_3CF_3)_2$ from chloride precursors.[130] The limited thermal stability of $BrSO_3CF_3$ makes it unsuitable for use in metal oxidation reactions, where elevated temperatures and long reaction times are often required. The anhydride of HSO_3CF_3, $(CF_3SO_2)_2O$, may be useful in reactions with metal alkoxides. So far, only partial substitution is reported resulting in the synthesis of $Ti(OCH_3)_2(SO_3CF_3)_2$,[131] as seen in Table 3.7.

For the iron, cobalt, and copper derivatives, $Fe(SO_3CF_3)_2$,[132] $Co(SO_3CF_3)_2$, and $Cu(SO_3CF_3)_2$,[133] a layered lattice structure involving hexacoordinated metal centers has been suggested on the basis of vibrational and electronic spectra, as well as magnetic and Mössbauer data. The SO_3CF_3 groups act as bridging tridentate ligands.

However, in contrast both $Ag(SO_3F)_2$ (Refs. 87 and 97) and $Ag(SO_3CF_3)_2$ (Ref. 127) indicate primarily bidentate bridging SO_3F and SO_3CF_3 groups,

TABLE 3.7. Transition Metal Trifluoromethylsulfates.[a]

Compound	Synthesis	Analysis	References
$TiCl_3SO_3CF_3$	$TiCl_4 + HSO_3CF_3$	M/A	175
$TiCl_2(SO_3CF_3)_2$	$TiCl_4 + HSO_3CF_3$ 60°C	M/A	175
$TiCl(SO_3CF_3)_3$	$TiCl_4 + HSO_3CF_3$, 100°C	M/A	175
$Zr(SO_3CF_3)_4$	$ZrCl_4 + HSO_3CF_3$	M/A	175
$V(SO_3CF_3)_3$ green solid	$V(CO_2CF_3)_3 +$ HSO_3CF_3 in HCO_2CF_3 25°C, 8 h	M/A, IR, DTG, μ_{eff}	176
$TaF_4(SO_3CF_3)$	$Ta(SO_3F)_5 + HSO_3CF_3$, 25°C, 7 d	M/A, IR	164
$Cr(SO_3CF_3)_3$ green solid	$CrO_2(SO_3CF_3) + (CF_3SO_2)_2O$, 8 d	M/A, IR, UV, X-ray powder	111, 177
$Mo_2(SO_3CF_3)_4$ yellow–red solid	$Mo_2(CO_2CH_3)_4 + HSO_3CF_3$, 100°C	IR, MS	140
$Fe(SO_3CF_3)_2$ white powder	$FeCl_2 + HSO_3CF_3$ 70°C, 40 h	M/A, IR, Elec spec $\mu_{eff}^{80-178K}$ Mössbauer	132
$Fe(SO_3CF_3)_3$ white powder	$FeCl_3 + HSO_3CF_3$ 70°C, 40 h	M/A, IR, $\mu_{eff}^{80-306K}$ Mössbauer	132
$Co(SO_3CF_3)_2$ pink powder	$Co[m\text{-}BrC_6H_4CO_2]_2 + HSO_3CF_3$, 56°C, 2 d, 8 torr	M/A, IR, Elec Spec, $\mu_{eff}^{84-310K}$	133
		M/A, elec spec, μ_{eff}	178
$Ni(SO_3CF_3)_2$ yellow powder	$NiCO_3 + HSO_3CF_3$	M/A, elec spec, μ_{eff}	178
$Pd(SO_3CF_3)_2$ purple powder	$Pd(SO_3F)_3 + HSO_3CF_3$, 4 d	IR	102
$Cu(SO_3CF_3)$	$Cu(SO_3CF_3)_2 + Cu + HSO_3CF_3$, in CH_3CN		124
$Cu(SO_3CF_3)_2$ white powder	$CuCO_3 + HSO_3CF_3$		124
	$Cu(CH_3CO_2)_2 + HSO_3CF_3$		124
	$Cu(C_6H_5CO_2)_2 + HSO_3CF_3$, 70°C, 20 torr	M/A, IR, elec spec, $\mu_{eff}^{127-311QK}$	133
$Ag(SO_3CF_3)$	$Ag_2CO_3 + Ba(SO_3CF_3)_2$	M/A, IR	125
	in H_2O	IR	126
$Ag(SO_3CF_3)_2$, dark brown powder dec pt 140°C	$Ag(SO_3F)_2 + HSO_3CF_3$, 2 d	M/A, IR, ESR	127
$Au(SO_3CF_3)_3$ yellow brown powder dec pt = 160°C	$Au(SO_3F)_3 + HSO_3CF_3$, 100°C, 2 d	M/A, IR	127
$Hg(SO_3CF_3)_2$			179

[a]See footnotes in Table 3.1 for definition of abbreviations.

respectively, suggesting approximately square planar or tetragonally elongated coordination of Ag^{2+}.

The compounds $Fe(SO_3CF_3)_2$, $Co(SO_3CF_3)_2$, and $Cu(SO_3CF_3)_2$ appear to be magnetically dilute down to 80, 84, and 127 K, respectively. For

$Fe(SO_3CF_3)_3$ magnetic moments decrease with decrease in temperature down to 80 K, possibly due to antiferromagnetic interactions. Antiferromagnetism is shown for $Ag(SO_3CF_3)_2$ between 298–80 K,[137] while its precursor $Ag(SO_3F)_2$ is ferromagnetic at low temperatures.[134]

Although the solvolysis of transition metal fluorosulfates in excess trifluoromethylsulfuric acid should allow the synthesis of a comparatively large number of SO_3CF_3 compounds since this method is generally applicable,[102] only limited use has been made of this synthetic possibility. The principal reasons for this are threefold: (a) As discussed in Section 3.1, the SO_3CF_3 group has a comparable basicity to that of the SO_3F group. An extended synthetic approach is not fruitful, where compounds with similar properties and molecular structures result, as is often the case. (b) With vibrational spectroscopy used as a principal method of structural analysis, the coincidence of SO_3 and CF_3 stretching modes in $SO_3CF_3^-$ causes a greater complexity, making vibrational assignments and structural conclusions frequently uncertain and ambiguous. (c) The S—C linkage in HSO_3CF_3 and its derivatives is sensitive to oxidative cleavage,[102] and its use in the synthesis of high-valent metal derivatives with a good oxidizing potential is not advisable.

There are nevertheless a number of interesting cases, where metal trifluoromethylsulfates display fundamentally different magnetic properties as the corresponding fluorosulfates do, as will be discussed in Section 3.3.

3.3. RECENT DEVELOPMENTS

In this section, a number of recent, only partly published developments will be summarized. They involve new synthetic initiatives, reveal unusual magnetic properties, and suggest synthetic pathways to unusual species. All studies involve either fluorosulfates or hexafluoroantimonates, which rank among the most weakly basic anionic ligands.

3.3.1. New Binary Fluorosulfates of the Early Transition Metals

It has recently been possible to extend the general route of metal oxidation in HSO_3F under relatively mild conditions to the Group 4 metals Ti, Zr, and Hf, as well as to Nb and Ta. The Group 4 metals are converted readily to the corresponding tetrakis(fluorosulfates), $M(SO_3F)_4$ (M = Ti, Zr, and Hf). Only $Zr(SO_3F)_4$ has been reported previously, obtained via solvolysis of the corresponding trifluoroacetate.[93] All three resulting compounds act as SO_3F^- acceptors, as is evidenced by their reaction with 2 mol of $CsSO_3F$ in HSO_3F solution. The resulting salts, containing the hexakis(fluorosulfato) metallate(IV) anion $[M(SO_3F)_6]^{2-}$ (M = Zr or Hf), show similar thermal stability, in excess of 250 °C. All give nearly identical vibrational spectra, similar to those reported for $Cs_2[Sn(SO_3F)_6]$,[17] where the ^{119}Sn Mössbauer spectrum suggests octahedral coordination for tin, $Cs_2[Ge(SO_3F)_6]$,[135] $Cs_2[Pt(SO_3F)_6]$,[109] and $Cs_2[Pd(SO_3F)_6]$.[114]

However, their use as Lewis acids in HSO_3F, to give viable superacids, for example, the HSO_3F—$Pt(SO_3F)_4$ system,[109] is doubtful, because both $Zr(SO_3F)_4$ and $Hf(SO_3F)_4$ are fluorosulfate bridged polymers, and like $Sn(SO_3F)_4$ (Ref. 136) lack any appreciable solubility in fluorosulfuric acid. Only $Ti(SO_3F)_4$, which is a viscous, oily liquid is somewhat soluble.

While this apparent lack of solubility in HSO_3F has permitted the facile isolation of the Group 4 fluorosulfates, the very high solubility of the binary fluorosulfates of niobium and tantalum, has on one hand prevented their isolation, but on the other hand, permitted solution studies of $M(SO_3F)_5$ (M = Ta or Nb), generated *in situ*, and the development of novel superacid systems[137] with $Ta(SO_3F)_5$—HSO_3F being the stronger of the two. This is probably due to a lower degree of association.[90] The fluorosulfate acceptor ability of $M(SO_3F)_5$ (M = Nb or Ta), generated *in situ*, is also evident from the reaction with $CsSO_3F$, resulting in two series of salts of the types $Cs[M(SO_3F)_6]$ and $Cs_2[M(SO_3F)_7]$ which are fully characterized.

More recently, it has become apparent that both niobium and tantalum may also be oxidized by bis(fluorosulfuryl)peroxide in the absence of HSO_3F, however, these reactions require prolonged times to reach completion (about 1–2 weeks). It becomes subsequently very difficult to remove the excess $S_2O_6F_2$ from the viscous, oily reaction products, which requires gentle heating, which in turn causes partial decomposition with the elimination of SO_3.

While analytically pure, undecomposed $M(SO_3F)_5$ has not been obtained, the oxidation reaction may be used to directly and conveniently form fluoride fluorosulfates of niobium and tantalum according to the general equation

$$nM + (5-n)MF_5 + n/2\,S_2O_6F_2 \xrightarrow{S_2O_6F_2} 5MF_{(5-n)}(SO_3F)_n \qquad (3.42)$$

where $n = 1$–4; M = Nb or Ta.

All the products obtained are clear colorless liquids, with the viscosity increasing with increasing SO_3F^- content. Of these materials, MF_4SO_3F and $MF_3(SO_3F)_2$ may be distilled undecomposed in vacuo. Their vibrational spectra are nearly identical to those obtained for the corresponding antimony(V) species[138] and suggest fluorosulfato bridged polymers. Like their antimony(V) analogues,[139] generated *in situ*, the solutions in HSO_3F increase in acidity with increasing fluorosulfate substitution on Ta or Nb.

All attempts to obtain binary fluorosulfates of molybdenum have failed, however, and resulted in the formation of $MoO(SO_3F)_4$ and the previously reported $MoO_2(SO_3F)_2$. But solvolysis of $Mo_2(CH_3CO_2)_4$ in HSO_3F has allowed isolation of the first, and so far only, binary binuclear fluorosulfate ($Mo_2(SO_3F)_4$) which is sufficiently volatile to allow analysis by mass spectrometry. The solvolysis reaction follows a precedent, the formation of $Mo_2(SO_3CF_3)_4$.[140]

3.3.2. Unusual Magnetic Properties of Electron Rich Transition Metal Derivatives

The magnetic properties reported for the transitional metal compounds covered in this survey follow, for the most part, in particular where 3*d*-block metals are involved, expected trends,[141,142] for complexes formed by weakly coordinating, weak field ligands, with commonly octahedral geometries about the metal and with layer structures in case of the MX_2 compounds ($X = SO_3F$ or SbF_6). Hence, it is not unexpected that palladium(II) fluoride[122] and fluorosulfate[105,108] are paramagnetic, suggesting octahedral, rather than square planar coordination, and a $^3A_{2g}$ ground state. It is also not surprising that the fluorosulfate ligands provide a magnetically dilute environment, and that the magnetic susceptibilities follow the Curie–Weiss law down to 80 K, while fluoro complexes of palladium(II) and PdF_2 are magnetically concentrated with relatively low magnetic moments.

The observed ferromagnetism at very low temperatures (below 40 K) found for $Pd(SO_3F)_2$, $Pd[Pd(SO_3F)_6]$, $Ag(SO_3F)_2$,[134] and also for $Ni(SO_3F)_2$ is unprecedented for coordination polymers with polyatomic, anionic ligands. There have been reports of weak ferromagnetism for both PdF_2 (Ref. 143) and AgF_2 (Refs. 141 and 142) and $Pd^{II}[Pd^{IV}F_6]$.[144] However, in these materials antiferromagnetic ordering appears to dominate, with ferromagnetism due to a canting of spins. Similar complex magnetic behavior termed "metamagnetism" is reported for $NiCl_2$, $CoCl_2$, and $FeCl_2$.[145] In all instances, small monoatomic ligands are involved.

Ferromagnetic ordering of Pd^{II}, Ag^{II}, and Ni^{II} fluorosulfates is not observed for the corresponding trifluoromethylsulfates, as already discussed for the Ag^{II} compounds, which show antiferromagnetic ordering in all instances. Interestingly, as seen in Fig. 3.1, where magnetic moments are plotted against temperature, the onset of magnetic ordering for $Ni(SO_3X)_2$ ($X = F$ or CF_3) becomes observable at about the same temperature, but takes different directions.

The corresponding hexafluoroantimonates, $M[SbF_6]_2$ (M = Ni, Pd, Cu, or Ag), show diverging complex magnetic behavior, with the absence of ferromagnetism. The nickel, copper, and palladium compounds obtained from the fluorosulfates by solvolysis in SbF_5 (Ref. 24) follow the Curie–Weiss law, and show antiferromagnetism at very low temperatures (≈ 10 K). The magnetic moments observed at temperatures above 10 K are invariant with temperature but unprecedentedly low,[141,142] and fall below the spin only values of 1.73 or 2.83 μ_B for $Cu[SbF_6]_2$ and $Ni[SbF_6]_2$, respectively.

The compound $Ag[SbF_6]_2$, obtained by solvolysis of $Ag(SO_3F)_2$, is diamagnetic and formulated as a mixed-valency Ag^IAg^{III} species with the metal in octahedrally compressed, near-linear, and elongated, near-square planar environments, respectively.[24] However, a true Ag^{II} compound is also known[19] and may be converted to the mixed-valency form. This appears to be a first case for valence isomerism for a silver(II) salt. It appears reasonable to assume a similar situation for $Cu[SbF_6]_2$ with Cu^I, Cu^{II}, and Cu^{III} present in an equilibrium of the

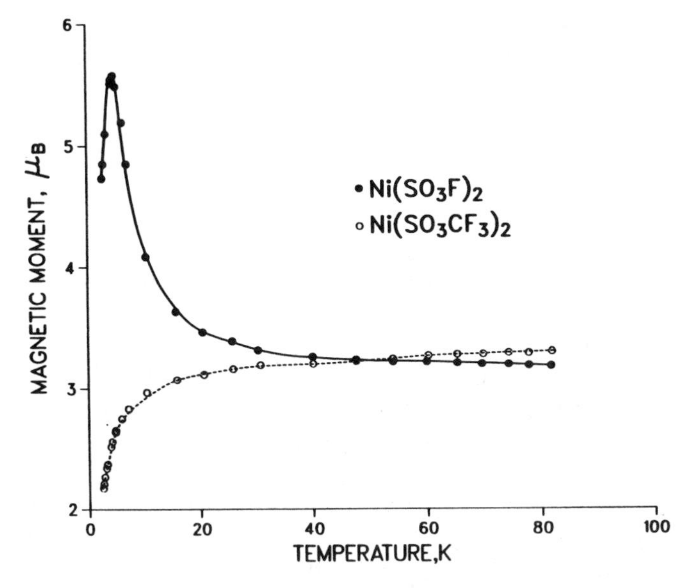

Figure 3.1. Magnetic moment of $Ni(SO_3F)_2$ and $Ni(SO_3CF_3)_2$ vs temperature.

type:

$$2Cu^{II} \rightleftharpoons Cu^{I}Cu^{III} \tag{3.43}$$

with the exact position differing with the experimental conditions of the formation reaction, and the subsequent treatment of the reaction product.

For $Ni[SbF_6]_2$, four different synthetic routes are reported (see Section 3.2), and magnetic moments μ_{eff}^{298} vary widely between 2.65 and 4.25 μ_B, with the highest values found for the product formed by the reaction of Ni with F_2 in the presence of SbF_5 (Ref. 34) and the lowest values for the $Ni[SbF_6]_2$ formed by solvolysis of the fluorosulfate in liquid SbF_5.[24] Since the spin–orbit coupling constant for Ni^{2+} is relatively small (630 cm^{-1}), its involvement in state mixing cannot account for μ_{eff} values as high as 4.25 μ_B,[141,142] and the presence of Ni^{3+} or even Ni^{4+} is suspected.

Many of the observations summarized here are not readily explained, and more detailed magnetic measurements as well as ESR studies into a wider range of compounds with layer structures formed by weakly basic fluoro anions are required. However, the examples cited here point to a very fruitful area of research into unusual magnetic behavior.

3.3.3. The Reduction of Gold(III) Fluorosulfate, $[Au(SO_3F)_3]_2$

Gold(III) fluorosulfate, $[Au(SO_3F)_3]_2$, one of the best characterized oxyacid derivatives of gold(III),[146,147] has a dimeric molecular structure and is soluble in fluorosulfuric acid, giving rise to a strong, monoprotonic superacid.[108] Attempts to protonate carbon monoxide (CO) resulted instead in the formation of gold(I) carbonyl derivatives according to:

$$3CO_{(g)} + Au(SO_3F)_{3(solv)} \longrightarrow [Au(CO)_2]^+_{(solv)} + S_2O_5F_2 + CO_{2(g)} + SO_3F^-_{(solv)} \quad (3.44)$$

The bis(carbonyl) cation of gold(I), $[Au(CO)_2]^+$, initially characterized in solution by vibrational spectroscopy[148] and ^{13}C NMR, has now also been obtained in solid form as $[Au(CO)_2][Sb_2F_{11}]$. From HSO_3F solution, $Au(CO)SO_3F$, only the second example of a mononuclear gold(I) carbonyl derivative, is isolated by sublimation. This species, as well as the five atomic linear $[Au(CO)_2]^+$ cation ($D_{\infty h}$ symmetry), reveal an unusual type of CO bonding to a transition metal, because the CO stretching frequency is raised to approximately 2200–2250 cm^{-1}, well beyond the 2143 cm^{-1} value reported for CO itself.[149] Similarly, high ν_{CO} values are found also for *cis*-$Pt(CO)_2(SO_3F)_2$ formed in the reduction of $Pt(SO_3F)_4$ in HSO_3F solution.

It may be recalled that the conventional model of synergic metal–CO bonding would require a lowering of the CO stretching frequencies, usually for terminally bonded CO groups to wavenumbers between 2100 and 1850 cm^{-1}, due to π back donation from filled metal d orbitals to antibonding π^* levels of the CO group. Such π-back donation must be absent in the instances discussed above, and also in $[Cu(CO)]^+[AsF_6]^-$ (Ref. 150) and Au(CO)Cl.[151] The compounds formed are thermally stable—for example, $Au(CO)SO_3F$ will decompose at approximately 190°C with CO evolution.

Similarly, the product of the reduction of gold(III) fluorosulfate dissolved in fluorosulfuric acid by gold powder is new. In addition to diamagnetic $Au(SO_3F)_2$, already discussed in the preceding section, which is insoluble in HSO_3F and interpreted as a mixed-valency Au^I—Au^{III} compound, an ESR active species is obtained in solution where $Au(SO_3F)_3$ is present in excess, and is characterized by an unusually large g_{iso} value, with the observation of a four line hyperfine splitting pattern due to the $\frac{3}{2}$ spin of ^{197}Au.

A nearly identical ESR spectrum (see Fig. 3.2) is obtained when solid $[Au(SO_3F)_3]_2$ is partly pyrolyzed, with the reductive elimination of $SO_3F^{\bullet}$ radicals which, depending on the reaction conditions, may remain trapped in the solid. Such samples have also permitted magnetic susceptibility measurements, and Curie–Weiss law is found to be obeyed.

All the evidence from the magnetic measurements, the ESR spectra, and the close similarity to the corresponding silver(II) fluorosulfate and its complexes,[87] suggests that the highly unusual Au^{2+} ion, with a $5d^9$ electron configuration, is the cause of the observed paramagnetism. This ion has not, so far, been unambiguously identified.

While it has not been possible to obtain paramagnetic $Au(SO_3F)_2$, presumably on account of disproportionation to a Au^I—Au^{III} species, the study briefly

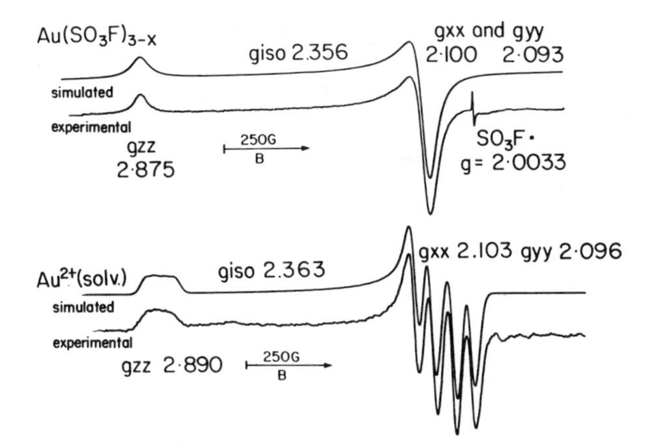

Figure 3.2. ESR Spectra of Au^{2+}

summarized here indicates under what conditions this disproportionation may be prevented and how unusual cations, stabilized by weakly basic anions, may be detected. It is expected that a wider range of metal ions with the metal in an unusual oxidation state can be generated and studied in this manner.

3.4. GENERAL SUMMARY

The survey undertaken here allows a few general conclusions to be made. The types of compounds formed by the various anions reflect the coordinating ability and thermal stability of these anions, while their number reflect the availability of versatile and useful synthetic reagents. There is not a single, universally applicable synthetic route available, but most of the groups mentioned have a principal synthetic reagent (and reaction type).

The Lewis acids AsF_5 and SbF_5 are the reagents of choice for the synthesis of hexafluoroarsenates and hexafluoroantimonates on account of their oxidizing power in SO_2 as a solvent, and their ability to abstract fluoride ions in HF solutions. In addition, the solvolysis of fluorosulfates in liquid antimoney(V) fluoride widens the scope of synthetic reactions. The coordinating ability and low basicity of the anions restricts the range of transition metal compounds to those having metals in the $+1$ or $+2$ oxidation state, and allows, in the case of mercury, the retention of Hg—Hg bonds.

The transfer reagent $B(OTeF_5)_3$, and to a lesser extent $Hg(OEF_5)_2$ (E = Te or Se), plays a similar important role. The range of compounds obtained is limited by the thermal stability of the anion, which is very apparent from the small number of binary $OSeF_5^-$ derivatives known, and, equally importantly, by the preference of the anion to function almost solely as a monodentate ligand.

In the case of the fluorophosphates, the lack of a single useful reagent has spawned a number of interesting, specific synthetic routes, but has limited the number of transition metal derivatives reported so far to compounds with the metal mainly in a +2 or +3 oxidation state.

As discussed already, the availability of fluorosulfuric acid and bis(fluorosulfuryl) peroxide as reagents on a large scale, together with the versatile coordinating ability and the reasonably high thermal stability of the fluorosulfate ion, have been responsible for the large number of transition metal compounds. The possibility of converting any fluorosulfate into a trifluoromethylsulfate by solvolysis of $M(SO_3F)_n$ in an excess of HSO_3CF_3 has great potential. This potential has so far not been widely realized.

ACKNOWLEDGEMENTS

The help of Sheri Harbour and Rita Margitay in producing this manuscript is gratefully acknowledged. We are indebted to Professor R. C. Thompson of this department and Professor H. Willner of the Universität Hannover, Germany, for many helpful discussions, and for the permission to include unpublished material. Finally, the members of our research group, past and present, are thanked for their assistance.

REFERENCES

1. Olah, G. A., Prakash, G. K. S., and Sommer, J., in *Superacids*, Wiley, New York, 1985.
2. Fabre, P. L., Devnyk, J., and Tremillon, B. *Chem. Rev.*, **82**, 591–614 (1982).
3. Dove, M. F. A. and Clifford, A. F., in *Chemistry in Non-Aqueous Ionizing Solvents* (J. Jander, H. Spandau, and C. C. Addison, Eds.), Vol. 2.1, Vieweg, Wiesbaden, 1971, pp. 121–300.
4. Lawrance, G. A., *Chem. Rev.*, **86**, 17–33 (1986).
5. Jache, A. W., *Adv. Inorg. Chem. Radiochem.*, **16**, 177–200 (1974).
6. Woolf, A. A., in *New Pathways in Inorganic Chemistry* (E. A. V. Ebsworth, A. G. Maddock, and A. G. Sharpe, Eds.), Cambridge University Press, London and New York, 1968, pp. 327–362.
7. Aubke, F. and DesMarteau, D. D., *Fluorine Chem. Rev.*, **8**, 73–118 (1977).
8. Howells, R. D. and McCown, J. D., *Chem. Rev.*, **77**, 69–92 (1977).
9. Hagenmüller, P., *Inorganic Solid Fluorides*, Academic, London and New York, 1985.
10. Dehnicke, K. anhd Shihada, A. F., in *Structure and Bonding* (J. D. Dunitz, P. Hemmerich, R. H. Holm, J. A. Ibers, C. K. Jorgensen, J. B. Neilands, D. Reinen, and R. J. P. Williams, Eds.), Vol. 28, Springer-Verlag, Berlin, 1976, pp. 51–82.
11. Seppelt, K., *Angew. Chem. Int. Ed. Engl.*, **21**, 877–956 (1982).
12. Engelbrecht, A. and Sladky, F., *Adv. Inorg. Chem. Radiochem.*, **24**, 189–223 (1981).
13. Wilkinson, G. Ed., in *Comprehensive Coordination Chemistry*, Vol. 3, Pergamon Press, Oxford, 1987.

14. Mallela, S. P., Yap, S., Sams, J. R., and Aubke, F., *Inorg. Chem.*, **25**, 4327–4329 (1986).
15. Pauling, L., in *The Nature of the Chemical Bond*, 3rd ed., Cornell University Press, New York, 1960.
16. Taft, R. W. Jr., in *Steric Effects in Organic Chemistry*, (M.S. Newman, Ed.), Wiley, New York, 1956, pp. 556–675.
17. Yeats, P. A., Sams, J. R., and Aubke, F., *Inorg. Chem.*, **12**, 328–331 (1973).
18. Yeats, P. A., Sams, J. R., and Aubke, F., *Inorg. Chem.*, **11**, 2634–2641 (1972).
19. Gantar, D., Leban, I., Frlec, B., and Holloway, J. H., *J. Chem. Soc. Dalton Trans.*, 2379–2383 (1987).
20. Felec, B., Gantar, D., and Holloway, J. H., *J. Fluorine Chem.*, **19**, 485–500 (1982).
21. Frlec, B., Gantar, D., and Holloway, J. H., *J. Fluorine Chem.*, **20**, 217–226 (1982).
22. Frlec, B., Gantar, D., and Holloway, J. H., *J. Fluorine Chem.*, **20**, 385–396 (1982).
23. Frlec, B., Gantar, D., and Holloway, J. H., *Vestn. Slov. Kem. Drus.*, **34**, 317–323 (1987).
24. Cader, M. S. R. and Aubke, F., *Can. J. Chem.*, **67**, 1700–1707 (1989).
25. Dean, P. A. W., *J. Fluorine Chem.*, **5**, 499–507 (1975).
26. Cutforth, B. D., Davies, C. G., Dean, P. A. W., Gillespie, R. J., Ireland, P. R., and Ummat, P. K., *Inorg. Chem.*, **12**, 1343–1347 (1973).
27. Court, T. L. and Dove, M. F. A., *J. Fluorine Chem.*, **6**, 491–498 (1975).
28. Desjardins, C. D. and Passmore, J., *J. Fluorine Chem.*, **6**, 379–388 (1975).
29. Brown, D., Cutforth, B. D., Davies, C. G., Gillespie, R. J., Ireland, P. R., and Vekris, J. E., *Can. J. Chem.*, **52**, 791–793 (1974).
30. Golic, L. and Leben, I., *Acta Cryst.*, **B33**, 232–234 (1977).
31. Gillespie, R. J. and Landa, B., *Inorg. Chem.*, **12**, 1383–1388 (1973).
32. Gantar, D., Frlec, B., Russell, D. R., and Holloway, J. H., *Acta Cryst.*, **C43**, 618–620 (1987).
33. Cutforth, B. D., Gillespie, R. J., Ireland, P. R., Sawyer, J. F., and Ummat, P. K., *Inorg. Chem.*, **22**, 1344–1347 (1983).
34. Christe, K. O., Wilson, W. W., Bougon, R. A., and Charpin, P., *J. Fluorine Chem.*, **34**, 287–298 (1987).
35. Meuwsen, A. and Mögling, H., *Z. Anorg. Allg. Chem.*, **285**, 262–270 (1956).
36. Baillie, M. J., Brown, D. H., Moss, K. C., and Sharp, D. W. A., *Proc. Int. Conf. Coord. Chem., 8th, Vienna,* 322–324 (1964).
37. Dukat, W. and Naumann, D., *Rev. Chim. Min.*, **23**, 589–603 (1986).
38. Gutman, V. and Emeléus, H. J., *J. Chem. Soc.*, 1046–1049 (1950).
39. Sharp. D. W. A. and Sharpe, A. G., *J. Chem. Soc.*, 1855–1858 (1956).
40. Woolf, A. A. and Emeléus, H. J., *J. Chem. Soc.*, 1050–1052 (1950).
41. Cox, B., *J. Chem. Soc.*, 876–878 (1956).
42. Brown, I. D., Gillespie, R. J., Morgan, K. R., Tun, Z., and Ummat, P. K., *Inorg. Chem.*, **23**, 4506–4508 (1984).
43. Brown, I. D., Gillespie, R. J., Morgan, K. R., Sawyer, J. F., Schmidt, K. J., Tun, Z., Ummat, P. K., and Vekris, J. E., *Inorg. Chem.*, **26**, 689–693 (1987).
44. Müller, B. G., *Angew. Chem. Int. Ed. Engl.*, **26**, 689–690 (1987).

45. Seppelt, K., *Angew. Chem. Int. Ed. Engl.*, **15**, 44–45 (1976).
46. Seppelt, K., *Angew. Chem. Int. Ed. Engl.*, **11**, 630 (1972).
47. Engelbrecht, A. and Sladky, F., *Angew. Chem. Int. Ed. Engl.*, **3**, 383 (1964).
48. Engelbrecht, A. and Sladky, F., *Monatsh. Chem.*, **96**, 159 (1968).
49. Engelbrecht, A. and Rode, R. M., *Monatsh. Chem.*, **103**, 1315–1319 (1972).
50. Seppelt, K., *Chem. Ber.*, **108**, 1823–1829 (1975).
51. Sladky, F., Kropschofer, H., and Leitzke, O., *J. Chem. Soc. Chem. Commun.*, 134–135 (1973).
52. Lentz, D. and Seppelt, K., *Angew. Chem. Int. Ed. Engl.*, **18**, 66–67 (1979).
53. Lentz, D., Pritzkow, H., and Seppelt, K., *Angew. Chem. Int. Ed. Engl.*, **18**, 729–730 (1977).
54. Lentz, D., Pritzkow, H., and Seppelt, K., *Inorg. Chem.*, **17**, 1926–1931 (1978).
55. Huppmann, P., Labischinski, H., Lentz, D., Pritzkow, H., and Seppelt, K., *Z. Anorg. Allg. Chem.*, **487**, 7–25 (1982).
56. Tempelton, L. K., Templeton, D. H. Bartlett, N., and Seppelt, K., *Inorg. Chem.*, **15**, 2720–2722 (1976).
57. Seppelt, K., *Chem. Ber.*, **109**, 1046–1052 (1976).
58. Huppmann, P., Hartl, H., and Seppelt, K., *Z. Anorg. Allg. Chem.*, **524**, 26–32 (1985).
59. Jones, P. G., *Gold Bull.*, **14**, 102–118, 159–166 (1981).
60. Jones, P. G., *Gold Bull.*, **16**, 114–124 (1983).
61. Jones, P. G., *Gold Bull.*, **19**, 46–57 (1986).
62. Schroeder, K. and Sladky, F., *Chem. Ber.*, **113**, 1414–1419 (1980).
63. De Marco, R. A. and Shreeve, J. M., *Adv. Inorg. Chem. Radiochem.*, **16**, 109–176 (1974).
64. Seppelt, K. and Nothe, D., *Inorg. Chem.*, **12**, 2727–2730 (1973).
65. Lange, W., in *Fluorine Chemistry* (J. H. Simons, Ed.), Vol. 1, Academic, New York, 1950, pp. 125–188.
66. Lange, W. and Livingston, R., *J. Am. Chem. Soc.*, **69**, 1073–1076 (1947).
67. Lange, W. and Livingston, R., *J. Am. Chem. Soc.*, **72**, 1280–1281 (1950).
68. Semmoud, A., Benghalem, A., and Addou, A., *J. Fluorine Chem.*, **46**, 1–6 (1990).
69. Singh, E. B. and Sinha, P. C., *J. Indian Chem. Soc.*, **41**, 407–414 (1964).
70. Singh, E. B. and Sinha, P. C., *J. Indian Chem. Soc.*, **47**, 491–492 (1970).
71. Hill, O. F. and Audrieth, L. F., *Inorg. Synth. III*, 106–110 (1950).
72. (a) Semoud, A., Vast, P., Sombret, B., and Legrand, P., *Rev. Chim. Miner.*, **21**, 28–33 (1984). (b) Chaigneau, M. and Santarromana, M., C.R. *Acad. Sci. Ser.*, **C278**, 1453–1455 (1974).
73. (a) Durand, J., Cot, L., Berraho, M., and Rafiq, M., *Acta Cryst.*, **C-43**, 611–613 (1987). (b) Durand, J., Larbot, A., Cot, L., Duprat, M., and Dabosi, F., *Z. Anorg. Allg. Chem.*, **504**, 163–172 (1983).
74. Tan, T. H. (1970), *M. Sc. Thesis*, University of British Columbia, Vancouver, Canada.
75. Dove, M. F. A., Hibbert, R. C., and Logan, N., *J. Chem. Soc. Dalton Trans.*, 707–710 (1985).

76. Begley, M. J., Dove, M. F. A., Hibbert, R. C., Logan, N., Nunn, M., and Sowerby, D. B., *J. Chem. Soc. Dalton Trans.*, 2433–2436 (1985).
77. Vast, P., Semmoud, A., Addou, A., and Palavit, G., *J. Fluorine Chem.*, **38**, 297–302 (1988).
78. Brown, S. D., Emme, L. M., and Gard, G. L., *J. Inorg. Nucl. Chem.*, **37**, 2557–2558 (1975).
79. Weidlein, J., *Z. Anorg. Allg. Chem.*, **358**, 13–20 (1968).
80. Pebler, J. and Dehnicke, K., *Z. Naturforsch.*, **B26**, 747–750 (1971).
81. Dalziel, J. R., Klett, R. D., Yeats, P. A., and Aubke, F., *Can. J. Chem.*, **52**, 231–239 (1974).
82. Thorpe, T. E. and Kirman, W., *J. Chem. Soc.*, 921–924 (1892).
83. Barr, J. Gillespie, R. J., and Thompson, R. C., *Inorg. Chem.*, **3**, 1149–1156 (1964).
84. Thompson, R. C., in *Inorganic Sulfur Chemistry*, (G. Nickless, Ed.), Elsevier, New York, 1968, pp. 587–606.
85. Gillespie, R. J., *Acc. Chem. Res.*, **1**, 202–209 (1968).
86. Dudley, F. B. and Cady, G. H., *J. Am. Chem. Soc.*, **79**, 513–514 (1957).
87. Leung, P. C. and Aubke, F., *Inorg. Chem.*, **17**, 1765–1772 (1978).
88. Noftle, R. E. and Cady, G. H., *Inorg. Chem.*, **4**, 1010–1012 (1965).
89. Cicha, W. V., Herring, F. G., and Aubke, F., *Can. J. Chem.*, **68**, 102–108 (1990).
90. Cicha, W. V., Lee, K. C., and Aubke, F., *J. Solution Chem.*, **19**, 609–622 (1990).
91. Edwards, D. A., Stiff, M. J., and Woolf, A. A., *Inorg. Nucl. Chem. Lett.*, **3**, 427–429 (1967).
92. Goubeau, J. and Milne, J. B., *Can. J. Chem.*, **45**, 2321–2326 (1967).
93. Singh, S., Bedi, M., and Verma, R. D., *J. Fluorine Chem.*, **20**, 107–119 (1982).
94. Hayek, E., Puschmann, J., Czaloun, A., *Monatsh. Chem.*, **85**, 359–364 (1954).
95. Brazier, J. N. and Woolf, A. A., *J. Chem. Soc. A*, 99–100 (1967).
96. Leung, P. C., Wong, G. B., and Aubke, F., *J. Fluorine Chem.*, **35**, 607–620 (1987).
97. Leung, P. C. and Aubke, F., *Inorg. Nucl. Chem. Lett.*, **13**, 263–266 (1977).
98. Alleyne, C. S., Mailer, K. O., and Thompson, R. C., *Can. J. Chem.*, **52**, 336–342 (1974).
99. Roberts, J. E. and Cady, G. H., *J. Am. Chem. Soc.*, **82**, 353–354 (1960).
100. Dev, R. and Cady, G. H., *Inorg. Chem.*, **11**, 1134–1135 (1972).
101. Brown, S. D. and Gard, G. L., *Inorg. Chem.*, **17**, 1363–1364 (1978).
102. Mallela, S. P., Sams, J. R., and Aubke, F., *Can. J. Chem.*, **63**, 3305–3312 (1985).
103. Dev, R. and Cady, G. H., *Inorg. Chem.*, **10**, 2354–2355 (1971).
104. Lee, K. C. and Aubke, F., *J. Fluorine Chem.*, **19**, 501–516 (1982).
105. Lee, K. C. and Aubke, F., *Can. J. Chem.*, **55**, 2473–2477 (1977).
106. Lee, K. C. and Aubke, F., *Can. J. Chem.*, **59**, 2835–2838 (1981).
107. Johnson, W. M., Dev, R., and Cady, G., *Inorg. Chem.*, **11**, 2260–2262 (1972).
108. Lee, K. C. and Aubke, F., *Inorg. Chem.*, **18**, 389–393 (1979).
109. Lee, K. C. and Aubke, F., *Inorg. Chem.*, **23**, 2124–2130 (1984).
110. Lee, K. C. (1976), B.Sc. Thesis, University of British Columbia, Vancouver, Canada.
111. Brown, S. D. and Gard, G. L., *Inorg. Nucl. Chem. Lett.*, **11**, 19–21 (1975).

112. Niyogi, D. G., Singh, S., and Verma, R. D., *Can. J. Chem.*, **67**, 1895–1897 (1989).
113. Leung, P. C. and Aubke, F., *Can. J. Chem.*, **62**, 2892–2897 (1984).
114. Lee, K. C. and Aubke, F., *Can. J. Chem.*, **57**, 2058–2064 (1979).
115. Mallela, S. P., Lee, K., Gehrs, P. F., Christensen, J. I., Sams, J. R., and Aubke, F., *Can. J. Chem.*, **65**, 2649–2655 (1987).
116. Willner, H., Rettig, S. J., Trotter, J., and Aubke, F., *Can. J. Chem* **69**, 391–396 (1991).
117. Clark, E. S., Templeton, D. H., MacGillavry, C. H., *Acta Cryst.*, **11**, 284–288 (1958).
118. Lörcher, K. P., Strähle, J., *Z. Naturforsch.*, **30b**, 662–664 (1975).
119. Einstein, F. W. B., Rao, P. R., Trotter, J., Bartlett, N., *J. Chem. Soc. A*, 478–482 (1967).
120. Willner, H., Mistry, F., Hwang, G., Herring, F. G., Cader, M. S. R., and Aubke, F., *J. Fluorine Chem.* **52**, 13–27 (1991).
121. Charbonnier, F., Faure, R., Loiseleur, H., *Acta Cryst.*, **B33**, 1478–1481 (1977).
122. Bartlett, N. and Rao, P. R., *Chem. Soc. Proc.*, 393–394 (1964).
123. Griffith, W. P., in *Chemistry of the Rarer Platinum Metals* (Os, Ru, Ir, Rh), Interscience, London, (1967).
124. Jenkins, C. L. and Kochi, J. K., *J. Am. Chem. Soc.*, **94**, 843–855 (1972).
125. Haszeldine, R. N. and Kidd, J. M., *J. Chem. Soc.*, 4228–4232 (1954).
126. Bürger, H., Burczyk, K., and Blaschette, A., *Monatsh. Chem.*, **101**, 102–119 (1970).
127. Leung, P. C., Lee, K. C., and Aubke, F., *Can. J. Chem.*, **57**, 326–329 (1979).
128. Olah, G. A. and Ohyama, T., *Synthesis*, **5**, 319 (1976).
129. Noftle, R. E., *Inorg. Nucl. Chem. Lett.*, **16**, 195–200 (1980).
130. Johri, K. K., Katsuhara, Y., and DesMarteau, D. D., *J. Fluorine Chem.*, **19**, 227–242 (1982).
131. Niyogi, D. G., Singh, S., Gill, S., and Verma, R. D., *J. Fluorine Chem.*, **48**, 421–428 (1990).
132. Haynes, J. S., Sams, J. R., and Thompson, R. C., *Can. J. Chem.*, **59**, 669–678 (1981).
133. Arduini, A. L., Garnett, M., Thompson, R. C., and Wong, T. C. T., *Can. J. Chem.*, **53**, 3812–3819 (1975).
134. Cader, M. S. R., Thompson, R. C., and Aubke, F., *Chem. Phys. Lett.*, **164**, 438–440 (1989).
135. Mallela, S. P., Lee, K. C., and Aubke, F., *Inorg. Chem.*, **23**, 653–659 (1984).
136. Yeats, P. A., Poh, B. L., Ford, B. F. E., Sams, J. R., and Aubke, F., *J. Chem. Soc. A*, 2188–2191 (1970).
137. Cicha, W. V. and Aubke, F., *J. Am. Chem. Soc.*, **111**, 4328–4331 (1989).
138. Wilson, W. W. and Aubke, F., *J. Fluorine Chem.*, **13**, 431–445 (1979).
139. Thompson, R. C., Barr, J., Gillespie, R. J., Milne, J. R., and Rothenbury, R. A., *Inorg. Chem.*, **4**, 1641–1649 (1965).
140. Abbott, E. H., Schoenewolf, F. Jr., and Backstrom, T., *J. Coord. Chem.*, **3**, 255–258 (1974).
141. Landolt-Bornstein, in *Magmetic Properties of Coordination and Organometallic Transition Metal Compounds, Numerical Data and Functional Relationships in Science and Technology*, Vol. 2, Springer-Verlag, Berlin, 1966.
142. Landolt-Bornstein, in *Magnetic Properties of Coordination and Organometallic*

Transition Metal Compounds, Numerical Data and Functional Relationships in Science and Technology, Vol. 8, *1st Supplement*, Springer-Verlag, Berlin, 1976.
143. Rao, R. P., Sherwood, R. C., and Bartlett, N., *J. Chem. Phys.*, **49**, 3728–3730, (1968).
144. Tressaud, A., Winterberger, N., Bartlett, N., and Hagenmuller, P., *Compt. Rend. Acad. Sci.* **C282**, 1069–1070 (1976).
145. Starr, C., Bitter, F., Kaufmann, A. R., *Phys. Rev.*, **58**, 977–983 (1940).
146. Puddephatt, R. J., *The Chemistry of Gold*, Elsevier, Amsterdam, 1978.
147. Puddephatt, R. J., in *Comprehensive Coordination Chemistry* (G. Wilkinson, Ed.), Vol. 7, Pergamon, Oxford, 1987, pp. 861–923.
148. Willner, H. and Aubke, F., *Inorg. Chem.*, **29**, 2195–2200 (1990).
149. Herzberg, G., in *Spectra of Diatomic Molecules*, Van Nostrand, New York, 1950.
150. Desjardins, C. D., Edwards, D. B., and Passmore, J., *Can. J. Chem.*, **57**, 2714–2715 (1979).
151. Browning, J., Goggin, P. L., Goodfellow, R. J., Norton, M. J., Rattray, A. J. M., Taylor, B. F., and Mink, J., *J. Chem. Soc. Dalton Trans.*, 2061–2067 (1977).
152. Schroeder, K. and Sladky, F., *Z. Anorg. Allg. Chem.*, **477**, 95–100 (1981).
153. Leitzke, O. and Sladky, F., *Z. Anorg. Allg. Chem.*, **480**, 7–12 (1981).
154. Mayer, E. and Sladky, F., *Inorg. Chem.*, **14**, 589–592 (1975).
155. Sladky, F., Kropshofer, H., Leitzke, O., and Peringer, P., *J. Inorg. Nucl. Chem., H. H. Hyman Memorial Volume*, 1976, 69–71.
156. Seppelt, K., *Chem. Ber.*, **105**, 2431–2436 (1972).
157. Ray, P. C., *Nature (London)*, **126**, 310–311 (1930).
158. Lange, W., *Chem. Ber.*, **B62**, 793–801 (1929).
159. Corbridge, D. E. C. and Lowe, E. J., *J. Chem. Soc.*, 4555–4564 (1954).
160. Grimmer, A. R., Mueller, D., and Neels, J., *Z. Chem.*, **23**, 140–142 (1983).
161. Shihada, A. F., Hassan, B. K., and Mohammed, A. T., *Z. Anorg. Allg. Chem.*, **466**, 139–144 (1980).
162. Paul, R. C., Kumar, R. C., and Verma, R. D., *J. Fluorine. Chem.*, **11**, 203–213 (1978).
163. Kleinkopf, G. L. and Shreeve, J. M., *Inorg. Chem.*, **3**, 607–609 (1964).
164. Cicha, W. V. and Aubke, F., *J. Fluorine Chem.*, **47**, 317–322 (1990).
165. Shreeve, J. M. and Cady, G. H., *J. Am. Chem. Soc.*, **83**, 4521–4525 (1961).
166. Clark, H. C. and Emeléus, H. J., *J. Chem. Soc.*, 4778–4781 (1957).
167. Woolf, A. A., *J. Chem. Soc. A*, 355–358 (1967).
168. Mallela, S. P. and Aubke, F., *Inorg. Chem.*, **24**, 2969–2975 (1985).
169. Sams, J. R., Thompson, R. C., and Tsin, T. B., *Can. J. Chem.*, **55**, 115–121 (1977).
170. Taylor, J. M. and Thompson, R. C., *Can. J. Chem.*, **49**, 511–515 (1971).
171. Woolf, A. A., *J. Chem. Soc.*, 1053–1056 (1950).
172. Sharp. D. W. A., *J. Chem. Soc.*, 3761–3764 (1957).
173. Woolf, A. A., *J. Fluorine Chem.*, **1**, 127–128 (1971).
174. Mallela, S. P. and Aubke, F., *Can. J. Chem.*, **62**, 382–385 (1984).
175. Schmeisser, M., Satori, P., and Lippsmeier, B., *Chem. Ber.*, **103**, 868–879 (1970).
176. Singh, S., Amita, Gill, M. S., and Verma, R. D., *J. Fluorine Chem.*, **27**, 133–142 (1985).

177. Brown, S. D. and Gard, G. L., *Inorg. Chem.*, **14**, 2273–2274 (1975).
178. Jansky, M. T. and Yoke, J. T., *J. Inorg. Nucl. Chem.*, **41**, 1707–1709 (1979).
179. Schmeisser, M., Sartori, P., and Lippsmeier, B., *Chem. Ber.*, **102**, 2150–2152 (1969).

CHAPTER 4

Fluorine Stabilized Carbon–Sulfur Multiple Bonding

KONRAD SEPPELT

4.1. INTRODUCTION

Multiple bonding between heavier nonmetal atoms have become common in the last few years. Mostly silicon and phosphorous multiple bonding have been investigated, and numerous reviews are already available. The method used to stabilize such Si—Si, Si—C, P—C, P—P multiple bonding in all cases is steric protection with hindered groups such as *tert*-butyl-2,4,6-(*tert*-$C_4H_9)_3C_6H_2^-$, $[(R_3Si)_3]_3Si^-$, and adamantyl. With this method many of the compounds containing multiple bonding gained remarkable thermal and chemical stability. A disadvantage of this method that cannot be avoided is the fact that the structure of such compounds may be distorted by steric interaction of the protecting groups, and important principal questions like planarity of hetero-olefinic systems are difficult to address.

In our work on fluorine stabilized carbon–sulfur multiple bonding we chose the way of electronic protection. This means that compounds such as $R_2C{=}SF_4$ or $R{-}C{\equiv}SF_3$ will possibly not exist with ligands other than F.

These compounds are often low molecular weight gases (e.g., $H_2C{=}SF_4$), and the full scale of physical measurements can be employed. The question was Do such molecules contain true double and triple bonds? The criteria used was (a) bond lengths, (b) planarity or linearity and static or dynamic deviations from

Synthetic Fluorine Chemistry,
Edited by George A. Olah, Richard D. Chambers, and G. K. Surya Prakash.
ISBN 0-471-54370-5

there, and (c) experimental electron densities. We rarely employ MO theory or other models since these can be easily adjusted to all findings. Due to the lack of space available the presentation is brief, and other enlighting similarities and dissimilarities with metal–carbon, or metal–metal multiple bonding, or similarities with the broad field of ylid chemistry is not included here. For further reading on this topic see Seppelt.[1]

4.2. ALKYLIDENE SULFUR TETRAFLUORIDES, $R_2C{=}SF_4$

The first alkylidene sulfur tetrafluoride happened to be the principal compound $H_2C{=}SF_4$, made in 1978 by a multiple step procedure.[2,3] The final step is the elimination of LiF from an unstable intermediate $LiCH_2{-}SF_5$. Similar reactions afforded the derivatives $H_3C{-}CH{=}SF_4$,[4] $F_3C{-}CH{=}SF_4$, and $F_3C(H_3C)C{=}SF_4$.[5]

$$\text{Br—CRR'—SF}_5 \xrightarrow{n\text{-C}_5\text{H}_{11}\text{Li}} [\text{Li—CRR'—SF}_5] \xrightarrow{-\text{LiF}} \text{RR'C=SF}_4$$

Because of the length of the reactions these materials were available only in small amounts. However, the 1,3 fluorine shift reaction of $F_2C{=}CH{-}SF_5$ and $O{=}C{=}CH{-}SF_5$ to $F_3C{-}CH{=}SF_4$ (Ref. 6) and $FCO{-}CH{=}SF_4$ (Ref. 7) made these two specimen easily available. Finally, $H_2C{=}SF_4$ became available by reaction of $ClCO{-}CH_2{-}SF_5$ with $Mn(CO)_5^-$ in one step.[8]

$$\text{O=C=CH—SF}_5 \xrightarrow{250\,^\circ\text{C, glass}} \text{O=CF—CH=SF}_4$$

$$\text{F}_2\text{C=CH—SF}_5 \xrightarrow{\text{CsF, }100\,^\circ\text{C}} \text{F}_3\text{C—CH=SF}_4$$

$$\text{O=CCl—CH}_2\text{—SF}_5 + \text{Mn(CO)}_5^- \xrightarrow{-\text{Cl}^-}$$

$$(\text{CO})_5\text{Mn—CO—CH}_2\text{—SF}_5 \xrightarrow{25\,^\circ\text{C, }-\text{CO}}$$

$$(\text{CO})_5\text{Mn—CH}_2\text{SF}_5 \xrightarrow{25\,^\circ\text{C}} \text{“(CO)}_5\text{MnF”} + \text{H}_2\text{C=SF}_4$$

The structure of all these compounds is based on simple bonding principles. The environment of sulfur is close to trigonal bipyramidal, with the double-bonded carbon substituent in an equatorial position. The carbon substituents occupy sites in the axial plane, thus allowing the π electron density space in the equatorial plane (see Fig. 4.1). These features have been established qualitatively on all alkylidene sulfur tetrafluorides by means of ^{19}F NMR spectroscopy, and quantitatively on $H_2C{=}SF_4$, $FCO{-}CH{=}SF_4$, and $F_3C(H_3C)C{=}SF_4$ with electron diffraction, microwave spectroscopy, and single-crystal structure methods.[7,9,10]

By far the most precise structure investigation is the crystal structure of $F_3C(H_3C)C{=}SF_4$.[10] It was obtained with a good R factor and deviations of

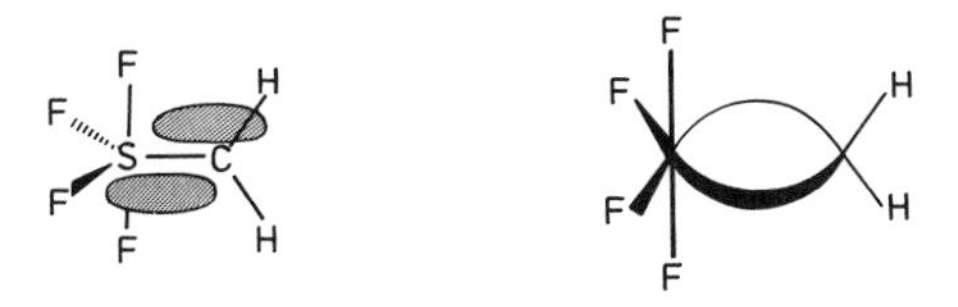

Figure 4.1. Structure of alkylidene sulfur tetrafluorides, $R_2C{=}SF_4$. Principles are a five-coordinated sulfur atom in a close to trigonal bipyramidal geometry, the double-bond carbon atom in an equatorial position, and the carbon substituents in the axial plane.

about 3° from planarity could be observed. These are probably caused by steric interaction of the CF_3 group with the axial fluorine atoms. Also, and more important, the deformation electron density not only established the different polarity of all bonds, but also the typical anisotropy of the CS double bond (see Fig. 4.2).

Chemical reactions of alkylidene sulfur tetrafluorides can be divided roughly into two categories. The first is the total cleavage of the CS double bond resulting in a carbene and SF_4. This reaction dictates the stabilities of the $R_2C{=}SF_4$ compounds. Carbenes are formed especially when these are stabilized (singlet), so the corresponding alkylidene sulfur tetrafluorides are not stable. This principle explains why substitution with a CF_3 group on carbon increases the stability. One very stable compound ought to be $(F_3C)_2C{=}SF_4$, which is not yet known, however.

The second and more productive typical chemical reaction is addition across the double bond.[3]

Polar agents (e.g., HF, HCl, or AsF_5) add very rapidly even at −78°C. This reaction is regioselective. Solely cis products with respect to the sulfur substitution are formed. Addition of HCl to prochiral $F_3C(H_9C)C{=}SF_4$ gives chiral $F_3C(H_3C)CH{-}SF_4Cl$ with four nonequivalent fluorine sulfur atoms.[10] Sulfur trioxide adds to give a four-membered ring.[11]

The intermolecular cycloaddition of carbonyl substituted alkylidene sulfur tetrafluoride, $R{-}CO{-}CR'{=}SF_4$ ($R{=}C_6H_5$ and $R' = H$ or Br), demand a special transition state, since one axial fluorine atom occupies the space for the oxygen atom. It can be expected that a *Berry* pseudorotation of the SF_4 group, combined with a C=S torsion, resolves this steric problem.[12]

$$R_2C{=}SF_4 \longrightarrow R_2C{:} + SF_4$$

$$\xrightarrow{+HX} \begin{matrix} R_2C{-}SF_4 \\ |\quad\;\; | \\ H\quad\; X \end{matrix}$$

$$\xrightarrow{+SO_3} \begin{matrix} R_2C{-}SF_4 \\ |\quad\;\; | \\ O_2S{-}O \end{matrix}$$

$$C_6H_5{-}\overset{\overset{O}{\|}}{C}{-}CH{=}SF_4 \longrightarrow \begin{matrix} O{-}SF_4 \\ |\quad\; | \\ C_6H_5{-}C{=}CH \end{matrix}$$

4.3. ALKYLIDENE SULFUR DIFLUORIDES, $R_2C{=}SF_2$

The first two members of this new class of material became known as recently as 1989.[13] This preparative pathway has so far not been generalized to other alkenes.

$$F_5S{-}CH{=}CF_2 \xrightarrow{CsF, SF_4} F_5S(F_3C)C{=}SF_2$$

$$F_3C{-}CH{=}CF_2 \xrightarrow{CsF, SF_4} (F_3C)_2C{=}SF_2$$

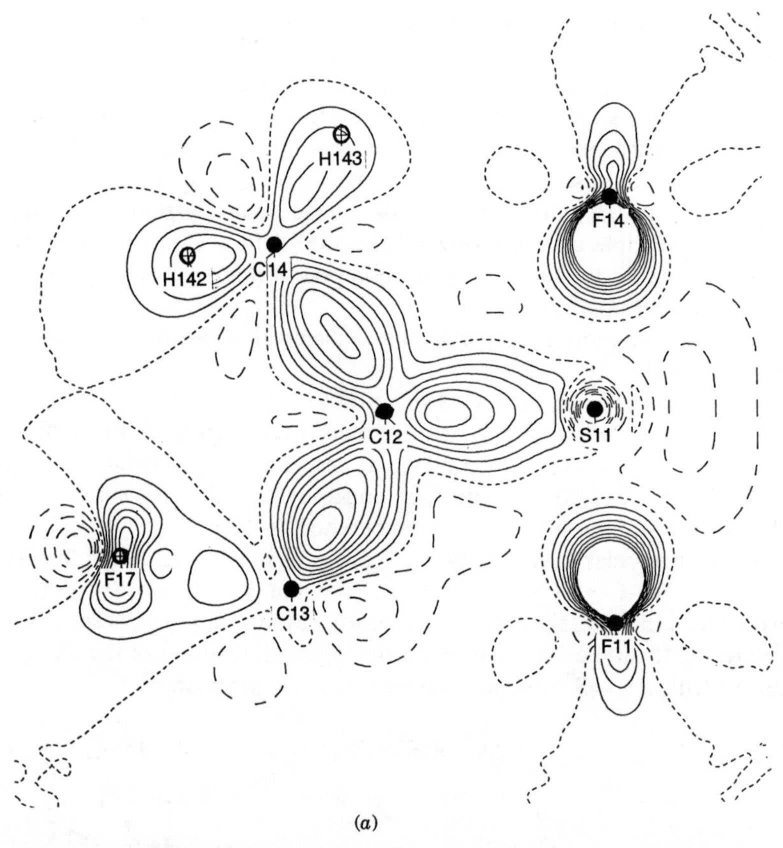

Figure 4.2. Deformation electron density of $F_3C(H_3C)C{=}SF_4$, (a) principal (axial) plane of the molecule. (*b*) Expected polarities of bonds are clearly visible. (*c*) The electron density of the CS double bond is highly anisotropic: it extends double as much out of plane as in plane. For the techniques and mathematical procedures of such high quality measurements see original. Reprinted with permission from[10]. Copyright © (1991) American Chemical Society.

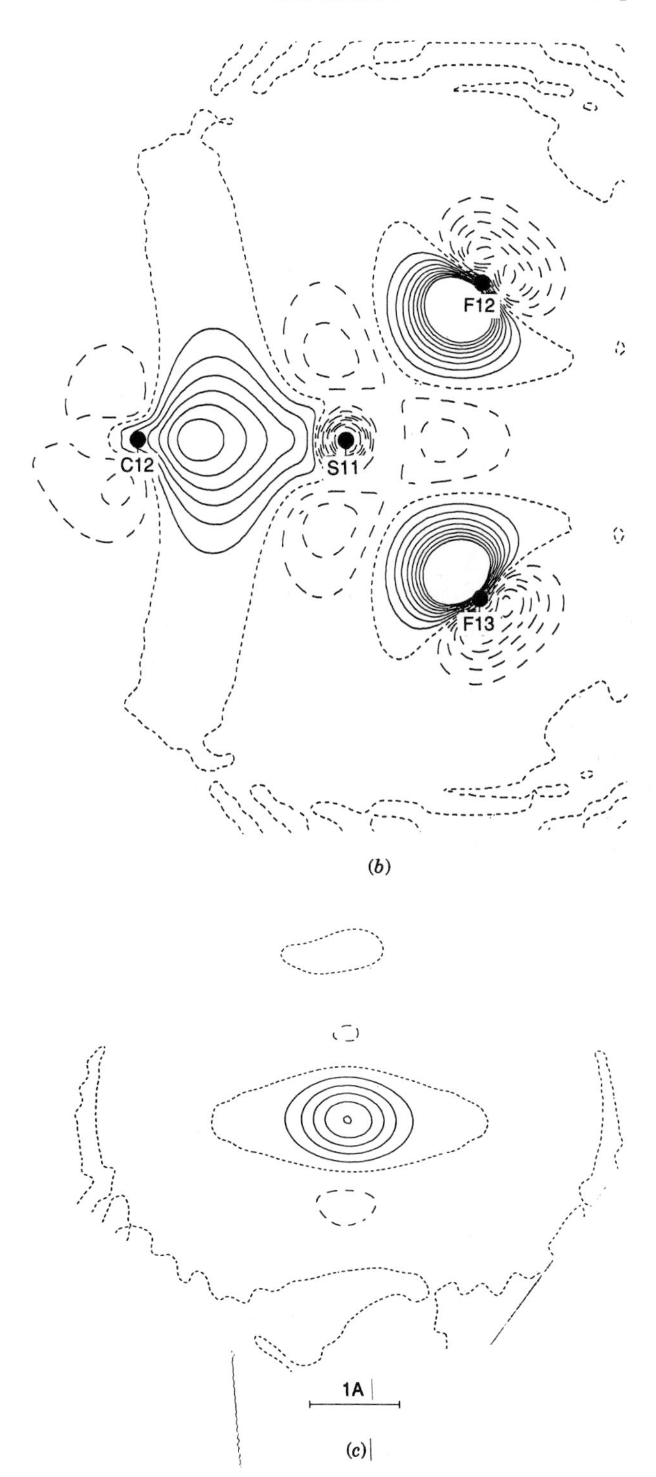

(b)

(c)

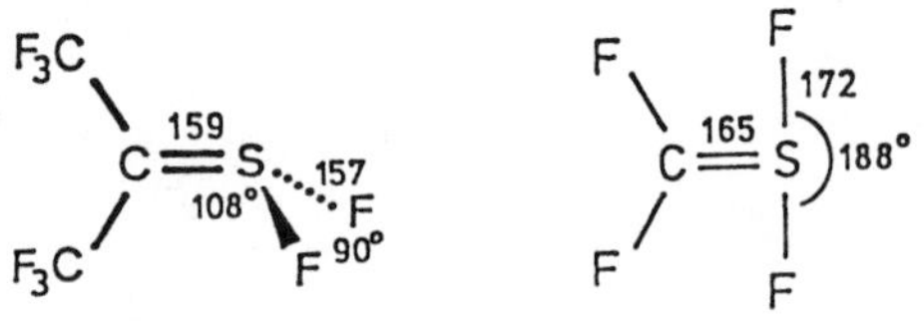

Figure 4.3. Structures of alkylidene sulfur difluorides. Left: Calculated structure of $(F_3C)_2C{=}SF_2$ and right: Calculated structure of $F_2C{=}SF_2$.

The structure of these two compounds, $(F_3C)_2C{=}SF_2$ and $F_5S(F_3C)C{=}SF_2$ is so obvious since the first shows nonequivalent CF_3 groups, and the latter exist as two isomers, while the fluorine atoms on sulfur are equivalent in both cases. This means qualitatively that the assumed plane formed by the carbon substituents bisects the SF_2 angle, and that the S environment is pyramidal. These experimental findings have been supported by ab initio calculations for $H_2C{=}SF_2$ and $(F_3C)_2C{=}SF_2$.[13] Surprisingly enough, the prediction for the yet unknown $F_2C{=}SF_2$ is different: Here the previously described geometry is not a minimum on the hyperface, the minimum is rather a total planar geometry with a T-shaped sulfur environment (see Fig. 4.3). (The global minimum, however, is of course $F_3C{-}S{-}F$.) This T-shaped geometry reminds us of trithiapentalenes and 1,6-dioxathiapentalenes.

Here a long standing discussion, whether the T-shaped sulfur geometry is a "bond tautomery" or "hypervalency," has finally been settled for the latter, however, with a very flat potential for the movement of the sulfur atoms between its two almost linear ligands.[14,15] If $F_2C{=}SF_2$ and its structure becomes known, it could be that the sulfur atom chose a "Non-Gillespie-structure" without steric strain by the five-membered rings.

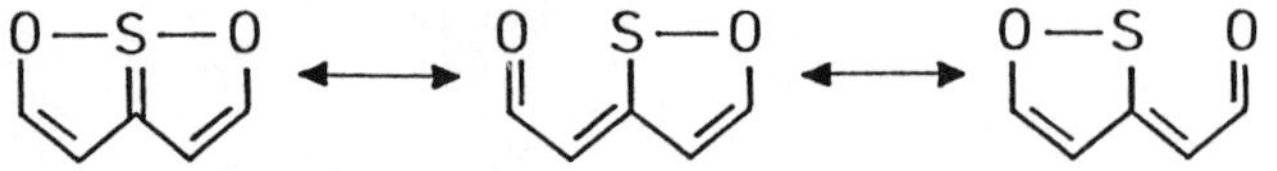

Little is known so far about the chemistry of alkylidene sulfur difluorides. Their general weakness is due to the 1,2 fluorine shift reaction to form sulfenylfluorides. The compound $F_5S(F_3C)C{=}SF_2$ serves as a precursor for the stable nonaromatic thioketone $F_5S(F_3C)C{=}S$, a deep violet liquid, that in itself may be an interesting precursor for further reactions.

4.4. ALKYLIDENE SULFUR DIFLUORIDE OXIDES, $R_2C{=}SF_2{=}O$

The first alkylidene sulfur difluoride oxide $FCO{-}CH{=}SF_2{=}O$ was made as recently as 1988.[16] The simple hydrolysis reaction of an alkylidene sulfur tetrafluoride worked only in one case, however.

$$O{=}CF{-}CH{=}SF_4 \xrightarrow{H_2O} O{=}CF{-}CH{=}SF_2{=}O$$

Soon after it was possible to generate $Hg[C(COF){=}SF_2{=}O]_2$ (Ref. 17) and $FSO_2(COF)C{=}SF_2{=}O$.[18] This class of compounds is similar to sulfuryl halides (SO_2X_2) or sulfur difluoride imide oxides ($R{-}N{=}SF_2{=}O$) and are structurally simple: that is, the sulfur geometry is derived from a tetrahedron.

The compound $COF{-}CH{=}SF_2{=}O$ shows, however, a complicated dynamic behavior in NMR experiments, which is so far not observed with the other two alkylidene sulfur difluoride oxides. On cooling to −70 °C two isomers are observed, and on further cooling to −135 °C three isomers are observed. Obviously, first the rotation around the CS double bond is frozen out, then the rotation around the CC single bond is seen. Of the four possible isomers, only three are observed. This may be a result of maximization of internal O ··· H bridges.[17] This idea is also supported by ab initio calculations.[19]

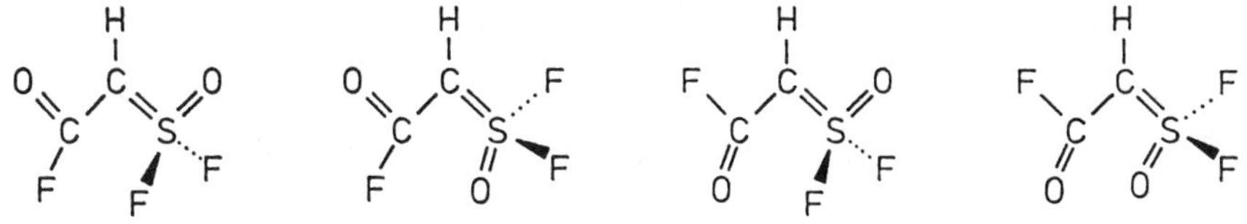

Int : 10	3	1	0
major isomer			not observed

The important question that remains is why in this compound the CS double bond is not rigid at least at room temperature, whereas the similar bond in compounds $R_2C{=}SF_4$ and $R_2C{=}SF_2$ is rigid at the highest temperatures measurable before they decompose (≈100 °C). In fact, this rigidity is one of the strongest arguments for the existence of a true double bond (the other arguments are the bond length and the anisotropic electron density distribution). We believe that in the alkylidene sulfur difluoride oxides there exists a double bond between C and S, since the bond distance of 155(3) pm in $Hg[C(COF){=}SF_2{=}O]_2$ is the same as in $R_2C{=}SF_4$ (experimental) and $R_2C{=}SF_2$ (calculated). The nonrigidity is possibly a result of the special geometry of alkylidene sulfur difluorides oxides: Since oxygen and fluorine are somewhat similar, the geometry with respect to the CS bond is approximately threefold on the sulfur side, but twofold (planar) on the carbon side. A full torsion around the C—S bond should exhibit a sixfold potential. Consequently, a sixfold potential cannot have as high rotational barriers as fourfold potentials ($R_2C{=}SF_4$) or twofold potentials ($R_2C{=}SF_2$ or $R_2C{=}CR_2$).

4.5. ALKYLIDYNE SULFUR TRIFLUORIDES, R—C≡SF$_3$

Alkylidyne sulfur trifluorides contain heteroatomic triple bonds, which are still rare among heavier main group elements. In fact two compounds are known,

$F_3C—C{\equiv}SF_3$ and $F_5S—C{\equiv}SF_3$.[20,21] These are formed by HF elimination from $F_3C—CH{=}SF_4$ and $F_5S—CH_2—SF_5$ with KOH.

$$F_3C—CH{=}SF_4 \xrightarrow{\text{KOH}} F_3C—C{\equiv}SF_3$$

$$F_5S—CH_2—SF_5 \xrightarrow{\text{KOH}} F_5S—C{\equiv}SF_3$$

These triple bonded species are unstable at room temperature, otherwise they are colorless solids, liquids, or gases. The critieria for triple bonds (very short bond length, linearity, and high electron density between the triply bonded atoms) are partly fulfilled. The bond length of 139–142 pm is approximately 20% shorter than that of a comparable single bond. This value fits well with the bond shrinkage between CC single and triple bonds.

Presently, little is known about the electron density of these compounds. A disk shaped electron density maximum between C and S is observed only in the crystal structure of $F_5S—C{\equiv}SF_3$, which now has a threefold symmetry, which is expected to occur because of the influence of the three fluorine atoms bonded to sulfur. Further work on this will be necessary to get better data. The criterion of linearity for such triple bonded systems raises a complicated problem. In the crystalline state at very low temperatures both molecules are linear or almost linear (172°).[21,22] However, in the gaseous state both are bent to approximately 155°.[23,24] The bending potential is obviously very soft. These compounds seem to be examples of "nonclassical behavior" of multiple bonding, as defined by Trinquier and Malrieu.[25] Such behavior in double-bonded species is expected to occur if the singlet–triplet gap for the cleavage products R_2X and R'_2Y from $R_2X{=}XR'_2$ becomes large, and similarly the doublet–quartet gap of RX and RY from $R—X{\equiv}Y—R'$. Heavier elements such as sulfur prefer electron pairing, and this is why such problems are expected with heavier nonmetal triple bonds. It is, however, possible to change the doublet–quartet gap a little by varying R on R—C. The smallest doublet–quartet gaps are expected with $R{=}CF_3$ and other strongly σ electron-withdrawing groups.

On the contrary, $F—C{\equiv}SF_3$ is predicted be bent down to 120° and having a ground state dicarbene structure $F—C—SF_3$ (see Fig. 4.4). The chemistry of these triple-bonded compounds is also influenced by these facts. It is possible to add HF twice across this bond in reversal of the formation reaction. But a carbene-type reaction is more prominent. The compound $F_3C—C{\equiv}SF_3$ dimerizes spontaneously, and this reaction is of first order, that means one molecule of $F_3C—C{\equiv}SF_3$ has to be thermally excited before it reacts. It is very likely to assume that the excited state is $F_3C—C—SF_3$, since the dimer contains the sulfur now in exactly the geometry of four valent sulfur.

$$2\,F_3C—C{\equiv}SF_3 \longrightarrow (F_3S)(CF_3)C{=}C(CF_3)(SF_3)$$

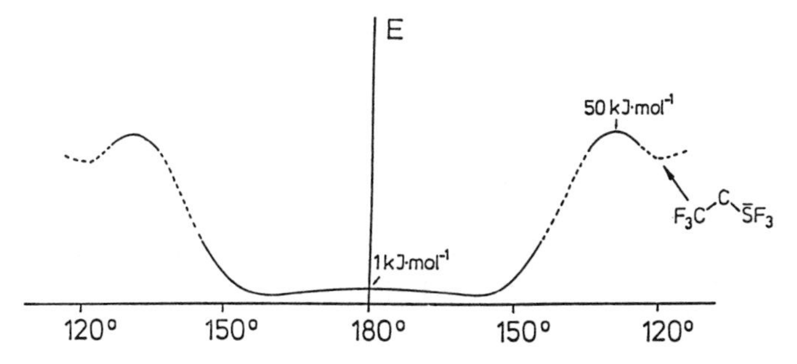

Figure 4.4. Bending potential of $F_3C—C{\equiv}SF_3$. Reprinted from[1] with permission of VCH Verlag, Weinheim, New York.

Very recently it was possible to trap both $F_5S—C{\equiv}SF_3$ and $F_3C—C{\equiv}SF_3$ with isonitriles. Also, it is noteworthy that $F_3C—C{\equiv}SF_3$ reacts with *all* hydrogen containing materials with the possible exception of saturated hydrocarbons resulting in very complex mixtures.

4.6. CONCLUSION

The novel compounds discussed in this chapter with CS double bonds have exciting structures and chemistry. But as more examples become available, chemists will view them as a normal extension of the double-bond principle. The compounds $F_3C—C{\equiv}SF_3$ and $F_5S—C{\equiv}SF_3$, however, show a limit of the triple-bond principle. The rearrangement of a triple bond $R—X{\equiv}Y—R'$ into a dicarbene-type species $R—X—Y—R'$ is so close in energy, that these materials exist only because of the careful choice of substituents. It can even be predicted that triple bonds between *two* heavier main group elements will always have a dicarbene as ground state, independent of substituents or steric protection.

REFERENCES

1. Seppelt, K., *Angew. Chem.* **103**, 399–413 (1991) *Angew. Chem. Int. Ed. Engl.*, **30**, 361–374 (1991).
2. Kleemann, G. and Seppelt, K., *Angew. Chem.*, **90**, 547–549 (1978); *Angew. Chem. Int. Ed. Engl.* **17**, 516–518 (1978).
3. Kleemann, G. and Seppelt, K. *Chem. Ber.*, **116**, 645–658 (1983).
4. Pötter, B. and Seppelt, K., *Inorg. Chem.*, **21**, 3147–3150 (1982).
5. Pötter, B., Kleemann, G., and Seppelt, K., *Chem. Ber.*, **117**, 3255–3264 (1984).
6. Grelbig, R., Pötter, B., and Seppelt, K., *Chem. Ber.*, **120**, 815–817 (1987).
7. Krügerke, T., Buschmann, J., Kleemann, G., Luger, P., and Seppelt, K., *Angew. Chem.*, **99**, 808–810 (1987); *Angew. Chem. Int. Ed. Engl.* **26**, 799–801 (1987).

8. Damerius, R., Leopold, D., Schulze, W., and Seppelt, K., *Z. Anorg. Allg. Chem.*, **578**, 110–118 (1989).
9. Simon, A., Peters, E.-M., Lentz, D., and Seppelt, K., *Z. Anorg. Allg. Chem.*, **468**, 7–14 (1980).
10. Buschmann, J., Koriatsanszky, T., Kuschel, R., Luger, P., and Seppelt, K., *J. Am. Chem. Soc.* (1991), **113**, 233–238 (1991).
11. Gerhardt, R., Kuschel, R., and Seppelt, K., *Chem. Ber.* (1991), in press.
12. Henkel, T., Krügerke, T., and Seppelt, K., *Angew. Chem.*, **102**, 1171–1172 (1990); *Angew. Chem. Int. Ed. Engl.*, **29**, 1128–1129 (1990).
13. Damerius, R., Seppelt, K., and Thrasher, J., *Angew. Chem.*, **101**, 783–785 (1989); *Angew. Chem. Int. Ed. Engl.*, **28**, 769–770 (1989).
14. Lozac'h, N., *Adv. Heterocycl. Chem.*, **13**, 161–234 (1971).
15. Gleiter, R. and Gygax, R., *Topics Curr. Chem.*, **63**, 49–88 (1976).
16. Krügerke, T. and Seppelt, K., *Chem. Ber.*, **121**, 1977–1981 (1988).
17. Bittner, J. and Seppelt, K., *Chem. Ber.* (1991), in press.
18. Winter, R. and Gard, G. L., *Inorg. Chem.*, **29**, 2386–2387 (1990).
19. Thrasher, J., personal communication.
20. Pötter, B. and Seppelt, K., *Angew. Chem.*, **96**, 138 (1984); *Angew. Chem. Int. Ed. Engl.*, **23**, 1504 (1984).
21. Gerhardt, R., Grelbig, T., Buschmann, J., Luger, P., and Seppelt, K., *Angew. Chem.*, **100**, 1592–1594 (1988); *Angew. Chem. Int. Ed. Engl.* **27**, 1534–1536 (1988).
22. Pötter, B., Seppelt, K., Simon, A., Peters, E.-K., and Hettich, B., *J. Am. Chem. Soc.*, **107**, 980–985 (1985).
23. Christen, D., Mack, H.-G., Marsden, C. J., Oberhammer, H., Schatte, G., Seppelt, K., and Willner, H., *J. Am. Chem. Soc.* , **109**, 4009–4018 (1987).
24. Weiß, J., Oberhammer, H., Gerhardt, R., and Seppelt, K., *J. Am. Chem. Soc.* **112**, 6839–6841 (1990).
25. Trinquier, G. and Malrieu, J.-P., *J. Am. Chem. Soc.*, **109**, 5303–5365 (1987).

CHAPTER 5

A NEW SYNTHETIC PROCEDURE FOR THE PREPARATION AND MANUFACTURE OF PERFLUOROPOLYETHERS

RICHARD J. LAGOW, THOMAS R. BIERSCHENK,
TIMOTHY J. JUHLKE, and HAJIMU KAWA

ABSTRACT

Perfluoropolyethers are an extraordinary class of high-performance fluids which are useful as high temperature lubricants and for many other applications. Lower molecular weight perfluoropolyethers are used for applications such as vapor phase soldering fluids, thermal shock fluids and inert fluids. While the first perfluoropolyethers appeared in the early 1970s, the technology that will dominate this field and the associated industry is emerging from our research laboratories at The University of Texas at Austin and at Exfluor Research Corporation. Our research group has already synthesized more new structures of perfluoropolyether lubricant fluids than all other sources combined. In our laboratories, controlled reactions of hydrocarbon starting materials and elemental fluorine have been pursued first on a small scale in the academic laboratory and scaled up very effectively at Exfluor Research Corporation. Currently, the highest molecular weight perfluoropolyether obtainable by others using conventional polymerization processes is 50,000—with a viscous

Synthetic Fluorine Chemistry,
Edited by George A. Olah, Richard D. Chambers, and G. K. Surya Prakash.
ISBN 0-471-54370-5

syruplike consistency. High molecular weight solids with a perfluoropolyether backbone have not been attained outside of our laboratory. The synthesis of hydrocarbon polymers as starting materials has many advantages and introduces great structural flexibility and capabilities not attainable using polymerization processes with various perfluorinated ethylene oxides.

5.1. INTRODUCTION

Perfluoropolyethers are the most extraordinary and exciting high-performance lubricants known today and have been perhaps the most important discovery in lubrication in the last 20 years.[1-3] For example, perfluoropolyether greases formulated from perfluoropolyether oils are used as the lubricants of choice for almost all civilian and military space activities (these greases often have fillers of either finely powdered Teflon or molybdenum disulfide) for lubricating bearings and other moving parts in satellites produced by all US and many foreign manufacturers. The satellite grease applications are very important because the satellites must operate for often 8–10 years in both high temperature and unusually cold environments, and failure marks a many million dollar loss. These lubricants are used on the space shuttle as well and will soon find their way into use as jet engine lubricants in aircraft.

Perfluoropolyethers have extraordinary properties and the materials are stable in some cases to 430°C [60°C more stable than poly(tetrafluoroethylene] thus giving very high thermal performance. This is true in part because carbon–oxygen bonds are thermochemically stronger than carbon–carbon bonds. In addition, the polymer backbones are protected by a sheath of fluorine [just as poly(tetrafluorethylene) is] protecting then chemically, as well as providing lubrication properties from the cloud of projecting nonbonding electrons associated with protruding fluorine atoms covering the chains.

On the other hand, perfluoropolyethers are the most extraordinary *low temperature* lubricants known because at very low temperatures the carbon–oxygen–carbon (ether linkage) acts as a hinge for storing vibrational energy. At even lower temperatures often exceeding −100°C, these materials still have *free rotation* around the carbon–oxygen ether bonds giving lower operating temperatures. In fact, in our laboratory we are working on a related project where elastomers having similar structures but higher molecular weights have glass transition temperatures as low as −110°C.

Perfluoropolyethers emerged on the market in the early 1970s; however, for the next 15 years there were only two basic structures known. The first perfluoropolyether was the homopolymer of hexafluoropropylene oxide produced by Du Pont having the structure:

$$-\!\!(CF_2-\underset{\displaystyle CF_3}{\underset{|}{CF}}-O)\!\!-_n$$

Du Pont called this new lubricant material Krytox[4] and initially it had such extraordinary properties that it sold for $100/lb (now $85/lb). Krytox was and is used in most of the vacuum pumps and diffusion oil pumps for the microelectronics industry in this country and in Japan because it produces no hydrocarbon (or fluorocarbon) vapor contamination. It has also found important applications in the lubrication of computer tapes and in other data processing applications as well as military and space applications.

The second material to emerge on the market was Montecatini Edison's polymer called Fomblin Z.[5] This polymer had better low-temperature properties than the Du Pont Krytox material yielding a selling price (initially for strategic defense applications) of $400/lb. The Montecatini material is made by photochemical polymerization of a mixture of oxygen and tetrafluoroethylene to prepare the random copolymer:

$$-\!\!(CF_2{-}O{-})_m({-}CF_2CF_2O)\!\!-_n$$

The methylene oxide, $-\!\!(CF_2O)\!\!-$, unit imparts even more extraordinary low-temperature properties than those derived from vibration and free rotation of other perfluoroether linkages. The $-\!\!(CF_2O)\!\!-$ unit is in that regard an ultimate.

Three years ago Daikin Kogyo of Japan[6] introduced the third structure to the market with a hybrid approach. The fluorooxetane

$$\begin{array}{ll} CF_2 & \!\!-\!\!-\,CF_2 \\ | & \quad | \\ O & \!\!-\!\!-\!\!-\,CH_2 \end{array}$$

was polymerized to give the hydrofluoro structure $-\!\!(CF_2{-}CF_2{-}CH_2{-}O)_n$, which was then perfluorinated using the direct fluorination technology previously developed in our laboratory to produce the perfluoropolyether structure $-\!\!(CF_2CF_2{-}CF_2{-}O)_n$. Daikin calls this material Demnum. Thus Demnum is the linear isomer of the original Krytox structure. Again no solid perfluoropolyether materials were attainable.

Consequently, there was a long period of time where the original two structures dominated the market and there have been no new entries in the market in the past 3 years since Demnum was discovered. The full potential of this new area had not been reached.

Work from our laboratory at the University of Texas on the synthesis of perfluoro ethers and perfluoropolyethers by direct fluorination (using elemental fluorine) is revolutionizing and in fact has revolutionized this field. Our laboratory has been responsible for putting at least 30 new perfluoropolyether structures into the literature and in some cases these perfluoropolyethers are now on the commercial chemical market. Exfluor Research Corporation prepared several hundred new polyether structures expanding upon the work from our laboratory. There is great flexibility in the elemental fluorine technique and substantially lower costs as well.

Therefore, we react fluorine with inexpensive hydrocarbon polyethers such as poly(ethylene oxide) to prepare perfluoropolyethers. In the simplest case, poly(ethylene oxide) is converted to the perfluoroethylene oxide polymer:

$$(CH_2CH_2O)_n \xrightarrow{F_2/He} R_f(OCF_2CF_2)_nOR_f$$

It was this simple reaction chemistry that we first reported in the literature.[2] As will later be seen, this direct fluorination technology as well as many new patents from Exfluor Research Corporation have been licensed to the 3M Corporation by the Lagow research group[7] and has been scaled up to a many gallons per hour scale.

The area of materials science and chemistry that we describe is that of *successfully* converting hydrocarbon polymers and hydrocarbon materials to fluorocarbon analogues without altering their structures. This is accomplished using elemental fluorine. Although the basic technology was developed in our laboratory at the University of Texas over the past 20 years and before that in our group at MIT, two very important breakthroughs have been made at the Exfluor Research Corporation in the last few years that have enabled across-the-board commercialization. These are the control of cross-linking during the direct fluorination process and the development of methods for processing, which assure that essentially every proton is replaced by a fluorine on the skeleton of the polymer or molecule. The latter has been accomplished so successfully by direct fluorination that the proton content in fluorocarbon fluids made by this new technology is less than three part per billion! This is two orders of magnitude lower than the proton content in, for example, commercial poly-(tetrafluoroethylene) and far too low to see with Fourier Transform IR or even high field NMR techniques. Previously, both of these potential problems with elemental fluorine synthesis had been thought by some to be insurmountable.

The key to removing the last protons lies in the fact that fluorine is soluble in fluorocarbons in the same manner as oxygen. It is now well known that up to 30% by volume oxygen can be dissolved in certain fluorocarbons (22% is the average amount). Such fluids have been used by Leland Clark, a frequent collaborator with Professor Lagow, in the liquid breathing area and the synthetic fluorocarbon blood area. Thus with proper activation, the last protons are easily removed from a liquid fluorocarbon that serves as a solvent for fluorine gas. With a solid fluorocarbon, pressurizing with fluorine usually enables complete fluorination; however, in difficult cases further reaction in a solvent may be used. Cross-linking is prevented using several types of solvent reactor techniques recently developed in our laboratory with appropriate fluorocarbon or chlorofluorocarbon solvents.

A review article concerning our early work appears in *Progress in Inorganic Chemistry*.[8] This work resulted in many new classes of organic compounds including perfluoroadamantanes,[9] perfluoro crown ethers,[10] perfluoro cryptands,[11] perfluoro inorganic compounds,[8,12] perfluoro organometallics,[13] and led to a general acceptance of this technique of very broad spectrum applicability toward the synthesis of novel fluoroorganic and inorganic compounds and structures.[8]

The Lagow group first entered the perfluoropolyether field in 1977[1] and again in 1978.[2] Thus, as in the previously mentioned reaction, the perfluoropolyethylene oxide structure that was not available by other synthetic routes was prepared:

$$\left(CF_2CF_2{-}O \right)_n$$

A

This structure is not amenable to manufacture by polymerizing tetrafluoroethylene oxide. Chemists at Du Pont attempted that reaction and encountered substantial explosion problems (from the very high heat of polymerization) dissuading them from preparing anything but small quantities of this material. Direct fluorination proceeded smoothly and produces fluids of almost any molecular weight desired. A version licenced to 3M by Exfluor will have a molecular weight of about 15,000, which is the molecular weight range of the original Krytox and Fomblin Z products.

New structures were important because the only differences in the physical properties among different Krytox materials and different Fomblin Z materials were produced by varying the degree of polymerization. Direct fluorination technology offers great flexibility in controlling molecular weights as well as new structures.

Subsequently, we were able to make perfluorinated analogues of Krytox from the *hydrocarbon* polypropylene oxide:[14]

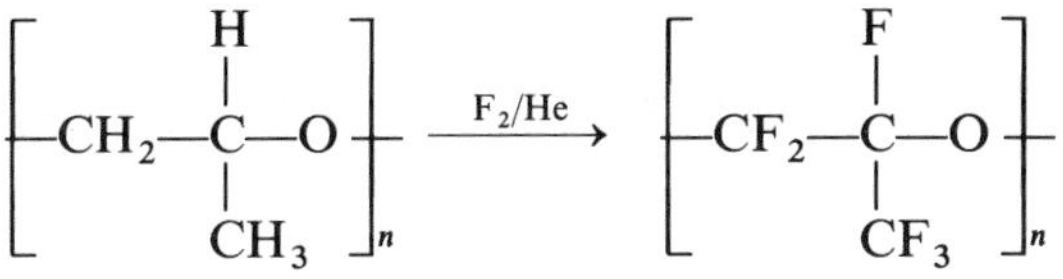

$$\left[CH_2{-}\underset{CH_3}{\overset{H}{C}}{-}O \right]_n \xrightarrow{F_2/He} \left[CF_2{-}\underset{CF_3}{\overset{F}{C}}{-}O \right]_n$$

B

We also reported an unusual perfluoropolyether structure prepared by the reaction of fluorine with the polyether:[15]

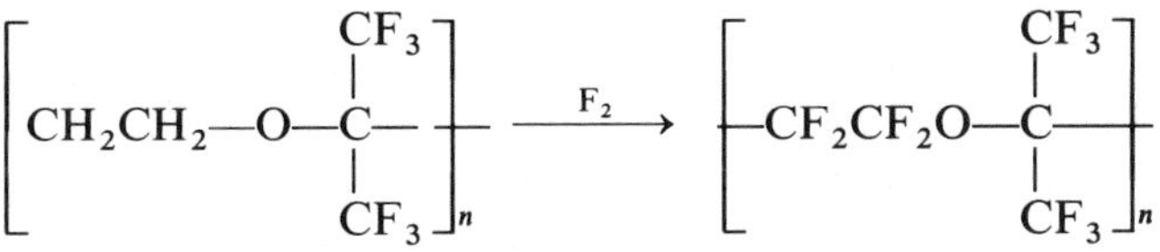

$$\left[CH_2CH_2{-}O{-}\underset{CF_3}{\overset{CF_3}{C}}{-} \right]_n \xrightarrow{F_2} \left[CF_2CF_2O{-}\underset{CF_3}{\overset{CF_3}{C}}{-} \right]_n$$

C

The precursor polymer was prepared by the method of Tabata et al.[16] using the radiation-induced low-temperature copolymerization of ethylene and hexafluoroacetone:

$$CH_2{=}CH_2 + CF_3{-}\overset{\overset{\displaystyle O}{\|}}{C}{-}CF_3 \xrightarrow[T<-9°C]{^{60}Co\,\gamma\text{-rays}} [CH_2CH_2OC(CF_3)_2]_n$$

Recently, we published three interesting new perfluoro ether structures. First we copolymerized hexafluoroacetone with ethylene oxide, propylene oxide, and trimethylene oxide. Subsequent fluorination yielded the following structures:[17]

Hexafluoroacetone–Ethylene Oxide Copolymer

$$\left(\underset{\displaystyle CF_3}{\overset{\displaystyle CF_3}{C}}OCH_2CH_2O\right)_x \xrightarrow[amb/\Delta]{He/F_2} \left(\underset{\displaystyle CF_3}{\overset{\displaystyle CF_3}{C}}OCF_2CF_2O\right)_y$$

D

Hexafluoroacetone–Propylene Oxide Copolymer

$$\left(\underset{\displaystyle F_3C}{\overset{\displaystyle F_3C}{C}}O\underset{\displaystyle H}{\overset{\displaystyle CH_3}{C}}CH_2O\right)_x \xrightarrow[amb/\Delta]{He/F_2} \left(\underset{\displaystyle F_3C}{\overset{\displaystyle F_3C}{C}}O\underset{\displaystyle F}{\overset{\displaystyle CF_3}{C}}CF_2O\right)_y$$

E

Hexafluoroacetone–Oxetane Copolymer

$$\left(\underset{\displaystyle CF_3}{\overset{\displaystyle CF_3}{C}}OCH_2CH_2CH_2O\right)_x \xrightarrow[amb/\Delta]{He/F_2} \left(\underset{\displaystyle CF_3}{\overset{\displaystyle CF_3}{C}}OCF_2CF_2CF_2O\right)_y$$

F

We discovered another synthesis that we consider to be truly general and to have many important ramifications, which we are now exploring. This has opened synthetic routes to literally hundreds of different perfluoroether and perfluoropolyether structures.[18] The synthetic technique involves the conversion by direct fluorination of hydrocarbon polyesters to perfluoropolyesters followed by treatment with sulfur tetrafluoride to produce new perfluoropolyethers. In some cases perfluoropolyethers esters are produced which can be hydrolyzed to produce functional end groups on the fluorocarbon polyethers.

General Synthetic Scheme

$$-\!\!\left(OCR_hCOR'_h\right)_{\!n}\ (\text{C=O, C=O}) \xrightarrow[\text{amb}]{F_2/He} -\!\!\left(OCR_fCOR'_f\right)_{\!n}\ (\text{C=O, C=O}) \tag{5.1}$$

$$-\!\!\left(O\overset{O}{\overset{\|}{C}}R_f\overset{O}{\overset{\|}{C}}OR'_f\right)_{\!n} \xrightarrow[\text{amb}]{SF_4/HF} -\!\!\left(OCF_2R_fCF_2OR'_f\right)_{\!n} \tag{5.2}$$

Our first paper in this area of research reported the conversion of poly(2,2-dimethyl-1,3-propylene succinate) (**G**) and poly(1,4-butylene adipate) (**H**) by using this procedure to novel branched and linear perfluoropolyether structures, respectively:

$$-\!\!\left(CH_2C(CH_3)_2CH_2O\overset{O}{\overset{\|}{C}}CH_2CH_2\overset{O}{\overset{\|}{C}}O\right)_{\!n} \xrightarrow[\text{2. } SF_4/HF]{\text{1. } F_2/He} -\!\!\left[CF_2C(CF_3)_2CF_2O(CF_2)_4O\right]_{n}$$

G

$$-\!\!\left(CH_2CH_2CH_2CH_2O\overset{O}{\overset{\|}{C}}CH_2CH_2CH_2\overset{O}{\overset{\|}{C}}O\right)_{\!n} \xrightarrow[\text{2. } SF_4/HF]{\text{1. } F_2/He} -\!\!\left[(CF_2)_4O(CF_2)_6O\right]_{n}$$

H

A second more recent paper[19] concerns the application of this same technology base directed toward oligomers, diacids, diesters, and surfactants.

Spectral Assignments of Compounds from 0% SF_4 Reaction

Compound	Highest m/z in mass spec	^{19}F NMR (relative $CFCl_3$)	^1H NMR (relative TMS)
$(CF_2CO_2CH_3)_2$	159 P—CO_2CH_3	−120.6	δ3.91
$(CO_2CH_3)_2$	59 P—CO_2CH_3		δ3.89
$(CF_2CO_2H)_2$	145 P—CO_2H	−120.3	δ11.43
$(CO_2H)_2$	45 P—CO_2H		δ9.00

Spectral Assignments of Compounds from 25% SF_4 Reaction

Compound	Highest m/z in mass spec	^{19}F NMR	^1H NMR[a]
$(CF_2CO_2CH_3)_2$	159 $P—CO_2CH_3$	−120.6	δ3.96
$(CO_2CH_3)_2$	59 $P—CO_2CH_3$		δ3.96
$H_3CO_2CCF_2OCF_2CF_2CF_2CO_2CH_3$ a b c d	275 $P—CO_2CH_3$	a −77.8 b −83.7 c −126.8 d −119.1	δ3.96

[a]Average chemical shift of CH_3 groups.

Spectral Assignments of Compounds from 50% SF_4 Reaction

Compound	Highest m/z in mass spectroscopy	^{19}F NMR	^1H NMR[a]
$H_3CO_2CCF_2OCF_2CF_2CF_2CO_2CH_3$ a b c d	275 $P—CO_2CH_3$	a −77.3 b −84.2 c −127.0 d −119.0	δ3.93
$H_3CO_2CCF_2CF_2CF_2O(CF_2)_2OCF_2$- a b c d c $CF_2CF_2CO_2CH_3$ b a	491 $P—CO_2CH_3$	a −119.0 b −127.0 c −84.2 d −88.7	δ3.93
$H_3CO_2CCF_2OCF_2CF_2CF_2OCF_2$- a b c c ba CO_2CH_3	391 $P—CO_2CH_3$	a −77.3 b −84.2 c −125.5	δ3.93
$H_3CO_2CCF_2OCF_2CF_2CF_2CF_2OCF_2$ a b c c b d $CF_2OCF_2CF_2CF_2CO_2CH_3$ d b e f	607 $P—CO_2CH_3$	a −77.3 b −84.2 c −125.5 d −88.7 e −127.0 f −119.0	δ3.93

[a]Average chemical shift of CH_3 groups.

These new methods offer a number of extremely significant and important advantages. The first advantage using this technique is that it is possible to prepare perfluoropolyethers containing *more than two* sequential carbon atoms in the perfluoropolyether backbone between adjacent oxygen sites. In the tetrafluoroethylene oxide and hexafluoropropylene oxide technology (i.e., polymerization of vinyl monomers and vinyl epoxides), one is limited to repeating two carbon–ether chains. A second important advantage is that perflu-

Spectral Assignments of Compounds from 100% SF_4 Reaction

Compound	Highest m/z in mass spectroscopy	^{19}F NMR	^{1}H NMR[a]
$(H_3CO_2CCF_2OCF_2CF_2CF_2CF_2OCF_2)_2$ (a b c c b d)	723 P—CO_2CH_3	a −78.0 b −83.3 c −125.3 d −88.6	δ3.94
$(H_3CO_2CCF_2CF_2CF_2OCF_2CF_2OCF_2CF_2)_2$ (a b c d d c e)	823 P—CO_2CH_3	a −119.3 b −126.6 c −83.3 d −88.6 e −125.3	δ3.94
$H_3CO_2CCF_2O(CF_2CF_2CF_2CF_2OCF_2CF_2O)_2$ (a b c c b d d) $CF_2CF_2CF_2CO_2CH_3$ (b e f)	939 P—CO_2CH_3	a −78.0 b −83.3 c −125.3 d −88.6 e −126.6 f −119.3	δ3.94

[a]Average chemical shift of CH_3 groups.

oropolyethers with unsymmetrical repeating units (alternating copolymers) are available (AOBOAOB), whereas with vinyl epoxides, other than random copolymers, one must have repeating AOA structures. A third advantage is that this technique is also capable of producing highly branched ethers and fluids of higher thermal stability.

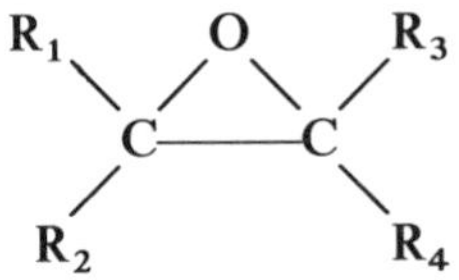

$$R_1R_2C\overset{O}{—}CR_3R_4$$

It is clear that if one considers as a class vinyl epoxides with the structure where R groups on either fluorine atoms or trifluoromethyl groups, even two trifluoromethyls hamper the polymerization leading only to low molecular weight materials. Thus substitution of larger, sterically bulky groups could limit even more severely the degree of polymerization. In many cases the synthesis of certain vinyl-substituted epoxide monomers would be extremely difficult or impossible even if polymerization were not a problem. A very important advantage to this method is the ease with which one can leave ester units in the high polymer and subsequently base hydrolyze to produce difunctional fluorocarbon polyesters of lower molecular weight.

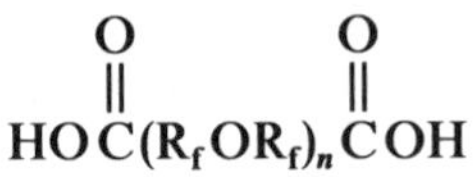

$$HO\overset{O}{\overset{\|}{C}}(R_fOR_f)_n\overset{O}{\overset{\|}{C}}OH$$

Functionalization of fluorocarbon polyethers is exceedingly difficult with conventional technology and often involves many steps and extremely high costs. An important effect observed when nonstoichiometric amounts of SF_4 are used is illustrated below. Normally, a twofold excess of SF_4 is necessary to obtain the high molecular weight nonfunctional polymer. As illustrated in Figure 5.1, if less than stoichiometric amounts of SF_4 are used, one obtains different Gaussian distributions of molecular weights, indicative of the average distance between the ester units left in the macromolecule. By varying the amount of SF_4 used, it is possible to shift the average molecular weight distribution at will. This technology makes available a very low-cost route to important new fluorocarbon surfactants and intermediates.[19]

In point of fact, high molecular weights are not obtained by polymerization of fluorocarbon epoxides with more than one trifluoromethyl group (i.e., Krytox). This is only one of many advantages associated with using the *hydrocarbon* polymer route for which there is much greater flexibility than using fluorocarbon monomers. The reasons for this are twofold: Organic hydrocarbon chemistry is now a mature science and polymerization of fluorocarbon monomers is very exothermic and usually lead to *only one molecular weight range* due to the large energy difference between the monomer and polymer. Direct fluorination technology can convert the hydrocarbon polymer into a fluorocarbon analogue with the same degree of polymerization as the starting material.

There are some 350 commercially obtainable linear hydrocarbon polyesters; thus the sulfur tetrafluoride–ether conversion technique is very broadly applicable to produce many novel perfluoroether structures. There are also over 750 hydrocarbon polyester structures prepared and characterized in the literature, allowing this technique almost total structural flexibility. Another important advantage is that cross-linkage can be accomplished in the hydrocarbon stages of the synthesis prior to converting the materials to fluorocarbon. Traditionally, cross-linking of fluorocarbons has been one of the most demanding and property-limiting problems in all of organofluorine polymer chemistry.

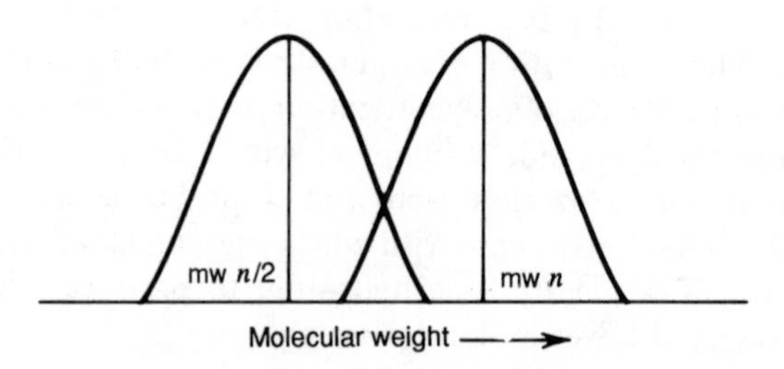

Figure 5.1. Gaussian distribution of difunctional perfluoropolyether molecular weights produced with n and $n/2$ mol of SF_4. Both samples are hydrolyzed to produce the diacids after treatment with SF_4.

Over 4 years ago the 3M Corporation licensed a number of patents from Exfluor Research Corporation and the R. J. Lagow laboratory at the University of Texas and more recently 3M put on line a new commercial production facility using our direct fluorination technology. A public announcement has recently been issued concerning 3M's licensing. Both Exfluor Research Corporation, an Austin corporation comprised primarily of a small group of former students from the Lagow laboratory, and 3M have now the capability of making commercial quantities of interesting new fluorocarbons using elemental fluorine.

Perfluoropolyether fluids are very dense clear liquids ($\rho = 1.80$), which range from an oil-like consistency at 5000 molecular weight to very thick "syruplike" viscosity at 20,000 molecular weight. Lower molecular weight perfluoro ethers make excellent vapor phase soldering fluids, thermal shock fluids, cooling fluids, and are useful in many applications in the microelectronics industry. We can now prepare any molecular weight up to and including solids. For example, in the polymerizations of Du Pont's hexafluoropropylene oxide with *only one* trifluoromethyl group, the maximum molecular weight obtainable is the syruplike 50,000 molecular weight and no solids are produced.

Two of the fluid structures that will be commercialized are very interesting. The first structure (I) is a strictly alternating copolymer of ethylene oxide and methylene oxide, which currently has the *longest liquid range of any molecule containing carbon.*[20]

$$+\!\!\left(CF_2O{-}CF_2CF_2O\right)\!\!_n$$

I

While it is a liquid down to as low as $-100°C$, its decomposition point is about 350°C.

The second perfluoropolyether structure (**J**) was long sought by a number of laboratories. Other laboratories had

$$+\!\!\left(CF_2O{-}CF_2{-}O\right)\!\!_n$$

J

attempted to polymerize carbonyl fluoride to synthesize material and used a number of other approaches as well. We succeeded in capping this structure with perfluoro ethyl groups to provide stability.[21] This is one of the lowest temperature high molecular weight liquids known. It remains a liquid lubricant down to $-112°C$ but unfortunately is only stable to about 250°C when it is in contact with metals or in an acidic or basic environment. This fluid may find use in solving exotic low temperature lubrication problems such as found in gyroscopes and for other uses where an inert fluid (lubricant) is needed with extremely low temperature properties. Our new perfluoropolyether lubricants were the subject of a paper recently published jointly with Dr. Bill Jones, of the Tribology Branch at NASAs Lewis Research Center.[22]

In addition to the linear perfluoropolyethers just described, branched structures can also be prepared. Fluorination of poly(propylene oxide) gives perfluoropoly(propylene oxide) in a nearly quantitative yield.[23] The fluid has properties identical to those of Du Pont's Krytox fluid, which is prepared by polymerizing hexafluoropropylene oxide. In a similar reaction, poly(butylene oxide) has been converted to its fluorocarbon analogue by reaction with elemental fluorine.

$$\left(CF_2\underset{\substack{|\\ CF_2CF_3}}{C}FO\right)_n$$

Unlike the linear fluids just described, perfluoropoly(butylene oxide) is very viscous even at room temperature. A fluid having an average molecular weight of about 3000 amu has a viscosity curve over 5500 cst at 20°C.[23] The high viscosity can be attributed to the presence of the pentafluoroethyl pendant groups.

The fluorination of chlorine-containing polyethers has given rise to a new class of materials that exhibited outstanding properties. For example, telomers of epichlorohydrin can be fluorinated to give perfluoropolyepichlorohydrin.[23b]

$$\left(CF_2\underset{\substack{|\\ CF_2Cl}}{C}FO\right)_n$$

We have shown that the oxidative stability of this fluid is essentially identical to that of perfluoropoly(propylene oxide). Fluids having molecular weights of approximately 3000 amu are excellent lubricants while lower molecular weight fluids, approximately 1500 amu, appear to be very good nonflammable hydraulic fluids. The presence of chlorine increases the bulk modulus (an important physical property for hydraulic fluids) and makes the fluids more compatible with existing additives. Finding suitable additives for perfluoropolyethers has been a difficult task since conventional additives are insoluble in perfluoropolyethers.

Functional perfluoropolyethers can also be prepared by direct fluorination in high yields. Difunctional perfluoropolyethers based on fluorinated poly(ethylene glycols) are of particular interest as possible precursors for elastomers, which should have both outstanding high-temperature and low-temperature properties.

$$CH_3\overset{\substack{O\\ \|}}{C}O(CH_2CH_2O)_n\overset{\substack{O\\ \|}}{C}CH_3 \xrightarrow{F_2/N} CF_3\overset{\substack{O\\ \|}}{C}O(CF_2CF_2O)_n\overset{\substack{O\\ \|}}{C}CF_3 \qquad n = 1\text{–}50$$

$$\xrightarrow{H_2O} HOOCCF_2O(CF_2CF_2O)_{n-2}CF_2COOH + 2CF_3COOH + 2HF$$

5.2. SPHERICAL PERFLUOROPOLYETHERS

We recently published work on the synthesis of perfluoro orthocarbonates.[24]

$$C(OCH_3)_4 \xrightarrow{F_2/He} C(OCF_3)_4$$

Tetramethyl orthocarbonate — 49.5%

$$C(OCH_2CH_3)_4 \xrightarrow{F_2/He} C(OCF_2CF_3)_4$$

Tetraethyl orthocarbonate — 56.5%

$$C(OCH_2CH_2CH_3)_4 \xrightarrow{F_2/He} C(OCF_2CF_2CF_3)_4$$

Tetra-*n*-propyl orthocarbonate — 48.6%

Comparison of Boiling Points

Compound	bp (°C)	Compound	bp (°C)
$C(OCH_3)_4$	114	$C(OC_2F_5)_4$	80
$C(OC_2H_5)_4$	160	$C(OC_3F_7)_4$	132
$C(OC_3H_7)_4$	224	$C(CF_2OCF_3)_4$	130
$C(OCF_3)_4$	20.8	$C(CF_2OC_2F_5)_4$	170

We should note that if we use the new solvent fluorination reaction techniques, the yields would be almost quantitative.

We are in the process of exploring a *new class of spherical perfluoropolyethers* based on pentaerythritol structures. Our idea is to make a very stable "molecular ball bearing" lubricant. This concept had been tried 30 years ago in the silicone field, however, the inherent stability limitation on the silicones is very substantial ($<250°C$). Some silicones are too rubbery to serve as lubricants.

Originally we entered the field of spherical perfluoroethers based on pentaerythritol by exploring the synthesis of perfluoro(pentaerythritol tetra-*t*-butyl ether) (K).

Perfluoro(pentaerythritol tetramethyl Ether)

$$C(CH_2-O-CH_3)_4 \xrightarrow{F_2} \underset{a}{C}(\underset{b}{CF_2}-O-\underset{c}{CF_3})_4 \qquad \text{MW } 552$$

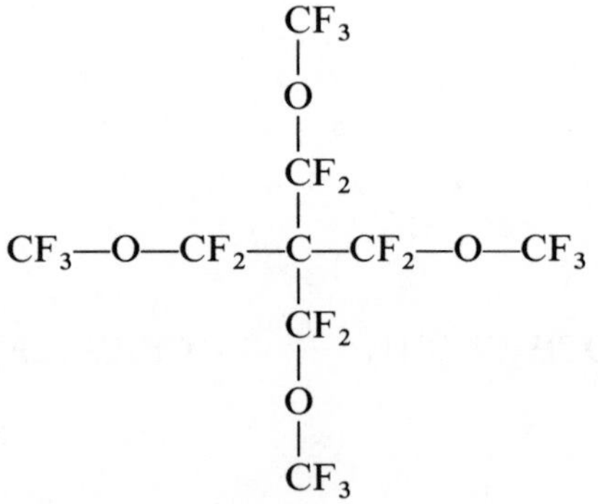

$$\begin{array}{c} CF_3 \\ | \\ O \\ | \\ CF_2 \\ | \\ CF_3-O-CF_2-C-CF_2-O-CF_3 \\ | \\ CF_2 \\ | \\ O \\ | \\ CF_3 \end{array}$$

13C{^{19}F} and ^{19}F NMR Data

(a) 66.556	
(b) 119.363	−70.0 (relative to $CFCl_3$)
(c) 117.282	−57.9

Mass Spectral Data

(P—F)	m/z 533
$C_7F_{13}O_3$	379
$C_7F_9O_2$	291
$C_5F_7O_2$	225
CF_3OCF_2	135 (base peak)
CF_3	69

Elemental Analysis

Calculated (%)	Element	Found
19.56	C	19.34
68.66	F	68.84
0.00	H	0.00
12.00	O	11.59

Boiling point 130.7 °C
Melting point −87.8 °C
Yield 68%

Perfluoro(pentaerythritol tetra-*t*-butyl Ether) (K)

$$C(CH_2-O-C(CH_3)_3)_4 \xrightarrow{F_2} C(CF_2-O-C(CF_3)_3)_4 \quad \text{MW } 1152$$

^{19}F NMR Data

NMR shows a sharp singlet at −70.0 for CF_3
The CF_2 signal is an envelope of peaks centered at −130

Mass Spectral Data

No (P—F) observed
Shows regularly spaced peaks characteristic of polymerized material
Material obtained is a viscous, very nonvolatile oil

$$C\left(CF_2-O-C(CF_3)_3\right)_4$$

Perfluoro(pentaerythritol tetraethyl Ether)

$$C{+}CH_2{-}O{-}CH_2{-}CH_3)_4 \xrightarrow{F_2} \underset{a}{C}{+}\underset{b}{C}F_2{-}O{-}\underset{c}{C}F_2{-}\underset{d}{C}F_3)_4 \quad \text{MW } 752$$

$$CF_3{-}CF_2{-}O{-}CF_2{-}C(CF_2{-}O{-}CF_2{-}CF_3)_2{-}CF_2{-}O{-}CF_2{-}CF_3$$

$^{13}C\{^{19}F\}$ and ^{19}F NMR Data

	^{13}C	^{19}F
(a)	67.012	
(b)	118.453	−90.3 (relative to $CFCl_3$)
(c)	116.047	−89.3
(d)	114.356	−67.2

Mass Spectral Data

(P—F)	*m/z* 733	$CF_3CF_2OCF_2$	*m/z* 185
$C_9F_{17}O_3$	479	CF_3CF_2	119 (base)
$C_7F_{11}O_3$	341	CF_3	69
$C_6F_9O_2$	275		

Elemental Analysis

Calculated	Element	Found
20.76	C	20.44
70.73	F	70.61
0.00	H	0.07
8.51	O	8.88

Boiling point 170°C
Melting point −60°C
Yield 25%

Perfluoro(pentaerythritol tetrapropyl Ether)

$$C(CH_2—O—CH_2—CH_2—CH_3)_4 \xrightarrow{F_2} \underset{a}{C}(\underset{b}{CF_2}—O—\underset{c}{CF_2}—\underset{d}{CF_2}—\underset{e}{CF_3})_4 \quad \text{MW 952}$$

$^{13}C(^{19}F)$ and ^{19}F NMR data

	$^{13}C(^{19}F)$	^{19}F
(a)	67.142	
(b)	118.648	−85.5 (relative to $CFCl_3$)
(c)	117.347	−83.3
(d)	106.747	−131.0
(e)	116.112	−66.5

Mass Spectral Data

No (P—F) observed			
$C_{13}F_{25}O_4$	*m/z* 695	C_4F_7	*m/z* 181
$C_{12}F_{23}O_4$	645	$CF_3CF_2CF_2$	169 (base peak)
$C_8F_{13}O_3$	391	CF_3CF_2	119
$CF_3CF_2CF_2OCF_2$	235	CF_2CF_2	100
CF_3	69		

Elemental Analysis

None to date

Boiling point 232 °C

Melting point −54 °C

31% yield

```
                              CF3
                              |
                              CF2
                              |
                              CF2
                              |
                              O
                              |
                              CF2
                              |
CF3—CF2—CF2—O—CF2—C—CF2—O—CF2—CF2—CF3
                              |
                              CF2
                              |
                              O
                              |
                              CF2
                              |
                              CF2
                              |
                              CF3
```

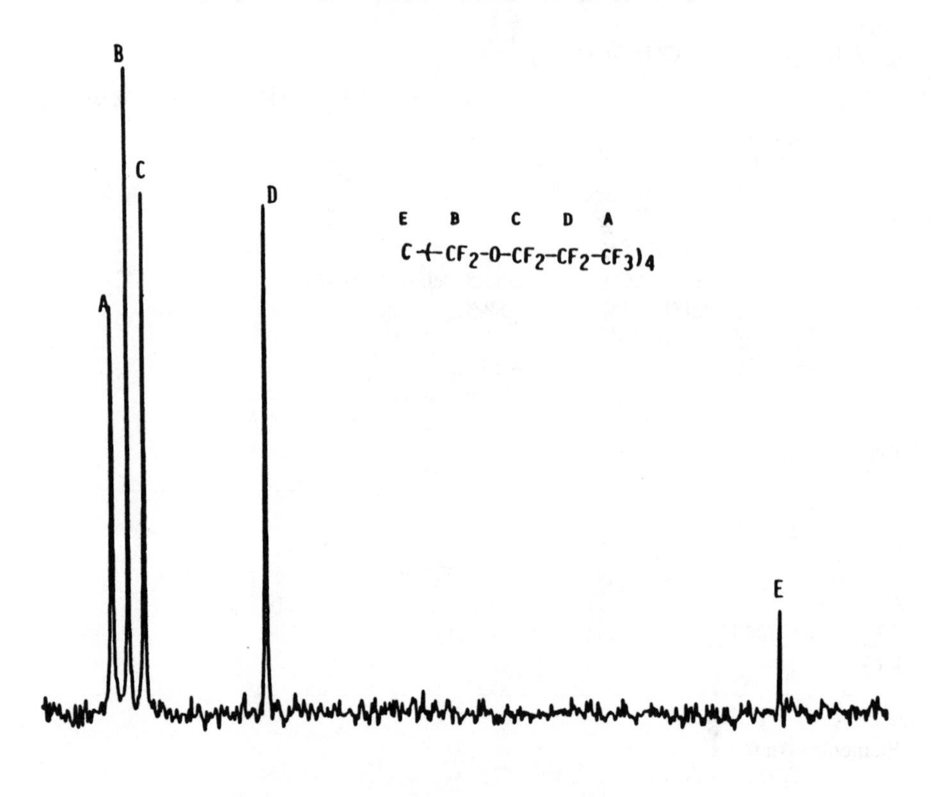

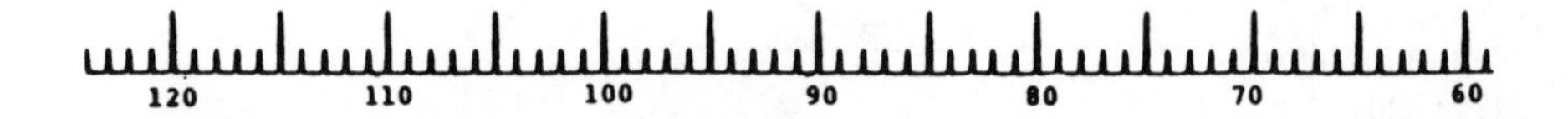

Gordon Rutherford, a graduate student in our research program, has vigorously pursued other pentaerythritol perfluoropolyethers and has raised the yields using the newer technology.[25] We found that we can synthesize many unusual perfluorinated pentaerythritol structures with the added capabilities. With our new synthetic techniques, we are in a position to prepare many spectacular new compounds. In most cases we have ^{13}C spectra in addition to the spectra reported.

Perfluoro(pentaerythritol tetrabutyl Ether)

$$C(CH_2-O-CH_2-CH_2-CH_2-CH_3)_4 \xrightarrow{F_2}$$

$$C(\underset{a}{CF_2}-O-\underset{b}{CF_2}-\underset{c}{CF_2}-\underset{d}{CF_2}-\underset{e}{CF_3})_4$$

MW 1152

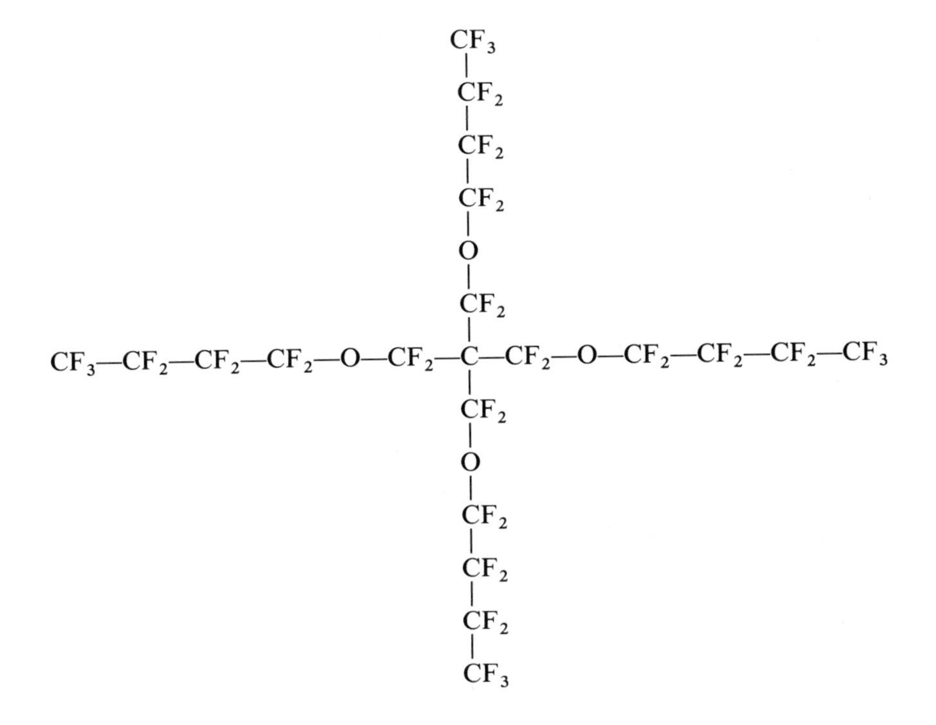

^{19}F NMR Data: (ppm) $\delta(CFCl_3)$
(a) −65.3
(b) −82.9
(c) −126.2
(d) −126.7
(e) −81.8

Mass Spectral Data

(P—F)	*m/z* 1133
(P—C_4F_9O)	917
$C_{17}F_{33}O_4$	895
$C_{13}F_{25}O_3$	679
$C_8F_{15}O$	397
C_4F_9	219
C_4F_7	181
C_3F_5	131 (base peak)
C_2F_5	119
C_2F_4	100

Elemental Analysis

Calculated	C 21.89%
	F 72.55%
Found	C 21.65%
	F 72.76%

$$C(CH_2—O—CH_2—CH_2—CH_2—CH_2—CH_2—CH_3)_4 \xrightarrow[29\%\text{ yield}]{F_2}$$

$$C(\underset{a}{CF_2}—O—\underset{b}{CF_2}—\underset{c}{CF_2}—\underset{d}{CF_2}—\underset{e}{CF_2}—\underset{f}{CF_2}—\underset{g}{CF_3})_4 \quad 7$$

MW 1552

```
                                          CF3
                                          |
                                          CF2
                                          |
                                          CF2
                                          |
                                          CF2
                                          |
                                          CF2
                                          |
                                          CF2
                                          |
                                          O
                                          |
                                          CF2
                                          |
CF3—CF2—CF2—CF2—CF2—CF2—O—CF2—C—CF2—O—CF2—CF2—CF2—CF2—CF2—CF3
                                          |
                                          CF2
                                          |
                                          O
                                          |
                                          CF2
                                          |
                                          CF2
                                          |
                                          CF2
                                          |
                                          CF2
                                          |
                                          CF2
                                          |
                                          CF3
```

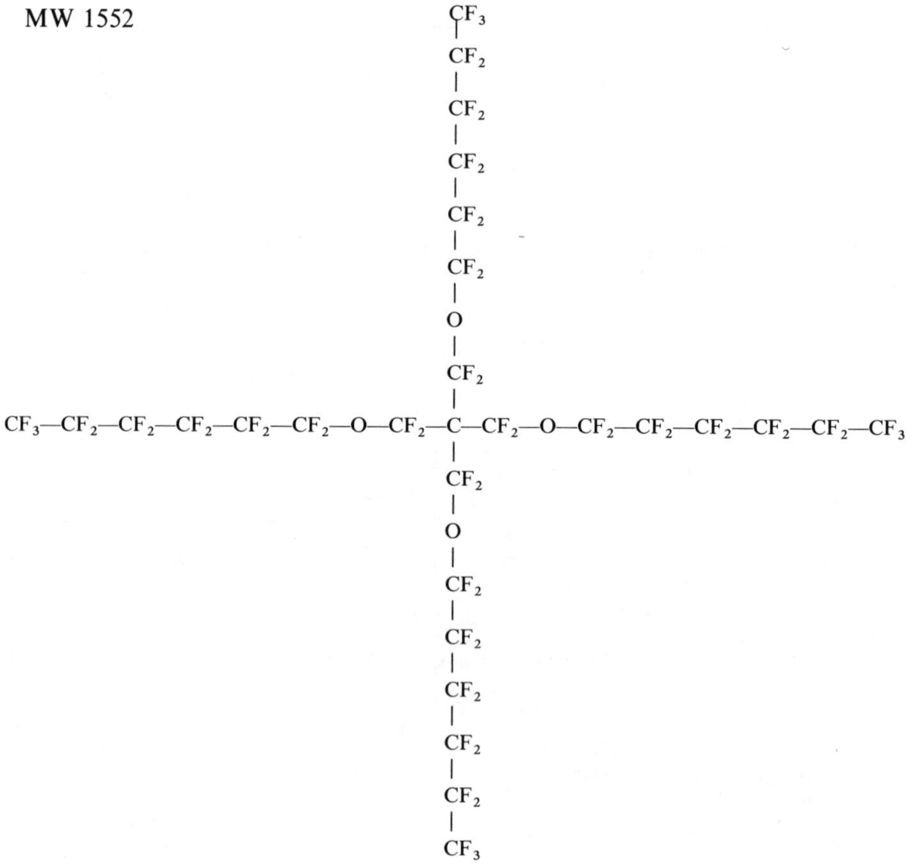

Mass Spectral Data

(P—F)	*m/z* 1533
$C_{29}F_{57}O_4$	1495
(P—C_4F_9)	1333
(P—$C_6F_{13}+1$)	1234
$C_{23}F_{45}O_4$	1195
$C_{20}F_{39}O_4$	1045
$C_{19}F_{37}O_4$	995
$C_{18}F_{35}O_4$	945
$C_{17}F_{33}O_4$	895
$C_{17}F_{33}O_3$	879 (base peak)
$C_{16}F_{31}O_3$	829
$C_{16}F_{31}O_2$	813
$C_{11}F_{21}O_3$	579

^{19}F NMR data (ppm) $\delta(CFCl_3)$

(a)	−65.3	(e)	−125.0
(b)	−83.0	(f)	−126.3
(c)	−122.3	(g)	−81.5
(d)	−122.8		

Elemental Analysis

Calculated	C 22.4%,	F 73.44%
Found	C 22.00%,	F 73.48%

bp 292 °C

$$[(CH_3CH_2{-}O{-}CH_2)_3C{-}CH_2]_2O \xrightarrow[58\%\,\text{yield}]{F_2} [(\underset{a}{CF_3}{-}\underset{b}{CF_2}{-}O{-}\underset{c}{CF_2})_3C{-}\underset{d}{CF_2}]_2O$$

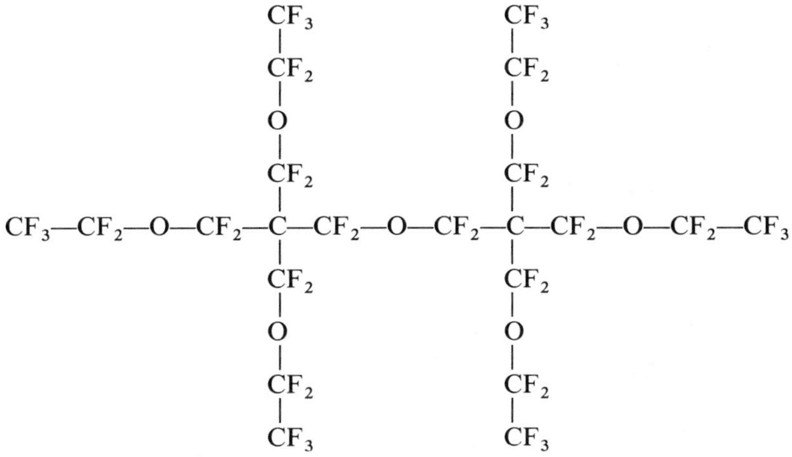

$$CF_3{-}CF_2{-}O{-}CF_2{-}C(CF_2{-}O{-}CF_2{-}CF_3)_2{-}CF_2{-}O{-}CF_2{-}C(CF_2{-}O{-}CF_2{-}CF_3)_2{-}CF_2{-}O{-}CF_2{-}CF_3$$

MW 1250
bp 243 °C

^{19}F NMR Data (ppm) $\delta(CFCl_3)$

(a)	−87.5
(b)	−88.9
(c)	−66.3
(d)	−65.3

Elemental Analysis:

Calculated	C 21.14%,	F 69.90%
Found	C 20.90%,	F 70.02%

Mass Spectral Data

(P—F)	*m/z* 1231 (base peak)
$C_{20}F_{39}O_7$	1093
$(P{-}C_3F_7O)$	1065
$C_{11}F_{23}O_3$	617
$C_9F_{17}O_3$	479

The new fluorination techniques invented at Exfluor were used for these syntheses. We would expect the yields to be nearly quantitative. These unusual materials are of definite interest for a number of applications.

Many novel small molecule perfluoropolyethers are available using this technology. It is perhaps true to say that one can prepare any ether structure that has a satisfactory hydrocarbon precursor. For example, the following branched ethers have recently been reported:[26]

CHART 5.1. $^{13}C\{^{19}F\}$ Assignments[a]

Bis(perfluoroisopropyl)ether

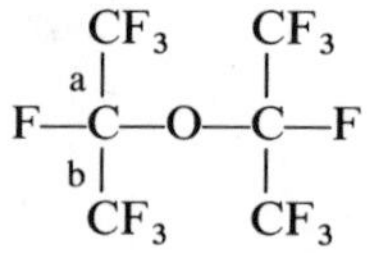

```
  CF3    CF3
 a|      |
F—C—O—C—F
 b|      |
  CF3    CF3
```

a 118.049
b 102.962

Bis(perfluoroisobutyl)ether

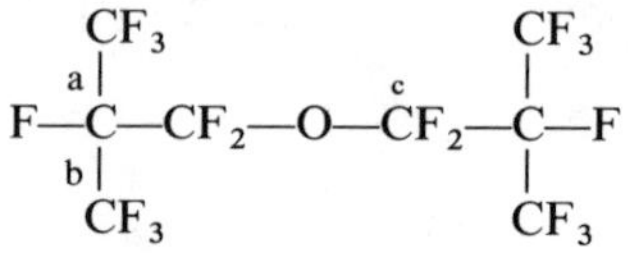

```
  CF3                 CF3
 a|          c        |
F—C—CF2—O—CF2—C—F
 b|                   |
  CF3                 CF3
```

a 118.960
b 88.980
c 117.659

Bis(perfluoroisopentyl)ether

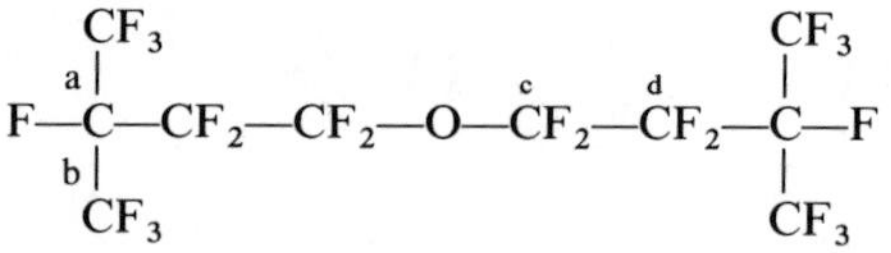

```
  CF3                            CF3
 a|               c      d       |
F—C—CF2—CF2—O—CF2—CF2—C—F
 b|                              |
  CF3                            CF3
```

a 119.285
b 90.215
c 116.684
d 110.831

Bis(perfluoroneopentyl)ether

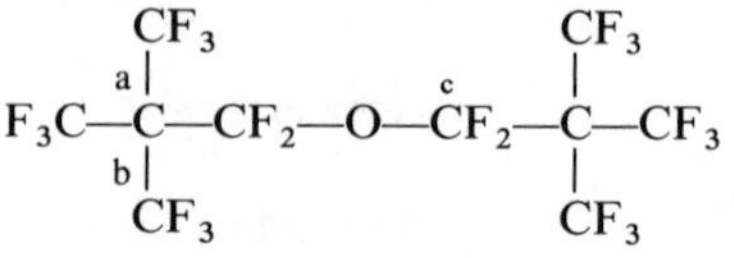

```
     CF3                 CF3
    a|          c        |
F3C—C—CF2—O—CF2—C—CF3
    b|                   |
     CF3                 CF3
```

a 119.805
b 65.112
c 118.895

[a]In parts per million relative to external Me_4Si.

Recently an even more comprehensive study of branched ethers has been reported.[27]

5.3. PERFLUORO CROWN ETHERS AND CRYPTANDS

A recent breakthrough in our laboratory has involved the synthesis of perfluorinated crown ethers and cryptands. We have previously reported the synthesis of the first perfluoro crown ethers, perfluoro 18-crown-6, perfluoro 15-crown-5, and perfluoro 12-crown-4.[9a]

Perfluoro crown ethers[9a,28] are becoming very important as the molecules of choice for many ^{19}F NMR imaging applications[29] in humans and is particularly

Properties and characterization of perfluoro 15-crown-5 and perfluoro 12-crown-4.[a,b]

	15-Crown-5	12-Crown-4
Boiling point (°C)	146	118
IR (vapor phase) (cm^{-1})	1250(s)	1260(vs)
	1228(vs)	1188(vs)
	1158(vs)	1160(vs)
	745(m)	1080(m)
		825(m)
		745(br)
NMR (neat liquid)	^{19}F—91.8(s) ppm	^{19}F—90.0(s) ppm
	(ext. $CFCl_3$)	(ext. $CFCl_3$)
	$^{13}C\delta 114.9$(s)	$^{13}C\delta 114.9$(s)
Mass spectrum (m/z)	580($C_{10}F_{20}O_5$, M^+)	445($C_8F_{15}O_4$, M^+—F)

[a]A table for the straight-chain fragmentation products listing mass spectral and ^{19}F data (two pages) is available from the authors.
[b]Satisfactory elemental analyses. (C, F) were obtained.

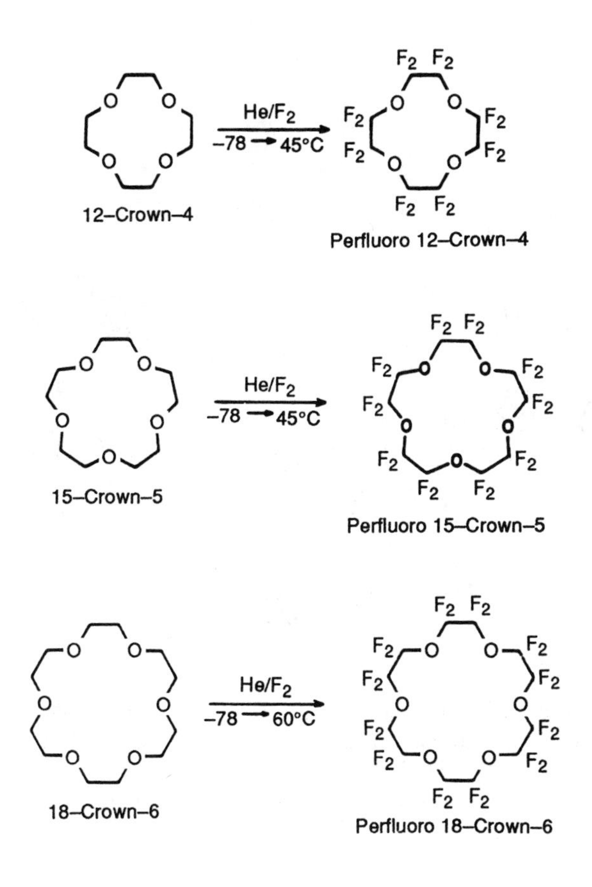

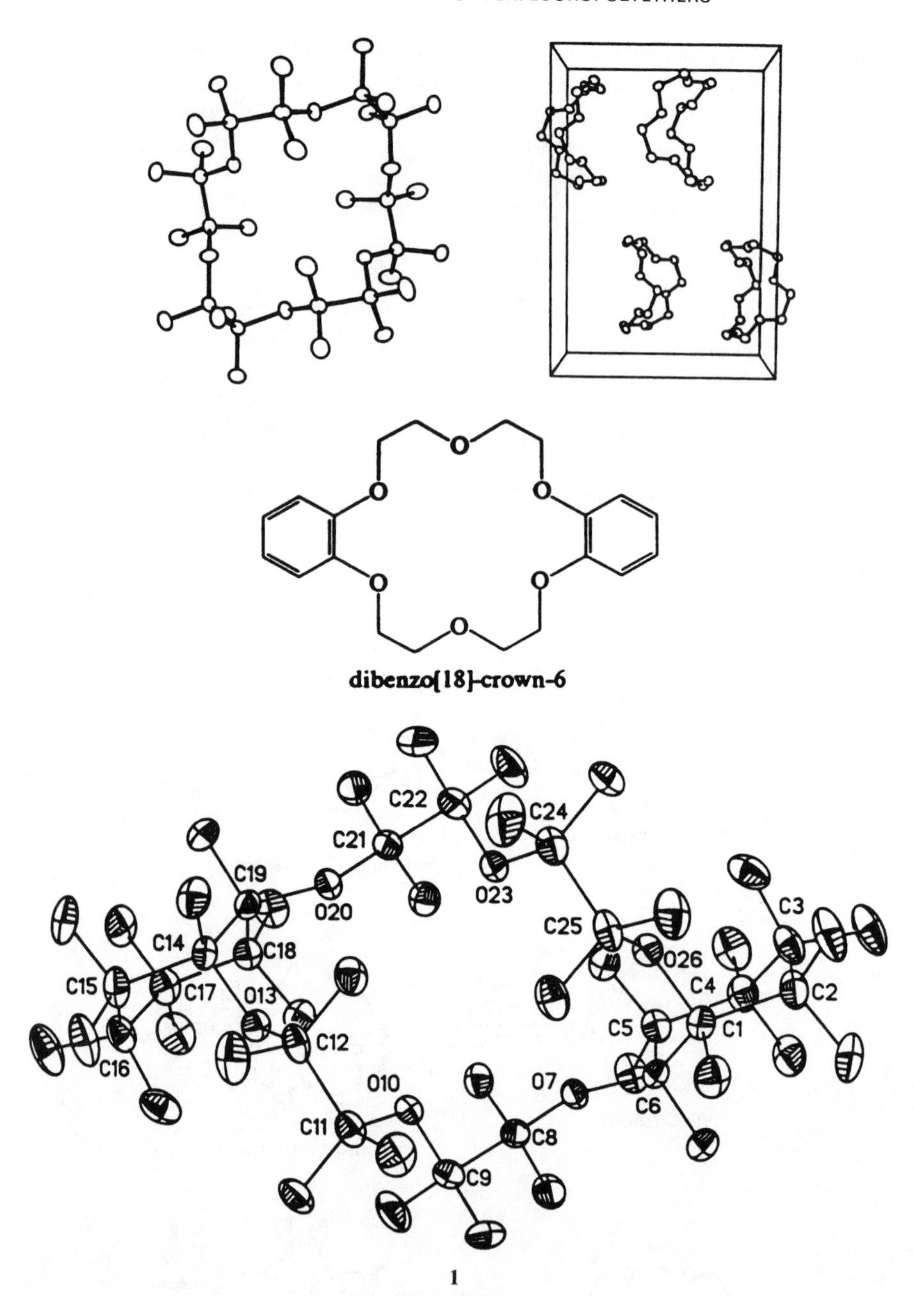

View (ORTEP of cis-syn-cis (**1**) showing the atom numbering scheme. Distances (Å) between adjacent oxygen atoms are as follows: O_{13}—O_{20} 2.662, O_{20}—O_{23} 2.713, O_{23}—O_{26} 2.665, O_{26}—O_7 2.667, O_7—O_{10} 2.707, O_{10}—O_{13} 2.667. Distances (Å) between oxygen atoms to the center of the molecule are as follows: O_7 2.536, O_{10} 2.129, O_{13} 3.177, O_{20} 2.543, O_{23} 2.141, O_{26} 3.181.

effective in brain and spinal diagnostics when administered to the cerebrospinal fluid compartment. Synthesis scale up of perfluoro 15-crown-5[9a,28] and plans for commercialization are underway while research is being conducted on other biological applications of these new compounds.[30] In collaboration with Air Products, very excellent brain imaging scans have been obtained by infusing the perfluoro 15-crown-5 in the spinal fluids. Toxicology reports on these are very favorable with essentially no toxic effects physiologically in several different animals. There are some pharmaceutical companies actively negotiating to obtain licensing on perfluoro 15-crown-5 on which we have a composition of matter patent.[28] Exfluor is preparing a 700-ml sample of perfluoro 15-crown-5 for one of these companies so that extensive testing may be done.

Very recently we have synthesized perfluoro crown ethers from the hydrocarbon dibenzo crown ethers.[31] We have prepared two interesting isomers of perfluorodicyclohexyl-18-crown-6 ethers[31], the cis-syn-cis and cis-anti-cis isomers. Their structures have also been established by X-ray crystallography.

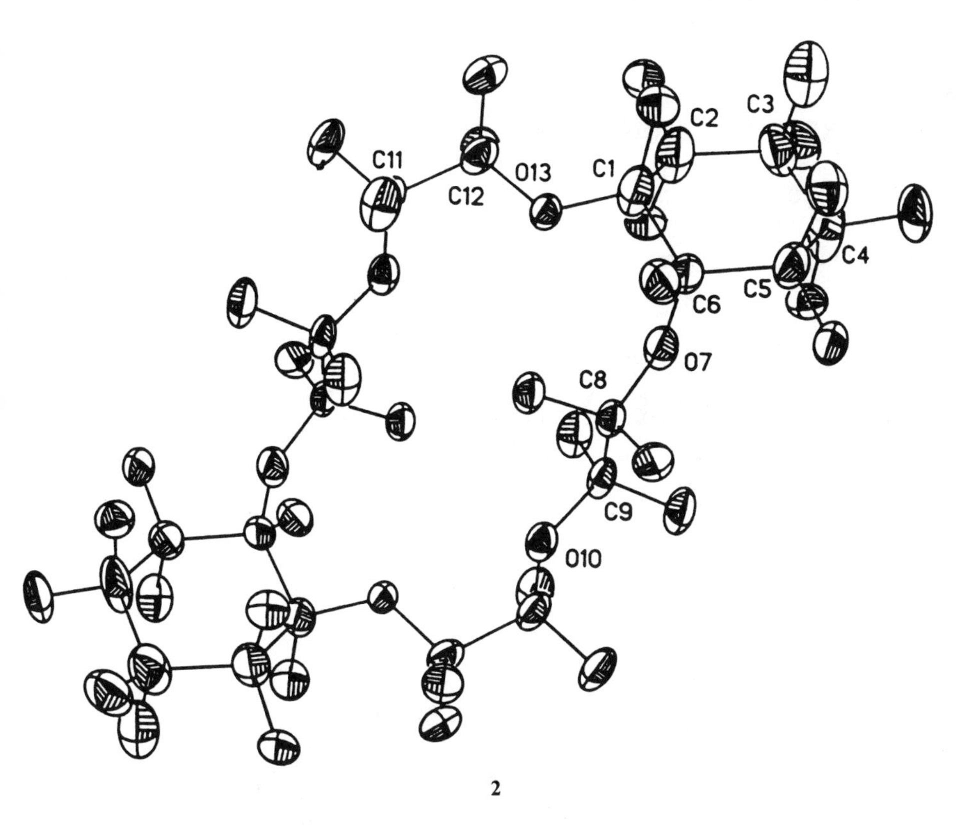

2

View (ORTEP) of cis-anti-cis (**2**) showing the atom numbering scheme. Distances (Å) between adjacent oxygen atoms are as follows: $O_{10'}$—O_{13} 2.727, O_{13}—O_7 2.702, O_7—O_{10} 3.525. Distances (Å) between oxygen atoms to the center of the molecule are as follows: O_7 3.156, O_{10} 2.403, O_{13} 3.332.

We have also reported the first perfluorocryptand molecule, namely, the perfluorocryptand [222].[11] The perfluorocryptand is a very stable, inert, high boiling clear oil. While hydrocarbon crown ethers coordinate cations, both the perfluoro crown ethers and the new perfluorocryptand coordinate anions. Two manuscripts have recently applied in colloboration with Professor J. S. Broodbelt in which both perfluorocrown ethers and perflurocryptands tenaciously encapsulate O_2^- and F^-.[32,33]

CF_2 O CF_2-CF_2 O CF_2
CF_2 CF_2
N CF_2-CF_2 O CF_2-CF_2 O CF_2-CF_2 N
CF_2 O CF_2-CF_2 O CF_2
CF_2 CF_2

5.4. PERFLUORO SPIRO COMPOUNDS

We have already synthesized several perfluoro spiro compounds in our laboratory. We have, for example, characterized and done the crystal structure[33] for perfluoro-1,4,9,12-tetraoxadispiro[4.2.4.2]tetradecane:

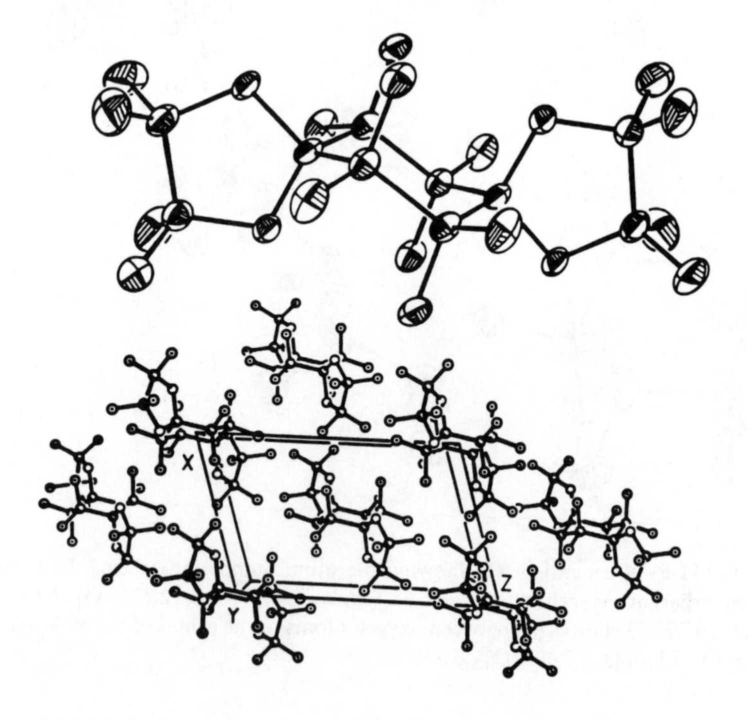

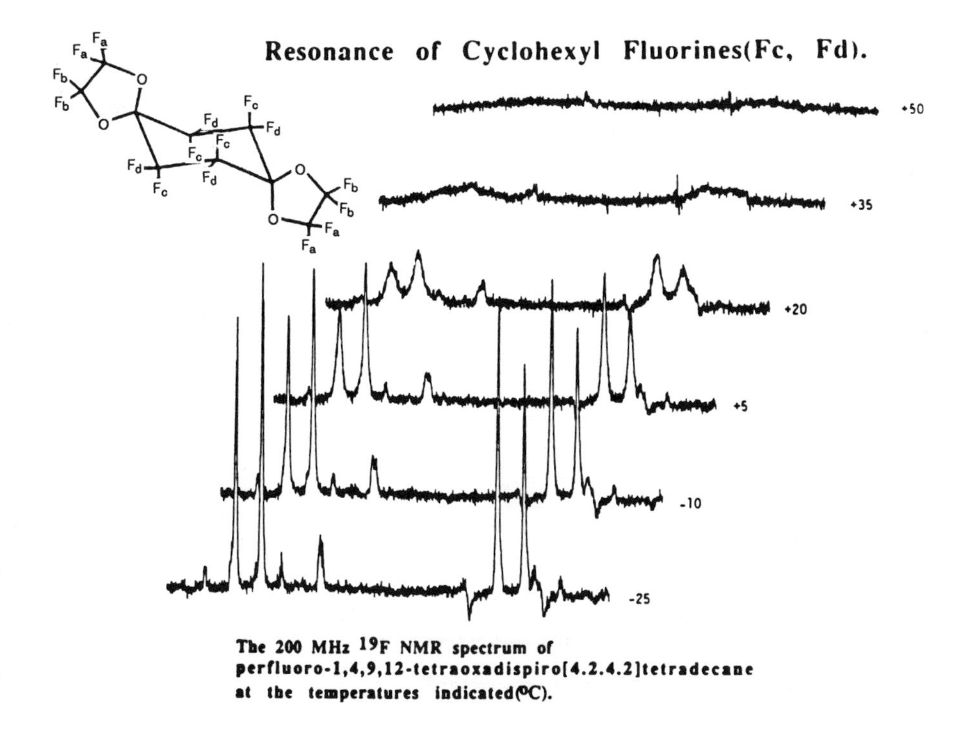

The 200 MHz ^{19}F NMR spectrum of perfluoro-1,4,9,12-tetraoxadispiro[4.2.4.2]tetradecane at the temperatures indicated(°C).

We have recently prepared ten other spiro perfluorinated compounds.[34]

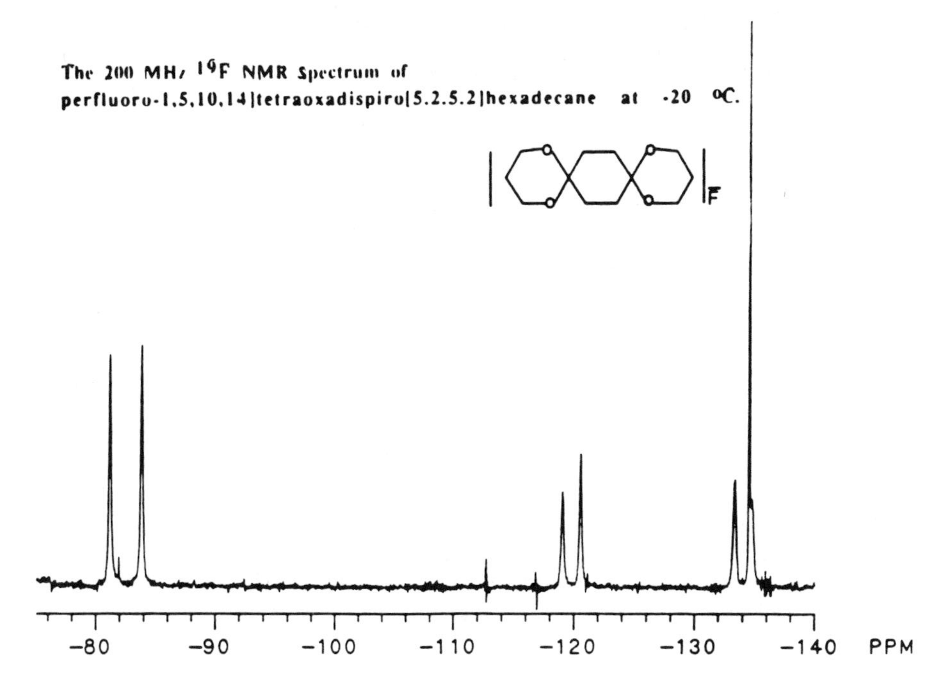

The 200 MHz ^{19}F NMR Spectrum of perfluoro-1,5,10,14]tetraoxadispiro[5.2.5.2]hexadecane at -20 °C.

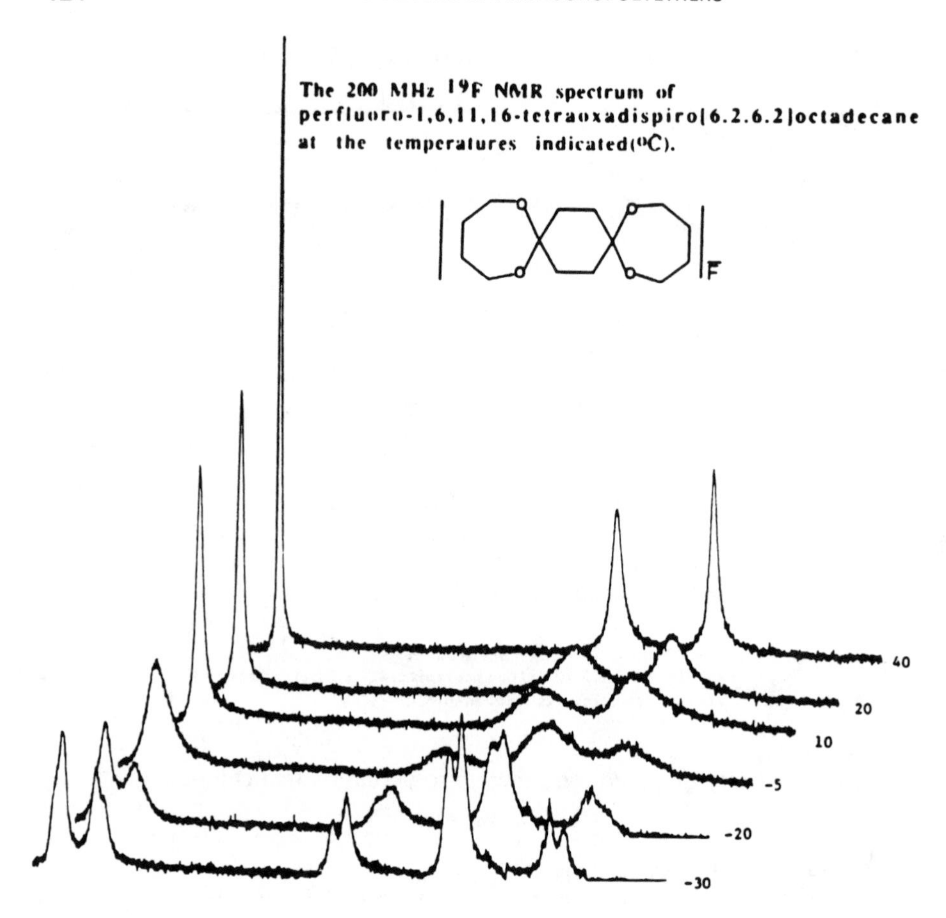
The 200 MHz 19F NMR spectrum of
perfluoro-1,6,11,16-tetraoxadispiro[6.2.6.2]octadecane
at the temperatures indicated(°C).
F
40
20
10
-5
-20
-30

Thus it is clear that many new perfluoropolyethers are available by the R. J. Lagow direct fluorination technology that are not available through other routes. The information in this chapter should serve as a strong indication that perhaps the best and ultimate synthetic methods on both a laboratory and manufacturing and industrial scale will be direct fluorination reactions in the future.

ACKNOWLEDGMENT

We are grateful for support of this work by the Air Force Office of Scientific Research (AFOSR-88-0084).

REFERENCES

1. Gerhardt, G. E. and Lagow, R. J., *J. Chem. Soc. Chem. Commun.*, 259 (1977).
2. Gerhardt, G. E. and Lagow, R. J., *J. Org. Chem.*, **43**, 4505 (1978).
3. (a) Tiers, G. V. D., *J. Am. Chem. Soc.*, **77**, 4837 (1955); (b) p. 6703; (c) p. 6704.
4. (a) Hill, J. T., *J. Macromol. Sci., Chem.*, **8**, 499 (1974); (b) Eleuterio, H. S., *J. Macromol. Sco. Chem.*, **6**, 1027 (1972).
5. (a) Sianesi, D., Benardi, G., and Moggi, F., Fr. Patent 1 531 902, 1968, to Montecatini Edison S.P.A.; (b) Sianesi, D. and Fontanelli, R., Br. Patent 1 226 566, 1971, to Montecatini Edison S.P.A.
6. Ohsaka, Y., Tohzuka, T., and Takaki, S., European Patent Application 84116003.9, 1984, to Daikin Industries, Ltd.
7. (a) Lagow, R. J. and Inoue, S., US Patent 4 113 772 (1972); (b) Gerhardt, G. E. and Lagow, R. J., US Patent 4 523 039 (1985); (c) Persico, D. F. and Lagow, R. J., US Patent 4 675 452 (1987); (d) Kagow, R. J., US Patent Application Serial No. 147,173 (1988).
8. Margrave, J. L. and Lagow, R. J., *Prog. Inorg. Chem.*, **26**, 161 (1979).
9. (a) Maraschin, N. J., Catsikis, B. D., Davis, L. H., Jarvinen, G., and Lagow, R. J., *J. Am. Chem. Soc.*, **97**, 513 (1975); (b) Adcock, J. L., Beh, R. A., and Lagow, R. J., *J. Org. Chem.*, **40**, 3271 (1975).
10. (a) Lin, W. H., Bailey, W. I., Jr., and Lagow, R. J., *J. Chem. Soc. Chem. Commun.*, 1550 (1985); (b) Lin, W. H., Bailey, W. I., Jr., and Lagow, R. J., *Pure Appl. Chem.* **60**, 473 (1988).
11. Clark, W. D. and Lagow, R. J., *J. Org. Chem.*, **55**, 5933 (1990).
12. Aikman, R. E. and Lagow, R. J., *Inorg. Chem.*, **21**, 524 (1982).
13. (a) Liu, E. K. S. and Lagow, R. J., *J. Am. Chem. Soc.*, **98**, 8270 (1976); (b) Liu, E. K. S. and Lagow, R. J., *J. Organomet. Chem.*, **145**, 167 (1978).
14. Gerhardt, G. E. and Lagow, R. J., *J. Chem. Soc. Perkin Trans. 1* 1321 (1981).
15. Gerhardt, G. E., Dumitru, E. T., and Lagow, R. J., *J. Polym. Sci. Polym. Chem. Ed.*, **18**, 157 (1979).
16. Tabata, Y., Ito, W., Matsbakashi, H., Chatani, Y., and Tadokoro, H., *Polym. J. (Japan)* **9**, 2, 145 (1977).
17. Persico, D. F. and Lagow, R. J., *Macromolecules*, **18**, 1383 (1985).
18. Persico, D. F., Gerhardt, G. E., and Lagow, R. J., *J. Am. Chem. Soc.*, **107**, 1197 (1985).
19. Persico, D. F. and Lagow, R. J., *J. Polym. Sci. Polym. Chem. Ed.*, **29**, 233, (1991).
20. Bierschenk, T. R., Juhlke, T. J., and Lagow, R. J., US Patent 4 760 198 (1988).
21. Bierschenk, T. R., Juhlke, T. J., and Lagow, R. J., US Patent 4 827 042 (1989).
22. Jones, W. R., Jr., Bierschenk, T. R., Juhlke, T. J., Kawa, H. and Lagow, R. J., *Ind. Eng. Chem. Res.*, **27**, 1497 (1988).
23. (a) Bierschenk, T. R., Juhlke, T. J., Kawa, H., and Lagow, R. J., US Patent Application Serial No. 07/414,134. (b) Lagow R. J., Bierschenk T. R., Juhlke, J. S., Kawa, H., US Patent 4,931,199 (1989).
24. Lin, W. H., Clark, W. D., and Lagow, R. J., *J. Org. Chem.*, **54**, 1990 (1989).

25. Rutherford, G. B., Clark, W. D., and Lagow, R. J., to be published.
26. Huang, H. N., Persico, D. F., Clark, L. C. Jr., and Lagow, R. J., *J. Org. Chem.*, **53**, 78 (1988).
27. Persico, D. F., Huang, H. N., Clark, L. C., Jr., and Lagow, R. J., *J. Org. Chem.*, **50**, 5156 (1985).
28. Lin, W. H. and Lagow, R. J., US Patent 4 570 005 (1986).
29. Schweighardt, F. K. and Rubertone, J. A., US Patent 4 838 274 (1989).
30. Lin, T. Y., Clark, L. C., Jr., and Lagow, R. J., to be publihed.
31. Lin, T. Y., and Lagow, R. J., *J. Chem. Soc. Chem. Commun.*, 12 (1991).
32. Brodbelt, J., Maleknia, S. D., Lin, T. Y., and Lagow, R. J., *J. Am. Chem. Soc.*, **113**, 5913 (1991).
33. Broodbelt, J., Maleknia, S. D., Lin, T. Y., and Lagow R. J., *J.C.S. Chem. Comm.*, 1705 (1991).
34. Lin, T. Y. and Lagow, R. J., to be published.

CHAPTER 6

Universal Synthesis of Perfluorinated Organic Compounds

JAMES L. ADCOCK

In the early 1970s Margrave and Lagow [1-4] showed that the direct fluorination reaction could be controlled by temporal separation of the individual propagation steps of the fluorine free radical chain reaction with organic compounds. If the abstraction of a hydrogen by atomic fluorine producing a carbon centered radical and hydrogen fluoride ($\Delta H = -141\ \mathrm{kJ\,mol^{-1}}$) does not occur simultaneously on the same molecule with a radical interaction with molecular fluorine producing a carbon–fluorine bond and regenerating the fluorine atom ($\Delta H = -289\ \mathrm{kJ\,mol^{-1}}$), then, with adequate heat dissipation, insufficient energy is released to dissociate a carbon–carbon bond ($\Delta H = -351\ \mathrm{kJ\,mol^{-1}}$) during a direct fluorination. In later work by Lagow [5-10] the benefits of high dilution, low temperatures, solid-state stabilization, high surface area, and long reaction times were recognized along with the need to conclude the reaction with higher temperatures and high fluorine concentrations. Yields obtained by the low-temperatures gradient (LTG) process were higher than had ever been previously obtained with elemental fluorine for a variety of functional organic compounds. The process was, however, slow, requiring several days to process a batch of material. As a participant in this work between 1971 and 1974, I saw this as a serious drawback and sought a novel solution to this problem upon receiving a faculty position at the University of Tennessee. In 1979 this effort came to fruition with the discovery and development of the aerosol direct fluorination process.[11]

Aerosol Fluorination, A Flow Process. Aerosol fluorination has been reviewed by Adcock and Cherry.[12] In this review the process was described as it was then understood, and a substantial body of organic functionalities has been fluorinated with exceptional success. As a flow process the reaction time was

Synthetic Fluorine Chemistry,
Edited by George A. Olah, Richard D. Chambers, and G. K. Surya Prakash.
ISBN 0-471-54370-5

reduced, 1–5 min, and it became possible to stoichiometrically control the fluorination reaction. The rate of reaction was no longer linked to the low temperature volatility of the material being fluorinated as in the LTG process. Throughput was now related to the aerosol density in the carrier gas, carrier flow rate, and the cross-sectional area of the reactor. Solid state stabilization of the organic molecule was preserved, and surface area exposure to fluorine gas was maximized by adsorption on sodium fluoride particulates, which reduced the acidity of the medium and protected molecules from backside attack. Sodium fluoride apparently exhibits a mild catalytic or antiinhibitory effect on the fluorination as well. The concept that direct fluorination had to be carried out in at least two stages, a reaction limiting stage followed by a reaction forcing stage, developed as a result of discussions during my work with Lagow. In LaMar and LTG technologies a solid phase batch was subjected to a high initial dilution of fluorine commonly coupled with a low reaction temperature which was gradually changed over time to a higher fluorine concentrations and a higher reaction temperature, a temporal separation of reaction limiting and reaction forcing conditions. In aerosol fluorination technology, these conditions were simultaneously operating in different regions of the reactor. A gradual increase in both fluorine concentration and reaction temperature was established over the length of the reactor. A gradient from reaction limiting conditions at the beginning of the reactor to reaction forcing conditions at the end of the reactor, a spatial separation of reaction limiting and reaction forcing conditions, was made functionally possible by a flowing solid (aerosol) reactant.

In an effort to achieve high fluorine efficiencies, a photochemical "finishing stage" was added that reduced the need for excess fluorine to near 20%. The photochemical stage produced two important findings. First, as the lamp power increased, the need for excess fluorine to achieve perfluorination was significantly reduced. Second, molecules so activated were cascaded to perfluorination even with a 25% deficit of fluorine.[11] Unactivated molecules passed into the product trap almost unchanged from their entry into the photochemical stage. The photochemical stage had almost no effect on product distribution and substitution if less than 50% of the stoichiometrically required fluorine was present, indicating that molecules were 50 + % fluorinated on leaving the prior reaction stage. The reactivity of fluorine with organic molecules in the dark at temperatures substantially below room temperature indirectly support the postulate that an associative initiation step is active in the free radical reactions of fluorine and organic compounds. This idea was first suggested by Miller.[13] Fluorinations to about 50–60% of perfluorination occur readily at fluorine concentrations of only 1–2% by volume. Fluorinations over this value are progressively more difficult in the dark. In early aerosol fluorination work fluorine concentrations of over 30% by volume produced 10% or less perfluorination during the 1–5-min reaction time with temperatures up to 40°C. The impracticality of achieving still higher fluorine concentrations by adding pure fluorine to the relatively high volume flow of dilute fluorine, forced us to seek fluorine efficiency and perfluorination by photochemical activation. The achievement of perfluorination with very short reactor residence times, minimal residual organic hydrogen, minimal fragmentation, high product yields, and high product effluent concentrations direct from the reactor at such low fluorine concentrations has been a unique achievement of aerosol technology. No other technology utilizing elemental fluorine has achieved such results. The significance of high yields in direct fluorination, a series of substitution reactions equal to the number of hydrogen atoms, is noteworthy. If one considers the yield required for each step in a 10-step reaction sequence to achieve an overall yield of 80–90% for the sequence, one can appreciate the success achieved by LaMar, LTG, and aerosol technologies.

Low Acidity Fluorinations. The conduct of fluorinations or any organic reaction in strong Lewis acid or protonic superacid media (anhydrous HF) should be expected to result in structural rearrangement of the carbon skeleton, bond cleavages, loss or alteration of functionality, and other side reactions. The introduction of alkali fluoride into the reaction medium significantly reduces this acidity and significantly reduces or eliminates such undesired results. Aerosol fluorination technology efficiently utilizes this increased basicity of the medium and the low reaction temperatures to preserve sensitive structures and functionalities.

The strong Lewis acidity and high operating temperatures of the cobalt trifluoride process produce significant skeletal rearrangement and cleavage of even the thermodynamically stable adamantane structures.[14] This is problem reduced but not completely solved by LaMar-type LTG direct fluorination used by Lagow et al.[15] The aerosol process on the other hand, due presumably to the intimate association between alkali fluoride aerosol particulates and the adsorbed organic substrate shows no rearrangements in any of the adamantanelike materials fluorinated. The most complex "diamond-oid" molecule fluorinated haks been diamantane ($C_{14}H_{20}$) prepared by the method of Olah et al.[16] The molecule was isolated in 8% yield and was identified by its negative chemical ionization mass spectrum, base peak 548 m/z (100%), 549 $^{13}C_1$ (16.1%), 550 $^{13}C_2$ (2.5%), and its ^{19}F NMR [$\phi_a = -117.4$ ppm (6); $\phi_b = -209.6$ ppm (3); $\phi_c = -223.6$ ppm (1)], Analysis for $C_{14}F_{20}$: calc. %C 30.68, %F 69.32; found %C 30.75, %F 69.78. 1,3-Dimethyladamantane was fluorinated to its perfluorinated analogue in 41.9% isolated yield. The effluent contained *F*-1,3-dimethyladamantane (90%) and *F*-1-methyladamantane (5%) by gas liquid chromatography. Mass spectra EI + for $C_{12}F_{20}$ showed a weak M—F ion at 524 m/z; ^{19}F NMR [$\phi_{CF_3} = -55.7$ ppm (6); $\phi_{CF_2} = -99.5$ ppm (2); $\phi_{CF_2} = -111.1$ ppm (8), $\phi_{CF} = -218.3$ ppm (2)] which compared well with an authentic sample.[14]

Fluorination of Solutions. Because the stoichiometry, the F_2/H ratio, is important in the control of the reaction and many of the compounds we have been asked to synthesize are solids, we have successfully introduced these solid substrates into the hydrocarbon evaporator unit of the aerosol fluorinator as solutions by syringe pump. The successful fluorination of solutions emphasizes the individual molecule's isolation as it is fluorinated. Choice of the solvent is important as it must not boil at a temperature below the melting point of the solute in order to prevent solid deposition in the tubes feeding the evaporator. It must also fluorinate to a material easily separable from the solid reactant after perfluorination. In most cases we have found that aliphatic hydrochlorocarbons are excellent choices but that carbon tetrachloride and chloroform are not. In every case where this technique has been used, production rates if not yields of perfluorinated solid reactant are higher than when temperature-controlled sublimation is used. Solvent perfluorinations are comparable to reactions where solvent alone is fluorinated. In all cases to date there has been no documented case of intermolecular reactions between solute and solvent. They fluorinate independently. The cases of carbon tetrachloride and chloroform are interesting in that these solvents interfere somehow in the fluorination and neither solvent nor solute fluorinate well. This may well be connected to the observation by Rozen[17] that chloroform moderates the reactivity of fluorine in his highly selective, low temperature, solution reactions. In our case this moderating influence is detrimental. Successful application of this solution injection procedure is commonplace and will be noted parenthetically henceforth.

Adamantane is the simplest diamondoid molecule and because of its symmetry is poorly soluble in most aliphatic solvents and only slightly more soluble in aromatic solvents, which are unsuitable in cofluorination. Since 1-bromoadamantane is considerably more soluble in aliphatic chloroalkanes and bromine does not survive direct fluorination, 1-bromoadamantane is a reasonable precursor for *F*-adamantane, which can be conveniently cofluorinated with 1,2-dichloroethane. The molecule is identical to that produced by direct fluorination of adamantane introduced by sublimation, but is substantially free of hydryl-*F*-adamantane materials and can be readily separated from tetrafluoro-1,2-dichloroethane (R-114) by simple vacuum line fractionation.[18] As we shall see in the following paragraph, bromines in aerosol direct fluorination are oxidized and generate a carbocation at the bromine substitution site. This is not a complication for 1-bromoadamantane since the 1-adamantyl cation is the most stable cation and does not rearrange under the conditions present.

Fluorine Generated Carbocations. Aerosol fluorinations of neopentyl chloride results in near quantitative yields of *F*-neopentyl chloride.[19] However, an analogous fluorination of neopentyl bromide produces a 90% yield of *F*-isopentane. The structural rearrangement observed is th classical 1,2-methyl shift commonly observed whenever the neopentyl cation could result as an intermediate.[20] The power of the aerosol technique to probe the physical organic chemistry of elemental fluorine to an extent never before practical can be appreciated by the deconvolution of the stepwise reactions observable by stoichiometric control of fluorine introduction. If 1 mol of fluorine per mole of neopentyl bromide is introduced the product observed exclusively is the alkene,

2-methyl-2-butene. The presumed intermediate is a neopentyl-BrF_2 derivative that disproportionates by BrF_2^- elimination to form neopentyl cation, which rearranges rapidly to the tertiary 2-methyl-2-yl cation. Interestingly, higher fluorine ratios begin to produce statistically anomalous products derived from 2-methyl-3,3-difluorobutane. This observation was explained by the presumed oxidation of bromine to a neopentyl-BrF_4, which underwent two disproportionations, the first resulting in an α-fluorinated *tert*-butyl-CHF—BrF_2. The second resulting in the α-fluorinated neopentyl carbocation, which after undergoing the 1,2-methyl shift stabilizes as the α-fluoro secondary cation, $(CH_3)_2CHC^+FCH_3$, rather than the tertiary cation formed in the earlier example. Fluoride ion capture by the cation results in 2-methyl-3,3-difluorobutane.

Free Radical Rearrangements, 1,2 Chloride Shifts. Aerosol fluorination of alkyl chlorides does not always produce the analogous perfluorinated alkyl chlorides. Although primary alkyl chlorides fluorinate without rearrangement, secondary and tertiary alkyl chlorides are significantly more prone to 1,2-chloride shifts. Secondary chlorides always produce mixtures of perfluorinated primary and secondary isomers unless the rearrangement is degenerate. Tertiary chlorides will rearrange completely to principally primary perfluoroalkyl chlorides although quantities of secondary perfluoroalkyl chlorides have been isolated when a statistically significant rearrangement can occur. For example, fluorination of 2-chloro-2-methylbutane produces 1-chloro-*F*-2-methylbutane (43.1%) and 1-chloro-*F*-3-methylbutane (17%) but also produces 2-chloro-*F*-3-methylbutane (2.7%). Even here a second 1,2-chloride shift occurred leading to a primary *F*-alkyl chloride, which predominated over the secondary intermediate species.[21] The rearrangements have been shown to occur early during the fluorination when the degree of fluorination is small. More extensive work has clearly related the product distribution outcomes to the stabilities of the free radicals formed by fluorine abstraction.[12] The only exceptions to these rules have been observed in the fluorination of 1-chloroadamantane (with 1,1,2-trichloroethane) to *F*-1-chloroadamantane, which was isolated in 40% yield (100% effluent concentration). Apparently, the rigidity of the adamantane structure impedes the rearrangement.[26].

F-1-Chloroadamantane is a reactive molecule and the addition of methyllithium in ethyl ether at −100 to −30°C for over 7 h followed by addition of iodine in ether at −30°C and allowed to warm to 22°C for over 2 h results in 27% conversion to 1-iodo-*F*-adamantane. The reaction of 1-chloro-*F*-adamantane in zinc–1,4-dioxane at 100°C for over 12 h followed by the addition of water results in a 42% conversion to 1-hydryl-*F*-adamantane. In contrast the isomeric 2-chloro-*F*-adamantane, which was cofluorinated with 1,1,2-trichloroethane and isolated in 49% yield, was much less reactive. It did not react with zinc–dioxane and required the substitution of the more vigorous free radical dechlorinating reagent tri(*n*-butyl)tin hydride–AIBN reacting at 80°C for 22 h to produce a 54% conversion to 2-hydryl-*F*-adamantane.

Aerosol fluorination of polychloroalkanes have been extensively investigated. Aerosol fluorination of methyl chloroform (CH_3CCl_3) results in a virtually quantitative yield of the rearranged 1,1,2-trichloro-*F*-ethane (R-113). Extensive analysis of the product of this reaction revealed only traces of the 1,1,1-trichloro-*F*-ethane.[22] Rearrangements of initially formed radicals by 1,2-chloride shifts can be rationalized by examination in Fig. 6.1 of the relative stabilities of radicals formed by these shifts. Figures 6.2 and 6.3 describe the observed radical rearrangements occurring when complete and partial rearrangement of reactants, respectively is observed during fluorination. The ease of this rearrangement can be seen in Fig. 6.3 for the rearrangement of 1,2-dichlorocyclopentanes to 1,3-dichlorocyclopentanes during fluorination. We believe the degree of rearrangement is kinetically based and not an equilibration, since secondary chlorides rearrange to primary chlorides during fluorination to produce mixtures but primary chlorides do not. This would indicate that complete rearrangements occur when the rate of rearrangement is fast relative to capture of the radical by molecular fluorine but slower in cases where only partial rearrangement is observed.

Free Radical Rearrangements, Cleavage. In the previous reactions we examined numerous rearrangements in which groups are not eliminated. Chlorine loss in most of the preceding reactions is minimal, even in early work it was usually less than 20%. Fluorination of a series of polychloroalkanes where each carbon atom is monochlorinated produced some interesting results. 1,2-Dichloroethane and 1,2,3-trichloropropane fluorinate to their perfluorinated analogues in 70% and greater than 40% yield, respectively. However, 1,2,3,4-tetrachlorobutane cofluorinates in 1,1,2-trichloroethane to 1,2-dichloro-*F*-ethane in 42% yield. This too has a ready explanation in radical chemistry. In Fig. 6.4 we see that elimination of 1,2-dichloroethylene from the radical(s) produced by hydrogen abstraction on 1,2,3-trichloropropane and 1,2,3,4-tetrachlorobutane requires that a chloromethyl radical be eliminated in the first

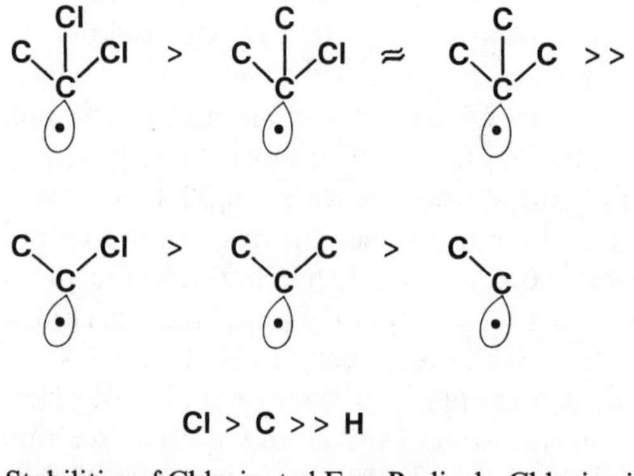

Cl > C >> H

Figure 6.1. Relative Stabilities of Chlorinated Free Radicals. Chlorine is more stabilizing than carbon, which is much more stabilizing than hydrogen.

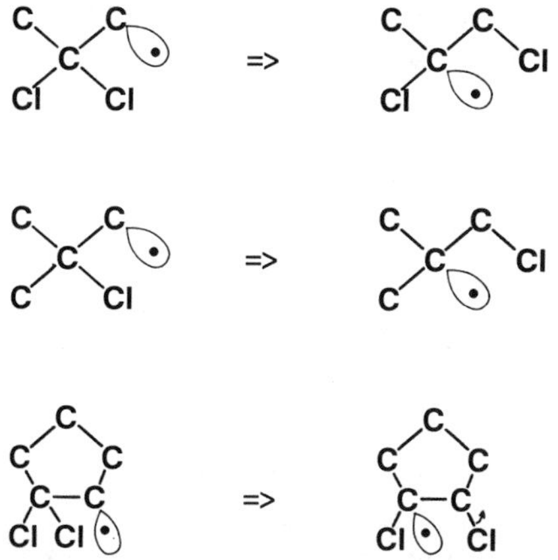

Figure 6.2. Complete Rearrangements Chloroalkane Radicals. Small arrow indicates cis or trans isomers.

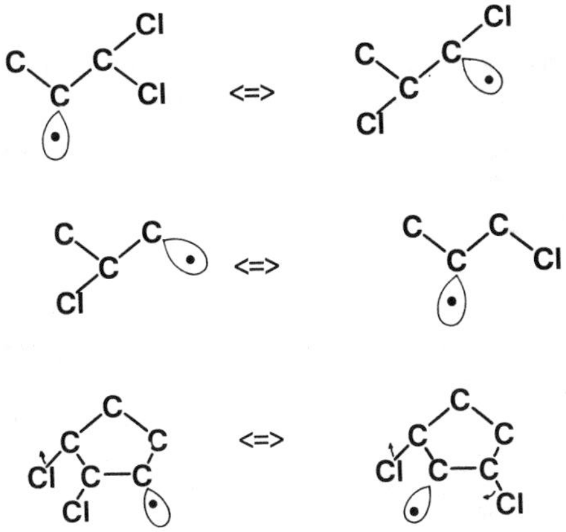

Figure 6.3. Incomplete Rearrangements Chloroalkane Radicals. Small arrows indicate cis or trans isomers.

$$\cdot CHCl\text{-}CHCl\text{-}CH_2Cl \Rightarrow CHCl{=}CHCl + \cdot CH_2Cl$$

$$\cdot CHCl\text{-}CHCl\text{-}CHCl\text{-}CH_2Cl \Rightarrow CHCl{=}CHCl + \cdot CHCl\text{-}CH_2Cl$$

Figure 6.4. Radical cleavage in polychloroalkanes.

case but a more stable 1,2-dichloroethyl radical be eliminated in the second case. Although mechanistic proof is not available, thus explanation does not seem particularly unrealistic.

In the radical reactions to follow we shall explore the radical chemistry of ethers. In general, aerosol fluorination of ethers results in high yields of the analogous perfluorinated ethers. A large number of ethylene glycol dimethyl ethers, cyclic and acyclic alkyl ethers, and polyethers have been successfully perfluorinated in good to excellent yields. It is instructive to examine the ethers that are prone to bond scissions during fluorination. Beta scission (bond scission one atom removed from the initial radical) occurs when two criteria are met. First, a good leaving group can be formed, preferably a stable molecule, and second, a more stable radical can be formed as a result of the scission (see Fig. 6.5). Although beta-scissions occur to a limited extent between methylene groups on ethylene glycol polyethers, the degree is only a few percent.[23] The effect becomes most serious in the special class of polyethers called orthoesters.[24]

Aerosol fluorination of tetramethyl orthocarbonate, $C(OCH_3)_4$, was subject to extensive β-scission. The products isolated $C(OCF_3)_4$ (7%), $FC(OCF_3)_3$ (14%), $F_2C(OCF_3)_2$ (6%), and CF_3OCF_3 (66%) plus large quantities of carbonyl fluoride indicate extensive beta-scission with a rate comparable to the rate of fluorination.

By comparison, aerosol fluorination of trimethylorthoformate, $HC(OCH_3)_3$, resulted in isolation of $FC(OCF_3)_3$ (5%), $F_2C(OCF_3)_2$ (29%), and CF_3OCF_3 (60%) again with large quantities of carbonyl fluoride. In this case as with the orthocarbonate the yield of the product resulting from one β-scission was higher than the intact perfluorinated molecule. The major cleavage product, *F*-dimethyl ether, was in both cases the result of exhaustive β-scission. The fact that different intermediate products predominate in the two closely related reactions is puzzling. This difference is repeatable and seems to be kinetic rather than thermodynamic in origin.

In contrast, a third related reaction, aerosol fluorination of the formal, dimethoxymethane, $H_2C(OCH_3)_2$, resulted in the synthesis of *F*-dimethoxymethane in 58% yield (yields that have since risen to 70+%) with only minor amounts of *F*-dimethyl ether. This would indicate that at least two oxygens are needed to stabilize the secondary radical formed by β-scission. It may also be that significant fluorination of the central carbon, which has the most reactive hydrogens, occurred first, resulting in reduced rearrangement.

Trimethylorthoacetate fluorinated to *F*-1,1-dimethoxyethane (24%) and *F*-methyl *F*-ethyl ether (<5%), no perfluorinated trimethyl orthoacetate could be

$$R_3C\text{-}O\text{-}CH_3 \;\text{-AF->}\; R_3C\text{-}O\text{-}\dot{C}H_2$$

$$R_3C\text{-}O\text{-}\dot{C}H_2 \;\text{--->}\; R_3C\cdot \;+\; O{=}CH_2$$

Figure 6.5. Beta-scission reactions in aerosol fluorinations.

isolated. One β-scission occurred fast with respect to the reaction with elemental fluorine. Large quantities of carbonyl fluoride were also isolated as well as smaller quantities of perfluorinated esters, which were removed from the perfluorinated ethers by hydrolysis.

In alkyl ethers β-scission is hardly noticeable, except when a tertiary radical can be formed (see Fig. 6.6). We found significantly greater β-scission when fluorinating these molecules.[25] Aerosol fluorination of *tert*-butyl ethyl ether results in isolation of *F*-*tert*-butyl *F*-ethyl ether in 42% yield. Significant quantities of *F*-isobutane and *F*-acetyl fluoride are also isolated. This is a much higher scission rate than is seen in the nontertiary alkyl ethers, which can be fluorinated in yields approaching 80–90%. The longer alkyl chains (ethyl and larger) appear to be slightly less prone to scission than the methyl group. Aerosol fluorination of *tert*-butyl methyl ether produces only a 36% yield of *F*-tert-butyl *F*-methyl ether under the same conditions. Apparently, *F*-acetyl fluoride is a poorer leaving group than carbonyl fluoride, a finding reinforced by our work on esters.[26]

Fluoride Ion Catalysis. It was mentioned earlier in the discussion of this process that sodium fluoride was either catalytic or antiinhibitory in the direct fluorination reaction. Our most substantial evidence for this the effect of eliminating the sodium fluoride from the reaction or substituting lithium fluoride for sodium fluoride.[22] It is well known that lithium does not form a stable bifluoride salt and that such salts increase in stability as the size of the alkali metal cation increases. This size effect is also though to underlie the increased catalytic activity of the larger alkali metal fluorides: CsF > RbF > KF > NaF > LiF. In experiments with larger organic molecules as starting materials, which were carried out under very similar conditions except that the sodium fluoride is eliminated or replaced with lithium fluoride, the degree of perfluorination is very low and substantial fragmentation occurred under conditions that would otherwise produce excellent results. Methyl chloride, on the other hand, fluorinates well under most conditions. Quantitative results for the fluorination of methyl chloride are published,[22] which show that when the sodium fluoride aerosol is present, even when a

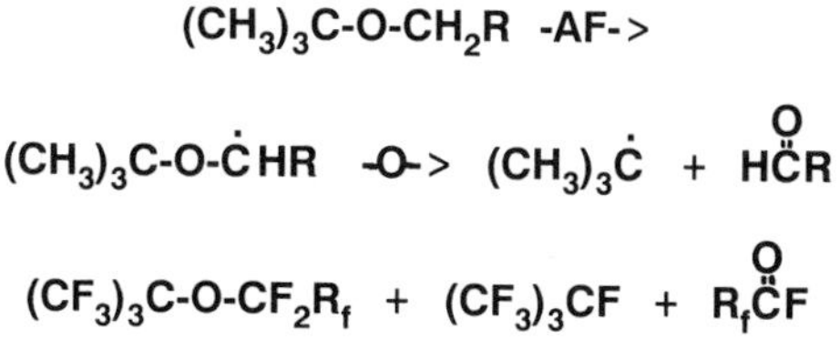

$$(CH_3)_3C\text{-}O\text{-}CH_2R \;\text{-AF->}$$

$$(CH_3)_3C\text{-}O\text{-}\dot{C}HR \;\text{-O->}\; (CH_3)_3\dot{C} + HC(=O)R$$

$$(CF_3)_3C\text{-}O\text{-}CF_2R_f + (CF_3)_3CF + R_fC(=O)F$$

$R = H$; $R_f = F$: yield = 36%

$R = CH_3$; $R_f = CF_3$: yield = 42%

Figure 6.6. Aerosol fluorination tertiary butyl ethers.

TABLE 6.1. Direct Fluorination of Methyl Chloride.

			Products (%)					
Run Number	F_2/CH_3Cl[a]	Temperature (°C)[b]	CF_4	CF_3Cl	CF_2HCl	CH_2FCl	CH_3Cl	CH_2F_2
1[c]	1.0	−50, −50	9.0	18.0	2.4	45.0	22.0	3.7
2[c]	2.0	−50, −50	30.5	46.7	3.0	12.0	0.0	7.8
3[c]	1.0	−50, −50, −50	6.6	9.6	1.2	46.0	32.0	5.0
4[c]	2.0	−50, −50, −50	24.0	44.0	2.5	18.0	1.0	10.0
5[d]	3.0	25	17.6	64.0	10.1	3.6	0.0	4.6
6[d,e]	2.7	85, 85	13.0	38.0	25.0	10.0	0.0	0.0
7[d,f,g]	2.7	85, 85	14.3	29.9	7.4	0.0	35.4	0.0

[a]Molar ratio of F_2 to CH_3Cl fed to the reactor.
[b]Reactor temperature; more than one entry indicates the temperatures of each module in a multistage reaction.
[c]Substrate deposited on an NaF aerosol; no photochemical finishing step was employed.
[d]Reaction run without NaF aerosol; CH_3Cl introduced as the vapor.
[e]Also obtained a 14% relative yield of unidentified material.
[f]HF injected into main carrier stream during this reaction.
[g]Also obtained a 14.6% relative yield of unidentified material.

deficiency of fluorine and low ($-50°C$) temperatures are employed, substantial amounts of perfluorination, and for the chloromethanes substantial chlorine substitution, are observed (see Table 6.1). The product distribution is anomalous with larger amounts of monofluorinated and trifluorinated material and small amounts fo difluorinated material isolated when the stoichiometry was 2:1 for F_2:CH_3Cl. A 3:1 stoichiometry at $+25°C$ without the sodium fluoride aerosol produced predominantly CF_3Cl and the distribution is near normal. The amount of chlorine substitution is significantly reduced. In the 3:1 stoichiometry at $+85°C$ in which hydrogen fluoride is injected the amount of CF_3Cl is halved and the amount of unreacted starting material is 30% of the total material collected. These observations, although qualitative, clearly indicate the inhibitory effect of added hydrogen fluoride on this fluorination and the catalytic–antiinhibitory effect of sodium fluoride, which promotes fluorination and in the case of methanes produces chlorine substitution.

The catalytic activities of the heavier alkali fluorides have not been extensively explored in the aerosol fluorinator. Although reactions to investigate potassium fluoride reactivity are planned, we have only recently devised means of controlling the rates and amounts of alkali fluorides evaporated. Sodium fluoride is considered to be a generally poor source of the catalytically active fluoride ion. It is, however, difficult to dismiss the possibility of very high activity for freshly generated aerosol particulate of sodium fluoride, especially, when we can readily demonstrate its essential contribution to the fluorination reaction process. In this regard we have examined the aerosol fluorination reactions of a number of functionalities, which in the form of their perfluorinated derivatives are sensitive to the fluoride ion.

Fluorination of Esters. Perfluorinated esters have been produced by the condensation of *F*-acid fluorides over cesium fluoride at low temperatures.[27] *F*-Acid fluorides or *F*-esters have been shown to equilibrate to a mixture of *F*-acid fluorides and *F*-esters (Fig. 6.7). We were concerned that the high fluoride availability in the aerosol reactor would make the direct fluorination synthesis of *F*-esters nearly impossible. We were surprised to learn that esters do in fact survive.[26] Their survival as an intact functional group is both a function of the fluorination conditions and the structure of the esters. Virtually all methyl esters, which are the most labile, can be fluorinated intact and isolated, if the fluorination conditions are optimized to provide the gentlest possible conditions consistent with perfluorination. For a given set of conditions it is found that the

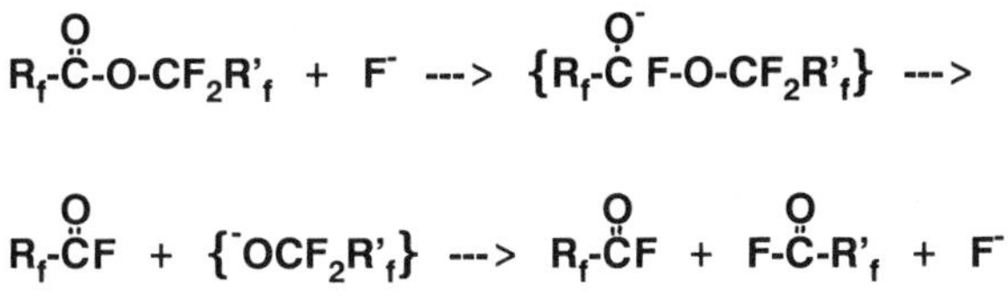

$$R_f\text{-}\overset{O}{\overset{\|}{C}}\text{-O-}CF_2R'_f + F^- \text{--->} \{R_f\text{-}\overset{O^-}{\overset{|}{C}}F\text{-O-}CF_2R'_f\} \text{--->}$$

$$R_f\text{-}\overset{O}{\overset{\|}{C}}F + \{^-OCF_2R'_f\} \text{--->} R_f\text{-}\overset{O}{\overset{\|}{C}}F + F\text{-}\overset{O}{\overset{\|}{C}}\text{-}R'_f + F^-$$

Figure 6.7. Fluoride ion catalyzed cleavage of *F*-esters.

steric bulk of R in the perfluoroalkyl group (R_f) in R_fCOOCF_3 is effective in protecting the carbonyl group from nucleophilic attack by the fluoride ion (Fig. 6.8). The perfluorination of methyl pivalate, methyl trimethylacetate, is effected without cleavage of the trifluoromethyl group under conditions in which methyl isobutyrate and methyl propanoate undergo significant cleavage to form the respective acid fluorides and carbonyl fluoride. The ready ejection of the trifluoromethoxide ion and its reversion to COF_2 and F^- is pronounced and can be difficult to reduce. A qualitative trend relating steric bulk and relative difficulty of ester cleavage is clear. It seemed logical that such a trend might be seen if one varies the steric bulk of the alcohol part of the ester. A series of propanoic acid esters of ethanol, isopropanol, and *tert*-butanol were synthesized (Fig. 6.9). The results were not as gratifying. In none of the reactions was simple cleavage observed although significant fragmentation of the *tert*-butyl propanoate occurred during fluorination. Although not obeying our simple steric hypothesis, the conclusion that only the methyl esters are susceptible to a significant cleavage due to fluoride ion (as sodium fluoride) is inescapable. The ready cleavage if the *tert*-butyl ester can likely be attributable to radical rearrangements as seen earlier. It is significant to note that all of the

$$(CH_3)_3CC(=O)\text{-O-}CH_3 \text{ -AF-> } (CF_3)_3CC(=O)\text{-X} \quad 86\%$$

$(X = OCF_3\%)/(X = F\%)$: 100/0

$$(CH_3)_2CHC(=O)\text{-O-}CH_3 \text{ -AF-> } (CF_3)_2CFC(=O)\text{-X} \quad 41\%$$

$(X = OCF_3\%)/(X = F\%)$: 22/55

$$CH_3CH_2C(=O)\text{-O-}CH_3 \text{ -AF-> } CF_3CF_2C(=O)\text{-X} \quad 58\%$$

$(X = OCF_3\%)/(X = F\%)$: 59/41

Figure 6.8. Fluoride ion cleavage of esters vs. acid structure.

$$CH_3CH_2C(=O)\text{-O-}CH_2CH_3 \text{ -AF-> } CF_3CF_2C(=O)\text{-O-}CF_2CF_3 \quad 23\%$$

$$CH_3CH_2C(=O)\text{-O-}CH(CH_3)_2 \text{ -AF-> } CF_3CF_2C(=O)\text{-O-}CF(CF_3)_2 \quad 65\%$$

$$CH_3CH_2C(=O)\text{-O-}C(CH_3)_3 \text{ -AF-> } CF_3CF_2C(=O)\text{-O-}C(CF_3)_3 \quad 33\%$$

In none of the above reactions was simple cleavage observed!

Figure 6.9. Fluoride ion cleavage of esters vs. alcohol structure. In none of these reactions was simple cleavage observed.

perfluorinated esters are readily cleaved to their respective acid fluoride pairs except that the isopropyl ester forms propanoyl fluoride plus hexafluoroacetone and the *tert*-butyl ester gave only propanoyl fluoride. The solid residue from the latter reaction was treated with 50% sulfuric acid and perfluoro-*tert*-butanol was identified.

Aerosol fluorination of methyl esters has proven to be a convenient route to perfluorinated acid derivatives without dealing with water sensitive and corrosive acid chlorides and fluorides. Although both acid fluorides and chlorides fluorinate to perfluorinated acid fluorides in excellent yields, these starting materials are unpleasant to handle and must be protected from atmospheric moisture. The sacrifice of the two carbon–fluorine bonds released by displacement of carbonyl fluoride would probably need to be avoided in an industrial process but is not consequential for a laboratory scale preparation. Fluorination of molecules with bifunctionality has not proved to complicated in any of the cases examined so far.[26]

Synthesis of methyl 4-(2-methoxy·ethoxy)butyrate was accomplished by reaction of sodium 2-methoxy·ethoxide with γ-butyrolactone (40% yield), and was converted to the methyl ester using dry methanol and catalytic amounts of concentrated sulfuric acid (80% yield). Aerosol fluorination of this material produced a mixture of 48% *F*-methyl *F*-4-(2-methoxy·ethoxy)butyrate and 6% *F*-4-(2-methoxy·ethoxy)butyryl fluoride. These materials were converted by sodium hydroxide to sodium *F*-4-(2-methoxy·ethoxy)butyrate. Careful drying of this material and vacuum pyrolysis at 270°C produced a 46% yield of the alkene, *F*-3-(2-methoxy·ethoxy)-1-propene. Conversion of the sodium salt to the silver salt followed by extraction with ether, drying of the salt protected from light, and pyrolysis with iodine at 100°C for 2 days resulted in a 50% yield of *F*-3-(2-methoxy·ethoxy)-1-iodopropane. We were surprised that the anion formed by decarboxylation of sodium *F*-4-(2-methoxy·ethoxy)butyrate did not eliminate the *F*-2-methoxy·ethoxide ion by an allylic mechanism.

F-Alkenes from F-Esters. Perfluorinated alkenes of unique structures are accessible by aerosol fluorination. A synthesis of perfluorinated tertiary butyl ethylene was carried out as follows. Ethylene was Friedel–Crafts alkylated with *tert*-butyl chloride in 45% yield. The chloride was converted to the Grignard reagent and carbonylated. The resulting acid was isolated in 35% yield and converted to the methyl ester as previously outlined. Aerosol fluorination of the methyl ester produced a mixture of *F*-methyl *F*-4,4-dimethylpentanoate (31%) and *F*-4,4-dimethylpentanoyl fluoride (26%). The mixture was hydrolyzed with sodium hydroxide to sodium *F*-4,4-dimethylpentanoate and after careful drying pyrolyzed under vacuum at 340°C to produce *F*-3,3-dimethylbutene in 68% yield.

Aerosol fluorination has been utilized to produce directly materials otherwise unavailable. The reaction of 7-oxabicyclo[2.2.1]heptane (1,4-epoxycyclohexane) resulted in a 39% yield of the fluorinated bicyclic ether. This molecule has been reported to fragment extensively under even "LaMar" "LTG" methodology.[28]

Aerosol fluorination of 3,3-dimethyloxetane produced this otherwise unavailable perfluorinated oxetane in 44% yield. The *F*-2,2-dimethyloxetane derivatives are available by photolysis of hexafluoroacetone and tetrafluoroethylene. Carbonyl fluoride generally fails to cyclize with alkenes and precludes the preceding method in the synthesis of the 3,3-derivative. For the same reason *F*-oxetane has never been available until it was synthesized by aerosol fluorination of oxetane in yields now in the 70% range.

In conclusion, it can be said with little exaggeration that the aerosol direct fluorination method is in fact a method for the Universal Synthesis of Perfluorinated Organic Compounds.

ACKNOWLEDGMENTS

Research in fluorine chemistry at the University of Tennessee has been supported in part by grants from, the Research Corporation, the Office of Naval Research, the 3M Company, the E. I. du Pont de Nemours, the Sun Oil Corporation, the Gas Research Institute, the Allied Signal Corporation, the US Environmental Protection Agency, and the Electric Power Research Institute, whose support is greatly appreciated.

REFERENCES

1. Margrave, J. L. and Lagow R. J., *Chem. Eng. News.*, 40 (1970).
2. Margrave, J. L. and Lagow R. J., *Proc. Natl. Acad. Sci.*, *69*, 8A (1970).
3. Margrave, J. L. and Lagow, R. J., *J. Polym. Sci.*, *12*, 177 (1974).
4. Margrave, J. L. and Lagow, R. J., *Prog. Inorg. Chem.*, *26*, 161 (1979).
5. Lagow, R. J. and Maraschin, N. J., *J. Am. Chem. Soc.*, *94*, 8601 (1972).
6. Lagow, R. J. and Maraschin, N. J., *Inorg. Chem.*, *12*, 1459 (1973).
7. Lagow, R. J. and Adcock, J. L., *J. Org. Chem.*, *38*, 3617 (1973).
8. Lagow, R. J. and Adcock, J. L., *J. Am. Chem. Soc.*, *96*, 7588 (1974).
9. Lagow, R. J. Adcock, J. L., and Beh, R. A., *J. Org. Chem.*, *40*, 3271–3275 (1975).
10. Lagow, R. J. and Adcock, J. L., *J. Fluorine Chem.*, *7*, 197 (1976).
11. Adcock, J. L., Horita, K., and Renk, E. B., *J. Am. Chem. Soc.*, *103*, 6937–6947 (1981).
12. Adcock, J. L. and Cherry, M. L., *Ind. Eng. Chem. Res.*, *26*, 208–215 (1987).
13. Miller, W. T. and Koch, S. D., *J. Am. Chem. Soc.*, *79*, 3084 (1957).
14. Moore, Robt. E. and Driscol, Gary, L., *J. Org. Chem.*, *43*, 4978–4980 (1978).
15. Lagow, R. J., Robertson, G., and Liu, E. K. S., *J. Org. Chem.*, *43*, 4981–4983 (1978).
16. Olah, G. A., Wu, A., Farooq, O., and Prakash, G. K. S., *J. Org. Chem.*, *54*, 1450–1451 (1989).
17. Rozen, S., *Acc. Chem. Res.*, *21*, 307 (1988).
18. Adcock, J. L. and Robin, M. L., *J. Org. Chem.*, *48*, 3128–3130 (1983).

19. Adcock, J. L., Evans, W. D., and Heller-Grossman, L., *J. Org. Chem.*, *48*, 4953–4957 (1983).
20. Skell, P. S., Storer, I., and Krapcho, A. P., *J. Am. Chem. Soc.*, *82*, 5257 (1960).
21. Adcock, J. L. and Evans, W. D., *J. Org. Chem.*, *49*, 2719–2723 (1984).
22. Adcock, J. L., Kunda, S. A., Taylor, D. R., Nappa, M. J., and Sievert, A. C., *Ind. Eng. Chem. Res.*, *28*, 1547–1549 (1989).
23. Adcock, J. L. and Cherry, M. L., *J. Fluorine Chem.*, *30*, 343–350 (1985).
24. Adcock, J. L., Robin, M. L., and Zuberi, S., *J. Fluorine Chem.*, *37*, 327–336 (1987).
25. Taylor, D. R. (1987), M. Sc. Thesis, The University of Tenessee, March 1987.
26. Adcock, J. L. et al. (1992) to be published.
27. Shreeve, J. M., De Marco, R. A., and Couch, D. A., *J. Org. Chem.*, *37*, 3332–3334 (1972).
28. Lagow, R. J., Huang, H. N., Persico, D. F., and Clarke, L. C., *J. Org. Chem.*, *53*, 78–85 (1988).

CHAPTER 7

Electrophilic Fluorination Reactions with F_2 and Some Reagents Directly Derived from It

SHLOMO ROZEN

Using elemental fluorine during the first 90 years since its isolation by Moissan in 1886,[1] was almost a "taboo" in organic chemistry. A few sporadic attempts to react certain perfluoro derivatives with F_2 during this period had been made,[2] but the results and the impact of these works did not encourage more comprehensive research projects nor did they inspire other researchers to join the pitifully small club of organic chemists who used fluorine gas.

Part of the "blame" for this neglect is due, of course, to the unusual reactivity and corrosiveness of the element. This created among most chemists the feeling, transformed with time into an unchallenged "wisdom," that it is pointless to burn and destroy valuable compounds while probably blowing yourself up in the process. Surprisingly, there are still many great chemists who yet have to free themselves from such legends and prejudice deeply rooted during their basic chemical education. Another reason for this early apathy was the fact that apart, from the rarely found fluoroacetic acid and some of its simple derivatives, nature has not created any fluoroorganic compounds of its own. Therefore, in the first 50 years, very few scientists were really interested in the chemistry of this element.

It was the Second World War, mostly through the famous Manhattan

Synthetic Fluorine Chemistry,
Edited by George A. Olah, Richard D. Chambers, and G. K. Surya Prakash.
ISBN 0-471-54370-5

Project, which started fluorine chemistry in grand scale. The need for compounds and lubricants that can withstand the awesome corrosiveness of F_2 and UF_6 was behind the massive research for new perfluorinated materials with outstanding stabilities and unique features. The rapid growth in the standard of living after the war brought with it a strong demand for cooling devices, which in its turn boosted more research and production of what was then considered to be the perfect family of chemicals—the halofluorocarbons.

While the above compounds, which have excellent physical and chemical stability, represented one aspect of the fluorine containing chemicals, Fried and his co-workers started to reveal in the 1950s the importance of this atom in biological processes.[3] Soon many fluorosteroids, fluoronucleosides and nucleotides, perfluorinated molecules serving as blood substitutes, fluoroantibiotics, fluoroanesthetics and much much more, captured the mind of everybody concerned with biological and health issues. Today, conservative estimates call for more than $50 billion/year to be associated with this element in organic chemistry alone.

With this boom in fluoro compounds one would imagine there to be a wide spectrum of reagents available to meet all the synthetic demands in the area. It was thus quite surprising to a young chemist 20 years ago to find out that the arsenal of fluorinating reagents was not quite as rich as one would imagine and that the primary source for the fluorine atom, namely, the element itself, was conspicuously missing from the list.

It was our good fortune at that period to spend some time in the laboratories of Barton and Hesse at Cambridge, Massachusetts, where the novel concept of electrophilic fluorination with the known, but yet unused, CF_3OF was shaping rapidly. It was demonstrated, for example, that reacting this gas under suitable conditions with electron-rich double bonds, such as enol acetates, resulted in high yield fluoroketones. The same pattern was observed with regular alkenes, where the oxygen bound fluorine always attached itself to the more electron-rich end of the double bond proving its electrophilic character (Eq. 7.1).[4] Thus, the chemistry of CF_3OF started its short-lived but intensive bloom.[5]

AcO ... $\xrightarrow{CF_3OF}$ F ... O

AcO ... AcO ... F ... OCF_3

(7.1)

After a few years CF_3OF ceased to be available and the stream of publications concerning this chemical had come almost to a halt.[6] This, along with the fact that the strength of the O—F and the F—F bonds are similar (42 vs 39 kcal mol^{-1}), turned our attention to the readily available and much cheaper elemental fluorine.

In the early 1970s Lagow had already used fluorine in reactions with hydrogen-containing compounds and had shown that the skeleton of the organic substrate could be preserved. His reactions were of a radical nature, however, as his main goal was replacing most or all hydrogens with fluorine atoms.[7] We, on the other hand, were more interested in the introduction of a single fluorine atom at specific sites, and the question was whether the most reactive element could be tamed to perform regio- and stereoselective reactions.

7.1. DIRECT ELECTROPHILIC FLUORINATION WITH F_2

7.1.1. Selective Substitution of Tertiary C—H Bonds

The chances are that whenever a chemist discusses electrophilic reactions he means reactions on some π center such as a double bond or an aromatic ring. Electrophilic substitutions on saturated carbon atoms are very rare but have a considerable theoretical interest and have largely been covered in an excellent book published by Olah.[8] Most of these reactions have not been practical and proceed with low yields, since the common electrophiles will not generate strong driving forces in favor of the substitution reaction. If fluorine is to be the electrophile, it should be one of the strongest, capable of reacting even with the electrons of a saturated C—H bond: $R_3CH + F_2 \rightarrow R_3CF + HF$. Thermodynamically such reactions are very favorable and should proceed exothermically by more than 100 kcal mol^{-1}. It was clear that, in order to minimize unselective radical processes, the reaction should be efficiently cooled, the fluorine and the substrates quite diluted and not much light should be allowed. These conditions alone, however, were found to be insufficient and when dilute fluorine in nitrogen[9] was bubbled through a solution of 4-methylcyclohexanol acetate in trichlorofluoromethane, a relatively fast radical reaction took place resulting in fluorinated tars (Eq. 7.2). Changing the solvent from $CFCl_3$ to hexane, pentane, or perfluoro 2-butyl THF (tetrahydrofuran) did not improve the outcome. A dramatic change was observed when the solvent system of the reaction was replaced by a 1:1 mixture of $CFCl_3$ and $CHCl_3$, which resulted in the formation of the simple but yet unknown 4-fluoro-4-methylcyclohexanol acetate in higher than 60% yield (Eq. 7.2).[10] A very interesting and important point is the full retention of configuration with which this type of reaction always proceeds. Thus the trans derivative was forming only the trans fluoro compound, while the cis isomer was transformed exclusively to the cis fluoro isomer. Obviously this rules out any radical mechanism, since a tertiary radical at C-4 would lead to a mixture of isomers (Eq. 7.2).

F_2/N_2, $CFCl_3$ → Tars

OAc —(F_2/N_2, $CHCl_3/CFCl_3$)→ OAc

H F

(7.2)

OAc —(F_2/N_2, $CHCl_3/CFCl_3$)→ OAc

H F

If indeed this reaction is ionic in nature and the positive pole of the fluorine molecule reacts as an electrophile, it should attack mainly the most electron-rich C—H bonds, which in the above case are the tertiary carbon–hydrogen bonds.[11] The ionic mechanism is further supported by using a mixture of 4-methyl- and 4-*t*-butylcyclohexanol acetates. Despite the steric shielding felt by the axial tertiary carbon–hydrogen bond in the latter compound, it is more electron rich than the corresponding one in the methyl compound and therefore reacts considerably faster (Eq. 7.3). It should be noted that if the reactions are not monitored and stopped at the right time, the small amount of fluorine radicals always present would eventually consume all the starting material as well as the products, resulting typically in fluorinated tars.

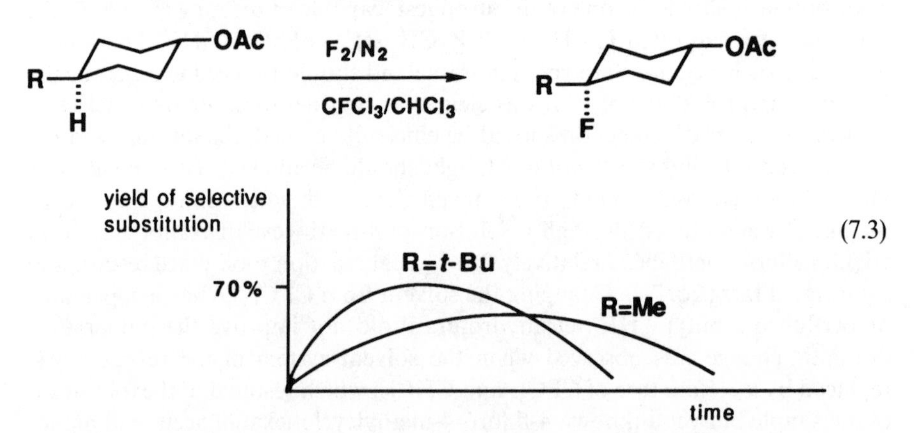

(7.3)

The mechanism of electrophilic substitutions on saturated carbon atoms was suggested by Olah[8] some years ago and involves a nonclassical three-center two-electron carbonium ion. Tertiary fluorination provides new unique support for this mechanism. Since the electrophile attacks neither the carbon nor the hydrogen but instead attacks the electrons bonding them, the resulting reaction should proceed with a full retention of configuration (a frontside attack) as demonstrated for the first time in the above examples. In order to encourage this transition state the solvent has to be polar, act as a radical scavenger, and provide an acceptor for the counterion of the electrophile, which in this case is the fluoride ion. Chloroform, acetic acid, and nitromethane are all polar solvents with somewhat acidic hydrogens and can act as such acceptors through hydrogen bonding with the fluoride ion (Eq. 7.4). Solvents like $CFCl_3$ or hexane cannot support the ionic intermediate and the alternative radical reactions become dominant. The nonclassical three-center two-electron carbocation does not encourage eliminations and rearrangements and indeed, with the exception of one case, we have not observed such processes in the large number of reactions we have examined. The one exception recorded well supports the idea of the unclassical carbocation. Reacting bicyclo[2.2.2]octane resulted in a high yield formation of only two compounds identified as 1-fluoro bicyclo[2.2.2]octane and 2-fluoro bicyclo[3.2.1]octane.[12] The pentacoordinated carbonium ion initially formed is stabilized by both the fluorine atom and the resonance with the nonclassical carbonium ion A keeping eight electrons around each carbon.[8] This ion eventually ejects wither the C—1 or the C—2 hydrogen atom with the appropriate rearrangement of the relevant C—C bonds forming the two fluoro derivatives (Eq. 7.4).

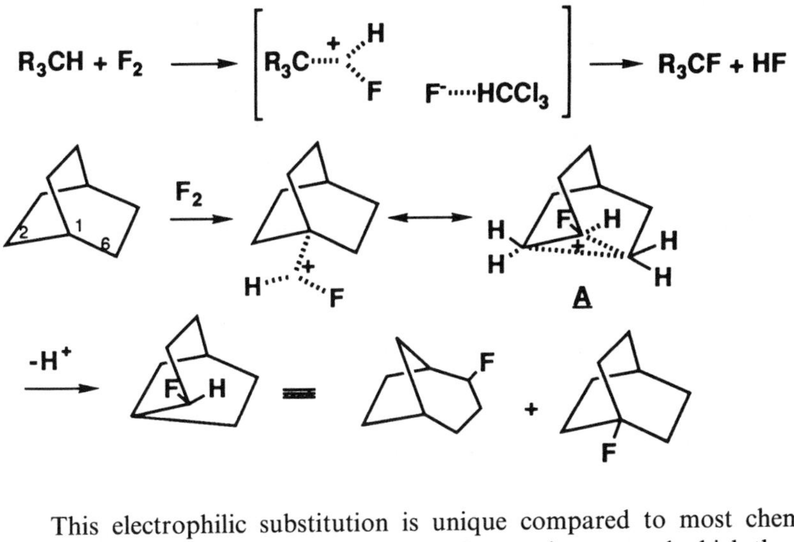

$$R_3CH + F_2 \longrightarrow \left[R_3\overset{+}{C}\cdots\begin{smallmatrix}H\\F\end{smallmatrix} \quad F^-\cdots HCCl_3 \right] \longrightarrow R_3CF + HF \qquad (7.4)$$

This electrophilic substitution is unique compared to most chemical reactions, which usually need some kind of an anchor around which the reagents can regroup. Thus, there are numerous reactions performed on carbonyls, double bonds, aromatic nuclei, heteroatoms, and so on. Extremely few processes, mainly of radical nature, involve an attack on a CH moiety remote from such activating handles. Examining the chemistry of the simple 4-methyl cyclohexanol derivatives, for example, reveals hundreds of reactions performed on the oxygen atom and the three adjacent carbon atoms, but practically none was reported on the inactivated C—4. Elemental fluorine reacts contrastingly with this carbon–hydrogen bond. Subsequent HF elimination results in a double bond thus creating an opening for many chemical transformations. Many similar reactions resulting in activation of "impossible sites" can be easily performed. For example, decalins, bicyclohexyls, and alkylcycloalkane derivatives can all be fluorinated at their tertiary position and subsequently dehydrofluorinated (Eq. 7.5).

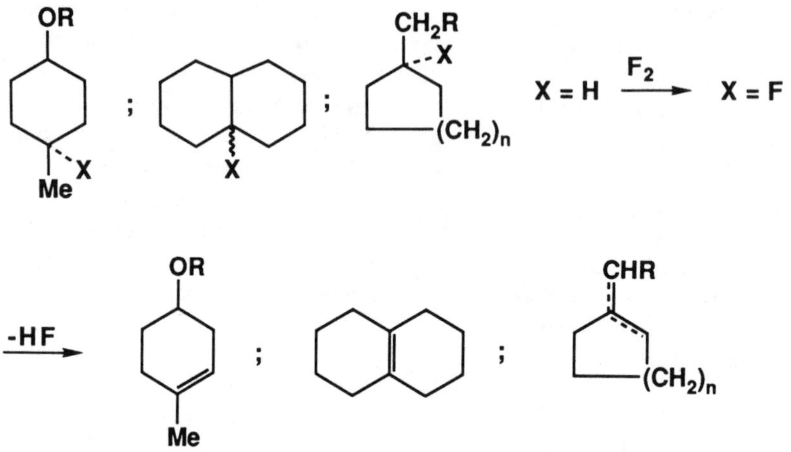

OR; ; CH_2R, X, $(CH_2)_n$; Me, X; X; X = H $\xrightarrow{F_2}$ X = F

$\xrightarrow{-HF}$ OR, Me; ; CHR, $(CH_2)_n$ (7.5)

Usually the dehydrofluorination process can be carried out either by employing Lewis acid reagents such as HF and $BF_3 \cdot OEt_2$ or basic ones like NaOH, or better still, MeMgBr. While the former elimination is a fast E1 process frequently associated with carbonium rearrangements, the basic dehydrofluorination proceeds with a slow syn-E2 mechanism, but in many cases results in cleaner not rearranged products.

Despite the fact that the reaction is performed by one of the most reactive reagents, it is still very stereo- (as already mentioned) and regioselective.[13] Fluorine will differentiate, for example, between the three tertiary hydrogen atoms in menthol acetate and substitute the one with the highest hybridization in *p*, the C-7 hydrogen atom, in 60% yield. Similarly, only the hydrogen further away from the electron-withdrawing oxygen atom in 2-acetoxybicyclopentyl will be substituted by fluorine (Eq. 7.6).

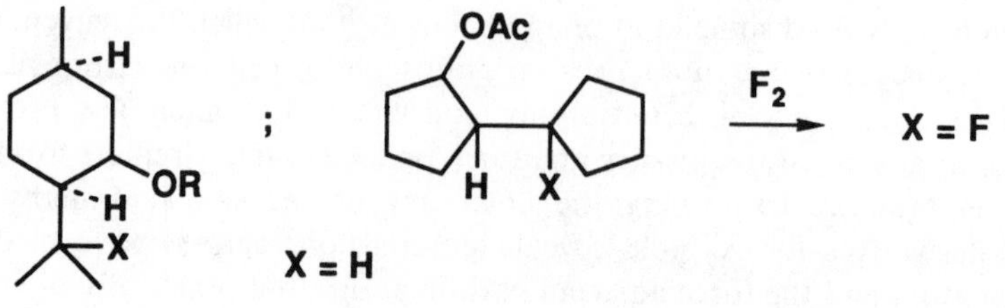

H, OR, H, X; OAc, H, X; X = H $\xrightarrow{F_2}$ X = F (7.6)

It should be mentioned that if a tertiary C—H bond is relatively electron poor, fluorine will not react with it under ionic conditions. Thus, when the tertiary hydrogen is too close to a strong electron-withdrawing group or otherwise has a low *p* orbital contribution as in the cyclopropane ring, the electrophilic substitution is no longer favorable and apart from some deterioration, due to minor radical reactions, the starting material can be largely recovered (Eq. 7.7).

O; CH_2R; CH_2OR; H; Me; R; H; F_2 → no reaction (7.7)

The tertiary fluorination reaction is quite general and steroids are very attractive substrates. By using the right electron-withdrawing substituent in various places, one can substitute almost any tertiary hydrogen of the skeleton. Consecutive dehydrofluorination produces double bonds serving as entries for constructing biologically important steroids.[14] Thus it is possible to introduce fluorine at positions 9 (offering an attractive route for the construction of cortisone-like 9-fluoro-11-hydroxy derivatives), at 14 (an entry to the important cardenolides), at 17 (for converting the inexpensive cholesterol to androstanes), and more. The only tertiary position that has never been attacked is the β hydrogen at C—8, which is effectively shielded by the two angular methyl groups at C-18 and C-19 (Eq. 7.8).

F_2 (R = alkyl), Δ^5, protected as dichloride → X=Y=H; Z=F

F_2 (R =OR'), Δ^5, protected as dichloride → X=Z=H; Y=F (7.8)

F_2 (R = OR') or corticosteroidal chain → Y=Z=H; X=F

Another steric factor becomes important when working with bile acids or other A/B *cis* (5β) steroids. Despite the fact that the C-9 hydrogen bond is the most electronically favored for electrophilic substitution, ring A, which is perpendicular to the steroidal plane, blocks the access to it. The fluorine, which has to approach with a solvent molecule, attacks the other tertiary sites at C-5, C-14, C-17 (Eq. 7.9).[15]

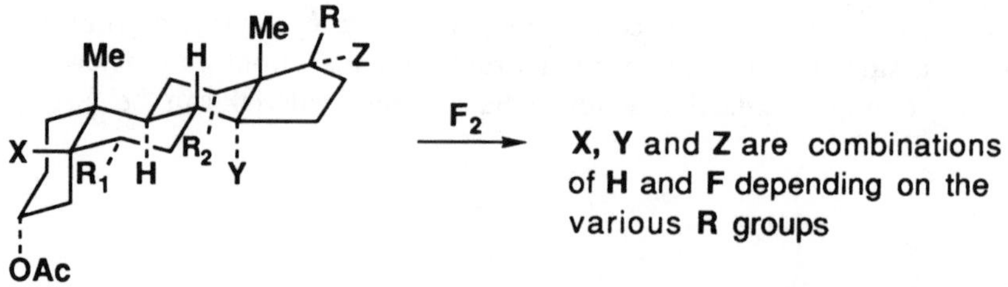

$\xrightarrow{F_2}$ **X, Y** and **Z** are combinations of **H** and **F** depending on the various **R** groups

(7.9)

These fluorosteroids, which were available for the first time, presented an excellent opportunity to learn more about the structure–^{13}C NMR relationship.[16] The observed changes between the parent steroids and the fluoro derivatives are sterically dependent and all γ-carbon atoms *gauche* to the fluorine atom regardless of the general shape of the steroid (A/B cis or trans) or the location of the halogen atom, are shielded by up to 8 ppm. In contrast, the γ-carbon atoms antiperiplanar to the fluorine atom exhibit a very characteristic downfield shift of 2–5 ppm. This difference is also reflected in the long-range $^3J(CF)$ coupling constants. In most cases the gauche γ-carbon atoms have very small couplings of up to 0.3 Hz, while anti carbon atoms frequently exhibit coupling constants of 6–8 Hz.

The steroidal skeleton being quite rigid, enabled the development of some other methods for the activation of unreactive sites. Breslow's "remote control" oxidation is probably one of the best known examples.[17] Such methods, however, are practically nonexistent in the field of the flexible aliphatic compounds. Since fluorine is little affected by steric factors the scope and limitations of the tertiary fluorination substitution are similar to those found in the cyclic and steroidal derivatives. The aliphatic substrate may be a paraffin, alcohol or carboxylic acid derivative, amide or ketone and as long as the electron-withdrawing group is far enough from the reacting center, electrophilic substitution can be carried out (Eq. 7.10).[18] The differences in yield and reaction rates are again easily understood qualitatively in terms of electron density of the reactive C—H bond.

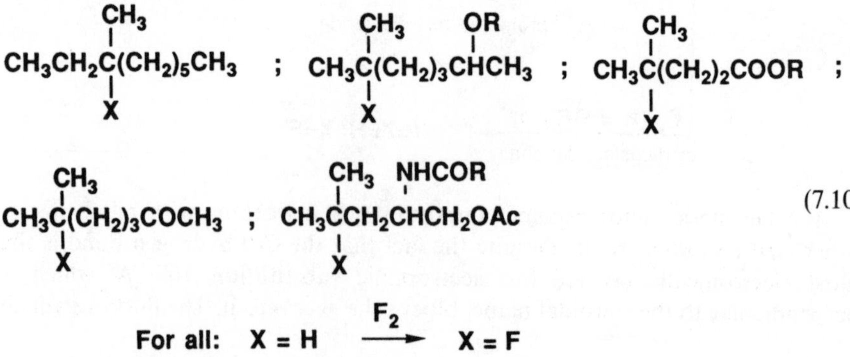

$$CH_3CH_2C(CH_3)(X)(CH_2)_5CH_3\ ;\ CH_3C(CH_3)(X)(CH_2)_3CH(OR)CH_3\ ;\ CH_3C(CH_3)(X)(CH_2)_2COOR\ ;$$

$$CH_3C(CH_3)(X)(CH_2)_3COCH_3\ ;\ CH_3C(CH_3)(X)CH_2CH(NHCOR)CH_2OAc \qquad (7.10)$$

$$\text{For all: } X = H \xrightarrow{F_2} X = F$$

When the electron-withdrawing group is too close to the tertiary CH bond the electrophilic attack is prevented and the starting material may be recovered. The presence of an aromatic ring should be avoided since when not strongly deactivated, it reacts with the fluorine faster than any tertiary center (Eq. 7.11).

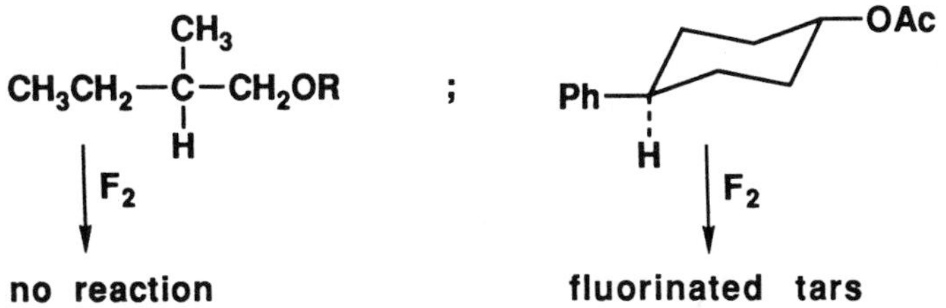

$CH_3CH_2-C(CH_3)(H)-CH_2OR \xrightarrow{F_2}$ no reaction ; Ph–(cyclohexyl, H)–OAc $\xrightarrow{F_2}$ fluorinated tars (7.11)

We have already seen that bicyclo[2.2.2]octane reacts readily with fluorine (Eq. 7.4). Other bicyclic derivatives also react and as long as the mode of the reaction can be of electrophilic nature, the results are predictable.[12] Thus derivatives of adamantane, bicylco[3.2.1]octane, bornane, and norbornane can all be fluorinated (Eq. 7.12). The yields and reaction rates are governed, as in the previous cases by the hybridization of the C—H bond in question. Thus compounds with high ring strain such as the tricyclic derivative A are completely resistant to F_2, while the pinane system B gives very low yields of the corresponding fluoro derivatives. The case of the endo-tricyclo[5.2.1.0^{2,6}]decane (C) is illuminating. This molecule has no electron-withdrawing elements, which may polarize or deactivate some of the potential reactive centers, and yet due to the relatively small difference in hybridization of the two types of the tertiary hydrogen atoms ($sp^{2.7}$ vs $sp^{2.8}$) only the 2-fluoro derivative was isolated in higher than 75% yield.

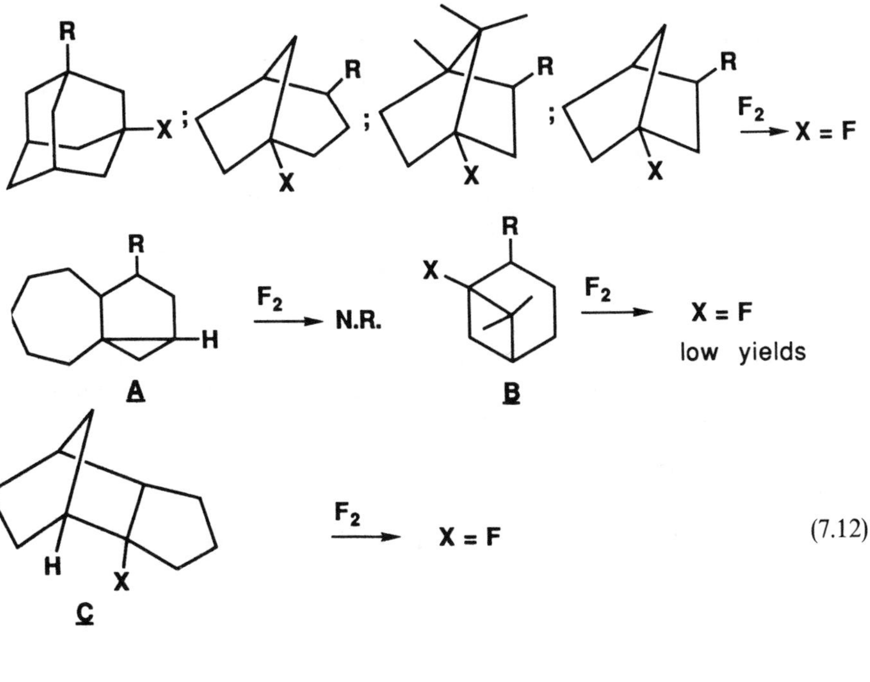

R, X ; R, X ; R, X ; R, X $\xrightarrow{F_2}$ X = F

A $\xrightarrow{F_2}$ N.R. B $\xrightarrow{F_2}$ X = F low yields

C $\xrightarrow{F_2}$ X = F (7.12)

7.1.2. Addition of Fluorine to Alkenes[19]

Adding chlorine and bromine to double bonds is one of the fundamental processes in chemistry and has been constantly investigated since the nineteenth century. Characteristically, however, not much has been done with F_2 in this field. It was Merritt et al.[20] in the mid-1960s who showed that such an addition is possible, but his technique was quite unusual and rather inconvenient. It was clear that in order to discourage unselective radical reactions and encourage the more controllable electrophilic ones, we had to provide a polar solvent, low temperature, and low fluorine concentration. Since F_2 reacts slower with ethanol than with alkenes, we decided to add some alcohol to the reaction solvent mixture of $CHCl_3$ and $CFCl_3$. The alcohol was also intended to serve as a temporary acceptor to the negative pole of the fluorine molecule through hydrogen bonding. Following the above assumptions we passed very dilute fluorine ($\sim 1\%$ F_2 in N_2) through a cold solution of trans-3-hexene-1-ol acetate and obtained the corresponding threo difluoro adduct in good yield. The cis-alkene was also reacted and stereospecifically converted to the erythro derivative (Eq. 7.13). The syn addition, contrary to that observed with Cl_2 and Br_2, was not surprising as this is the preferred mode for every addition involving electrophilic fluorine. The rationale behind this steroselectivity was first presented by Barton and Hesse[21] for reactions of stilbenes with CF_3OF and is based on the formation of a tight ion pair constituting the very short living α-fluoro carbocation A. This ion pair collapses before any rotation around the C—C bond in question can take place (Eq. 7.13).

$$CH_3CH_2CH{=}CH(CH_2)_2OAc \xrightarrow{F_2} CH_3CH_2CHF{-}CHF(CH_2)_2OAc$$

trans → *threo*
cis → *erythro*

$$>C{=}C< \xrightarrow{FX} \left[\text{F–C–C}^{+}\ \ X^{-}\cdots\text{H-solvent} \right]_{\mathbf{A}} \longrightarrow \text{F–C–C–X} \qquad (7.13)$$

The addition of fluorine to alkenes is not confined to a single type of double bonds and was also demonstrated with cyclic alkenes, steroidal ones,[22] terminal alkenes, and relatively deactivated carbonyls and lactones (Eq. 7.14).

$$\text{cycloheptene} \xrightarrow{F_2} \text{1,2-difluorocycloheptane} \quad ; \quad RCH{=}CH_2 \xrightarrow{F_2} RCHF{-}CH_2F$$

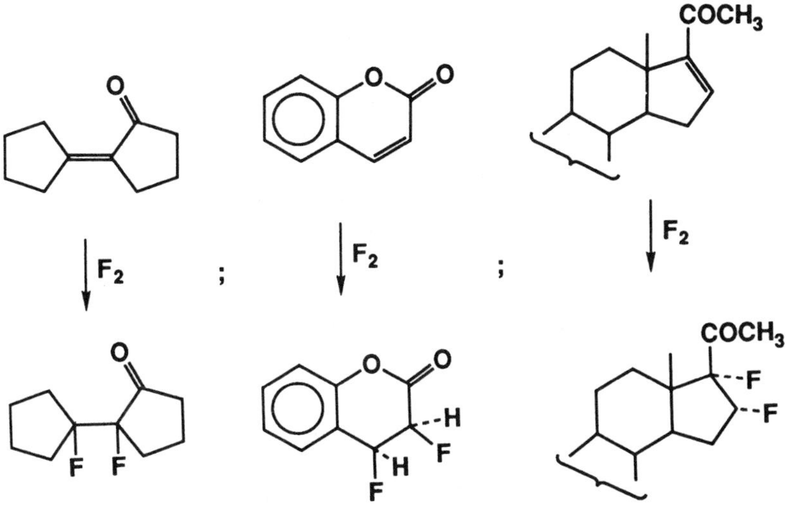

(7.14)

7.2. INDIRECT ELECTROPHILIC FLUORINATIONS WITH F_2

7.2.1. Perfluoroalkyl and Acyl Hypofluorites

As we have seen there are reactions where F_2, can be very selective. There are other cases, however, where applying elemental fluorine results in inseparable mixtures of many compounds. Such was the case when F_2 was reacted with enol acetate solutions in an attempt to make the important α-fluorocarbonyl moiety.[23] This was quite disappointing since CF_3OF executes these reactions very well indeed.[4] As in many cases, the solution came while working on a different project. For a long time we were trying to find some media, liquid or even solid (such as molecular sieves), in which we could "dissolve" fluorine. We did not have much success in this area, but in several experiments we passed fluorine through $CFCl_3$ in which dry sodium trifluoroacetate was suspended. This resulted in the formation of an oxidizing material and disappearance of the CF_3COONa salt (Eq. 7.15). A short investigation showed that fluoroxy pentafluoroethane (CF_3CF_2OF) was produced. It was added then to various alkenes and the regioselectivity obtained clearly demonstrated the electrophilicity of the oxygen bound fluorine.[24]

The formation of the CF_3CF_2OF, however, is a low yield process and considerable quantities of other fluoroxy derivatives were also formed. This is partly because it is technically difficult to keep a uniform fluorine flow during the reaction (higher fluorine concentrations produce higher proportions of polyfluoroxy derivatives). It is also important to keep the reaction dry and HF free in order to enhance the nucleophilicity of the NaF, which participates in the first step of the reaction.

$$CF_3COONa + F_2 \longrightarrow [CF_3C(=O)OF \cdot NaF] \xrightarrow{H_2O \text{ or } HF} CF_3COOF$$

$$[CF_3C(=O)OF \cdot NaF] \longrightarrow CF_3CF(ONa)OF \xrightarrow{F_2} CF_3CF(OF)_2$$

$$CF_3CF(ONa)OF \xrightarrow{-NaOF} NaF + \tfrac{1}{2}O_2$$

$$[CF_3C(=O)F \cdot NaF] \longrightarrow CF_3CF_2ONa \xrightarrow{F_2} CF_3CF_2OF \qquad (7.15)$$

$$CF_3CF_2OF + R \rightarrow C_6H_4\text{–}CH{=}CHPh \longrightarrow R \rightarrow C_6H_4\text{–}CHF\text{–}CH(OCF_2CF_3)Ph$$

$$CF_3CF_2OF + R \rightarrow C_6H_4\text{–}CH{=}CHPh \longrightarrow R \rightarrow C_6H_4\text{–}CH(OCF_2CF_3)\text{–}CHFPh$$

The proposed mechanism suggested that if the fluoride ion were to lose its nucleophilicity due to solvatation, the reaction would stop with the formation of the trifluoroacetyl hypofluorite, CF_3COOF. Indeed, when instead of dry sodium trifluoroacetate we used the hydrated salt, or did not exclude the HF always found in commercial fluorine, a much higher yield of oxidizing material was obtained.[25] It was easy to prove that this oxidant was the expected CF_3COOF by its addition to several stilbenes with high regio- and stereoselectivity (again syn addition, as in Section 7.1.2).[26] One of the most valuable features of this reaction is the easy hydrolysis of the adduct resulting in the important fluorohydrin moiety (Eq. 7.16).

$$MeOOC\text{–}C_6H_4\text{–}CH{=}CHPh \xrightarrow{CF_3COOF} MeOOC\text{–}C_6H_4\text{–}CHF\text{–}CH(OCOCF_3)Ph$$

$$\xrightarrow{H_2O/py} MeOOC\text{–}C_6H_4\text{–}CHF\text{–}CH(OH)Ph \qquad (7.16)$$

Independent of the exact reaction conditions, the fluoroxy group is common to all the oxydants produced. This allows us to consider the whole mixture as a single homogeneous reagent ("F^+") when the formation of the α-fluorocarbonyl moiety from an enol acetate is the aim. The reaction is general and cyclic and bicyclic ketones, aliphatic and aromatic ones, steroidal ketones, and enones were all successfully converted to the corresponding α-fluorocarbonyl derivatives via their enol acetates (Eq. 7.17).[27] Thus fluorine, although in a somewhat indirect way, is after all fully capable of replacing the expensive and not easy available CF_3OF.

$$CF_3COONa + F_2 \longrightarrow \text{A mixture of fluoroxy derivatives} \equiv \text{"}F^+\text{"}$$

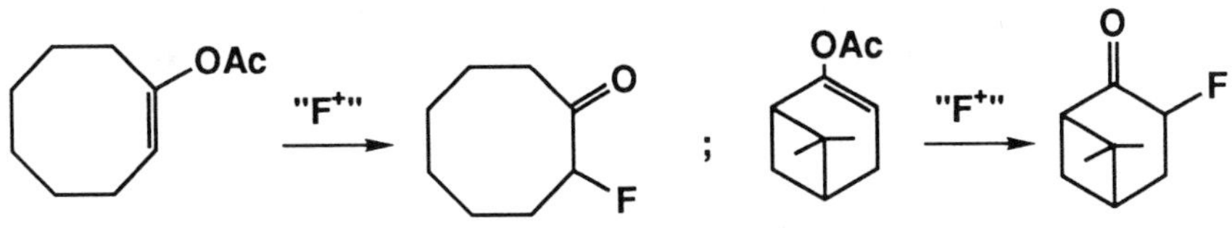

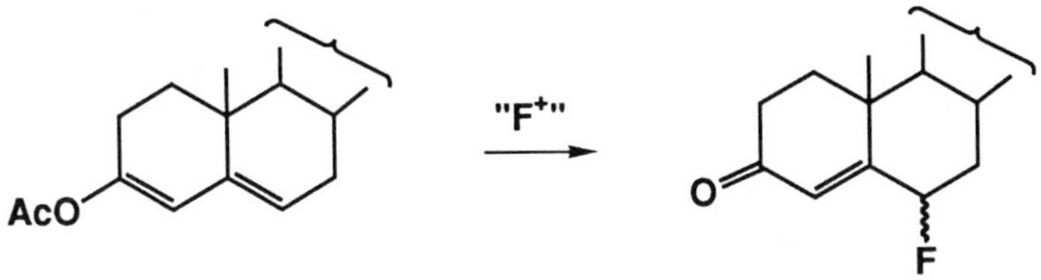

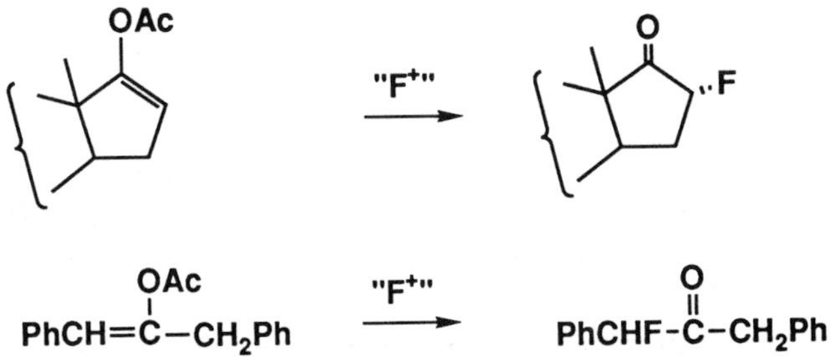

$$PhCH{=}C(OAc){-}CH_2Ph \xrightarrow{\text{"}F^+\text{"}} PhCHF{-}C(=O){-}CH_2Ph \quad (7.17)$$

Higher homologues of trifluoroacetyl hypofluorites have not been described prior to their synthesis by us. The homologues turned out to be indefinitely stable in $CFCl_3$ solution at 0°C.[28] These solutions can therefore serve as "off the shelf" electrophilic sources and react with enol acetates to form α-fluoro ketones in the same manner as CF_3COOF does. Their uniqueness, however, lies in their ability to serve as radical initiators for polymerization of fluoro-alkenes. Thus, unlike the commonly used persulfate initiation, which creates a sizable proportion of undesired oxygen containing "end groups," catalytic amounts of $C_8H_{17}COOF$, for example, will efficiently polymerize tetrafluoroethylene producing high molecular weights polymers with practically no detectable "end groups" (Eq. 7.18).

$$R_fCOOK \xrightarrow{F_2} R_fCOOF \xrightarrow{\Delta} R_f\cdot + F\cdot + CO_2 \xrightarrow{CF_2{=}CF_2} R_f{-}(CF_2{-}CF_2)_n{-}F \quad (7.18)$$

7.2.2. Acetyl Hypofluorire and Higher Homologues

Until some years ago it was assumed that all fluoroxy compounds needed to have a perfluorinated alkyl chain, otherwise HF elimination would occur immediately. While this line of reasoning is valid to some extent, as described later, it is by no means a general rule. The most notable exception is acetyl hypofluorite, CH_3COOF.

When fluorine was passed through a mixture of acetic acid and $CFCl_3$ no oxidative compounds were formed. When, however, a sodium salt was added to the AcOH, forming and equilibrium with AcONa, or when just plain sodium acetate was dispersed in trichlorofluoromethane, the fluorine formed an oxidative solution. Reactions with some stilbenes suggested that the previously unknown AcOF was formed.[29] After a few experiments it was concluded that adding water or AcOH to the reaction mixture enhances the efficiency of formation of AcOF and increases its stability at room temperatures for up to several hours (Eq. 7.19).

$$CH_3COONa.nHOR \xrightarrow[CFCl_3]{F_2/N_2} CH_3COOF \xrightarrow{PhCH=CHPh} PhCH(F)-CH(OAc)Ph \quad (7.19)$$

R = H, Ac; syn addition

Although for all synthetic purposes the *in situ* prepared AcOF solution is sufficient, Appelman[30] did manage to isolate pure AcOF and together with our independent studies[31] its spectral properties and thermal behavior[32] were recorded.

As with other electrophilic fluorination reagents, AcOF adds itself across double bonds in a syn mode. Being more tamed than R_fOF or R_fCOOF it was added successfully to many types of double bonds including benzylic, cyclic, straight chain, and steroidal ones. The relative mildness of AcOF is evident from the fact that alkynes and certain types of flexible α,β-unsaturated carbonyls fail to react. On the other hand electron-rich alkenes such as enol acetates, react instantaneously to produce the expected α-fluoro ketones offering a simple and very inexpensive route for this transformation (Eq. 7.20).

$$\text{cyclohexene} \xrightarrow{AcOF} \text{cis-1-F-2-OAc-cyclohexane} \;;\; RCH=CH_2 \xrightarrow{AcOF} RCH(OAc)-CH_2F$$

$$\text{steroidal enone} \xrightarrow{AcOF} \text{4-F, 5-OAc adduct} \xrightarrow{-AcOH} \text{4-fluoro enone}$$

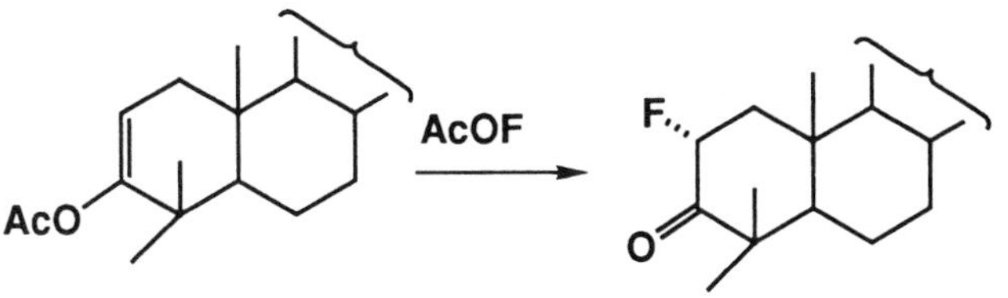

$$CH_3CH{=}CHCOOR;\ PhCH{\equiv}CH \xrightarrow[-78\ ^\circ C]{AcOF} \text{no reaction} \tag{7.20}$$

Most electrophilic fluorinations of double bonds are very fast and acetyl hypofluorite is of no exception. This prompted many radiochemists to employ the reaction for the synthesis of ^{18}F containing compounds suitable for positron emitting tomography (PET). Soon ^{18}F—F was prepared from neon, passed through a suspension of sodium acetate in $CFCl_3$ (Ref. 33) and the resulting $AcO^{18}F$ was immediately reacted with tri-O-acetyl-D-glucal. Thus the formation and the necessary workup of the 2-deoxy-2-(^{18}F)fluoro-D-glucose was completed in a less than one-half lifetime of the ^{18}F isotope (110 min) and used extensively in PET for various brain studies (Eq. 7.21).[34] The $AcO^{18}F$ has been since used for the construction of many other biologically important compounds designed for this modern type of tomography.[35]

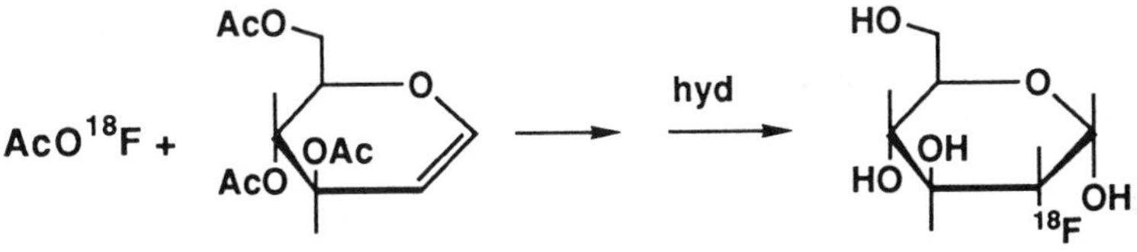

(7.21)

The mildness of AcOF is evident also from its smooth reactions either with 1,3-dicarbonyl compounds or their enolates.[36] Even lithium enolates of monocarbonyl derivatives can be directly fluorinated,[37] eliminating the need to prepare the corresponding enol acetates, silyl enol ethers, and the like. The parallel reactions with F_2, CF_3OF, or CF_3COOF resulted only in tars or in very low yields of the expected monofluoro derivatives (Eq. 7.22).

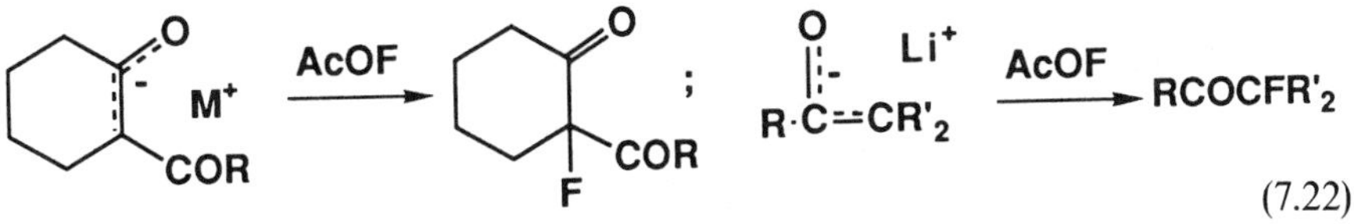

$$R\text{-}C(O^-Li^+){=}CR'_2 \xrightarrow{AcOF} RCOCFR'_2 \tag{7.22}$$

Since AcOF has an electrophilic fluorine we have tried using it for the aromatic fluorination achieved usually through the Balz–Schiemann reaction. Indeed, AcOF could be successfully reacted with many activated aromatic compounds to produce mainly the ortho fluoro derivatives in very good conversions and yields of up to 85%.[38] The dominant ortho substitution was a result of the addition of AcOF across the most electron-rich region of the aromatic ring. A subsequent spontaneous elimination of AcOH restored the aromaticity, but in cases where this was not possible, the resulting cyclohexadiene reacted very rapidly with the reagent and tars were obtained. Only in certain cases and with careful monitoring could the corresponding adducts be isolated (Eq. 7.23).

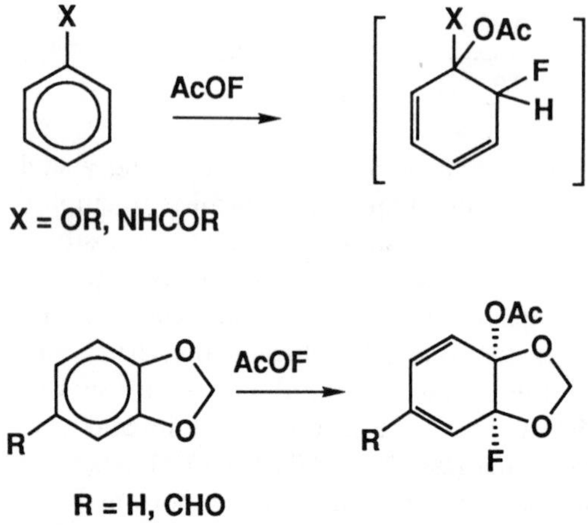

X = OR, NHCOR

R = H, CHO

(7.23)

We have used this reaction to prepare many types of biologically interesting derivatives including fluoro hexestrols, fluorotyrosine, various 4-fluoro steroidal enones, 6-fluorosteroids,[39] and fluorobimans, which are important fluorescent compounds (Eq. 7.24).[40]

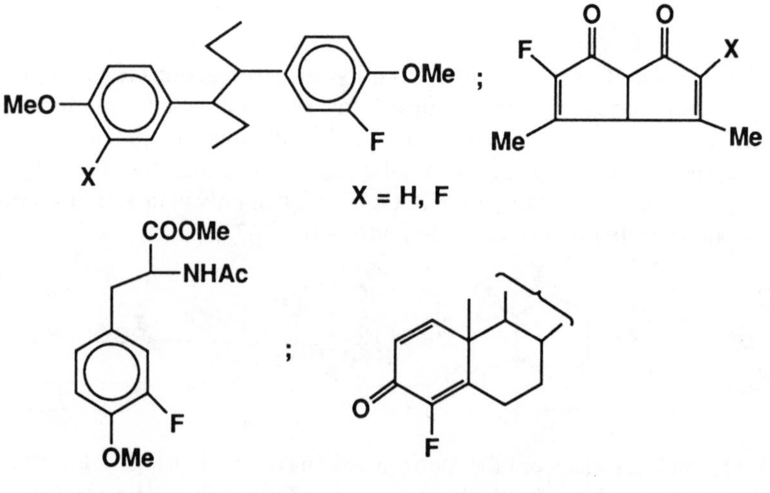

X = H, F

(7.24)

Acetyl hypofluorite also cleaves the carbon–mercury bond (Eq. 7.25) and this provides an easy entry to many fluoroethers. Since the electrophilic fluorine attacks the electrons of the C—Hg bond, the reaction proceeds with a full retention of configuration. We have thus prepared several fluoro-methoxy derivatives from the corresponding alkenes,[41] formally accomplishing the addition of the elements of "MeOF" across a double bond (Eq. 7.25).[42]

$$R_3C\text{-}HgCl + F\text{-}OAc \longrightarrow R_3CF + AcO^- + [HgCl]^+$$

$$\text{cyclohexene} \xrightarrow[\text{NaCl}]{\text{Hg(OAc)}_2/\text{MeOH}} \text{1-OMe-2-HgCl-cyclohexane} \xrightarrow{\text{AcOF}} \text{1-OMe-2-F-cyclohexane} \quad (7.25)$$

Recently, we addressed the question of whether AcOF is the only fluoroxy compound with fluorine-free alkyl residue. Although we have shown that there are some homologues of this type,[43] their number seems to be limited. If a chain is too long there are many conformations where an alkyl hydrogen may be near the oxygen bound fluorine, apparently triggering HF and CO_2 elimination followed by uncontrollable fragmentation of the chain. The higher acyl hypofluorites, possessing an electrophilic fluorine, react with enol acetates to form α-fluoro ketones (Eq. 7.26).

$$CH_3(CH_2)_n\text{-}CH(X)\cdot COONa \xrightarrow{F_2} CH_3(CH_2)_n\text{-}CH(X)\cdot COOF \equiv RCOOF$$

$$n = 1, 2;\ X = Cl, NO_2, COOR \quad ; \quad n > 3 \Rightarrow \text{decomposition}$$

$$RCOOF + \text{(OAc-dihydronaphthalene)} \longrightarrow \text{(2-fluoro-1-tetralone)} \quad \text{good yields} \quad (7.26)$$

We believe that the reagents and the reactions outlined here are only the tip of the iceberg, but nevertheless they demonstrate the surprising ability of fluorine to react mildly and cleanly in an electrophilic and in other modes and perform many novel and useful transformations in organic chemistry.

ACKNOWLEDGMENT

We thank the CR&D Department of Du Pont, Wilmington, DE, for partial financial support.

REFERENCES

1. Moissan, H., *Ann. Chim. Phys.*, *19*, 272 (1891).
2. Bockemullar, W., *Justus Liebigs Ann. Chem.*, *506*, 20 (1933); Miller, W. T., Calfee, J. D., and Bigelow, L. A., *J. Am. Chem. Soc.*, *59*, 2072 (1937); Fukuhara, N. and Bigelow, L. A., *J. Am. Chem. Soc.*, *60*, 427 (1938).
3. Fried, J. and Sabo, E. F., *J. Am. Chem. Soc.*, *76*, 1455 (1954).
4. Barton, D. H. R., Godhino, L. S., Hesse, R. H., and Pechet, M. M., *J. Chem. Soc. Chem. Commun.*, 804 (1968). Barton, D. H. R., Danks, L. J., Ganguly, A. K., Hesse, R. H., Tarzia, G. and Pechet, M. M., *J. Chem. Soc. Chem. Commun.*, 227 (1969).
5. Hesse, R. H., *Isr. J. Chem.*, *17*, 60 (1978).
6. CF_3OF is again commercially available although quite expensive.
7. Adcock, J. L. and Lagow, R. J., *J. Am. Chem. Soc.*, *96*, 7588 (1974).
8. Olah, G. A., Prakash, G. K. S., Williams, R. E., Field, L. D., and Wade, K., in *Hypercarbon Chemistry;* Wiley: New York, 1987.
9. If not otherwise stated, reaction with fluorine means reaction with 1–10% prediluted fluorine in nitrogen.
10. Rozen, S., Gal, C., *J. Org. Chem.*, *52*, 2769 (1987).
11. By "electron rich" we mean that the molecular orbital of the corresponding C—H bond has high *p* character, or in other words, its hybridization in *p* is higher and thus it is more susceptible to electrophilic attack. The experimental results are supported by hybridization calculations using the semiempirical PRDDO program.
12. Rozen, S., Gal, C., *J. Org. Chem.*, *53*, 2803 (1988).
13. The link between reactivity and selectivity has been reevaluated: Pross, A., *Isr. J. Chem.*, *26*, 390 (1985) and references cited therein.
14. Alker, D., Barton, D. H. R., Hesse, R. H., James, J. L., Markwell, R. E., Pechet, M. M., Rozen, S., Takashita, T., and Toh, H. T., *Nouv. J. Chim.*, *4*, 239 (1980).
15. Rozen, S. and Ben-Shoshan, G., *J. Org. Chem.*, *51*, 3522 (1986).
16. Rozen, S. and Ben-Shoshan, G., *Org. Magn. Res.*, *23*, 116 (1985).
17. See, for example: Breslow, R., Brandl, M., Hunger, J., Adams, A. D., *J. Am. Chem. Soc.*, *109*, 3799 (1987).
18. Rozen, S. and Gal. C., *J. Org. Chem.*, *52,* 4928 (1987).
19. Rozen, S. and Brand, M., *J. Org. Chem.*, *51*, 3607 (1986).
20. Merritt, R. F. and Johnson, F. A., *J. Org. Chem.*, *31*, 1859 (1966); Merritt, R. F. and Steven, T. E., *J. Am. Chem. Soc.*, *88*, 1822 (1966); Merritt, R. F., *J. Org. Chem.*, *31*, 3871 (1966); Merritt, R. F., *J. Am. Chem. Soc.*, *89*, 609 (1967).
21. Barton, D. H. R., Hesse, R. H., Jackmann, G. P., Ogunkoya, L., and Pechet, M. M., *J. Chem. Soc. Perkin Trans. 1*, 739 (1974).
22. Barton, D. H. R., James, J. L., Hesse, R. H., Pechet, M. M., and Rozen, S., *J. Chem. Soc. Perkin Trans. 1*, 1105 (1982).
23. Rozen, S. and Filler, R., *Tetrahedron*, *41*, 1111 (1985).
24. Lerman, O. and Rozen, S., *J. Org. Chem.*, *45*, 4122 (1980).
25. Rozen, S. and Lerman, O., *J. Org. Chem.*, *45*, 672 (1980).

26. This hypofluorite was prepared first by Cady by a different and difficult route. Gard, G. L. and Cady, G. H., *Inorg. Chem.*, *4*, 594 (1965).
27. Rozen, S. and Menahem, Y., *J. Fluorine Chem.*, *16*, 19 (1980).
28. Barnette, W. E., Wheland, R. C., Middleton, W. J., and Rozen, S., *J. Org. Chem.*, *50*, 3698 (1985).
29. Rozen, S., Lerman, O., and Kol, M., *J. Chem. Soc. Chem. Commun.*, 443 (1981).
30. Appelman, E. H., Mendelsohn, M. H., and Kim, H., *J. Am. Chem. Soc.*, *107*, 6515 (1985).
31. Hebel, D., Lerman, O., and Rozen, S., *J. Fluorine Chem.*, *30*, 141 (1985).
32. Note that pure or highly concentrated AcOF may explode. See Ref. 30 and Adam's letter to the Chem. Eng. *News* Feb. 8, p. 2 (1985).
33. Some other variations for the preparation of AcOF were also published: Jewett, D. M., Potocki, J. F., and Ehrenkaufer, R. E., *Synth. Commun.*, *14*, 45 (1984); Jewett, D. M., Potocki, J. F., and Ehrenkaufer, R. E., *J. Fluorine Chem.*, *24*, 477 (1984).
34. Shieu, C. Y., Salvadori, P. A., and Wolf, A. P., *J. Nucl. Med.*, *23*, 108 (1982); Adam, M. J., *J. Chem. Soc. Chem. Commun.*, 730 (1982); Ehrenkaufer, R. E., Potocki, J. F., and Jewett, D. M., *J. Nucl. Med.*, *25*, 333 (1984).
35. Luxen, A., Barrio, J. R., Satyamurthy, N., Bida, G. T., and Phelps, M. E., *J. Fluorine Chem.*, *36*, 83 (1987); Mislankar, S. G., Gildersleeve, D. L., Wieland, D. M., Massin, C. C., Mulholland, G. K., and Toorongian, S. A., *J. Med. Chem.*, *31*, 362 (1988); Diksic, M. and Diraddo, P., *Int. J. Appl. Radiat. Isot.*, *36*, 643 (1985); Chirakal, R., Firnau, G., Couse, J., and Garnett, E. S., *Int. J. Appl. Radiat. Isot.*, *35*, 651 (1984).
36. Lerman, O. and Rozen, S., *J. Org. Chem.*, *48*, 724 (1983).
37. Rozen, S. and Brand, M., *Synthesis*, 665 (1985).
38. Lerman, O., Tor, Y., Hebel, D., and Rozen, S., *J. Org. Chem.*, *49*, 806 (1984).
39. Hebel, D., Lerman, O., and Rozen, S., *Bull. Soc. Chim. France*, *861*, (1986).
40. Kosower, E. M., Hebel, D., Rozen, S., and Radkowski, A. E., *J. Org. Chem.*, *50*, 4152 (1985).
41. Hebel, D. and Rozen, S., *J. Org. Chem.*, *52*, 2588 (1987).
42. Recently, and for the first time, we have synthesized and characterized MeOF itself. Kol, M., Rozen, S., and Appelman, E., *J. Am. Chem. Soc.*, **113**, 2646 (1991).
43. Rozen, S. and Hebel, D., *J. Org. Chem.*, *55*, 2621 (1990).

CHAPTER 8

FLUORINATION WITH ONIUM POLY(HYDROGEN FLUORIDES): THE TAMING OF ANHYDROUS HYDROGEN FLUORIDE FOR SYNTHESIS

GEORGE A. OLAH and XING-YA LI

Synthetic Fluorine Chemistry,
Edited by George A. Olah, Richard D. Chambers, and G. K. Surya Prakash.
ISBN 0-471-54370-5

8.1. INTRODUCTION

Various nonbonded electron-pair donor bases form with excess anhydrous hydrogen fluoride remarkably stable complexes having the general formula $BH^+(HF)_xF^-$, which can be generally named as "onium poly(hydrogen fluorides)." Examples are

$R_2OH^+ (HX)_xF^-$

Oxonium
Poly(hydrogen fluoride)

$R_3NH^+ (HF)_xF^-$

Ammonium
Poly(hydrogen fluoride)

$C_5H_5NH^+ (HF)_xF^-$

Pyridinium
Poly(hydrogen fluoride)

[structure: poly(4-vinylpyridine) repeat unit, subscript m, pyridinium N^+–H $(HF)_xF^-$]

Polyvinylpyridinium
Poly(hydrogen fluoride)

$R_3PH^+ (HF)_xF^-$

Phosphonium
Poly(hydrogen fluoride)

Hydrogen fluoride, having a history of more than 200 years,[1] is the principal manufactured fluorine compound and was once called the life blood of the modern fluorochemical industry.[2] As a classical and inexpensive fluorinating agent, anhydrous hydrogen fluoride (AHF) has been widely used for the preparation of organic as well as inorganic fluoro compounds.[3] However, the low boiling point of AHF (19.5°C), its high degree of hazardness and corrosiveness make its handling very difficult and dangerous, especially for laboratory purposes. Pressure equipments made of special materials like nickel or Monel alloy are usually needed. Besides the difficulties in handling and the need of special equipments, the use of AHF as fluorinating agent for organic compounds often bring some complications caused by undesired side reactions, such as rearrangement, dehydration, and polymerization, which are caused by the strong acidity of AHF.

To overcome these difficulties, both the volatility and acidity of AHF should be reduced. Hirschmann et al.[4] in 1956 and Taub et al.[5] in 1957 reported the use of solutions of hydrogen fluoride in organic bases, such as tetrahydrofuran (THF), pyridine, and alcohol to improve the yield of the desired fluorohydrin from its steroid epoxide precursor. In 1963 Bergstrom et al. reported[6] the successful use of 70% HF solution in pyridine for transferring hydroxyl groups of steroids into fluorine. A variety of HF–base complexes were subsequently reported, for example, solutions of hydrogen fluoride in amines,[7] amides,[8] carbamic acids and esters,[9] trialkyl phosphines,[10] and alcohols.[11] These reports, however, remained basically limited to fluorinations of specific organic compounds (mostly steroids). The use of HF–base complexes had not achieved wider acceptance until Olah et al. published their first report[12] and subsequent extensive studies on pyridinium poly(hydrogen fluoride) (PPHF) as fluorinating agent in a wide variety of fluorination reactions.[13,14] Pyridinium poly(hydrogen fluoride) (30:70), known as Olah's reagent,[15] and some of the subsequently developed systems, for example, melamine–HF complexes,[16] are commercially available and can be used as common reagents in the laboratory. These reagents are not only less volatile equivalents of AHF, but, in their own right, are more efficient fluorinating agents. Recently developed poly(vinylpyridinium) poly(hydrogen fluoride) (PVPHF)[17] provides a solid hydrogen fluoride reagent, which bears similar reactivity compared to the liquid reagents, but is much easier to handle and extremely simplifies workup.

8.2. PREPARATION, PROPERTIES, AND GENERAL FLUORINATING ABILITY OF ONIUM POLY(HYDROGEN FLUORIDES)

8.2.1. Preparation

Generally, the onium poly(hydrogen fluoride), both liquid and polymer resin based, can be conveniently prepared in the laboratory by condensing at low temperature (usually at $-78°C$) the needed amount of anhydrous hydrogen fluoride into a polyethylene or Teflon vessel containing the selected base. Since at the beginning the reaction is very exothermic, the vessel containing the base should be well cooled and the flow rate of hydrogen fluoride should be adjusted to keep the reaction smooth. After the desired amount of hydrogen fluoride is condensed, the mixture is allowed to warm gradually to room temperature. For the preparation of PPHF, since pyridine solidifies at $-78°C$ and the dissolution of the solid in liquid hydrogen fluoride can be violent if the bath temperature is not well controlled, it was suggested[18] to keep pyridine as cold as possible without freezing ($\sim -40°C$), then slowly condense hydrogen fluoride into the vessel so that the mixture always remain liquid during the preparation. The ratio of HF and base is readily adjustable by varying the amount of the hydrogen fluoride used.

Caution: Proper precautions must be used when handling anhydrous hydrogen fluoride.[19]

8.2.2. Properties

Most onium poly(hydrogen fluoride) reagents are liquid at room temperature, except the polymer-based ones. The stability, corrosiveness, and hazard of the reagents largely depend on the molar ratio of hydrogen fluoride and base. Pyridinium poly(hydrogen fluoride) with 30% pyridine and 70% hydrogen fluoride (molar ratio 1:9) is a colorless liquid stable up to 55°C.[12,13] The complex of 40% pyridine and 60% HF (molar ratio 1:6) is stable up to 90°C.[20] Poly(vinylpyridinium) poly(hydrogen fluoride) with 50–60 wt% of hydrogen fluoride are dry beads stable up to 50°C.[17] Ammonium and trialkylammonium trihydrogen tetrafluorides were reported to be stable liquids that resist the loss of hydrogen fluoride even at elevated temperature and can be distilled intact.[21] However, a later report pointed out that the 1:4 complexes of tertiary amines and hydrogen fluoride can lose a molecule of hydrogen fluoride during distillation under vacuum to give distillates, which, according to elementary analysis, contain 2.6–3.0 hydrogen fluoride per amine molecule.[22] The compounds $(Et)_3N \cdot (HF)_3$ and $(n\text{-}Bu)_3N \cdot (HF)_3$ have very similar boiling points, that is around 78°C 1.5 $mbar^{-1}$, whereas the distilled pyridine–HF complex, $C_6H_5N \cdot (HF)_{2.7}$, distills at 50°C 1 $mbar^{-1}$. Interestingly, it was recently reported[23] that when the solid $Et_3N \cdot 2HF$ was heated in a dry pistol (15 torr, 65°C) it lost Et_3N, not HF, to give $Et_3N \cdot 3HF$. It is significant that these amine bis- or tris-hydrogen fluorides, convenient agents in nucleophilic replacement of chlorine or bromine by fluorine, can be handled without hazard and do not corrode borosilicate glass.

It must be pointed out, however, for the most frequently used fluorinating reagents, which usually contain more than 50 wt% of HF, that these poly(hydrogen fluorides) are always in equilibrium with a small amount of free hydrogen fluoride (as indicated by some fuming of the reagents when exposed to the atmosphere). Thus the reagents are somewhat corrosive and hazardous, although to a much lesser extent than anhydrous hydrogen fluoride itself, and require similar safety precautions to prevent any skin contact or inhalation of HF vapor. Reactions using onium poly(hydrogen fluoride) reagents should be carried out preferably in polyethylene or Teflon vessels.

8.2.3. The Role of Bases

Nonbonded electron pair donor bases can change the property and reactivity of hydrogen fluoride in two ways: (1) reducing the volatility and hazard of the reagents, and (2) reducing the acidity of the media and enhancing the nucleophilicity of the fluoride anion.

It is well known that liquid hydrogen fluoride is highly associated owing to strong hydrogen bonding. Even in the gas phase at temperatures below 80°C, hydrogen fluoride forms polymers, mostly the hexamer.[3,24] Liquid hydrogen fluoride, being a two-dimensional hydrogen-bonded zigzag chain polymer,[24,25] is, however, much more volatile than water, which is a three-dimensional hydrogen-bonded polymer. Formation of onium poly(hydrogen fluoride) complexes could result in a three-dimensional structure through Coulombic interaction between the onium center and fluorides involving enhanced hydrogen bonding. The ^{19}F NMR spectrum of PPHF (30:70) at −60°C showed a substantially deshielded quintet with $J_{HF} = 120$ Hz, indicating that each fluorine atom is somewhat bonded to four hydrogen atoms.[12] Subsequent low-temperature IR structural analysis[21] and X-ray diffraction[26,27] of amine–HF complexes further support this notion. It was found that the "inside" hydrogen of the complexes $R_3NH^+F^-(HF)_x$ ($x = 1$–3) is shared between N and the central fluorine. In the case of trimethylammonium trihydrogen tetrafluoride, the highest adducts stable in vacuo at room temperature, the IR spectra suggest a tetrahedronlike structure consisting of a trimethylammonium cation hydrogen bonded to the central fluorine of a C_{3v} trihydrogen tetrafluoride anion.[21] Ammonium and methylammonium trihydrogen tetrafluorides have similar structures. The cation and anion in these complexes are tightly bound together.

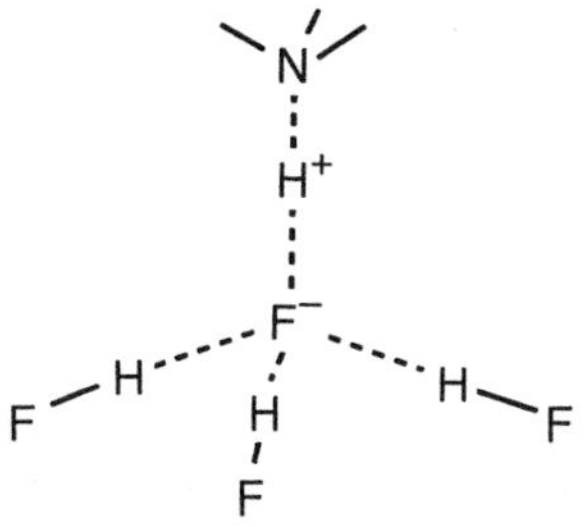

The solid complexes of pyridine–hydrogen fluoride were recently prepared at low temperature and studied by difference thermal analysis and X-ray powder diffraction.[28] The triclinic $C_5H_5 \cdot 3HF$ melts at −17°C and the monoclinic $C_5H_5 \cdot 4HF$ melts at −39°C. The X-ray structural study confirms that the complexes contain pyridinium cations and complex $[H_{n-1}F_n]^-$ anions. In the crystal structure of the 1:4 complex, there are unlimited ribbons along the *b*-axis direction. It is evident that in liquid complexes with higher HF content such as PPHF (30:70), hydrogen bonding by excess hydrogen fluoride molecules to the end fluorines of the trihydrogen tetrafluoride will result in a three-dimensional network and, consequently, reduce volatility.

Densities of HF–pyridine (py) binary systems have been studied.[29] The 1:3 complex, py · 3HF, has the highest density (1.24), while the density of py · 9HF (about 30:70 by weight) is 1.09.

The reactivity of the onium poly(hydrogen fluorides) differ from that of anhydrous hydrogen fluoride because the formation of onium complexes significantly reduces the acidity of hydrogen fluoride and enhances the nucleophilicity of the poly(hydrogen fluoride) anions. Fluorination reactions of organic compounds by using HF reagents are usually multistep reactions, which involve (1) formation of a carbocationic or related ionic intermediate and (2) subsequent nucleophilic attack of poly(hydrogen fluoride) anions on the intermediate. For example:

$$R^1R^2C{=}CR^3R^4 \xrightarrow{H^+} R^1R^2CH{-}\overset{+}{C}R^3R^4 \xrightarrow{(HF)_xF^-} R^1R^2CH{-}CFR^3R^4$$

$$R^1R^2R^3C{-}OH \xrightarrow{H^+} R^1R^2R^3C{-}\overset{+}{O}H_2 \xrightarrow{-H_2O} R^1R^2R^3C^+ \xrightarrow{(HF)_xF^-} R^1R^2R^3C{-}F$$

The high acidity of the hydrogen fluoride reagent may facilitate the formation of persistent carbocationic or other ionic intermediates. However, high acidity may also cause undesired side reactions such as rearrangement, ionic hydrogenation, dehydration, oligomerization, or polymerization.

Anhydrous hydrogen fluoride is a strong acid with $-H_0 > 10$. Recently, Gillespie and Liang found[30] that the $-H_0$ of 100% HF is as high as 15 but trace amounts of water rapidly lower the acidity . These acidities are too high for the purposes of many fluorination reactions. Lowering the acidity of HF by forming onium complexes may lessen the tendency for formation of persistent carbocations and thus suppress side reactions. Furthermore, formation of poly(hydrogen fluoride) complexes will enhance nucleophilic fluoride reactivity, allowing more ready nucleophilic attack on the carbocationic intermediate. To achieve optimum results, the composition of onium poly(hydrogen fluorides) should be adjusted to the requirements of specific reactions and substrates. Thus, although PPHF (30:70) is most frequently used for fluorination, reagents with a lower concentration of HF are preferred for certain systems, particularly when decreased acidity is needed.

Solvents used in reactions with PPHF include THF, which frequently enhance the reactivity. Halogenated solvents such as methylene chloride, chloroform, carbon tetrachloride, and 1,1,2,-trichlorotrifluoroethane are also often used. In many cases, however, no solvent is necessary.

8.3. FLUORINATION OF ALKENES AND ALKYNES

8.3.1. Hydrofluorination

Addition of hydrogen fluoride or halogen monofluoride to carbon–carbon multiple bonds is probably the simplest way to obtain secondary or tertiary alkyl fluorides. The complexes PPHF (30:70),[12,13] PVPHF (40:60),[17] and melamine–HF (14:86)[31] have been successfully used for hydrofluorination of alkenes and alkynes to obtain alkyl fluorides and gem-difluorides, respectively, in typical Markovnikov-type additions.

$$R^1R^2C{=}CR^3R^4 \xrightarrow[\text{or PVPHF(40:60)/CH}_2\text{Cl}_2]{\text{PPHF(30:70)/THF}} R^1R^2HC{-}CFR^3R^4$$

$$R^1{-}C{\equiv}C{-}R^2 \xrightarrow[\text{or PVPHF(40:60)/CH}_2\text{Cl}_2]{\text{PPHF(30:70)/THF}} R^1H_2C{-}CF_2R^2$$

Branched alkenes and cycloalkenes react more readily. Accompanying polymerization is generally minimal. Tetrahydrofuran was found to be a suitable solvent when using PPHF, while methylene chloride is preferred as the medium for PVPHF. The melamine–HF reagent was reported less affected by solvents[32] and moisture.[16]

Albert and Cousseau[33] recently reported that tetrabutylammonium dihydrogentrifluoride can fluorinate electrophilic alkynes containing up to two electron-withdrawing substituents, whereas Olah's reagent does not add HF to dimethylacetylenedicarboxylate. In all cases the reaction stops at the alkene stage giving a mixture of (*Z*) and (*E*) isomers in good yield. This is in contrast to the reactions of nonactivated alkynes with PPHF or PVPHF, in which case monofluoroalkenes were observed only as intermediate products and the reactions went further to give difluoroalkanes.

$$\begin{array}{c} A{-}C{\equiv}C{-}A \\ \text{or} \\ R{-}C{\equiv}C{-}A \end{array} \xrightarrow{n\text{-}Bu_4N^+H_2F_3^-} \begin{array}{c} A{-}CF{=}CH{-}A \\ \text{or} \\ R{-}CF{=}CH{-}A \end{array}$$

R = alkyl, phenyl; A = CN, COOR', COR'

8.3.2. Halofluorination

Halogen monofluorides add more readily to carbon–carbon double bonds than does hydrogen fluoride. Since the reaction conditions are mild and β-halo substituents of the products are useful groups for elimination or substitution reactions, this reaction is considered one of the most useful methods to introduce a single fluorine atom into organic molecules.

$$>C{=}C< \xrightarrow{\text{"XF"}} >C{-}C< \;(\overset{+}{X}\cdots F^-\ \text{bridged}) \longrightarrow >C(Br){-}C(F)<$$

Following early reports[34] of the addition of bromine monofluoride to double bonds to synthesize fluorinated steroids, the reaction was extend to include addition of IF and ClF to varied alkenes.[35] Iodine fluoride and bromine fluoride are not sufficiently stable for isolation and chlorine fluoride is too reactive to be directly used for reactions with organic molecules. Consequently, the halogen monofluorides are generally in situ prepared by the reaction of anhydrous hydrogen fluoride or silver fluoride with a source of positive halogen, usually *N*-haloacetamides or *N*-halosuccinimides.

The successful use of a combination of pyridinium or trialkylammonium poly(hydrogen fluoride) and *N*-halosuccinimide for halofluorination of alkenes and alkynes greatly simplified the reactions.[12,13] The recently developed poly(vinylpyridinium) poly(hydrogen fluoride)[17] and a vinylpyridine–styrene copolymer reagent[36] are also effective and even further simplify the workup procedure.

$$R^1R^2C{=}CR^3R^4 \xrightarrow[\text{PPHF or PVPHF}]{\text{N-Halosuccinimides}} R^1R^2CF{-}CXR^3R^4$$

The halofluorination with onium poly(hydrogen fluoride) and *N*-halosuccinimide (or other positive halogen sources) is a general reaction for varied unsaturated compounds and gives good yield of regiospecific Markovnikov-type adducts. In the case of alkynes, in contrast to hydrofluorination giving gem-difluorides, bromofluorination, and iodofluorination of disubstituted alkynes usually give monofluorinated products, that is, vinylic bromofluoro and iodofluoro alkenes, which are suitable precursors to a wide variety of interesting fluorine compounds.

$$R^1\text{—C}\equiv\text{C—}R^2 \xrightarrow[\text{PPHF}]{\text{N-Halosuccinimides}} (R^1)(F)\text{C=C}(X)(R^2)$$

R^1,R^2 = Alkyl, Aryl

Electron-deficient alkenes are generally less reactive. However, substituted ethyl cinnamates (e.g., $YC_6H_4CX{=}CHCO_2Et$)[37] and some halo-substituted alkenes, such as *cis*- and *trans*-α-fluoro, chloro, and bromostilbenes, can also be halofluorinated by using PPHF(30:60) and N-halosuccinimides[38]

$$YC_6H_4CX{=}CHCO_2Et \xrightarrow{\text{NBS/PPHF}} YC_6H_4CXF\text{—}CHBrCO_2Et$$

$$(Ph)(F)C{=}C(H)(Ph) \xrightarrow{\text{NBS/PPHF}} Ph(F)_2C\text{—}CBr(H)(Ph)$$

The HF–base complexes with lower concentration of hydrogen fluoride were also found effective for halofluorination. The combination of triethylamine tri(hydrogen fluoride), $(C_2H_5)_3N{-}3HF$, and *N*-halosuccinimides can halofluorinate cycloalkenes of different ring size as well as α-methylstyrene, which is prone to carbocationic oligomerization, to give anti-addition products.[39] The reactions were carried out in normal glass apparatus with dichloromethane as solvent.

$$C_6H_5C(CH_3){=}CH_2 \xrightarrow[\text{NBS}]{(C_2H_5)_3N/3HF} C_6H_5CF(CH_3)CH_2Br \quad 82\%$$

The less acidic $Et_3N{-}3HF$ reagent has different selectivity, compared to the more acidic PPHF (30:70) reagent, in the bromofluorination of cycloocta-1,5-diene. The reaction of (*Z*,*Z*)-cycloocta-1,5-diene with *N*-bromosuccinimide (NBS) and $Et_3N{:}3HF$ gave mainly trans-5-bromo-6-fluorocyclooctene, whereas the reaction with PPHF (30:70) yielded predominantly isomers of 2-bromo-6-

fluorobicyclo[3.3.0]octane.[40] Apparently, the more acidic PPHF (30:70) reagent favors the cross-transannular π cyclization. However, the bromofluorination of norbornadiene gave similar product distributions with either Et_3N-3HF or PPHF (30:70).[41]

NBS +

$(C_2H_5)_3N/3HF$	92	2	6
or			
Pyridine/10HF	0	26	66

NBS +

$(C_2H_5)_3N/3HF$	53	38	5
or			
Pyridine/10HF	66	34	0

Stereochemical studies showed that, by using PPHF–N-halosuccinimide reagent, the bromofluorination of phenyl-substituted alkenes, for example, 1,1-diphenylethylenes, β-alkylstyrenes, and stilbene follows the Markovnikov-type regioselectivity, and is sterospecific anti addition for trans alkenes and nonstereospecific for cis alkenes.[42] This can be explained by the large degree of isomerization of the cis alkenes under the reaction conditions. Bromofluorination of phenylcyclohexene is stereospecifically anti.[43] Bromofluorination and chlorofluorination of 1-phenyl-4-*tert*-butylcyclohexene[44] proceeded stereospecifically anti, but formed two pairs of vicinal halofluorocyclohexanes. The ratio of the products containing a phenyl group in the equatorial and axial position is around 60:40.

NBS / PPHF → *anti*-Adducts

Combined with substitution of the β-halo substituents or elimination of HX from the halofluorination products, various fluorine-containing compounds have been prepared. vic-Difluoroalkenes can be prepared by a modification of the halofluorination method adding, after the normal halofluorination procedure but without isolation of the products, silver fluoride to the reaction mixture to exchange the intermediately formed bromo- or iodo-fluoroalkanes to the corresponding difluorides.[12] This procedure is particularly well applicable to disubstituted alkenes and formally represents an efficient way to add F_2 to a double bond without employing elementary fluorine.

$$R^1R^2C{=}CR^3R^4 \xrightarrow[\text{PPHF}]{\text{NBS or NIS}} R^1R^2CF{-}CXR^3R^4 \xrightarrow[\text{PPHF}]{\text{AgF}} R^1R^2CF{-}CFR^3R^4$$

X = Br, I

Bromofluorination of styrene was recently achieved by using NBS and PPHF containing 55% of HF. Pyridinium poly(hydrogen fluoride) with 70% of HF causes polymerization of styrene. The obtained 2-fluoro-2-phenyl-1-bromoethane was transformed via its azide to 2-fluoro-2-phenyl-1-aminoethane, which can be used as a chiral derivatizing agent.[45]

$$PhCH{=}CH_2 \xrightarrow[\text{NBS}]{\text{PPHF(45/55)}} PhCHFCH_2Br \xrightarrow{\text{NaN}_3,\ \text{TBAB}} PhCHFCH_2N_3 \xrightarrow[\text{H}_2\text{O}]{\text{Ph}_3\text{P}} PhCHFCH_2NH_2$$

Highly stereospecific bromofluorination was combined with elimination of HBr to prepare stereospecifically (*Z*) and (*E*) isomers of vinyl fluorides.[46] The complex PPHF–NBS reacted with (*Z*) alkenes to give threo BrF adducts, which on elimination of HBr, afforded stereospecifically (*Z*) vinyl fluorides, whereas with (*E*) alkenes, via erythro-bromofluorides, (*E*)-vinyl fluorides were obtained.

$$(Z)\text{-}R^1HC{=}CHR^2 \xrightarrow[\text{PPHF}]{\text{NBS}} \text{threo-}R^1CHBr{-}CHFR^2 \xrightarrow{\text{KOH}} (Z)\text{-}R^1HC{=}CFR^2$$

threo (Z)

$$(E)\text{-}R^1HC{=}CHR^2 \xrightarrow[\text{PPHF}]{\text{NBS}} \text{erythro-}R^1CHBr{-}CHFR^2 \xrightarrow{\text{KOH}} (E)\text{-}R^1HC{=}CFR^2$$

erythro (E)

Halofluorination using onium poly(hydrogen fluoride) reagents were frequently utilized to introduce fluorine into bioactive organic molecules and to solve otherwise difficult synthetic tasks. In order to synthesize monofluoroethenyl γ-aminobutyric acid (GABA) derivatives, 5-ethenyl-2-pyrrolidinone was reacted with HF–pyridine and NBS in ether to afford a 1:3 mixture of positional isomers of (bromofluoroethyl)pyrrolidinones. Elimination of HBr followed by acid hydrolysis gave the target molecules, 4-amino-5-fluoro-5-hexenoic acid and (*Z*)-4-amino-6-fluoro-5-hexenoic acid.[47]

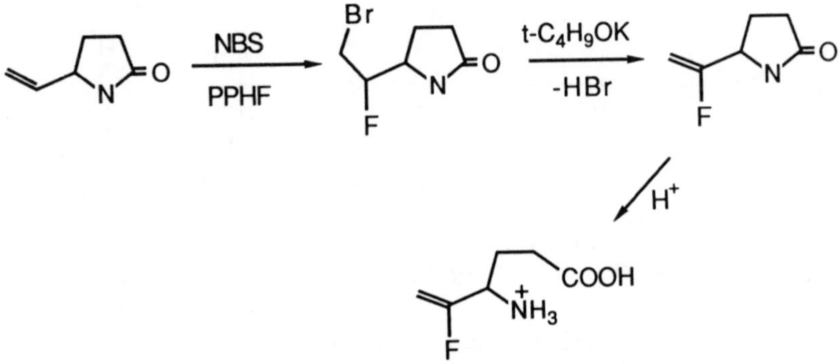

NBS / PPHF; t-C_4H_9OK / -HBr; H^+; Br; F; N; O; COOH; $\overset{+}{N}H_3$

The synthesis of radiopharmaceuticals labeled with short-lived ^{18}F requires rapid, convenient, and efficient methods. Bromofluorination using equivalent or substoichiometric quantities of HF–pyridine and 1,3-dibromo-5,5-dimethylhydantoin (DBH) was found suitable for this purpose.[48] Halofluorination proceeds faster with DBH than that with *N*-iodosuccinimide and much faster than that with NBS. A recently reported synthetic strategy for rapid and efficient preparation of fluoroalkylated amines and amides also involves bromofluorination of terminal alkenes with PPHF and DBH.[49]

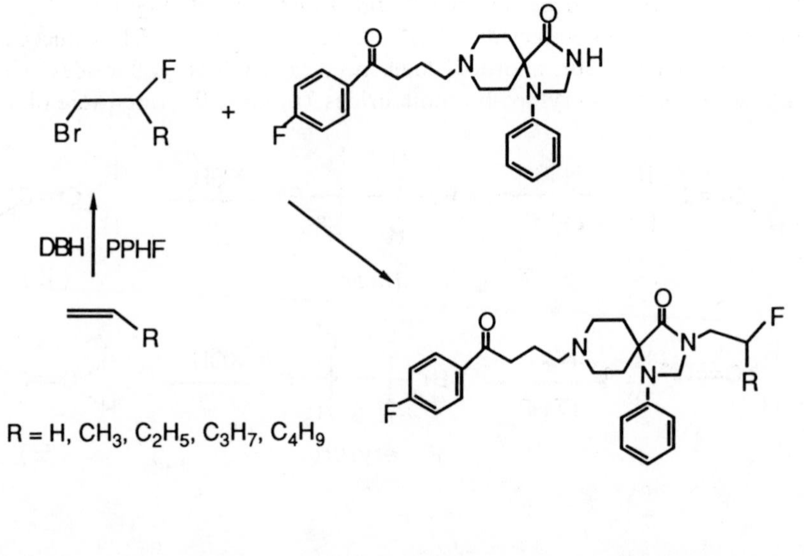

DBH | PPHF; Br; F; R; +; NH; N; O

R = H, CH_3, C_2H_5, C_3H_7, C_4H_9

8.3.3. Nitrofluorination

Alkenes react with nitronium tetrafluoroborate dissolved in PPHF (30:70) to afford β-nitrofluoroalkenes,[50] which are useful intermediates for the preparation of β-fluoroalkylamines.

$$R^1R^2C{=}CR^3R^4 \xrightarrow[\text{PPHF}]{NO_2^+BF_4^-} R^1R^2C(NO_2){-}CF R^3R^4$$

8.3.4. Fluorosulfenylation

It has been known that dimethyl(methylthio)sulfonium salts can be directly added to alkenes to form episulfonium salts.[51] The episulfonium salts can be *in situ* opened with triethylammonium dihydrogentrifluoride to give β-fluoroalkylmethylthioethers.[52] Treatment of alkenes such as cyclohexene, 1,5-cyclohexadiene, trans- and cis-cyclododecene, norbornadiene, trans-β-methylstyrene, and 2-cholestene in CH_2Cl_2 with $MeSS^+Me_2$ followed by addition of Et_3N–3HF gives 2-fluoroalkyl methyl sulfides in high yield. This fluorosulfenylation was reported to be diastereoselective.

$$R^1R^2C{=}CR^3R^4 \xrightarrow[\text{Et}_3\text{N/3HF, CH}_2\text{Cl}_2\text{, r t}]{Me_2\overset{+}{S}SMe\ BF_4^-} R^1R^2CF{-}C(SMe)R^3R^4$$

$$\text{indene} \xrightarrow[\text{Et}_3\text{N/3HF, CH}_2\text{Cl}_2\text{, r t}]{Me_2\overset{+}{S}SMe\ BF_4^-} \text{1-fluoro-2-(methylthio)indane}\quad 90\%$$

Another approach to obtain β-fluoroalkyl phenyl thioethers was also reported. This procedure[53] uses triethylamine tris(hydrogen fluoride) for exchanging the chlorine atom by fluorine in β-chloroalkyl phenyl thioethers, which can be easily prepared by the addition of sulfenyl chlorides to alkenes.

RSCl; [Cl⁻ episulfonium (S+, R)]; fast ⇌ SR/Cl adduct; $(HF)_2F^-$, slow → SR/F adduct

8.4. FLUORINATION OF SECONDARY AND TERTIARY ALCOHOLS

8.4.1. Alcohols

Pyridinium poly(hydrogen fluoride) (30:70)[13,54] and PVPHF[17] are suitable reagents for transferring secondary and tertiary alcohols into the corresponding fluorides. While most alcohols are highly soluble in the PPHF reagent, the resulting fluoride products are much less soluble and can be easily separated by extraction with cyclohexane or other solvents.[18] In the case of solid PVPHF reagent, separation is by simple filtration.

$$R^1R^2CH\text{-}OH \xrightarrow[\text{or PVPHF}]{\text{PPHF}} R^1R^2CH\text{-}F$$

$$R^1R^2R^3C\text{-}OH \xrightarrow[\text{or PVPHF}]{\text{PPHF}} R^1R^2R^3C\text{-}F$$

Secondary alcohols react with PPHF, usually 70% HF–30% pyridine or with lower (to 50%) hydrogen fluoride content, generally at room temperature. Tertiary alcohols react much faster and the reactions can be best carried out at temperatures as low as −70°C. Reagents with lower HF content may provide higher regioselectivity. In the fluorination reactions of benzyl alcohols, it was found that only ipso fluorides were formed when using low HF content reagents (HF/py mole ratio <5), whereas reactions with high HF content reagents (HF/py mole ratio >6.5) gave predominantly the rearranged fluorides, that is, fluorination products at the β carbon.[55]

$$\text{Ph-CH(OH)-CHRR'} \longrightarrow \text{Ph-}\overset{+}{\text{C}}\text{H-CHRR'} \xrightarrow{\text{1,2-migration}} \text{Ph-CH}_2\overset{+}{\text{C}}\text{RR'}$$

$$\text{Ph-}\overset{+}{\text{C}}\text{H-CHRR'} \xrightarrow{(HF)_xF^-} \text{Ph-CHF-CHRR'} \qquad \text{Ph-CH}_2\overset{+}{\text{C}}\text{RR'} \xrightarrow{(HF)_xF^-} \text{Ph-CH}_2\text{CFRR'}$$

The fluorination reaction of amino alcohols with PPHF (30:70) was found to be stereoselective, but not stereospecific. From either erythro or threo β-aminobenzyl alcohols, threo-fluoroamines were preferentially formed.[56]

Ph-CH(OH)-CH(CH_3)-N(piperidino) —PPHF(30:70)→ Ph-CH(F)-CH(CH_3)-N(piperidino)

		threo + erythro
threo	→	70 : 30
erythro	→	80 : 20

Primary alcohols show little reactivity. However, in the presence of sodium fluoride, primary alcohols such as *n*-hexanol, *n*-octanol, and neopentyl alcohol also react with PPHF (30:70) giving 30–55% yields of the corresponding 1-fluoroalkanes.[57] The formation of neopentyl fluoride(1-fluoro-2,2-dimethylpropane) from neopentyl alcohol without rearrangement indicates the S_N2 nature of the reaction.

$$n\text{-}C_8H_{17}OH \xrightarrow{\text{NaF/PPHF(30:70)}} n\text{-}C_8H_{17}F$$

$$t\text{-}C_4H_9CH_2OH \xrightarrow{\text{NaF/PPHF(30:70)}} t\text{-}C_4H_9CH_2F$$

The significant difference in reactivity of tertiary and secondary hydroxyl groups with PPHF allows regioselective monofluorination of steroid alcohols. 9α-Fluoro-3β-hydroxy-5α-cholest-8(14)-en-15-one has been prepared in 90% yield by treatment of 3β,9α-dihydroxy-5α-cholest-8(14)-en-15-one with PPHF (30:70) at −35°C.[58] Steroids with 3,5-dihydroxyl groups were converted to 5-monofluorides by a similar procedure.[59] In both cases, only the tertiary hydroxyl group (on the 5 or 9 position) was converted to fluoride with configurational retention. The reactions with PPHF reagent showed much higher stereospecificity than with AHF,[59] which often cause partial racemization.

[Steroid scheme: HO, OH, O —PPHF, -35°C, CH_2Cl_2→ HO, F, O]

[Steroid scheme: HO, OH —PPHF, -30°C, 15 min→ HO, F]

An interesting S_N2'-type fluorination of 2-hydroxy-androst-4-ene-3,17-dione to 6-fluoroandrost-4-ene-3,17-dione was observed with PPHF reagent:[60]

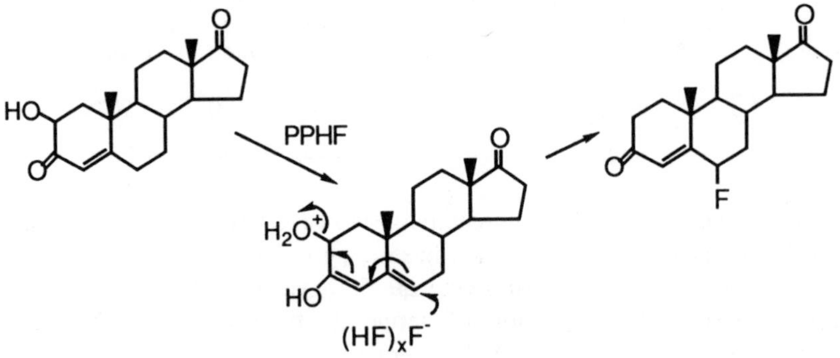

PPHF

$(HF)_xF^-$

8.4.2. Glycosyl Fluorides

Glycosyl fluorides, important building blocks in the synthesis of fluorine-containing carbohydrates and for the preparation of glycosides, have been effectively prepared using PPHF reagent. Appropriately protected sugars react with 50–70% HF–pyridine complexes to give 1-fluoroderivatives in good to excellent yields.[61,62] Both 1-unprotected[61] and 1-*O*-acetylated[62] sugars are suitable starting materials, but it was found that the acetylated sugars reacted more smoothly and gave better yields.[62]

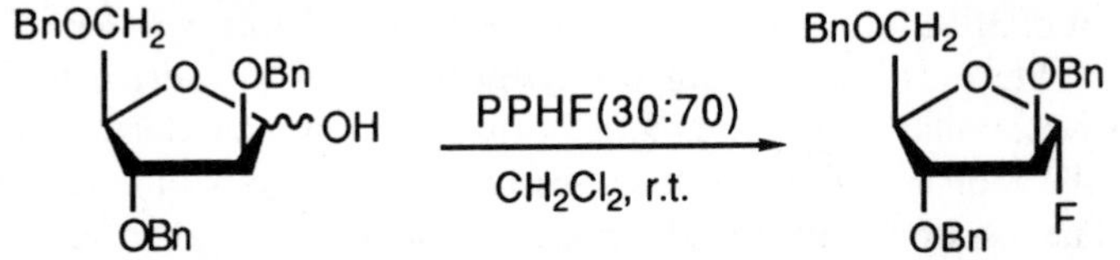

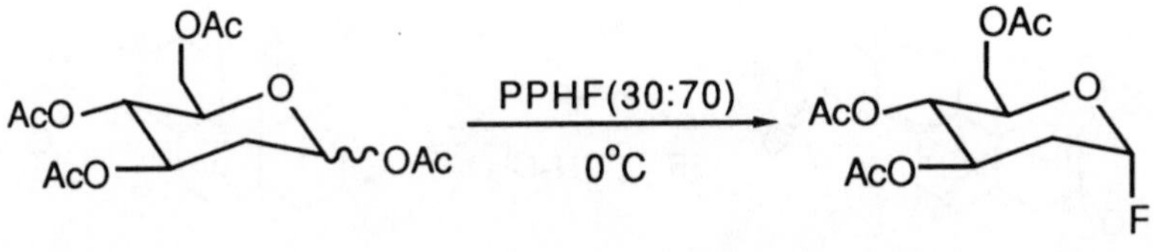

PPHF(30:70), CH_2Cl_2, r.t.

PPHF(30:70), 0°C

Regioselectivity of fluorination was also observed in these reactions. With fully acetylated sugars only the anomeric hydroxyl group were replaced by fluorine to give predominantly the thermodynamically more stable α isomers. Attempts to fluorinate methyl hexopyranosides at sites other than the anomeric carbon failed.[61] This can be explained by stereoelectronic or, more specifically, anomeric effects,[63] that is, back donation from the nonbonded electron pairs of the ring oxygen into the empty *p* orbital of the intermediate carbocation or the antibonding orbital of the anomeric C—F bond in the products:

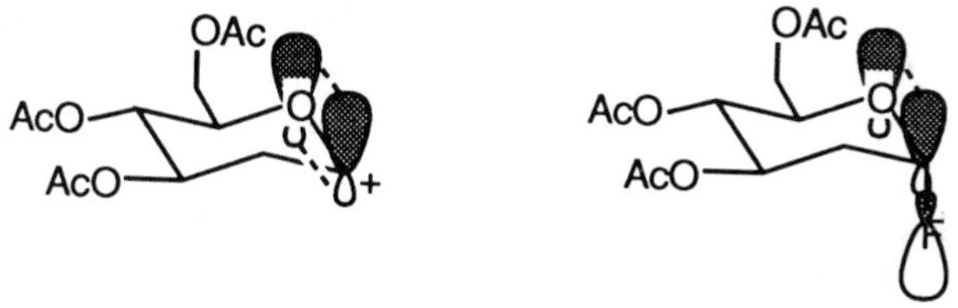

Glycosyl fluorides can also be efficiently prepared from phenyl thioglycosides by treatment with NBS and PPHF.[64]

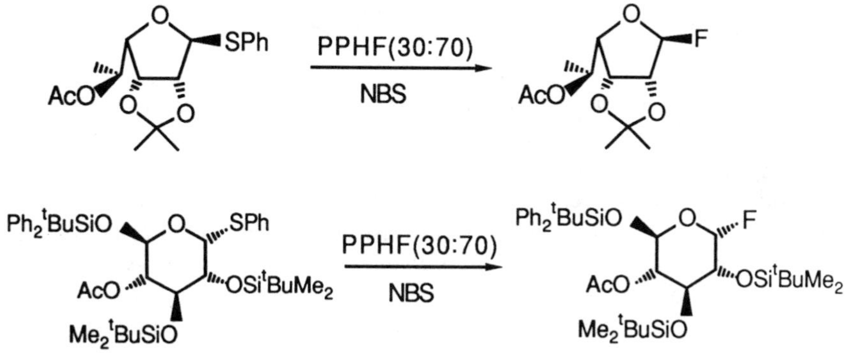

8.5. TRANSFORMATION OF AMINO, DIAZO, TRIAZENO, AND ISOCYANO GROUPS INTO FLUORINE

8.5.1. $ArNH_2$ to ArF

Transformation of arylamines to aryl fluorides via diazotization and fluorinative dediazoniation (Balz–Schieman reaction and its variations)[65] is the classical method for regiospecific introduction of fluorine into aromatic rings. A more practical method was developed by carrying out the diazotization and dediazoniation of arylamines in anhydrous hydrogen fluoride.[66,67] However, the low boiling point of AHF and difficulty in controlling side reactions limited the application of this method to industrial use. The use of PPHF for deaminative fluorination of amino arenes[13,68,69] greatly simplified the method. It not only overcomes the difficulty in handling anhydrous hydrogen fluoride but allows ready *in situ* fluorodediazoniation at atmospheric pressure with minimal by-products and offers the advantage of not requiring isolation of the intermediate diazonium salt.

$$ArNH_2 \xrightarrow[PPHF]{NaNO_2} \left[\begin{array}{c} ArN_2^+ \\ (HF)_xF^- \end{array} \right] \xrightarrow{in\ situ} ArF$$

The fluoroarenes obtained from substituted amino arenes are generally isomerically pure expected products. However, mixtures of isomers were obtained in some cases.[13,68] This could be attributed to the ambident nature of the arenediazonium ions (or the possibility of intermediate aryne formation).[68] A nucleophile can attack either at the ipso carbon or at different ring positions.

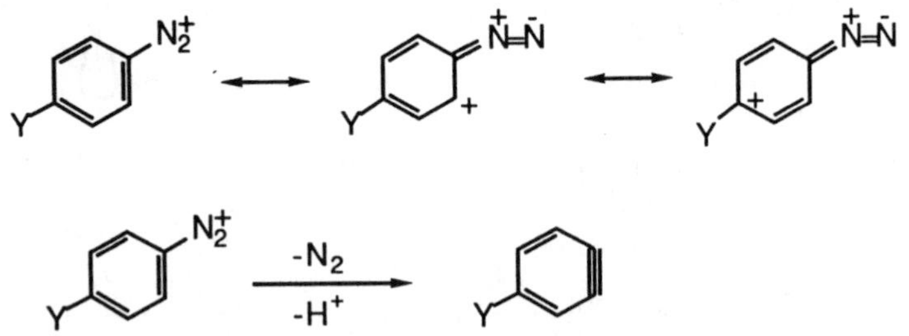

Using PPHF (40:60) instead of PPHF (30:70) for the transformation of arylamines to fluoroarenes afforded good to excellent yields (up to 98% by GC) of regiospecific products.[20,70] It is noteworthy that even *meta*- and *para*-nitroanilines can be converted to the corresponding *meta*- or *para*-nitrofluorobenzene in higher than 90% yield. In contrast, when using anhydrous hydrogen fluoride, aniline was converted to fluorobenzene in only 15% GC-yield and the reaction has poor reproducibility.[70] It has long been known[71] that in intermediate acidity media ($-H_0 = -1$ to 3), the rate of diazotization increases with acidity, but in highly acidic media ($-H_0 > 4$) the deprotonation of the *N*-nitrosoanilinium ion becomes rate limiting and the overall rate decreases with acidity.

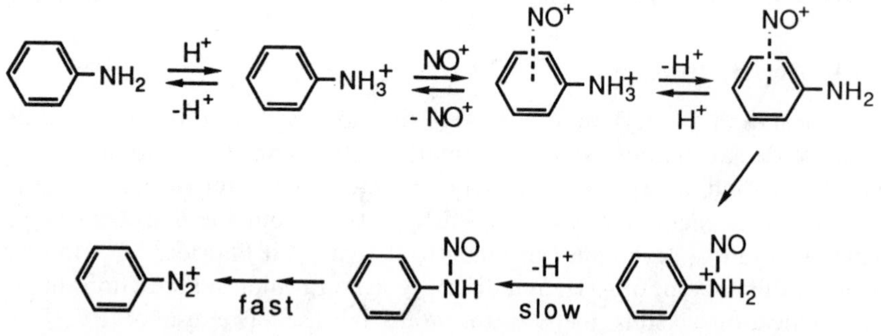

A recent study showed that the rate of diazotization of aniline in PPHF with a HF–py molar ratio of 9 (~70 wt% HF) is much slower than that with a HF/py ratio of 6 (~60% HF).[20] It was also found that it is the diazotization step, not dediazoniation, which plays the most important role in determining the yield of fluoroarenes. The higher thermal stability (up to 90 °C) of the 40:60 PPHF system also allowed us to carry out dediazoniation at higher temperatures without loss of HF or using pressure equipment. However, further lowering the concentration of hydrogen fluoride remarkably decreases the yield of fluorobenzene.

A PPHF reagent with 40% of HF was used to transfer *p*-aminophenol via diazotization to *p*-fluorophenol.[72] 4-Fluoro-5-(trifluoromethyl)phenol, a useful herbicide, was prepared from 4-amino-5-(trifluoromethyl)phenol via diazotization and dediazoniation in PPHF under UV irradiation in a yield of 55%.[73]

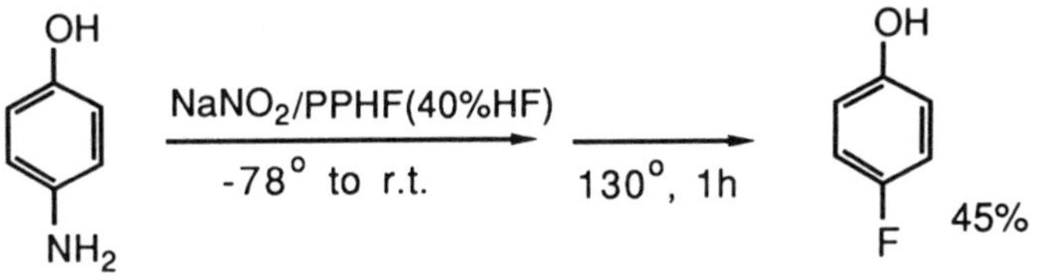

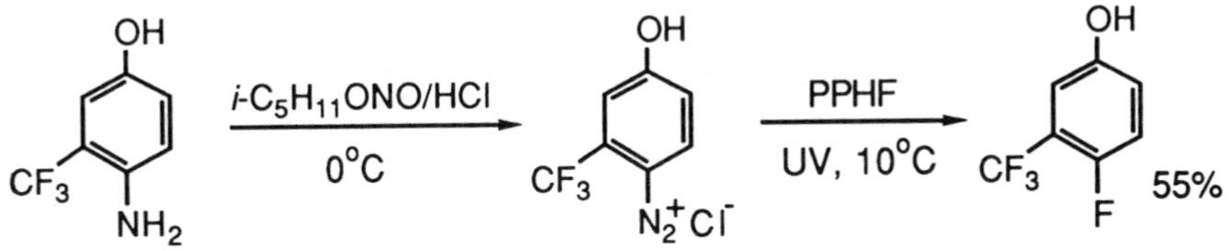

The discussed procedures can be readily adopted for efficient preparation of aromatic heterocyclic fluorides from the corresponding amino compounds.[69,74,75] By using PPHF (40:60), 2-, 3-, or 4-aminopyridines and substituted aminopyridines were converted to the corresponding fluoropyridines in high yield (>90%).[74] Notably, even *ortho*-nitroaminopyridines gave high yield of *ortho*-nitrofluoropyridines, in contrast to the aniline analogue, *ortho*-nitroaniline, which gave only 4% of *o*-fluoronitrobenzene under the same reaction conditions.

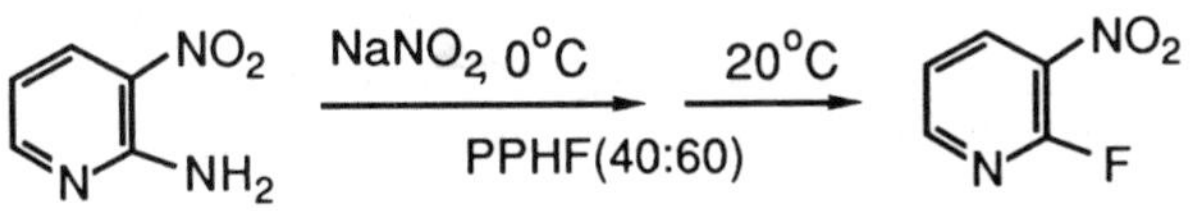

Protected aminopurine nucleosides can also be converted in good yield to their fluoro derivatives when reacted with *tert*-butyl nitrite (TBN) in PPHF (40:60) at −20 °C for 10 min.[75] It is interesting that when 2,6-diamino-9-(2,3,5-tri-*O*-acetyl-*β*-D-ribofuranosyl)purine was treated with TBN–PPHF only the amino group at the 2 position was converted to fluorine. It was observed that the fluoro substitution was inhibited by concentrations of HF–py greater than 65% HF. Reagents with 45–65% HF are effective.

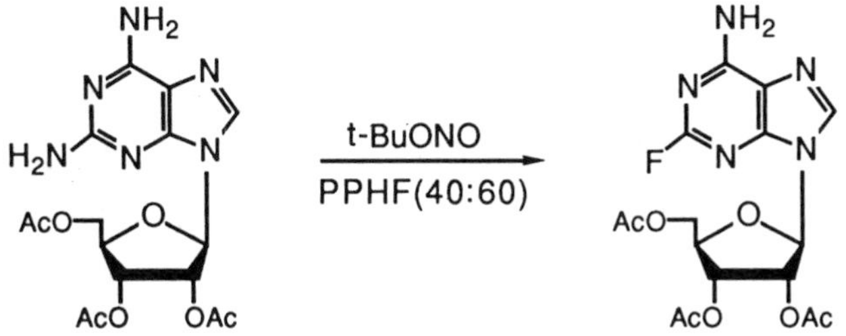

8.5.2. Amino Acids to Fluorocarboxylic Acids

α-Fluorocarboxylic acids are biologically interesting molecules and synthetically useful intermediates. They can be readily prepared from easily available α-amino acids via diazotization in PPHF.[76] The reactions were carried out at room temperature and the yields were generally good except for glutamic acid.

$$\underset{\displaystyle NH_2}{RCHCOOH} \xrightarrow[\text{PPHF}]{NaNO_2} \underset{\displaystyle N_2^+}{RCHCOOH} \xrightarrow[\text{in situ}]{} \underset{\displaystyle F}{RCHCOOH}$$

Stereochemical studies revealed that the substitution of the amino group in most cases takes place with retention of configuration.[77,78] This retention of configuration can be ascribed to the anchimeric assistance of the carboxylate group.

R, COOH, H_2N, H → R, OH, O, N_2^+, H → R, H, OH^+, $(HF)_xF^-$, O → R, COOH, F, H

However, rearrangement to yield β-fluorocarboxylic acids was observed in the reactions of some amino acids such as phenylalanine, tyrosine, threonine (complete rearrangement), and valine and isoleucine (partial rearrangement).[77,78] The mechanism of this rearrangement may involve the anchimeric assistance of phenyl, alkyl, or hydroxyl groups in the β position. For example, phenyl ring participation leads to formation of a stabilized phenonium ion, which is attacked by fluoride ion leading to the β-fluoro product:[78]

$PhCH_2$, COOH, H_2N, H $\xrightarrow[\text{PPHF(30:70)}]{NaNO_2}$ H, OH^+, O → $(HF)_xF^-$, COOH, +, H → FH_2C, COOH, H

Such anchimerically assisted rearrangement can be suppressed by carrying out the reaction in reagents of lower HF concentration, that is, 48% HF and 52% pyridine (by weight).[79,80] Rearrangement was completely prevented in the case of phenylalanine, valine, and isoleucine, whereas in the case of tyrosine and threonine the ratio of α- and β-fluoro product is about 80:20. Decreasing further the HF concentration of the reagent, however, drastically affects the yield of the α-fluoro product.

8.5.3. Alkyl Carbamates to Alkyl Fluoroformates

Treatment of alkyl carbamates in PPHF with sodium nitrite at room temperature yields, via diazotization and dediazoniation, the corresponding fluoroformates.[81]

$$R\text{-}O\text{-}\overset{O}{\overset{\|}{C}}\text{-}NH_2 \xrightarrow[\text{PPHF}]{NaNO_2} R\text{-}O\text{-}\overset{O}{\overset{\|}{C}}\text{-}F$$

8.5.4. Diazo Compounds to Fluoro Compounds

Diazoalkanes, diazoketones, and diazoacetates react readily with PPHF at 0 °C to give the corresponding fluoro derivatives.[13,82]

$$R\text{-}\overset{-}{C}H\text{-}\overset{+}{N}\equiv N \xrightarrow[\text{PPHF}]{H^+} R\text{-}\underset{H}{\underset{|}{C}}H\text{-}\overset{+}{N}\equiv N \xrightarrow{-N_2} R\text{-}\underset{H}{\underset{|}{C}}H^+ \xrightarrow{F^-} R\text{-}\underset{H}{\underset{|}{C}}HF$$

R = Ph, PhCO, $c\text{-}C_6H_{11}CO$, C_2H_5CO

In the presence of *N*-halosuccinimides, diazoketones and diazoacetates were converted to α-halo-α-fluoroketones and esters, respectively. The reaction was considered to involve initial electrophilic attack by positive halogen on the diazoalkanes

$$R\text{-}\overset{O}{\overset{\|}{C}}\text{-}\overset{-}{C}H\text{-}\overset{+}{N}\equiv N \xrightarrow[\text{PPHF}]{\text{N-Halosuccinimide}} R\text{-}\overset{O}{\overset{\|}{C}}\text{-}CHFX$$

R = Ph, C_2H_5, $c\text{-}C_6H_{11}$ X = Cl, Br, I

8.5.5. Aryltriazenes to Arylfluorides

Aryltriazenes were found a suitable source of aryldiazonium ion under controlled, mild acid conditions. Treatment of aryltriazenes with PPHF (30:70) at room temperature or above gave arylfluoride in moderate to good yields.[83]

$$Ar\text{-}N{=}N\text{-}NR_2 \xrightarrow[\text{PPHF}]{2H^+} ArN_2^+ \xrightarrow{(HF)_xF^-} ArF$$

The triazene route is considered to be most suitable for introducing ^{18}F into an aromatic ring of bioactive organic compounds, rather than the classical Balz–Schiemann reaction, because it is a rapid, mild reaction and has the potential of giving high radiochemical yields. A series of piperidinotriazenes[1-aryl-3,3-(1,5-pentanediyl)triazenes] bearing alkyl and/or oxygen functions in the aryl ring as models for estrogen systems were treated with PPHF in benzene. All triazenes bearing only alkyl groups underwent fluorination in moderate yields, as did those bearing alkoxy groups oriented meta with respect to the triazene group. None of the triazenes bearing an *ortho*-methoxy group, however, gave any fluorination products. Hexestrol triazene was converted, by reacting with PPHF (30:70) in benzene, to 2′-fluorohexestrol dimethyl ether in 43% yield, which could be demethylated to give the fluorinated nonsteroidal estrogen, 2′-fluorohexestrol.[84] The conversion of aminotamoxifen into fluorotamoxifen is another example of the application of this method.[85]

OMe PPHF(30:70) OMe MeO N N N MeO F 43%

2- Fluorohexestrol dimethyl ether

N-O PPHF N-O N-N=N F 28%

Fluorotamoxifen

8.5.6. Reactions of Isocyanates and Thioisocyanates

Aliphatic and aromatic isocyanates react with PPHF at room temperature to give the corresponding carbamyl fluorides.[13]

$$R\text{-}N{=}C{=}O \xrightarrow[0^\circ C \text{ to r t}]{\text{PPHF}(30{:}70)} R\text{-}NH\text{-}\overset{O}{\overset{\|}{C}}\text{-}F \quad 40\text{-}58\%$$

R = aliphatic or aromatic

In the presence of nitrosonium borontetrafluoride or sodium nitrite, isocyanoarenes and thioisocyanoarenes react with PPHF to give fluoroarenes in moderate yield.[86]

$$\text{Ar-N=C=O} \text{ or } \text{Ar-N=C=S} \xrightarrow[\text{PPHF(30:70)}]{NO^+BF_4^- \text{ or } NaNO_2} \text{ArF} \quad 30\text{-}50\%$$

8.6. RING-OPENING FLUORINATION

8.6.1. Cyclopropanes

Pyridinium poly(hydrogen fluoride) (30:70) cleaves cyclopropane readily at room temperature to give 75% of *n*-propyl fluoride.[12,13]

$$\triangle \xrightarrow{\text{PPHF(30:70)}} CH_3CH_2CH_2F$$

This ring-opening fluorination was recently used to prepare homoallylic fluorides from cyclopropylmethanols.[87] It was found that treatment of substituted cyclopropylmethanols in chlorobenzene with PPHF (30:70) modified by addition of diisopropylamine (30% to PPHF) and KHF_2 (4% to PPHF) at 0°C for 5 min yielded the corresponding homoallylic fluorides in good yield, whereas PPHF (30:70) alone afforded poor yield and by-products. This method provides a new facile pathway leading to synthons fluorinated at the homoallylic position under mild conditions, while most of susceptible functionalities, for example, double and triple bonds, remain intact. Treatment of 2,2-dihalogenocyclopropylmethanols with PPHF in the presence of diisopropyamine and NaF, followed by dehydrohalogenation with potassium *t*-butoxide, afforded 1-fluoro-1-halogenoalka-1,3-dienes in good yield.[88]

Ph, OH; PPHF(30:70); $i\text{-}Pr_2NH/KHF_2$; $(HF)_xF^-$; Ph; +; Ph; F

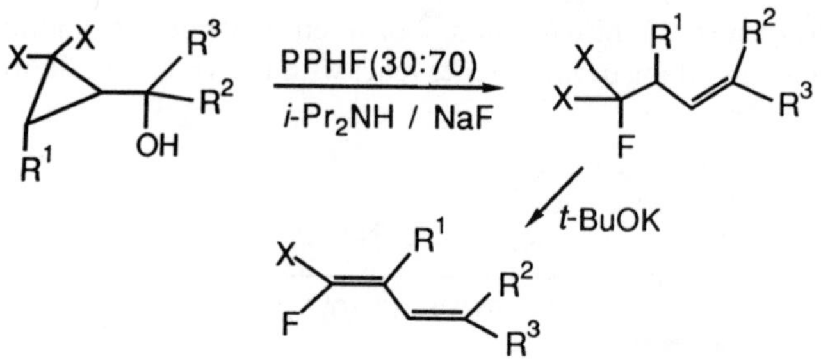

PPHF(30:70)
i-Pr$_2$NH / NaF
t-BuOK

An alkyl or aryl substituent at the 1-position of the cyclopropyl ring will change the pattern of the reaction Upon treatment with the modified reagent, PPHF (30:70)/*i*-Pr$_2$NH—KHF$_2$, phenyl 1-alkyl(or aryl)cyclopropylmethanols undergo ring-expansion fluorination giving exclusively 1-fluoro-1-alkyl(or aryl)cyclobutanes, whereas 1-alkyl(or aryl) cyclopropylethylene oxides gave fluorocyclobutanes containing hydroxymethyl group.[89] In these cases, generation of a tertiary cyclobutyl cation (via ring expansion) would be preferred to formation of a primary homoallylic cation, although the cyclobutane ring still has some ring strain.

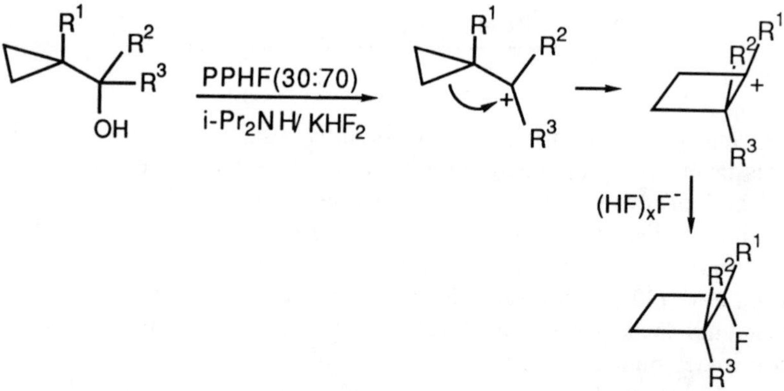

PPHF(30:70)
i-Pr$_2$NH/ KHF$_2$
(HF)$_x$F$^-$

However, no ring expansion occurred when substrates containing an electron-withdrawing group, such as acyl or cyano group attached to the cyclopropane ring, were reacted.[90]

Methoxyallenyl cyclopropylmethanols, when treated with PPHF (30:70) and NaF or KHF$_2$, can be converted to 3-fluoroethylcyclopentenones,[91] which are essential cores for a number of important bioactive molecules such as prostaglandin B, Jasmonoids, and pyrethrins.

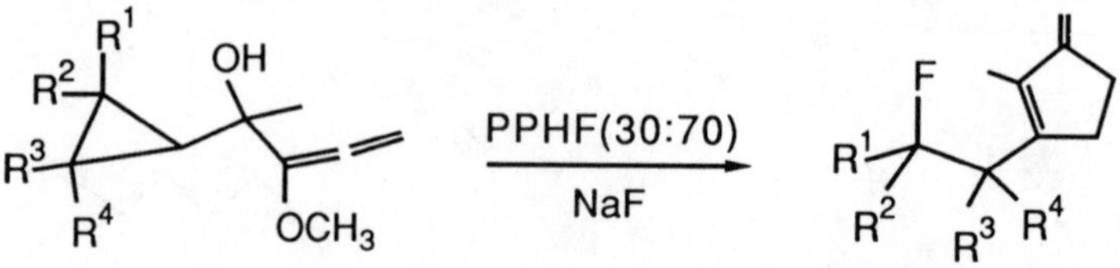

PPHF(30:70)
NaF

It was found that the yield of the desired products was sensitive to solvent and the procedures for the discussed ring-opening and ring-expansion fluorination. Generally, addition of PPHF into a mixture of the substrate, KHF_2, and diisopropylamine in dichloromethane or chlorobenzene at or below 0°C gave good results.[90] Stereochemistry of the products is highly dependent on the steric factors of the substituent at the α carbon, but the reaction is generally highly stereoselective.

8.6.2. Oxiranes

Hydrogen fluoride opens the oxirane ring to give fluorohydrins.[29b] However, use of anhydrous hydrogen fluoride also leads to polymerization. Yield of the HF-addition product can be improved by diluting HF in an electron-donor solvent such as THF or organic base[4,7] or using sodium bifluoride.[92]

Pyridinium poly(hydrogen fluorides) with various concentrations of HF (from 30 to 70 wt%) were used to hydrofluorinate both aromatic and aliphatic epoxides.[93] Since the ring opening in acidic medium occurs through initial formation of an epoxonium ion, the electron-donating or -withdrawing character of the substituents has a major influence on the direction of the reaction and the conditions needed for the ring opening. While styrene oxide reacts readily with PPHF containing 42% HF to give 2-fluoro-2-phenylethanol in 75% yield, the reaction of 3,3,3-trichloromethyl-1,2-propene oxide requires a PPHF containing 60% of HF and gives 1-fluoro-3,3,3-trichloromethyl-2-propanol.[93] The regioselectivity of the HF addition might be explained by different mechanisms. A carbenium ion-stabilizing substituent, such as phenyl, favors an S_N1 reaction, whereas a destabilizing substituent favors an S_N2 pathway.

The ring-opening hydrofluorination of oxiranes is generally sensitive to the acidity of the reagent, solvent, and reaction temperature. Moderate to good yields can be obtained if the reaction conditions are properly adjusted to the specific substrates. Higher concentration of HF causes polymerization, but too low a concentration of HF will inhibit the reaction. Decreasing reaction temperatures sometimes lead to increased polymerization. The milder diisopropyl- or triethylamine tris(hydrogen fluoride) reagents[94–96] are much less reactive, but more selective than PPHF (30:70). The ring opening of 1-benzyloxy-2,3-epoxypropane with the former reagent required 7 h at 110 °C giving predominantly 1-benzyloxy-2-fluoropropan-1-ol, whereas the reaction with PPHF (30:70) was completed in less than 15 min at −5 °C affording two isomers in comparable amount.[94]

PPHF(30/70)	53	:	47
i-$Pr_2NH/3HF$	93	:	7

A series of alkylphenyl 2,3-epoxycarboxylates from the well-known Darzens glycidic esters synthesis has been allowed to react with PPHF (30:70) at room temperature to give the corresponding 3-fluoro-3-phenyllactates in almost quantitative yield.[97] The reaction is highly regioselective and, in most cases, also stereoselective. Cleavage of glycidic esters with PPHF (30:70) followed by Jones oxidation gives fluoropyruvic esters in 60–85% yield.[98]

PPHF(30:70), r.t. — 80-95%

R = Ph, -$(CH_2)_5$-   Y = CO_2R', CONHR', CN

R^1, R^2 = alkyl, aryl

When 9-oxabicyclo[6.1.0]non-4-ene was reacted with triethylamine tris(hydrogen fluoride) or PPHF in the presence of NBS, transannular oxygen participation was observed in the bromofluorination instead of oxirane ring opening.[99]

NBS, $Et_3N/3HF$ or PPHF; $(HF)_xF^-$

An unusual ring-opening fluorination of 6β-fluoroandrost-4,5-epoxy-3,17-dione with PPHF in dichloromethane led to a 2,6-difluoro derivative:[60]

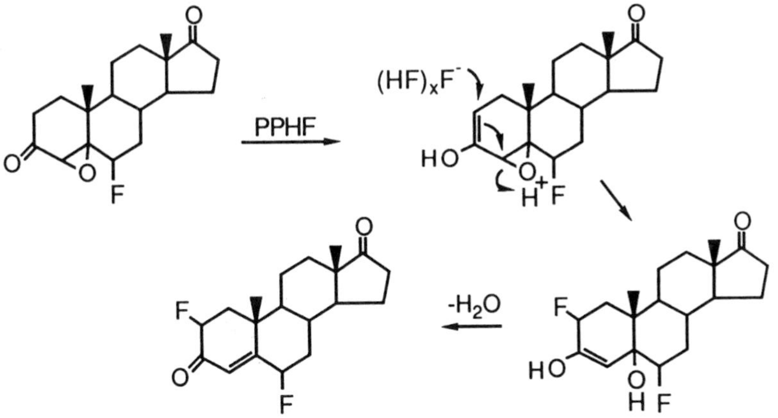

PPHF
$(HF)_xF^-$
$-H_2O$

Aziridines

Ring-opening hydrofluorination of aziridines by treatment with PPHF was shown to be an effective method for regiospecifically preparing β-fluoro amines.[100–105]

PPHF

Aziridines bearing a phenyl substituent are more reactive and the reaction can be carried out at room temperature, whereas those with only alkyl substituents need heating to give the corresponding ring-opening products. *N*-Alkyl substitution significantly decreases the reaction rates.[101] However, an electron-withdrawing group on nitrogen activates the aziridines, for example, *N*-(*tert*-butoxycarbonyl)-2-phenyl-3-methylaziridine needs a milder "partially neutralized Olah's reagent," that is, PPHF (30:70) plus triethylamine, giving *trans*-fluoroamine. In contrast, the corresponding unactivated aziridine reacts with PPHF to yield *cis*-fluoroamine.[104] Consequently, it is possible to selectively prepare each diastereomeric fluoroamine from the same aziridine.

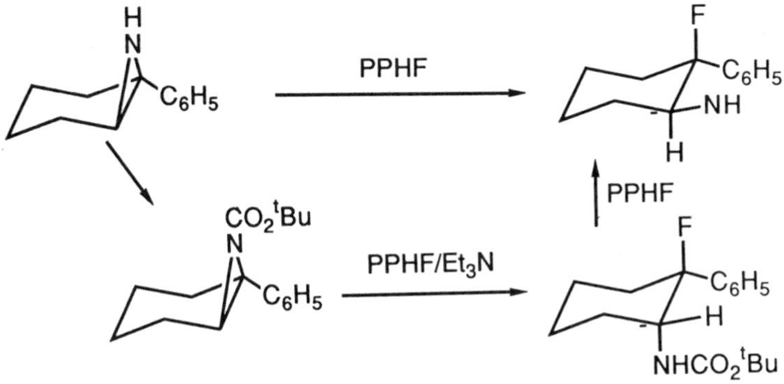

PPHF
PPHF
$PPHF/Et_3N$

Hydrogen fluoride addition to the aziridine ring is regioselective. Fluoride attack is generally directed to the carbon most capable of stabilizing a positive charge, that is, to a benzylic or the most substituted ring carbon.[100–104] Some complications in regioselectivity have been observed with bicyclic,[106] tricyclic[107] and α,β-ethylenic[108] aziridines.

The stereochemistry of the reaction is highly dependent on the structure of the substrates and the fluorinating agent used. However, in most cases one diastereoisomer is predominant. It has been observed that both cis and trans isomers of the starting aziridines afford the same fluoro amines while different reagents often lead to different stereoisomers.[101,104]

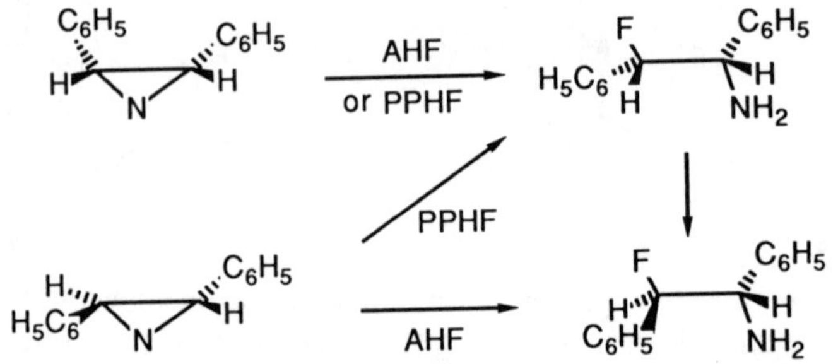

An S_N1 mechanism involving the formation of carbocationic intermediates and complexation of HF with the amino group was proposed based on the stereochemical results and on the fact that neither the reaction rates nor the isomer distribution was much affected by addition of sodium fluoride.[101,104]

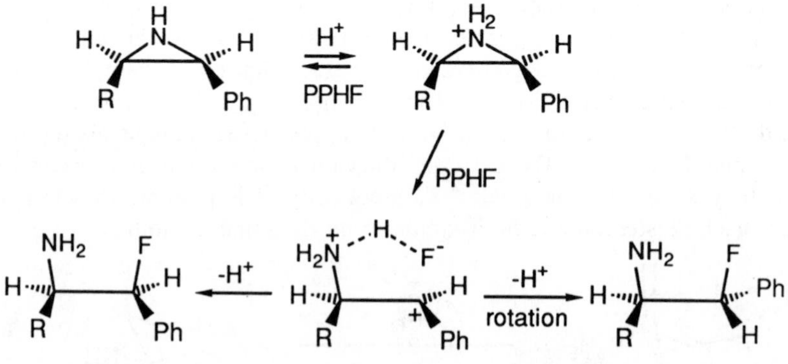

The stereochemistry of the ring-opening hydrofluorination of methyl-substituted bicyclic aziridines has been studied[109] and a dynamic conformational analysis of the steric course, using the torsion angle notation, was reported.[110]

8.6.4. Azirines

Substituted 1-azirines reacted under mild conditions with PPHF to give β,β-difluoroamines, β,β-difluoro-α-amino acid alkyl esters or,[111,112] in some cases, α-fluoroketones.[112,113] The yields of the products are moderate and can be improved by using a more nucleophilic reagent, PPHF (30:70) with triethylamine.[113]

The reaction is likely to proceed by one of the following two main pathways:

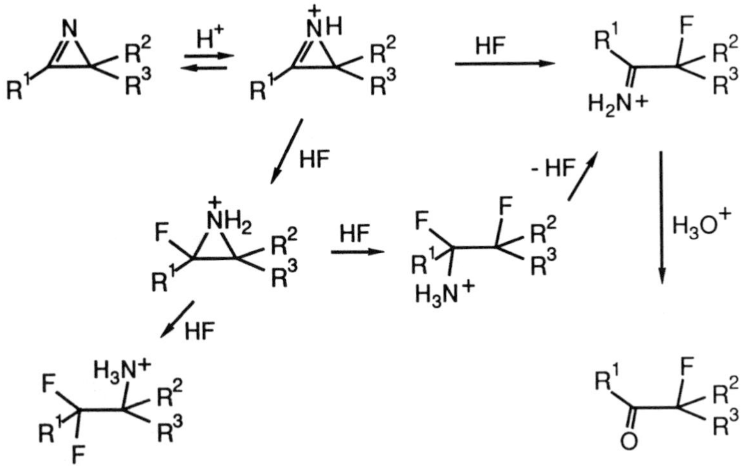

The relative importance of the two pathways is dependent on the substituents. When R^1 = Ph, R^2, R^3 = H, β-difluoroamine formed in 67% yield, whereas when R^1 = Ph, R^2, R^3 = CH_3, phenyl 2-fluoro-2-methylethyl ketone was obtained in 90% yield.[113]

8.7. *gem*-DIFLUOROINATION AND TRIFLUOROINATION

8.7.1. gem-Difluorination via 1,3-Dithiolanes

Fluorodesulfurization of thiols in liquid hydrogen fluoride with CF_3OF or NCS is an interesting new approach for C—F bond formation.[114]

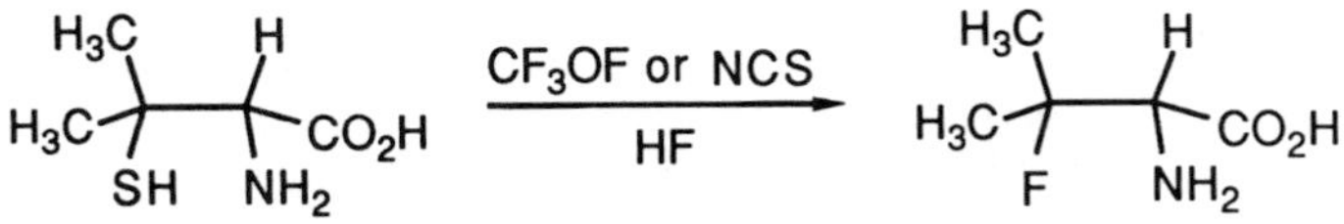

A modified procedure using PPHF (30:70)–NBS was used for preparation of glycosyl fluorides from phenyl thioglycosides.[64] Based on the fluorodesulfurization reaction, a facile method has been developed for the preparation of *gem*-difluoro compounds from ketones and aldehydes via formation of the corresponding 1,3-dithiolanes, followed by reaction with an equivalent BrF source, that is, 1,3-dibromo-5,5-dimethylhydantoin (DBH) or *N*-halosuccinimides and PPHF, in methylene chloride.[115] This two-step procedure is a convenient alternative to the use of sulfur tetrafluoride and DAST.

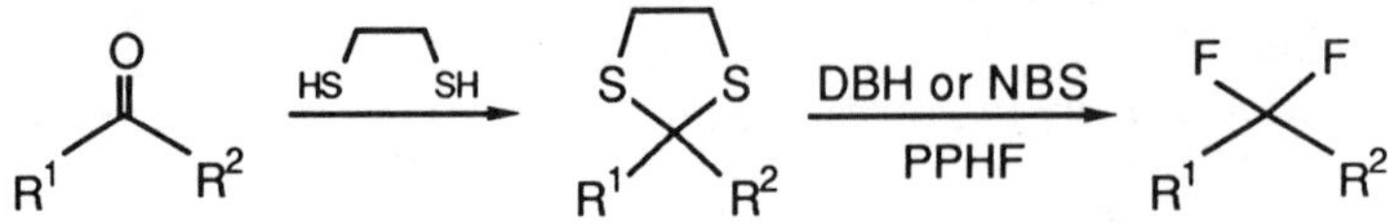

The fluorodesulfurization requires 2 eq of Br^+ and appears to proceed through a sequence of two bromosulfonium ions that open and cleave, respectively, to sulfur- and fluorine-stabilized carbocations, which are then trapped by fluoride ion.

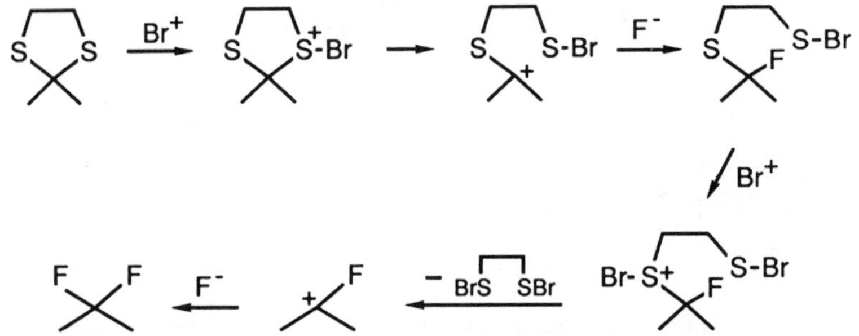

1,3-Dibromo-5,5-dimethylhydantoin proved to be the most effective positive halogen source. Formation of *gem*-difluoro compounds from dithiolanes derived from ketones is efficient and fast, even at $-78\,^{\circ}$C, whereas reaction of dithiolanes derived from aldehydes proceeds only at $0\,^{\circ}$C. While the reaction proceeds successfully with a variety of ketones and aldehydes, brominated byproducts are observed with electron-rich aromatic systems, and side reactions take place in systems prone to carbocationic rearrangements.

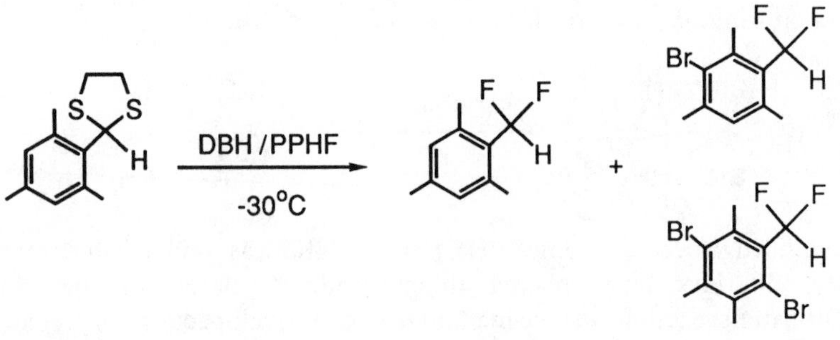

8.7.2. *gem*-Difluorination via Hydrazones

Carbonyl compounds can be transformed via their hydrazone into *gem*-difluoro compounds by reacting with combinations of NBS and PPHF (40:60) or poly-4-vinylpyridinium poly(hydrogen fluoride) (41:59) in satisfactory yields.[116]

$$R^1C(=O)R^2 \xrightarrow{H_2NNHR} R^1C(=NNHR)R^2 \xrightarrow[\text{PPHF or PVPHF}]{\text{NBS}} R^1CF_2R^2$$

This method is an improvement of that of Rozen et al.[117] for preparation of *gem*-difluoro compounds from carbonyl compounds by treatment of their hydrazones with IF, which in turn was prepared by reacting iodine with elemental fluorine in $CFCl_3$. Since formation of hydrazones is facile and there is no need to use elementary fluorine, the modified method has obvious advantages, even over methods using DAST or SeF_4. The major limitation in the method comes from the stability of the hydrazones, as some of them may be readily further oxidized to the corresponding azines.

8.7.3. Direct *gem*-Difluorination of Carbonyl Group

Ketones react with SCl_2 in PPHF (50:50) giving *gem*-difluoro products.[118]

$$\text{2-adamantanone} \xrightarrow{SCl_2\ /\ \text{PPHF(50:50)}} \text{2,2-difluoroadamantane}$$

It is known that SCl_2 reacts with PPHF giving SF_4 (see below), the method is a simplified, *in situ* modification of the SF_4 fluorination of ketones. The reactions are carried out at atmospheric pressure, a further improvement over the use of SF_4. However, in the cases of aromatic ketones, chlorination by-products are also formed in significant amounts.

8.7.4. *gem*-Trifluorination via Orthothio Esters

Aromatic orthothio esters react with DBH or NBS and PPHF (30:70) to give corresponding trifluoromethyl derivatives in fair yield.[119] Both NBS and DBH gave similar yields. The reaction can be carried out with both electron-rich and electron-poor aromatics and is compatible with thioethers. However, concomitant bromination of electron-rich aromatic rings was observed.

R = Me, Et

DBH or NBS, PPHF(30:70)

34-67%

DBH or NBS, PPHF(30:70)

40%

8.7.5. *gem*-Trifluorination via Dithioacids

Trifluoromethyl arenes can also be prepared by treatment of aromatic dithioacids with NBS–PPHF.[120] Similar to the reaction of orthothio esters, ring bromination was observed for substrates having electron-donating substituents.

CS_2 ; NBS, PPHF ; CF_3

8.8. TRANSFORMATION OF C–H AND C–X INTO C–F BONDS

8.8.1. C–H to C–F

Direct fluorination of a saturated hydrocarbon without the use of elementary fluorine is difficult. However, some reactive alkanes having tertiary C—H bonds, such as adamantane, diadamantane, and triphenylmethane, readily react with $NO^+BF_4^-$ or $NO^+PF_6^-$ in PPHF (30:70) to give the corresponding tertiary fluorohydrocarbons in high yield.[121] The needed NO^+ can also be *in situ* generated by inexpensive $NaNO_2$ in PPHF.

$NO^+ BF_4^-$; PPHF

The method is limited mainly to such tertiary hydrocarbons where the intermediate carbocations cannot undergo proton elimination to form alkenes that may readily polymerize under the reaction condition.

The C—H bonds β to a C—Br bond in secondary alkyl and cycloalkyl bromides can also be transformed to a C—F bond when reacted with $NO_2^+BF_4^-$ in PPHF solution.[122]

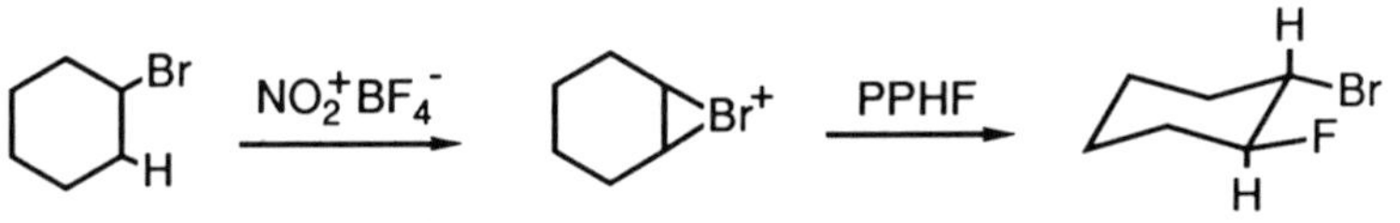

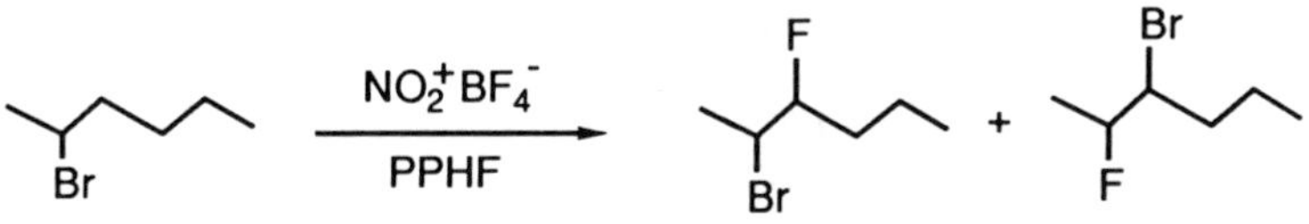

The β-fluorination seems to take place only with secondary bromides. Primary bromides and secondary chlorides do not react, whereas iodocyclohexane gives fluorocyclohexane under similar condition.

8.8.2. C–X to C–F

Halogen atoms located particularly on tertiary, allylic, and acyl carbon atoms or on a carbon α to a carbonyl group can be replaced with fluorine by using PPHF or its combination with other reagents.

Bridgehead adamantyl and diadamantyl halides undergo halogen exchange fluorination in the presence of $NO_2^+BF_4^-$ with PPHF.[123,124]

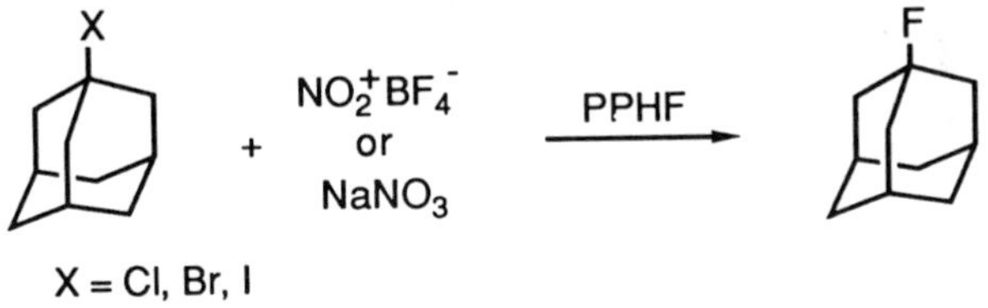

X = Cl, Br, I

Acyl fluorides can be prepared by reaction of acyl chlorides or anhydrides with PPHF[13] or melamine–HF complexes[125] in high yields.

$$R{-}C(=O){-}Cl \xrightarrow{\text{PPHF}} R{-}C(=O){-}F$$

$$Cl{-}C(=O){-}C(=O){-}Cl \xrightarrow{\text{HF/Melamine}} F{-}C(=O){-}C(=O){-}F$$

Teriary amine tris(hydrogen fluoride) was found to be very effective to fluorinate phosgene giving carbonyl difluoride in quantitative yield.[22] Similarly, cyanuric chloride was transformed into cyanuric fluoride in 90% yield.

$$COCl_2 \xrightarrow{Et_3N/3HF} COF_2 \quad 100\%$$

$$C_3N_3Cl_3 \xrightarrow[\text{N-methylpyrrolidone}]{Et_3N/3HF} C_3N_3F_3 \quad 90\%$$

When α-halo ketone or geminal dihalides were heated at 50°C with a suspension of yellow mercury oxide in PPHF (30:70) for approximately 15 h, the halides were changed for fluorides in good to moderate yield.[13]

$$C_6H_5{-}C(=O){-}CH_2Br \xrightarrow[50^\circ C]{HgO\ /\ PPHF} C_6H_5{-}C(=O){-}CH_2F$$

$$CH_3CH_2CHCl_2 \xrightarrow[50^\circ C]{HgO\ /\ PPHF} CH_3CH_2CHF_2$$

The HgO–PPHF combination was used to convert 6β-bromoandrost-4-ene-3,17-dione into 6β-fluoroandrost-4-ene-3,17-dione in 40% yield.[126] In this case an allylic bromide was replaced by fluoride.

$$\text{HO}\ \ldots\ \text{Br (steroid)} \xrightarrow[\text{rt, 4h}]{HgO\ /\ PPHF} \text{F (steroid)}$$

8.9. OXIDATIVE AND REDUCTIVE FLUORINATION

8.9.1. Oxidative Fluorination

Phenols are oxidatively fluorinated with PbO_2 and PPHF (30:70) into difluorodienones.[127] When anhydrous hydrogen fluoride and PbO_2 are used, no dienones are formed and only polymers are obtained.[128]

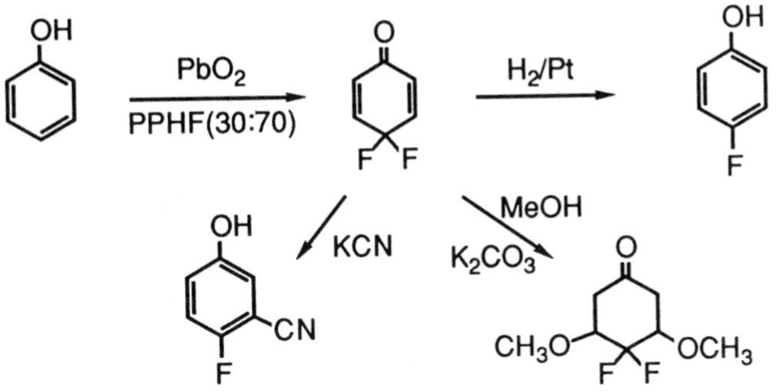

1,1-Diphenylethene, when treated with lead tetraacetate and PPHF, was oxidatively fluorinated with concurrent rearrangement to 1,1-difluoro-1,2-diphenylethane:[129]

Onium poly(hydrogen fluorides) such as PPHF and Et_3N-3HF were found to have several advantages as electrolytes and fluorine sources in electrochemical fluorinations. These include good stability towards oxidation, good electrical conductivity, and satisfactory solubility of the substrates.[127,130–132] Electrochemical oxidation of benzene in Et_3N-3HF gave several fluoro derivatives.[127]

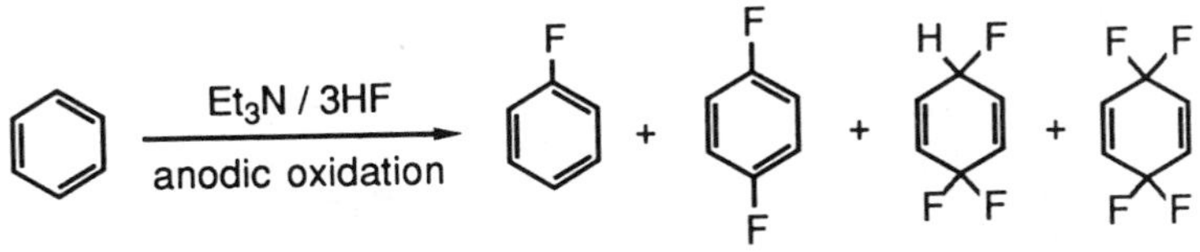

Benzylic ketones, esters, and nitriles[131,132] as well as sulfides substituted by an electron-withdrawing group[133] are readily converted to the corresponding α-monofluoro derivatives by electrochemical oxidation in the presence of Et_3N-3HF. Continued electrochemical oxidative fluorination gives α-difluoro compounds (as well as other by-products). Chemical oxidative fluorination using DBH–Et_3N-3HF is less efficient.

MeO–C$_6$H$_4$–CH$_2$–E → (Et_3N / 3HF, anodic oxidation, 1.2 to 1.3 v) → MeO–C$_6$H$_4$–CHF–E, 65-70%

E = COR, COOEt, CN

→ (Et_3N / 3HF, anodic oxidation, 1.45 to 1.6 v) → MeO–C$_6$H$_4$–CF$_2$–E, 50-60%

R^1R^2CH–S–R → (Et_3N / 3HF, anodic oxidation) → R^1R^2CF–S–R

A combination of PPHF (30:70) and triethylamine (1:0.6 by volume) was found to be the best electrolyte for the electrochemical fluorination of ethyl 1-methylpyrazole-4-carboxylate, whereas using either PPHF or Et_3N–3HF gave unsatisfactory results.[134]

Conditions (anodic oxidation)	5-F, N-CH$_3$	:	5-F, N-CH$_2$F	Yield
PPHF(30:70) / MeCN	25	:	75	16%
Et_3N / 3HF / MeCN	93	:	7	29%
PPHF / Et_3N / MeCN	83	:	17	40%

8.9.2. Reductive Fluorination

Ketones react with triethylsilane in PPHF to give the corresponding monofluoride:[135]

2-Adamantanone → (Et_3SiH / PPHF) → 2-fluoroadamantane

8.10. FORMATION OF S–F AND P–F BONDS

8.10.1. SF$_4$

Sulfur tetrafluoride can be conveniently prepared from SCl_2 and PPHF (30:70).[136] The reaction proceeds smoothly at 45°C at atmospheric pressure.

When the reaction was carried out with PPHF (py:HF = 1:3) at 30 °C, SF_4 was obtained in 90% yield.[137] Tertiary amine tris-hydrogen fluorides can also be used for the preparation of SF_4.[22]

$$SCl_2 \xrightarrow[30\text{-}45^\circ C]{PPHF} \underset{80\text{-}90\%}{SF_4} + S_2Cl_2$$

8.10.2. SF_6

Pyridine tris(hydrogen fluoride) reacted with S_2Cl_2 and Cl_2 at 35 °C to give a product mixture containing SF_5Cl (90%), SF_4 (9.5%), and SOF_2 (0.5%). This product mixture was pyrolyzed at 300–350 °C to form SF_6.[138]

$$S_2Cl_2 + Cl_2 \xrightarrow[35^\circ C]{Pyridine/3HF} SF_5Cl + SF_4 \xrightarrow{300\text{-}350^\circ C} SF_6$$

8.10.3. SO_2ClF

Sulfuryl chloride fluoride has been extensively used as a solvent for the generation of stable carbocations[139] and is also useful as a synthetic reagent. The preparation of sulfuryl chloride fluoride by the reaction of sulfuryl chloride with PPHF in 80% yield is by far the most convenient method.[136]

$$SO_2Cl_2 \xrightarrow{PPHF} SO_2ClF$$

8.10.4. P–N to P–F Bonds

Diastereomers of 2-(*N*-ethyleneimino)-2-oxo-3-(α-methylbenzyl)-1,3,2-oxazaphoshporinane have been found to react with PPHF (C_5H_5N–HF ~80:20 w/w) at 5 °C giving the corresponding cyclic phosphoramidofluoridates in 50% yield.[140] Displacement of the ethyleneimino ligand by fluoride is fully stereospecific and occurs with inversion of configuration at the P atom. Interestingly, not even traces of products resulting from the opening of the oxazaphosphorinane ring were found in the reaction mixture.

N, P=O, O, N, CH_3–C–H, Ph $\xrightarrow[5^\circ C]{PPHF(80:20)}$ O, P, O, F, N, CH_3–C–H, Ph

8.10.5. Hexafluorophosphates

Conventional methods for the preparation of hexafluorophophates require the use of anhydrous hydrogen fluoride.[141] Pyridinium poly(hydrogen fluoride) has been found to be a more convenient and efficient reagent for the preparation of pyridinium hexafluorophosphate from several phosphorous(V) halides.[142]

$$PCl_5 \text{ or } POCl_3 \xrightarrow{\text{PPHF(60:40)}} C_5H_5NH^+ PF_6^- \begin{cases} \xrightarrow{\text{NaOH}} Na^+PF_6^- \\ \xrightarrow{\text{NH}_4\text{OH}} NH_4^+PF_6^- \\ \xrightarrow{\text{KOH}} K^+PF_6^- \end{cases}$$

The reaction of phosphorus(V) halides($POCl_3$, $POBr_3$, $PSCl_3$, PCl_5, or PBr_5) and PPHF (60:40, py: HF = 3:8) afforded pyridinium hexafluorophosphate in 90% yield, which in turn was used to prepare ammonium, sodium, potassium, rubidium, and cesium hexafluorophosphates in good yield (>90%) and high purity (>99%).

REFERENCES

1. Glemser, O., in *Fluorine the First Hundred Years* (R. E. Banks, D. W. A. Sharp, and J. C. Tatlow, Eds.), Elsevier Sequoia, Lausanne and New York, 1986, pp. 45–70.
2. Barbour, A. K., in *Organofluorine Chemicals and Their Industrial Application* (R. E. Banks, Ed.), Ellis Horwood, Chichester, 1979, p. 45.
3. Kilpatrick, M. and Jones, J. G., in *The Chemistry of Non-Aqueous Solvents*, (J. J. Lagowski, Ed.), Academic, New York, London, 1967, pp. 43–99.
4. Hirschmann, R. F., Miller, R., Wood, J., and Jones, R. E., *J. Am. Chem. Soc.*, *78*, 4956 (1956).
5. Taub, D., Hoffsommer, R. D., and Wendler, W. L., *J. Am. Chem. Soc.*, *79*, 452 (1957).
6. Bergstrom, C. G., Nicholson, R. T., and Dodson, R. M., *J. Org. Chem.*, *28*, 2633 (1963).
7. (a) Aranda, G., Jullien, J., and Martin, J. A., *Bull. Soc. Chim. Fr.*, 1890 (1965). (b) *Bull. Soc. Chim. Fr.*, 2850 (1966).
8. (a) French Patent, FR 1370827, Chem. Abstr., *62*, P2712h (1965). (b) Hecht, S. S. and Rothman, E. S., *J. Org. Chem.*, *38*, 395 (1973).
9. French Patent, FR 1374591, Chem. Abstr., *62*, P9212a (1965).
10. van der Akken, M. and Jellinek, F., *Recl. Trav. Chim. Pays-Bas*, *86*, 275 (1967).
11. (a) Thomas, R. K., *Proc. R. Soc. London, Ser. A*, *322*, 137 (1971). (b) Politanskii, S. F., Ivanyk, G. D., Sarancha, V. N., and Shevchuk, V. U., *Z. Org. Khim.*, *10*, 693 (1974).
12. Olah, G. A., Nojima, M., and Kerekes, I., *Synthesis*, 779 and 781 (1973).

13. Olah, G. A., Welch, J. T., Vankar, Y. D., Nojima, M., Kerekes, I., and Ohal, J. A., *J. Org. Chem.*, *44*, 3872 (1979), and the references cited therein.
14. Olah, G. A., Shih, J. G., and Prakash, G. K. S., *J. Fluorine Chem.*, *33*, 377 (1986).
15. Fieser, M. and Fieser, L., *Reagents Org. Synth.*, *11*, 453 (1984).
16. Fukuhara, T., Yoneda, N., Abe, T., Nagata, S., and Suzuki, A., *Nippon Kagaku Kaishi*, *166*, 1951 (1985).
17. Olah, G. A. and Li, X. Y., Synlett, 267 (1990).
18. Olah, G. A. and Watkins, M., *Org. Synth.*, *Coll. Vol. 6*, 628 (1988).
19. For safety precaution and first aid, see: (a) Finkel, A. J., *Adv. Fluorine Chem.*, *7*, 199 (1973). (b) *Org. Synth.*, *Coll. Vol. 5*, 66 (1973). (c) Trevino, M. A., Herrman, G. H., and Sprout, W. L., *J. Occup. Med.*, *25*, 861 (1983).
20. Fukuhara, T., Sasaki, S., Yoneda, N., and Suzuki, A., *Bull. Chem. Soc. Jpn.* *63*, 2058 (1990).
21. (a) Gennick, I., Harmon, K. M., and Potvin, M. M., *Inorg. Chem.*, *16*, 2033 (1977); (b) Andrews, L., Davis, S. R., and Johnson, G. L., *J. Phys. Chem.*, *90*, 4273 (1986).
22. Franz, R., *J. Fluorine Chem.*, *15*, 423 (1980).
23. Giudicelli, M. B., Picq, D., and Veyron, B., *Tetrahedron Lett.*, *31*, 6527 (1990).
24. Jache, A. W., in *Fluorine Containing Molecules*, (J. F. Liebman, A. Greenberg, and W. R. Dolbier, Jr., Ed.), VCH, New York, pp. 165–197 (1988).
25. Desbat, B. and Houng, P. V., *J. Chem. Phys.*, *78*, 6377 (1983).
26. Mootz, D. and Poll, W., *Z. Naturforsch.*, *39b*, 290 (1984).
27. Mootz, D. and Boenigk, D., *Z. Anorg. Allg. Chem.*, *544*, 159 (1987).
28. Boenigk, D. and Mootz, D., *J. Am. Chem. Soc.*, *110*, 2135 (1988).
29. Carre, J., and Barberi, P., *J. Fluorine Chem.*, *50*, 1 (1990).
30. Gillespie, R. J. and Liang, J., *J. Am. Chem. Soc.*, *110*, 6053 (1988).
31. Yoneda, N., Abe, T., Fukuhara, T., and Suzuki, A., *Chem. Lett.*, 1135 (1983).
32. Yoneda, N., Nagata, S., Fukuhara, T., and Suzuki, A., *Chem. Lett.*, 1241 (1984).
33. Albert, P. and Cousseau, J., *J. Chem. Soc. Chem. Commun.*, 961 (1985).
34. (a) Bowers, A., *J. Am. Chem. Soc.*, *81*, 4107 (1959). (b) Robinson, C. H., Finckenor, L., Oliveto, E. P., and Gould, D., *J. Am. Chem. Soc.*, *81*, 2191 (1959).
35. For reviews see: (a) Sheppard, W. A. and Sharts, C. M., *Organic Fluorine Chemistry*, Benjamin, New York, 1969. (b) Sharts, C. M. and Sheppard, W. A., *Org. React.*, *21*, 125 (1974).
36. (a) Zupan, M., Sket, B., and Johar, Y., *J. Macromol. Sci. Chem.*, *A17*, 759 (1982). (b) Gregorcic, A. and Zupan, M., *J. Fluorine Chem.*, *24*, 291 (1984). (c) Gregorcic, A. and Zupan, M., *Bull. Chem. Soc. Jpn.*, *60*, 3083 (1987).
37. Hamman, S. and Beguin, C. G., *J. Fluorine Chem.*, *23*, 515 (1983).
38. Gregoric, A. and Zupan, M., *J. Org. Chem.*, *44*, 1255 (1979).
39. Alvernhe, G., Laurent, A., and Haufe, G., *Synthesis*, 562 (1987).
40. Haufe, G., Alvernhe, G., and Laurent, A., *Tetrahedron Lett.*, *27*, 4449 (1986).
41. Alvernhe, G., Anker, D., and Laurent, A., *Tetrahedron*, *44*, 3551 (1988).
42. Zupan, M. and Pollak, A., *J. Chem. Soc. Perkin Trans.* I, 971 (1976).
43. Zupan, M., *J. Fluorine Chem.*, *9*, 177 (1977).
44. Gregorcic, A. and Zupan, M., *J. Org. Chem.*, *49*, 333 (1984).

45. Hamman, S., *J. Fluorine Chem.*, *45*, 377 (1989).
46. Boche, G. and Fahrmann, U., *Berichte*, *114*, 4005 (1981).
47. Kolb, M., Barth, J., Heydt, J.-G., and Jung, M. J., *J. Med. Chem.*, *30*, 267 (1987).
48. Chi, D. Y., Kiesewetter, D. O., Katzenellenbogen, J. A., Kilbourne, M. R., and Welch, M. J., *J. Fluorine Chem.*, *31*, 99 (1986).
49. Chi, D. Y., Kilbourn, M. R., Katzenellenbogen, J. A., and Welch, M. J., *J. Org. Chem.*, *52*, 658 (1987).
50. Olah, G. A. and Nojima, M., *Synthesis*, 785 (1973).
51. Capozzi, G., De Lucchi, O., Lucchini, V., and Modena, G., *Tetrahedron Lett.*, 2603 (1975).
52. (a) Haufe, G., Alhernhe, G., Anker, D., Laurent, A., and Saluzzo, C., *Tetrahedron Lett.*, *29*, 2311 (1988). (b) Laurent, A., Haufe, G. and Alvernhe, G., FR 2616145, (1988) *Chem. Abstr.*, *112*, 20544 (1990).
53. Saluzzo, C., Alvernhe, G., and Anker, D., *J. Fluorine Chem.*, *47*, 467 (1990).
54. Olah, G. A., Nojima, M., and Kerekes, I., *Synthesis*, 786 (1973).
55. Dahbi, A., Hamman, S., and Beguin, C. G., *J. Chem. Research(s)*, 128 (1989).
56. Hamman, S. and Beguin, C. G., *J. Fluorine Chem.*, *37*, 343 (1987).
57. Olah, G. A., and Welch, J., *Synthesis*, 653 (1974).
58. Parish, E. J. and Schroepfer, Jr., G. J., *J. Org. Chem.*, *45*, 4034 (1980).
59. Ambles, A. and Jacquesy, R., *Tetrahedron Lett.*, 1083 (1976).
60. Mann, J. and Pietrzak, B., *J. Chem. Soc. Perkin Trans. 1*, 2681 (1983).
61. Szarek, W. A., Grynkiewicz, G., Doboszewski, B., and Hay, G. W., *Chem. Lett.*, 1751 (1984).
62. Hayashi, M., Hashimoto, S., and Noyori, R., *Chem. Lett.*, 1747 (1984).
63. Kirby, A., *The Anomeric Effect and Related Stereoelectronic Effects at Oxygen*, Springer Verlag, New York, 1983.
64. Nicolaou, K. C., Dolle, R. E., Papahatjis, D. P., and Randall, J. L., *J. Am. Chem. Soc.*, *106*, 4189 (1984).
65. For reviews see: (a) Suschitzky, H., *Adv. Fluorine Chem.*, *4*, 1 (1965). (b) Hewitt, C. D. and Silvester, M. J., *Aldrichimica Acta*, *21*, 3 (1988).
66. Osswald, X. and Scherer, X., Ger. Patent 600,706,1934, *Chem. Abstr.*, *23*, 7260 (1934).
67. Ferm, R. L. and VanderWerf, C. A., *J. Am. Chem. Soc.*, *72*, 4809 (1950).
68. Olah, G. A. and Welch, J., *J. Am. Chem. Soc.*, *97*, 208 (1975).
69. Boudakian, M. M., U.S. 4,096,196 (1978), *Chem. Abstr.*, *90*, 103597 (1979).
70. Fukuhara, T., Yoneda, N., Sawada, T., and Suzuki, A., *Synth. Commun.*, *17*, 685 (1987).
71. (a) Ridd, J. H., *Quart. Rev.*, *15*, 418 (1961). (b) Zollinger, H., *Helv. Chim. Acta*, *71*, 1661 (1988).
72. Yoneda, N. and Fukuhara, T., JP 01233232 (1989), *Chem. Abstr.*, *112*, 178326 (1990).
73. Yoneda, N., Fukuhara, T., Harada, K., and Matsushita, A., JP 01316336 (1989), *Chem. Abstr.*, *113*, 23358 (1990).
74. Fukuhara, T., Yoneda, N., and Suzuki, A., *J. Fluorine Chem.*, *38*, 435 (1988).
75. Robins, M. J. and Uznanski, B., *Can. J. Chem.*, *59*, 2608 (1981).

76. Olah, G. A., and Welch, J., *Synthesis*, 652 (1974).

77. Keck, R. and Retey, J., *Helv. Chim. Acta.*, *63*, 769 (1980).

78. Faustini, F., De Munari, S., Panzeri, A., Villa, V., and Gandolfi, C. A., *Tetrahedron Lett.*, 4533 (1981).

79. Olah, G. A., Prakash, G. K. S., and Chao, Y. L., *Helv. Chim. Acta.*, 2528 (1981).

80. (a) Barber, J., Keck, R., and Retey, J., *Tetrahedron Lett.*, *23*, 1549 (1982). (b) Hamman, S. and Beguin, C. G., *Tetrahedron Lett.*, *24*, 57 (1983).

81. Olah, G. A. and Welch, J., *Synthesis*, 654 (1974).

82. Olah, G. A. and Welch, J., *Synthesis*, 896 (1974).

83. Rosenfeld, M. N. and Widdowson, D. A., *J. Chem. Soc. Chem. Commun.*, 914 (1979).

84. Ng, J. S., Katzenellenbogen, J. A., and Kilbourn, M. R., *J. Org. Chem.*, *46*, 2520 (1981).

85. Shani, J., Gazit, A., Livshitz, T., and Biran, S., *J. Med. Chem.*, *28*, 1504 (1985).

86. Olah, G. A. and Ramos, M. T., unpublished results.

87. Kanemoto, S., Shimizu, M., and Yoshioka, H., *Tetrahedron Lett.*, *28*, 663 (1987).

88. Shimizu, M., Cheng, G.-H., and Yoshioka, H., *J. Fluorine Chem.*, *41*, 425 (1988).

89. Kanemoto, S., Shimizu, M., and Yoshioka, H., *Tetrahedron Lett.*, *28*, 6313 (1987).

90. Kanemoto, S., Shimizu, M., and Yoshioka, H., *Bull. Chem. Soc. Jpn.*, *62*, 2024 (1989).

91. Shimizu, M. and Yoshioka, H., *Tetrahedron Lett.*, *28*, 3119 (1987).

92. Karabinos, J. V. and Hazdra, J. J., *J. Org. Chem.*, *27*, 3308 (1962).

93. Olah, G. A. and Meidar, D., *Isr. J. Chem.*, *17*, 148 (1978).

94. Muehlbacher, M. and Poulter, C. D., *J. Org. Chem.*, *53*, 1026 (1988).

95. Alvernhe, G., Laurent, A., and Haufe, G., *J. Fluorine Chem.*, *34*, 147 (1986).

96. Amri, H. and El Gaied, M. M., *J. Fluorine Chem.*, *46*, 75 (1990).

97. Ayi, A. I., Remli, M., Condom, R., and Guedji, R., *J. Fluorine Chem.*, *17*, 565 (1981).

98. Ourari, A., Condom, R., and Guedj, R., *Can. J. Chem.*, *60*, 2707 (1982).

99. Haufe, G., Alvernhe, G., and Laurent, A., *J. Fluorine Chem.*, *46* 83 (1990).

100. Wade, T. N. and Guedj, R., *Tetrahedron Lett.*, 3247 (1987).

101. Wade, T. N., *J. Org. Chem.*, *45*, 5328 (1980).

102. Alvernhe, G., Kozlowska-Gramsz, Z., Lacombe-Bar, S., and Laurent, A., *Tetrahedron Lett.*, 5203 (1978).

103. Alvernhe, G., Lacombe, S., and Laurent, A., *Tetrahedron Lett.*, 289 (1980).

104. Alvernhe, G. M., Ennakoua, C. M., Lacombe, S. M., and Laurent, A. J., *J. Org. Chem.*, *46*, 4938 (1981).

105. Lacombe, S., Laurent, A., and Rousset, C., *Nouv. J. Chim.*, *7*, 219 (1983).

106. Haufe, G., Lacombe, S., Laurent, A., Rousset, C., *Tetrahedron Lett.*, *24*, 5877 (1983).

107. Girault, Y., Rouillard, M., Decouzon, M., and Geribaldi, S., *J. Fluorine Chem.*, *49*, 231 (1990).

108. Girault, Y., Rouillard, M., Decouzon, M., and Azzarro, M., *J. Fluorine Chem.*, *25*, 465 (1984).

109. Girault, Y., Geribaldi, S., Rouillard, M., and Azzaro, M., *J. Fluorine Chem.*, *22*, 253 (1983).

110. Girault, Y., Geribaldi, S., Rouillard, M., and Azzaro, M., *Tetrahedron*, *43*, 2485 (1987).

111. Wade, T. N. and Guedj, R., *Tetrahedron Lett.*, 3953 (1979).
112. Wade, T. N. and Kheribet, R., *J. Org. Chem.*, *45*, 5333 (1980).
113. Alvernhe, G., Lacombe, S., and Laurent, A., *Tetrahedron Lett.*, 1437 (1980).
114. Kollonitsch, J., Marburg, S., and Perkins, L. M., *J. Org. Chem.*, *41*, 3107 (1976).
115. Sondej, S. C. and Katzenellenbogen, J. A., *J. Org. Chem.*, *51*, 3508 (1986).
116. Prakash, G. K. S., Reddy, V. P., Li, X.-Y., and Olah, G. A., *Synlett*, 594 (1990).
117. Rozen, S., Brand, M., Zamir, D., and Hebel, D., *J. Am. Chem. Soc.*, *109*, 896 (1987).
118. Olah, G. A., Wang, Q., Li, X.-Y., and Prakash, G. K. S., unpublished result.
119. Matthews, D. P., Whitten, J. P., and McCarthy, J. R., *Tetrahedron Lett.*, *27*, 4861 (1986).
120. Olah, G. A., Wang, Q., Li, X.-Y., and Prakash, G. K. S., unpublished result.
121. Olah, G. A., Shih, J. G., Singh, B. P., and Gupta, B. G. B., *J. Org. Chem.*, *48*, 3356 (1983).
122. Hashimoto, T., Prakash, G. K. S., Shih, J. G., and Olah, G. A., *J. Org. Chem.*, *52*, 931 (1987).
123. Olah, G. A., Shih, J. G., Singh, B. P., and Gupta, B. G. B., *Synthesis*, 713 (1983).
124. Krishnamurthy, V. V., Shih, J. G., Singh, B. P., and Olah, G. A., *J. Org. Chem.*, *51*, 1354 (1986).
125. Nishimura, M. and Hirai, Y., JP 60260534 (1985), *Chem. Abstr.*, 104:206729.
126. Drew, M. G. B., Mann, J., and Pietrzak, B., *J. Chem. Soc. Perkin Trans.* I, 1049 (1985).
127. Meurs, J. H. H., Sopher, D. W., and Eilenberg, W., *Angew. Chem. Int. Ed. Engl.*, *28*, 927 (1989).
128. Feiring, A. E., *J. Org. Chem.*, *44*, 1252 (1979).
129. Olah, G. A., Wang, Q., Li, X.-Y., and Prakash, G. K. S., unpublished result.
130. Huba, F., Yeager, E. B., and Olah, G. A., *Electrochim. Acta*, *24*, 489 (1979).
131. Laurent, E., Marquet, B., Tardivel, R., and Thiebault, H., *Tetrahedron Lett.*, *28*, 2359 (1987).
132. Laurent, E., Marquet, B., and Tardivel, R., *J. Fluorine Chem.*, *49*, 115 (1990).
133. Brigaud, T. and Laurent, E., *Tetrahedron Lett.*, *31*, 2287 (1990).
134. Makino, K. and Yoshioka, H., *J. Fluorine Chem.*, *39*, 435 (1988).
135. Olah, G. A., Wang, Q., Li, X.-Y., and Prakash, G. K. S., unpublished result.
136. Olah, G. A., Bruce, M. R., and Welch, J., *Inorg. Chem.*, *16*, 2637 (1977).
137. Oda, Y., Otouma, H., Uchida, K., Morikawa, S., and Ikemura, M., US 4372938 (1983), *Chem. Abstr.*, 98, 145996.
138. Oda, Y., et al. GB 2081694 (1982), *Chem. Abstr.*, 96: 183678.
139. Olah, G. A., *Carbocations and Electrophilic Reactions*, Wiley, New York, 1974.
140. Misiura, K., Silverton, J. V., and Stec, W. J., *J. Org. Chem.*, *50*, 1815 (1985).
141. Woyski, M. M., *Inorg. Synth.*, **III**, 111 (1950).
142. Mohamed, K. S., Radma, D. K., Kalbandkeri, B. G., and Murthy, A. R. V., *J. Fluorine Chem.*, *23*, 509 (1983).

CHAPTER 9

ORGANOMETALLICS IN SYNTHETIC ORGANOFLUORINE CHEMISTRY

DONALD J. BURTON

9.1. INTRODUCTION

In the past 20 years there has been a rapid increase in the use of organometallic reagents for regio- and stereoselective control in organic synthesis. In contrast to these rapid advances in natural product chemistry, however, the use of organometallic reagents in organofluorine chemistry has proceeded at a much slower pace. In part, this reduced application of organometallics in organofluorine chemistry is due to the less stable nature of many fluorinated

Synthetic Fluorine Chemistry,
Edited by George A. Olah, Richard D. Chambers, and G. K. Surya Prakash.
ISBN 0-471-54370-5

organometallic reagents. For example, CF_3Li is as yet an unknown reagent due to its rapid α elimination of [:CF_2], even at low temperatures. The corresponding Grignard reagent, CF_3MgX, has been prepared at low temperatures but its rapid extrusion of [:CF_2] has also hampered its utility as a synthetic reagent. Higher homologues, $CF_3(CF_2)_nLi$ and $CF_3(CF_2)_nMgX$, exhibit greater stability than the trifluoromethyl reagents, but even these longer chain analogues must be formed and utilized at low temperatures. Similarly, $F_2C{=}CFLi$ rapidly undergoes β elimination and must therefore be formed and used at temperatures less than $-100\,°C$, thus hampering scale up processes with this reagent. The corresponding magnesium reagent, $F_2C{=}CFMgX$, is more thermally stable but still requires formation and use at low temperature.

The difficulty of dealing with these thermally labile reagents prompted us to investigate the preparation of *thermally stable* fluorinated organometallic reagents, which could be readily utilized in synthetic applications. This chapter details our work in the preparation of *stable* trifluoromethyl and perfluorovinyl organometallic reagents and the utility of these reagents in the regio- and stereospecific preparation of perfluoroalkyl and perfluoroalkenyl compounds.

9.2. TRIFLUOROMETHYL ORGANOMETALLICS

Agricultural and pharmaceutical chemicals that contain a trifluoromethyl group have been of increased research interest in recent years.[1-3] Concomitant with these applications have been renewed efforts to develop a cheaper and more efficient synthetic methodology for the introduction of the trifluoromethyl group into organic compounds. An adjunct of this strategy has been the incorporation of the trifluoromethyl group directly into the molecule via *in situ* generation and coupling of trifluoromethyl copper with aryl halides.[4-15] The reported methods have achieved some modest success; however, these methods have either utilized expensive reagents, required high temperatures or multistep processes, given low yields or low conversions, or have often been plagued with competing Ullmann coupling or reduction of the aryl halide.

In some of the early reports on the generation and capture of CF_3Cu (from a suitable trifluoromethyl precursor) it was noted that preformed [CF_3Cu] underwent coupling with aryl halides at lower temperatures and minimized these side reactions, which plagued the high temperature processes.[8,11] Unfortunately, no experimental data accompanied these pregeneration reports, and it was difficult to ascertain whether the same [CF_3Cu] species had been produced in these quite different pregenerative approaches.

Previous work in our laboratory had utilized dibromodifluoromethane (**1**) or bromodifluoromethyl phosphonium salts (**2**) as useful precursors to difluoromethylene Wittig reagents.[16]

$$\underset{}{R_3P} + \underset{\mathbf{1}}{CF_2Br_2} \longrightarrow \underset{\mathbf{2}}{[R_3\overset{+}{P}CF_2Br]Br^-} \xrightarrow{R_3P} R_3\overset{+}{P}\overset{-}{C}F_2 + R_3PBr_2$$

The phosphonium salt (**2**) could also be employed in a mild nonbasic route to difluorocarbene[17] and this application has been effectively utilized by others in both mechanistic and synthetic work.

$$\mathbf{2} + \text{KF (excess)} + \text{Me}_2\text{C=CMe}_2 \xrightarrow{\text{RT}} \text{(1,1-difluoro-2,2,3,3-tetramethylcyclopropane)} + \text{KBr} + \text{R}_3\text{PFBr}$$

The formation of difluorocarbene can be rationalized via the following scheme:

$$[\text{R}_3\overset{+}{\text{P}}\text{CF}_2\text{Br}]\text{Br}^- + \text{KF} \longrightarrow [\text{R}_3\overset{+}{\text{P}}\text{F}]\text{Br}^- + \text{K}^+[\text{CF}_2\text{Br}]^-$$

$$\text{K}^+[\text{CF}_2\text{Br}]^- \longrightarrow \text{KBr} + [:\text{CF}_2]$$

Alternatively, the formation of $[:CF_2]$ could be via a concerted process. In an attempt to provide mechanistic support for $[CF_2Br]^-$ formation, we attempted to capture this halodifluoromethide ion *in situ* with either phenyl mercuric chloride or chlorotrimethyl tin. Surprisingly, neither $PhHgCF_2Br$ nor Me_3SnCF_2Br were formed! The only products isolated were the trifluoromethyl

$$\mathbf{2} + \text{PhHgCl} + \text{KF (excess)} \longrightarrow \text{PhHgCF}_3 + \text{R}_3\text{PFBr} + \text{KBr} + \text{KCl}$$

$$\mathbf{2} + \text{Me}_3\text{SnCl} + \text{KF (excess)} \longrightarrow \text{Me}_3\text{SnCF}_3 + \text{KCl} + \text{R}_3\text{PFBr} + \text{KBr}$$

mercurial and stannane.[18] Mechanistically, the formation of the CF_3 analogue can be rationalized via the following scheme:

$$\mathbf{2} + \text{KF} \longrightarrow [\text{R}_3\overset{+}{\text{P}}\text{F}]\text{Br}^- + \text{K}^+[\text{CF}_2\text{Br}]^-$$

$$\text{K}^+[\text{CF}_2\text{Br}]^- \rightleftharpoons [:\text{CF}_2] + \text{KBr}$$

$$[:\text{CF}_2] + \text{KF} \rightleftharpoons \text{K}^+[\text{CF}_3]^-$$

$$\text{K}^+[\text{CF}_3]^- + \text{PhHgCl} \longrightarrow \text{PhHgCF}_3 + \text{KCl}$$

$$\underline{\text{or}}\quad \text{K}^+[\text{CF}_3]^- + \text{Me}_3\text{SnCl} \longrightarrow \text{Me}_3\text{SnCF}_3 + \text{KCl}$$

This work suggested that trifluoromethide ion could be captured *in situ* by metal salts to give the trifluoromethylated organometallic compound. Salt **2** merely served as the source of $[:CF_2]$ and presumably other sources of $[:CF_2]$ could be employed (in conjunction with fluoride ion) to produce the trifluoromethide ion. Thus, our attention turned to a more direct route to $[:CF_2]$; namely, dihalodifluoromethanes (precursor to **2**). Our goal was to strip the halogens (other

$$CF_2XY + M^\circ \text{ (metal)} \longrightarrow MXY + [:CF_2]$$

X = Br, Cl

Y = Br,Cl

than fluorine) from the dihalodifluoromethane with a metal (M^0) to produce the carbene and the necessary metal salt (MXY) *in situ*. If fluoride ion could be produced *in situ* or added as a metal fluoride, then perhaps the carbene would be converted *in situ* to $[CF_3]^-$, which could subsequently be captured by the salt, MXY, formed in the carbene generation step.

$$F^- + [:CF_2] + MXY \longrightarrow CF_3MX \text{ and/or } (CF_3)_2M$$

9.2.1. Trifluoromethyl Cadmium and Zinc Reagents

When CF_2Br_2 and CF_2BrCl are reacted at room temperature with acid-washed cadmium or zinc powder in dry *N*,*N*-dimethylformamide (DMF), an exothermic reaction occurs to produce the trifluoromethyl cadmium and zinc reagents in high yield.[19]

$$2\,M + 2\,CF_2X_2 \longrightarrow CF_3MX\ [\text{or } 1/2\,(CF_3)_2M + 1/2\,MX_2] + MX_2 + CO + [Me_2N{=}CFH]^+X^-$$

X = Br or Cl

M = Cd, 80-95%

M = Zn, 80-85%

On a large scale, sequential addition of the methane (CF_2Br_2 or CF_2BrCl) allows the zinc reagent to be routinely made in 0.5-mol preparations and the cadmium reagent in 0.1–0.2-mol preparations. In the absence of moisture, the CF_3CdX and CF_3ZnX reagents can be stored indefinitely without decomposition. When CF_2Cl_2 is used in this preparation, the reaction is carried out at 80°C in a sealed ampoule.

One perhaps might anticipate that a CF_2XMY organometallic had been formed via oxidative addition of the methane to the metal. However, both the cadmium and zinc reagents are readily hydrolyzed to CF_3H or CF_3D. The cadmium reagent can also be identified via its ^{113}Cd NMR spectrum. The CF_3CdX reagent exhibits the expected quartet at 459 ppm and the $(CF_3)_2Cd$ reagent the expected heptet at 410 ppm, respectively, in good agreement with reported values. The observed multiplicities are those expected for CF_3 groups attached to cadmium and *not* the multiplicities expected for one or two CF_2X

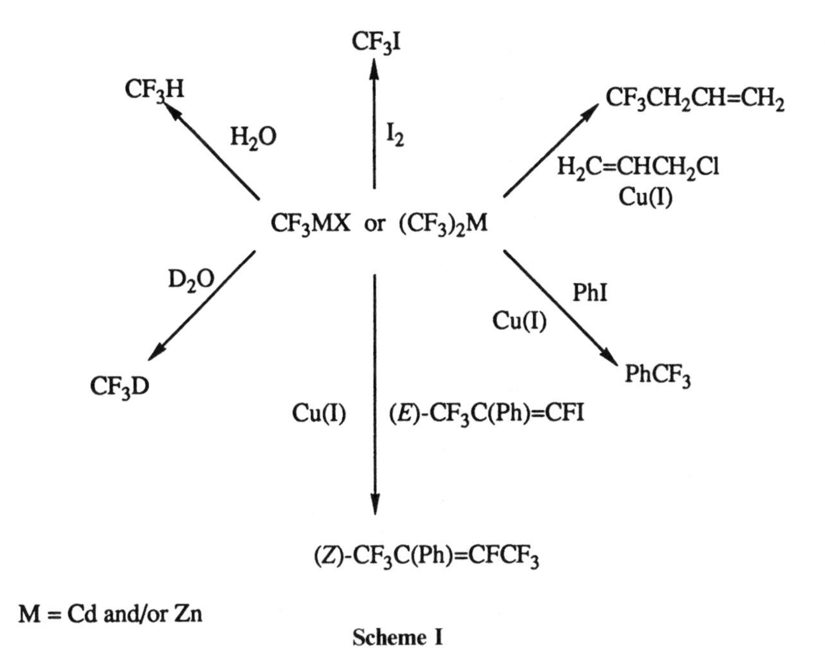

Scheme I

(X = halogen) groups bonded to cadmium (triplet and pentet, respectively). In addition subsequent chemical transformations are in agreement with the formation of the CF_3 group as outlined in Scheme **I**. Consequently, an unequivocal, novel, remarkable transformation of the difluoromethylene carbon had occurred. It is not intuitively obvious how this unique conversion takes place and how the additional fluoride required to convert CF_2 to CF_3 is produced. We have investigated some of the mechanistic features of this transformation and the mechanism consistent with our data is outlined in Scheme **II**.

Electron transfer from the metal to the dihalodifluoromethane to form the radical anion is consistent with our observation that the reaction between CF_2Br_2 and cadmium is totally quenched when the reaction is carried out in the presence of *p*-dinitrobenzene, a known radical–anion inhibitor. Evidence for carbene formation is consistent with the detection of 60% 1,1-difluoro-2,2,3,3-tetramethylcyclopropane (via ^{19}F NMR) when the reaction between CF_2Br_2 and cadmium is carried out in the presence of tetramethylethylene. The proposal for Me_2NCHF_2 formation and this amine as the source of fluoride ion is consistent with two experimental observations. First, Me_2NCHF_2 was detected spectroscopically via ^{19}F NMR analysis of the reaction mixture and enhancement of the NMR signal attributed to Me_2NCHF_2 on addition of an authentic sample of this amine to the reaction mixture. Second, the exit gas that is continuously produced during the course of the reaction was shown by FTIR to be identical to the IR spectrum of carbon monoxide. In addition, the volume of

Part 1: Carbene and Metal Halide Formation:

$$CF_2XY + M \longrightarrow [CF_2XY]^{\cdot -} + M^+$$

$$[CF_2XY]^{\cdot -} \longrightarrow Y^- + [CF_2X]^{\bullet}$$

$$M^+ + [CF_2X]^{\bullet} \longrightarrow [CF_2X]^- + M^{2+}$$

$$[CF_2X]^- \longrightarrow [:CF_2] + X^-$$

Part 2: Fluoride Ion Formation:

$$CF_2XY + M \longrightarrow [:CF_2] + MXY$$

$$[:CF_2] + Me_2NCH{=}O \longrightarrow CO + Me_2NCHF_2$$

$$Me_2NCHF_2 \rightleftharpoons [Me_2N{=}CHF]^+F^-$$

Part 3: Formation of Trifluoromethyl Organometallic

$$[:CF_2] + [Me_2N{=}CHF]^+F^- \longrightarrow [CF_3]^-[Me_2N{=}CHF]^+$$

$$[CF_3]^- + MXY \longrightarrow CF_3MX \text{ and/or } (CF_3)_2M + Y^-$$

Overall Reaction:

$$2\ CF_2XY + 2\ M \xrightarrow{DMF} CF_3MX + (CF_3)_2M + CO + MXY + [Me_2N{=}CHF]^+X^-$$

Scheme II

CO formed can be equated with the amount of trifluoromethyl groups obtained in the organometallic reagent. It should be noted that Me_2NCHF_2 has been demonstrated to be an excellent soluble source of fluoride ion.[20] Finally, independent evidence that $[CF_3^-]$ can displace halide from metal salts was obtained via the reaction of **2** with KF and cadmium halide, which produced CF_3CdX. Thus, the key features in this mechanism; namely, radical–anion formation, carbene formation, fluoride ion source, and displacement of halide ion from metal halides by trifluoromethide ion have experimental support, and this methodology is applicable to the preparation of many trifluoromethyl organometallics stable under these reaction conditions.

9.2.2. Preparation and Applications of Trifluoromethyl Copper

Metathesis of trifluoromethyl cadmium (generated as outlined in **I**) with Cu(I) salts gives a solution of trifluoromethyl copper.[21] Since the requisite cadmium reagent can be produced *in situ* from dihalodifluoromethanes, the overall process constitutes a simple, one-pot preparation of trifluoromethyl copper from cheap Freon precursors. The exchange reaction with $[CF_3CdX]$ is rapid even at low temperatures (-30 to $-50\,^\circ C$), whereas the analogous metathesis of the corresponding $[CF_3ZnX]$ is slow even at room temperature.

$$2\,Cd + CF_2XY \xrightarrow{DMF} [CF_3CdX + (CF_3)_2Cd] \xrightarrow[Y\,=\,Cl,\ Br,\ I,\ CN]{CuY/\text{-}50^\circ C\ \text{to RT}} [CF_3Cu]$$

X = Br, Cl 80-95% 90-100%

Y = Br, Cl

The metathesis process of $[CF_3CdX]$ with Cu(I) salts gave an intriguing and unexpected observation. We had anticipated a singlet for $[CF_3Cu]$ in the ^{19}F NMR spectrum for this reagent. However, when $[CF_3CdX]$ was exchanged with CuBr, the ^{19}F NMR spectrum of the resultant solution exhibited three signals. When the metathesis reaction was carried out carefully under degassed conditions, only the ^{19}F NMR signals at -28 and -32 ppm, respectively, were observed, indicating that CF_3Cu [C] is an oxidation product. Recent work in our laboratory has unequivocally demonstrated that CF_3Cu [C] is a *stable* Cu(III) reagent, $[(CF_3)_4Cu^-]$, and the crystal structure of an analogue of this reagent has been reported.[22] Preliminary work has also indicated that CF_3Cu[A] is $CF_3Cu\cdot L$ and CF_3Cu[B] is $[(CF_3)_2Cu^-]Cd^+X$. Thus, not only does the metathesis of the CF_3CdX reagent provide a synthetic route to a preformed solution of $[CF_3Cu]$, but this work has demonstrated that these copper reagents are not the simplistic entities as previously proposed. In addition, this route has demonstrated the first example of a stable perfluoroalkyl Cu(III) complex.

$$CF_3CdX + CuBr \xrightarrow{DMF} CF_3Cu\,[A]\ ; \quad CF_3Cu\,[B]\ ; \quad CF_3Cu\,[C]$$

$\delta = -28$ ppm $\quad \delta = -32$ ppm $\quad \delta = -35$ ppm

When $[CF_3Cu]$ is formed via the metathesis reaction noted above and the solution of this reagent is allowed to stand at room temperature (RT) overnight, the trifluoromethyl copper is converted to pentafluoroethyl copper. Heating

$$CF_3Cu \xrightarrow[DMF,\ 24\,h]{RT} CF_3CF_2Cu$$

(RT–50 °C) accelerates this chain extension. However, if hexamethylphosphoramide (HMPA) is added to a solution of $CF_3Cu[A]$ the chain extension to CF_3CF_2Cu is suppressed and reactions of this reagent can be accomplished at higher temperatures without significant decomposition or chain extension of the $[CF_3Cu]$.[21] Table 9.1 summarizes several typical examples of the trifluoromethylation of aromatic iodides.

$$CF_3Cu\,[A] + Y\text{-}C_6H_4\text{-}I \xrightarrow[\text{HMPA, } 70^\circ C]{\text{DMF}} Y\text{-}C_6H_4\text{-}CF_3 + CuI$$

>70%

Heterocyclic compounds also are trifluoromethylated well and multiple trifluoromethylations are readily accomplished with polyiodo aryl iodides. Thus, the pregeneration of CF_3Cu, stabilization of this reagent with HMPA, and trifluoromethylation of aryl iodides with this stable reagent can be easily accomplished and provides a facile, straightforward, mild, controlled route to trifluoromethylated aromatic and heterocyclic compounds.

TABLE 9.1. Trifluoromethylation of Aromatic Halides.

$$CF_3Cu + Y\text{-}C_6H_4\text{-}X \xrightarrow[\text{HMPA, } 70^\circ C]{\text{DMF}} Y\text{-}C_6H_4\text{-}CF_3 + CuX$$

Y	X	Position	Isolated Yield (%)
H	Br		0
H	I		(100)[a]
NO_2	I	*o*	75
NO_2	I	*p*	75
Cl	I	*o*	84
CH_3	I	*m*	72
CH_3	I	*p*	71
OCH_3	I	*p*	(78)[a]

[a] ^{19}F NMR yield vs. $PhCF_3$ standard

$$\text{2-iodothiophene} + CF_3Cu \xrightarrow[\text{HMPA, 70°C}]{\text{DMF}} \text{2-(trifluoromethyl)thiophene} \quad 60\%$$

$$\text{tetraiodothiophene} + 4\,CF_3Cu \xrightarrow[\text{HMPA, 70°C}]{\text{DMF}} \text{tetrakis(trifluoromethyl)thiophene} \quad 70\%$$

The success achieved in the trifluoromethylation of aryl iodides prompted us to examine the coupling of this reagent with vinyl iodides. The following examples demonstrate the applicability of this reagent for the stereospecific introduction of a trifluoromethyl group onto a vinylic carbon.[23]

$$C_6F_5CH{=}CFI + CF_3Cu \xrightarrow[\text{-50°C to RT, 2-3 h}]{\text{DMF}} C_6F_5CH{=}CFCF_3,\ 67\%$$

$$(E/Z) = 69/31 \qquad\qquad (E/Z) = 30/70$$

$$C_6H_5C(CF_3){=}CFI + CF_3Cu \xrightarrow[\text{-50°C to RT, 2-3 h}]{\text{DMF}} C_6H_5C(CF_3){=}CFCF_3,\ 72\%$$

$$(E/Z) = 89/11 \qquad\qquad (E/Z) = 10/90$$

$$(Z)\text{-}CF_3CF_2CF_2CF{=}CFI + CF_3Cu \xrightarrow[\text{-50°C to RT, 2-3 h}]{\text{DMF}} (E)\text{-}CF_3CF_2CF_2CF{=}CFCF_3,\ 52\%$$

$$100\%\ (E)$$

With vinylic iodides the coupling reaction is essentially complete by the time the reaction mixture is warmed to room temperature. Consequently, chain extension of CF_3Cu is not a problem and HMPA is not required in these trifluoromethyl vinylation reactions.

Another application of $[CF_3Cu]$ that proved successful was the regiospecific trifluoromethylation of allylic halides. Table 9.2 summarizes these results.

Trifluoromethylation occurs only at the least hindered carbon of the allyl functionality and only one isomer was obtained in the cases when either α or γ attack could have occurred. Again, the trifluoromethylation of allyl halides is rapid and no HMPA is required in the allylation reactions.

The success achieved in the allylation reactions of trifluoromethyl copper prompted us to reexamine the utility of perfluoroalkyl copper reagents in an

$$CF_3CdX + (CH_3)_2C(Cl)CH{=}CH_2 \xrightarrow[\text{-50°C to RT/2-3 h}]{\text{CuBr, DMF; 91\%}} (CH_3)_2C{=}CHCH_2CF_3$$

$$CF_3CdX + (CH_3)_2C{=}CHCH_2Cl \xrightarrow[\text{-50°C to RT/2-3 h}]{\text{CuBr, DMF; 83\%}} (CH_3)_2C{=}CHCH_2CF_3$$

TABLE 9.2. Trifluoromethylation of Allyl Halides.

$$CF_3Cu + RC{=}C{-}C{-}X \longrightarrow CF_3{-}C(R){-}C{=}C \text{ or } RC{=}C{-}C{-}CF_3$$

Allyl Halide	Product	Isolated Yield (%)
$H_2C{=}CHCH_2Br$	$CF_3CH_2CH{=}CH_2$	70
$H_2C{=}C(CH_3)CH_2Cl$	$CF_3CH_2C(CH_3){=}CH_2$	61
$CH_3CH{=}CHCH_2Cl$	$CF_3CH_2CH{=}CHCH_3$	53
$CH_3CH(Cl)CH{=}CH_2$	$CF_3CH_2CH{=}CHCH_3$	62
$(CH_3)_2C{=}CHCH_2Cl$	$CF_3CH_2CH{=}C(CH_3)_2$	83
$(CH_3)_2C(Cl)CH{=}CH_2$	$CF_3CH_2CH{=}C(CH_3)_2$	91
$PhCH{=}CHCH_2Br$	$CF_3CH_2CH{=}CHPh$	68
$CH_3O_2CCH{=}CHCH_2Br$	$CF_3CH_2CH{=}CHCO_2CH_3$	57
3-Bromocyclohexene (Br on ring)	3-(Trifluoromethyl)cyclohexene (CF_3 on ring)	49

S_N2' reaction with propargyl halides. Coe and Milner[24] reported that perfluoroheptyl copper reacted with propargyl bromide to give <10% of the desired propadiene and that experimental difficulties were encountered with this reaction.

$$C_7F_{15}Cu + HC{\equiv}CCH_2Br \xrightarrow{DMSO} C_7F_{15}CH{=}C{=}CH_2 \quad (<10\%)$$

However, with the analogous propargyl chloride or tosylate we have found that trifluoromethyl copper reacts readily at 0°C–RT to give the corresponding allene *without* any safety problems.[25] Subsequently, we extended this study to other perfluoroalkyl copper reagents with related propargyl chlorides or tosylates and these results are summarized in Table 9.3. In addition, Hung

TABLE 9.3. Preparation of Perfluoroalkyl Substituted Allenes.

$$R_fCu + HC{\equiv}CCR^1R^2X \xrightarrow[\text{0°C-RT, overnight}]{\text{solvent}} R_fCH{=}C{=}CR^1R^2$$

R_f	R^1	R^2	X	Solvent	Isolated Yield (%)
CF_3	H	H	OTs	DMF	68
CF_3	H	CH_3	OTs	DMF	65
CF_3	CH_3	CH_3	Cl	DMF	73
n-C_3F_7	H	H	OTs	DMSO	41
n-C_3F_7	H	CH_3	OTs	DMSO	67
n-C_3F_7	CH_3	CH_3	Cl	DMSO	68
n-C_6F_{13}	H	H	OTs	DMSO	49
n-C_6F_{13}	H	CH_3	OTs	DMSO	58
n-C_6F_{13}	CH_3	CH_3	Cl	DMSO	66
n-C_8F_{17}	H	H	OTs	DMSO	30
n-C_8F_{17}	H	CH_3	OTs	DMSO	55
n-C_8F_{17}	CH_3	CH_3	Cl	DMSO	60

$$CF_3Cu + HC{\equiv}CCH_2OTs \xrightarrow[0° - RT]{DMF} CF_3CH{=}C{=}CH_2 \quad 68\%$$

$$CF_3Cu + HC{\equiv}CC(R^1R^2)Cl \xrightarrow[0° - RT]{DMF} CF_3CH{=}C{=}CR^1R^2 \quad 73\%$$

extended this methodology to bis(copper) reagents as a preparative route to bis(allenes).[26]

Functionalized trifluoromethyl substituted allenes can also be prepared by application of this methodology to the appropriately substituted propargyl halide.

$$CF_3Cu + EtO_2CC{\equiv}CC(CH_3)_2Cl \xrightarrow[-20°C \text{ to RT}]{DMF} (CF_3)(EtO_2C)C{=}C{=}C(CH_3)_2$$

$$CF_3Cu + Cl(CH_3)_2CC{\equiv}CSi(CH_3)_3 \xrightarrow[-78°C \text{ to RT}]{DMF} (CF_3)((CH_3)_3Si)C{=}C{=}C(CH_3)_2$$

9.2.3. Conclusions

The reaction of dihalodifluoromethanes with cadmium or zinc powders provides a novel transformation of a simple, cheap, commercially available fluorinated precursor *directly* to the thermally stable trifluoromethyl cadmium and zinc reagents. Metathesis of the trifluoromethyl cadmium reagent with Cu(I) salts gives a stable solution of the trifluoromethyl copper reagent, which for the first time is observable spectroscopically. This [CF_3Cu] reagent can be utilized in the synthesis of trifluoromethyl aromatic and heterocyclic compounds, in the regio- and stereospecific preparation of trifluoromethyl-containing alkenes, in the synthesis of trifluoromethyl allenes and in the regiospecific trifluoromethylation of substituted allyl halides. It offers much more flexibility in its synthetic applicability compared to the high temperature *in situ* approaches reported previously, since the trapping agent for the [CF_3Cu] need not be present in the generation step of the reagent and competitive coupling or reduction of the trapping agent can be avoided. The mechanistic studies in the generation of these reagents from the dihalodifluoromethanes has suggested a wide scope for the preparation of trifluoromethyl organometallics and a recent report has demonstrated the applicability of our mechanistic ideas with an alternative precursor for the formation of difluorocarbene.[27] Undoubtedly, other precursors will be investigated for the generation of $[CF_3Cu]$ but the overall mechanistic concepts (carbene formation, capture of carbene to form trifluoromethide ion, and attack of trifluoromethide ion on the metal halide) formulated from our work will presage these approaches.

9.3. PERFLUOROALKENYL ORGANOMETALLICS

Although the preparation of poly(fluoroalkenyl)lithium and Grignard reagents has been described in the literature, the application of these reagents in synthesis has been impeded by the low thermal stability of the organometallic compounds. The preparations must be carried out at low temperatures and serious difficulties often arise when scale up processes are attempted. In addition, only reactive substrates can be employed, since it is not possible to force sluggish substrates via the use of higher temperatures.

Therefore, our initial attention concentrated on a general route to thermally *stable* perfluoroalkenyl organometallic compounds that satisfy the following criteria: (1) the reagents must be thermally stable at room temperature or above, so that scale up procedures can be readily accomplished; (2) the stereochemical integrity with (*E*)/(*Z*) precursors must be retained in the formation of the organometallic reagent to ensure stereochemical control throughout subsequent functionalization of these reagents; (3) these reagents must be capable of functionalization or exchange (metathesis) processes so that they can be utilized in the synthesis of polyfunctionalized compounds; and (4) these reagents should be capable of synthesis in a one-step (or one-pot) procedure from accessible precursors.

The utility of nonfluorinated alkenyl copper or alkenyl cuprates for regio- and stereochemical control in synthesis is well known. A notable omission, however, from the list of known copper species was the presence of fluorinated alkenyl copper reagents. At the 9th International Symposium on Fluorine Chemistry, Miller reported the first preparation of a fluorinated vinyl copper reagent via metathesis of the corresponding perfluoroalkenyl silver reagent. This vinyl copper reagent exhibited thermal stability, thus indicating that perfluoroalkenyl copper reagents might satisfy our criteria outlined above. However, Miller's approach presented some serious difficulties as a general approach to these reagents; (1) it required a perfluoroalkyne as a precursor; (2) with unsymmetrical perfluoroalkynes, a regiospecificity problem was introduced; and (3) only the (*E*) isomer was obtained via this approach.

$$CF_3C{\equiv}CCF_3 + AgF \longrightarrow (E)\text{-}CF_3CF{=}C(CF_3)Ag \xrightarrow{Cu^{\circ}} (E)\text{-}CF_3CF{=}C(CF_3)Cu$$

Therefore, we decided to utilize a strategy that employed vinylic halides as precursors with the expectation that the formation of the organometallic from the vinyl halide precursor would proceed with stereochemical control. Thus, by proper choice of the appropriate (*E*) or (*Z*) precursor, both regio- and stereochemical selectivity would be achieved.

Our initial studies involved the direct formation of a perfluorinated alkenyl copper reagent via oxidative addition of a perfluorovinyl iodide and copper metal. However, only perfluoro-1,3-butadiene was obtained and the expected vinyl copper reagent could not be detected. Presumably, the slow step in the reaction is the formation of the copper reagent and the subsequent coupling of the vinyl copper reagent with the perfluorovinyl iodide is the fast step.

$$2\,CF_2{=}CFI + Cu^{\circ} \longrightarrow F_2C{=}CFCF{=}CF_2 + 2\,CuI$$

Therefore, if one wishes to pregenerate the perfluorovinyl copper reagent, it is necessary to form this reagent in the *absence* of the vinyl halide in order to preclude any coupling processes.

$$F_2C{=}CFI + 2\,Cu^{\circ} \xrightarrow{\text{slow}} F_2C{=}CFCu + CuI$$

$$F_2C{=}CFCu \xrightarrow[\text{fast}]{F_2C{=}CFI} F_2C{=}CFCF{=}CF_2 + CuI$$

9.3.1. Perfluorovinyl Cadmium and Zinc Reagents

In order to achieve the formation of the perfluorovinyl copper reagent in the absence of perfluorovinyl halide, a metathesis approach was utilized. For this approach to be successful, it was required that (1) the formation of a stable perfluoroalkenyl organometallic that could be exchanged with Cu(I) salts be achieved; and (2) that this alternative stable perfluoroalkenyl organometallic precursor be capable of formation from a perfluorovinyl halide *without* dimerization of the vinyl halide to the symmetrical diene.

Subsequently, we discovered that perfluorovinyl cadmium and zinc reagents could be readily formed from polyfluorovinyl iodides and bromides.[28,29] The formation of the cadmium and zinc reagents proceeded with total stereochemical control and *no* symmetrical diene was observed.

$$R_fCF{=}CFX + M \xrightarrow{\text{RT-60°C}} R_fCF{=}CFMX + (R_fCF{=}CF)_2M$$

X = I, Br M = Zn, Cd

R_f = F, perfluoroalkyl, perfluoroalkenyl

In general, the vinyl iodides reacted at room temperatures; the vinyl bromides required mild heating (RT–60 °C). The cadmium reagents were best formed in

TABLE 9.4. Preparation of Perfluoroalkenyl Cadmium and Zinc Reagents.

$$R_fCF{=}CFX + M \xrightarrow{RT-60°C} R_fCF{=}CFMX + (R_fCF{=}CF)_2M$$

Alkene	Solvent	Organometallic Reagent[a]	Yield (%)[b]
$F_2C{=}CFI$	DMF	$F_2C{=}CFCdX$	99
(*E*)- $CF_3CF{=}CFI$	DMF	(*E*)- $CF_3CF{=}CFCdX$	92
(*Z*)- $CF_3CF{=}CFI$	DMF	(*Z*)- $CF_3CF{=}CFCdX$	96
(*Z*)-$CF_3(CF_2)_4CF{=}CFI$	DMF	(*Z*)- $CF_3(CF_2)_4CF{=}CFCdX$	95
$CF_3CF{=}CICF_3$[c]	DMF	$CF_3CF{=}C(CF_3)CdX$[d]	91
(*E*)- $PhC(CF_3){=}CFI$	DMF	(*E*)- $PhC(CF_3){=}CFCdX$	77
$CF_3CF{=}C(Ph)CF{=}CFBr$[e]	DMF	$CF_3CF{=}C(Ph)CF{=}CFCDX$[e]	61
$F_2C{=}CFI$	DMAC[f]	$F_2C{=}CFZnX$	97
$F_2C{=}CFBr$	DMF	$F_2C{=}CFZnX$	72
(*Z*)-$CF_3CF{=}CFI$	THF	(*Z*)- $CF_3CF{=}CFZnX$	98
(*E*)- $CF_3CF{=}CFI$	TG[g]	(*E*)- $CF_3CF{=}CFZnX$	100
(*Z*) -$CF_3CF_2CF{=}CFI$	TG	(*Z*)- $CF_3CF_2CF{=}CFZnX$	90
(*Z*)- $CF_3(CF_2)_4CF{=}CFBr$	DMF	(*Z*)- $CF_3(CF_2)CF{=}CFZnX$	77
$CF_3C(Ph){=}CFBr$[h]	DMF	$CF_3C(Ph){=}CFZnX$[h]	94
(*Z*)-$CF_3(C_6F_5)C{=}CFI$	THF	(*Z*)- $CF_3(C_6F_5)C{=}CFZnX$	86
(*E*)- $CF_3CH{=}CFI$	TG	(*E*)- $CF_3CH{=}CFZnX$	89
$CF_3CF{=}C(Ph)CF{=}CFBr$[i]	DMF	$CF_3CF{=}C(Ph)CF{=}CFZnX$[i]	71

[a] Mixture of mono and bis reagent
[b] ^{19}F NMR Yield vs. $PhCF_3$
[c] (*E*/*Z*) = 39:61
[d] (*E*/*Z*) = 37:63
[e] (*E*,*E*:*Z*,*Z*) = 90:10
[f] DMAC is N,N-Dimethylacetamide
[g] TG is Triglyme
[h] (*E*/*Z*) = 59:41
[i] (*E*,*Z*:*Z*,*Z*) = 90:10

DMF or related amide solvents, whereas the corresponding zinc reagents were readily formed in a variety of solvents. The zinc and cadmium reagents can be stored indefinitely at 0 °C–RT in the absence of moisture, which hydrolyzes both the zinc and cadmium reagents to $R_fCF{=}CFH$. Table 9.4 summarizes the stereospecific preparation of perfluorovinyl cadmium and zinc reagents directly from the perfluorovinyl halide.

The perfluoroalkenyl cadmium and zinc reagents exhibit remarkable thermal stability. Even at temperatures of 100 °C, these reagents only slowly lose activity. Therefore, reaction mixtures with these reagents can be heated without significant loss or decomposition of the perfluorovinyl cadmium or zinc reagents.

9.3.2. Perfluorovinyl Copper Reagents

The perfluorovinyl copper reagents could be stereospecifically prepared via metathesis of either the perfluorovinyl cadmium or zinc reagents with soluble Cu(I) salts.[30] Consequently, the overall transformation is a one-pot stereospecific synthesis of the perfluoroalkenyl copper reagent from the perfluorovinyl halide. For example:

$$R_f(F)C{=}C(F)X + M \xrightarrow{DMF} [R_f(F)C{=}C(F)MX] \xrightarrow[Y = Cl, Br, I]{CuY \text{ at RT}} R_f(F)C{=}C(F)Cu + MXY$$

X = Br, I M = Zn or Cd

Table 9.5 summarizes the preparation of perfluoroalkenyl copper reagents.

9.3.3. Applications of Perfluoroalkenyl Zinc and Copper Reagents

9.3.3.A. Palladium-Catalyzed Coupling of Perfluorovinyl Zinc Reagents

9.3.3.A.1. With Aryl Iodides. Literature reports on the preparation of α,β,β-trifluorostyrenes have either given low yields or have utilized multistep procedures. However, the perfluorovinyl zinc reagents readily react with aryl

TABLE 9.5. Preparation of Perfluoroalkenyl Copper Reagents.

$$R_fCF{=}CFX + M \longrightarrow R_fCF{=}CFMX \xrightarrow[RT]{CuBr,\ DMF} R_fCF{=}CFCu$$

Alkene	M	$R_fCF{=}CFCu$	Yield (%)[a]
$CF_2{=}CFBr$	Zn	$F_2C{=}CFCu$	100
$CF_2{=}CFI$	Cd	$F_2C{=}CFCu$	100
(*Z*)-$CF_3CF{=}CFI$	Cd	(*Z*)-$CF_3CF{=}CFCu$	90
(*Z*)-$CF_3CF{=}CFI$	Zn	(*Z*)-$CF_3CF{=}CFCu$	78
(*Z*)-$CF_3CCl{=}CFI$	Cd	(*Z*)-$CF_3CCl{=}CFCu$	78
(*Z*)-$CF_3(CF_2)_4CF{=}CFI$	Cd	(*Z*)-$CF_3(CF_2)_4CF{=}CFCu$	92
(*E*)-$CF_3CF{=}CFI$	Cd	(*E*)-$CF_3CF{=}CFCu$	96

[a] Yield of the copper reagent is based on the starting cadmium or zinc reagent; determined by ^{19}F NMR integration via internal $C_6H_5CF_3$ standard.

iodides in the presence of 1–2 mol % of a Pd^0 catalyst to give high yields of these styrenes under mild reaction conditions.[31,32]

$$F_2C{=}CFX + Zn \xrightarrow[THF\ or\ TG]{DMF} F_2C{=}CFZnX \xrightarrow[(Ph_3P)_4Pd,\ 60\text{-}80°C,\ 1\text{-}8\ h]{ArI} F_2C{=}CFAr$$

X = Br,I

The corresponding cadmium reagent will also undergo Pd^0 catalyzed coupling with aryl iodides, but the cadmium reagent offers no advantage. The zinc reagent avoids the formation and disposal of toxic cadmium salts and is the reagent of choice in these Pd^0 coupling reactions.

Similarly, the (*E*)- and (*Z*)-perfluoropropenyl zinc reagents[32] undergo stereoselective Pd^0 coupling with aryl halides and provide a superior entry into these 1-substituted propenes in comparison to the classical Grignard approach.[33] A variety of substituents are tolerated in the aryl iodide. Table 9.6 summarizes typical examples of the Pd^0 coupling of substituted aryl iodides with perfluoroalkenyl zinc reagents.

$$CF_3(F)C{=}C(F)ZnX + ArI \xrightarrow[60\text{-}80°C]{(Ph_3P)_4Pd} CF_3(F)C{=}C(F)Ar$$

TABLE 9.6. Pd⁰-Catalyzed Coupling of Perfluoroalkenyl Zinc Reagents with Aryl Iodides.

$$R_fCF{=}CFZnX + Y{-}Ar{-}I \xrightarrow[60-80^\circ C]{(Ph_3P)_4Pd} R_fCF{=}CFAr{-}Y$$

R_f	Stereochemistry	Y	Solvent	Product	Yield (%)
F		H	DMF	$C_6H_5CF{=}CF_2$	74
F		*o* - NO_2	DMF	*o* - $NO_2C_6H_4CF{=}CF_2$	73
F		*p* - MeO	THF	*p* - $MeOC_6H_4CF{=}CF_2$	61
F		*o* - CF_3	TG	*o* - $CF_3C_6H_4CF{=}CF_2$	73
F		*m* -NO_2	TG	*m* -$NO_2C_6H_4CF{=}CF_2$	81
F		*p* - Cl	DMF	*p* - $ClC_6H_4CF{=}CF_2$	77
CF_3	(*Z*)	H	TG	(*E*) - $C_6H_5CF{=}CFCF_3$	80
CF_3	(*Z*)	*p* -CH_3	TG	(*E*)-*p* - $CH_3C_6H_4CF{=}CFCF_3$	65
CF_3	(*Z*)	*p* -Cl	TG	(*E*)-*p* - $ClC_6H_4CF{=}CFCF_3$	61
CF_3	(*Z*)	*m*-NO_2	TG	(*E*)-*m* -$NO_2C_6H_4CF{=}CFCF_3$	80
CF_3	(*E*)	H	TG	(*Z*)-$C_6H_5CF{=}CFCF_3$	82[a]
CF_3	(*E*)	*p* -CH_3	TG	(*Z*)-*p* - $CH_3C_6H_4CF{=}CFCF_3$	70[b]
CF_3	(*E*)	*p* -Cl	TG	(*Z*)-*p* - $ClC_6H_4CF{=}CFCF_3$	74[a]

[a] (*Z*/*E*) = 97:3

[b] (*Z*/*E*) = 92:8

$$(CF_3)(F)C{=}C(F)(ZnX) + ArI \xrightarrow[60-80^\circ C]{(Ph_3P)_4Pd} (CF_3)(F)C{=}C(F)(Ar)$$

9.3.3.A.2. With Perfluorovinyl Iodides. The perfluoroalkenyl zinc reagents also couple readily with iodotrifluoroethylene in the presence of Pd^0 and provided the first stereospecific route to both (*E*)- and (*Z*)-perfluoro-1,3-pentadiene.[34] Additional work in our laboratory has demonstrated this methodology to be a general route to the stereospecific preparation of perfluoro dienes.

TABLE 9.7. Preparation of Perfluoroalkenyl Ketones.

$$R_fCF{=}CFZnX + RC(O)Cl \xrightarrow[\text{RT, glyme}]{\text{CuBr}} R_fCF{=}CFC(O)R$$

R_f	R	$R_fCF{=}CFC(O)R$	Yield Isolated (%)
F	CH_3	$F_2C{=}CFC(O)CH_3$	76
F	CH_2CH_3	$F_2C{=}CFC(O)CH_2CH_3$	83
F	$CH(CH_3)_2$	$F_2C{=}CFC(O)CH(CH_3)_2$	87
F	$C(CH_3)_3$	$F_2C{=}CFC(O)C(CH_3)_3$	81
F	$(CH_2)_4CH_3$	$F_2C{=}CFC(O)(CH_2)_4CH_3$	67
F	$C(O)OCH_2CH_3$	$F_2C{=}CFC(O)C(O)OCH_2CH_3$	50
F	CH_2Cl	$F_2C{=}CFC(O)CH_2Cl$	27
F	$(CH_2)_2C(O)OCH_3$	$F_2C{=}CFC(O)(CH_2)_2C(O)OCH_3$	44
CF_3[a]	CF_3[b]	$CF_3CF{=}CFC(O)CF_3$[c]	71
CF_3[a]	CF_2Cl[d]	$CF_3CF{=}CFC(O)CF_2Cl$[c]	77
CF_3[a]	CH_2CH_3[e]	$CF_3CF{=}CFC(O)CH_2CH_3$[c]	73
CF_3[f]	CH_3[g]	$CF_3CF{=}CFC(O)CH_3$[h]	69

[a] (*Z*) - Zinc reagent
[b] PhCN at -15°C
[c] 100% (*E*)-Isomer
[d] PhCN/0°C
[e] PhCN/25°C
[f] (*E*)-Zinc reagent
[g] TG + PhCN/0°C
[h] 100% (*Z*)-isomer

$$CF_3CF{=}CFZnX + F_2C{=}CFI \xrightarrow[\text{TG, 40°C, 18 h}]{(Ph_3P)_4Pd} CF_3CF{=}CF{-}CF{=}CF_2 \quad 80\%$$

$$CF_3CF{=}CFZnX + F_2C{=}CFI \xrightarrow[\text{TG, 75°C, 3 h}]{(Ph_3P)_4Pd^\circ} CF_3CF{=}CF{-}CF{=}CF_2 \quad 50\%$$

9.3.3.A.3. With Acyl Halides. The perfluorovinyl zinc reagents can be smoothly acylated and provide a practical route to α,β-unsaturated ketones.[35] When $R_f = F$ the product α,β-unsaturated ketones are extremely reactive and the choice of solvent is critical to success. When the zinc reagent is prepared in DMF, the acylation reaction proceeds readily; however, the resultant α,β-unsaturated ketone is destroyed by subsequent reaction with DMF. However, when (TG) or tetraglyme (TetG) are employed in the formation of the zinc reagent and in the acylation reaction, the α,β-unsaturated ketone can be isolated in good to excellent yields. Table 9.7 summarizes some typical α,β-unsaturated ketones that have been prepared by this methodology.

$$R_fCF{=}CFZnX + RC(O)Cl \longrightarrow R_fCF{=}CFC(O)R$$

9.3.3.B. Stereospecific Addition of Perfluoroalkenyl Copper Reagents to Perfluoroalkynes. Recent work in our laboratory has demonstrated that perfluoroalkenyl copper reagents add stereospecifically syn to perfluoroalkynes.[36] For example:

$$F_2C{=}CFCu + CF_3C{\equiv}CCF_3 \xrightarrow[\text{DMF}]{\text{RT}} [F_2C{=}CF(CF_3)C{=}C(Cu)CF_3]$$

$$CF_3CF{=}CFCu + CF_3C{\equiv}CCF_3 \xrightarrow[\text{DMF}]{\text{RT}} [CF_3CF{=}CF(CF_3)C{=}C(Cu)CF_3]$$

This addition process provides an alternative route to the preparation of perfluorinated unsaturated copper reagents with total stereochemical control. Future applications of these reagents will provide useful stereospecific routes to substituted fluorinated diene derivatives.

9.4. CONCLUSIONS

We have demonstrated that *stable* perfluoroalkenyl cadmium and zinc reagents can be stereospecifically prepared via direct reaction of perfluorovinyl bromides or iodides with cadmium and zinc powders. Metathesis of these perfluorovinyl cadmium and zinc reagents with Cu(I) salts provides the first general stereospecific route to *stable* perfluoroalkenyl copper reagents.

The perfluorovinyl zinc and copper reagents find practical application in the preparation of α,β,β-trifluorostyrenes, 1-arylperfluoropropenes, perfluoroalkenyl ketones, perfluoro dienes, and are valuable stereospecific building blocks to perfluoro dienyl copper reagents. These reagents will play an important role in future regio- and stereochemical controlled synthesis of a wide variety of functionalized fluorinated vinyl derivatives.

ACKNOWLEDGMENTS

We thank the National Science Foundation and the Air Force Office of Scientific Research for support of this work.

REFERENCES

1. Banks, R. E. (Ed.), *Organofluorine Compounds and Their Industrial Applications*, Ellis Horwood, Chichester 1979.
2. Filler, R. and Kobayashi, Y. (Eds.), *Biomedicinal Aspects of Fluorine Chemistry*, Kodasha/Elsevier, New York, 1982.
3. Filler, R. (Ed.), *Biochemistry Involving Carbon–Fluorine Bonds*, American Chemical Society, ACS Symposium Series, No. 28, Washington, DC, 1976.
4. McLoughlin, V. C. R. and Thrower, J., *Tetrahedron*, *25*, 5921–5940 (1969).
5. Kobayashi, Y. and Kumadaki, I., *Tetrahedron Lett.*, 4095–4096 (1969).
6. Kobayashi, Y., Kumadaki, I., Sato, S., Hara, N., and Chidami, E., *Chem. Pharm. Bull.*, *18*, 2334–2339 (1970).
7. Kobayashi, Y., Kumadaki, I., and Hanzawa, Y., *Chem. Pharm. Bull.*, *25*, 3009–3012 (1977).
8. Kobayashi, Y., Yamamoto, K., and Kumadaki, I., *Tetrahedron Lett.*, 4071–4072 (1979).
9. Kobayashi, Y., Kumadaki, I., Ohsawa, A., and Yamada, T., *Chem. Pharm. Bull.*, *20*, 1839 (1972).
10. Kobayashi, Y., Yamamoto, K., Asai, T., Nakano, M., and Kumadaki, I. *J. Chem. Soc. Perkin Trans. 1*, 2755–2761 (1980).
11. Kondratenko, N. V., Verirko, E. P., and Yagupolskii, L. M., *Synthesis*, 932–933 (1980).
12. Suzuki, H., Yoshida, Y., and Osuka, A., *Chem. Lett.*, 135–136 (1982).

13. Matsui, K., Tobita, E., Anso, M., and Kondo, K., *Chem. Lett.*, 1719–1720 (1981).
14. Leroy, J., Rubinstein, M., and Wakselman, C., *J. Fluorine Chem.*, *27*, 291–298 (1985).
15. Carr, G. E., Chambers, R. D., Holmes, T. F., and Parker, D. G., *J. Chem. Soc. Perkin Trans. 1*, 921–926 (1988).
16. Burton, D. J., *J. Fluorine Chem.*, *23*, 339–357 (1983).
17. Burton, D. J. and Naae, D. G., *J. Am. Chem. Soc.*, *95*, 8467–8468 (1973).
18. Kesling, H. S., Ph.D. Thesis, University of Iowa.
19. Burton, D. J. and Wiemers, D. M., *J. Am. Chem. Soc.*, *107*, 5014–5015 (1985).
20. Knunyants, I. L., Delyagina, N. I., and Igumnov, S. M., *Izv. Akad. Nauk SSSR, Ser. Khim.*, *4*, 857–859 (1981).
21. Wiemers, D. M. and Burton, D. J., *J. Am. Chem. Soc.*, *108*, 832–834 (1986).
22. Willert-Porada, M. A., Burton, D. J., and Baenziger, N. C., *J. Chem. Soc. Chem. Commun.*, 1633–1634 (1989).
23. Wiemers, D. M., Ph.D. Thesis, University of Iowa.
24. Coe, P. L. and Milner, N. E., *J. Organomet. Chem.*, *70*, 147–152 (1974).
25. Burton, D. J., Hartgraves, G. A., and Hsu, J., *Tetrahedron Lett.*, *31*, 3699–3702 (1990).
26. Hung, M. H., *Tetrahedron Lett.*, *31*, 3703–3706 (1990).
27. Chen, Q. Y. and Wu, S. W., *J. Chem. Soc. Perkin Trans. 1*, 2385–2387 (1989).
28. Burton, D. J. and Hansen, S. W., *J. Fluorine Chem.*, *31*, 461–465 (1986).
29. Hansen, S. W., Spawn, T. D., and Burton, D. J., *J. Fluorine Chem.*, *35*, 415–420 (1987).
30. Burton, D. J. and Hansen, S. W., *J. Am. Chem. Soc.*, *108*, 4229–4230 (1986).
31. Heinze, P. L. and Burton, D. J., *J. Fluorine Chem.*, *31*, 115–119 (1986).
32. Heinze, P. L. and Burton, D. J., *J. Organic Chem.*, *53*, 2714–2720 (1988).
33. Dmowski, W., *J. Fluorine Chem.*, *18*, 25–30 (1981).
34. Dolbier, W. R., Koroniak, H., Burton, D. J., Heinze, P. L., Bailey, A. R., Shaw, G. S., and Hansen, S. W., *J. Am. Chem. Soc.*, *109*, 219–225 (1987).
35. Spawn, T. D. and Burton, D. J., *Bull. Chim. Soc. Fr.*, 876–880 (1987).
36. Hansen, S. W., Ph.D. Thesis, University, University of Iowa.

CHAPTER 10

Nucleophilic Perfluoroalkylation of Organic Compounds Using Perfluoroalkyltrialkylsilanes

G. K. SURYA PRAKASH

10.1. INTRODUCTION

Organofluorine compounds have found rapidly increasing use in the areas of agrochemicals, pharmaceuticals, and fluoropolymers. A number of antiviral, antitumor, and antifungal agents have been developed in which fluorine substitution has been a key to their biological activity. Many organofluorine

Synthetic Fluorine Chemistry,
Edited by George A. Olah, Richard D. Chambers, and G. K. Surya Prakash.
ISBN 0-471-54370-5

applications stem from the fact that the introduction of fluorine into a molecule often leads to a significant change in its physical and chemical properties.[1] First, fluorine and hydrogen are comparable in size (the van der Waal's radii of F and H are 1.35 and 1.1 Å, respectively). Thus, whereas a molecule and its fluoro analogues would be sterically almost indistinguishable to a guest molecule, their chemical behavior could be different from one another. Second, the high C—F bond energy, which averages about 116 kcal mol^{-1}, leads to enhanced thermal stability. Third, due to its high electronegativity, fluorine-containing molecules often show different chemical properties. In fact, the difluoromethyl (CF_2H) and difluoromethylene (CF_2) groups are considered isopolar and isosteric with the hydroxy (OH) and ether oxygen (O) groups, respectively.[2] Finally, fluorine substitution can increase lipid solubility, and this increases the rate of transport of biologically active compounds across lipid membranes. Thus, many perfluoroalkyl, especially trifluoromethyl-substituted compounds have been examined for their ability to enhance transport rate *in vivo*.

Concomitant with an increased understanding of the behavior of organofluorine compounds,[1,3] considerable progress has also been made in the development of new synthetic methodologies.[4] A number of methods exist for the preparation of fluorinated and polyfluoroalkylated aromatic compounds.[5]

Trifluoromethyl-substituted compounds have been examined for their potential as biologically active drugs and agrochemicals.[6] Consequently, much effort has been put into developing more economical and efficient trifluoromethylating reagents using organometallic derivatives.

Trifluoromethylation of aromatics is readily achieved with a variety of methods most notably using trifluoromethyl copper,[7] sodium trifluoroacetate,[8] trifluoromethyl triflate,[9] bis(trifluoromethyl)mercury,[10] and other related reagents.[5,11] Although the literature abounds with examples of introducing perfluoroalkyl groups into carbonyl compounds through organometallic reagents of zinc,[12] calcium,[13] manganese,[14] magnesium,[15] silver,[14] and lithium (Eq. 10.1),[16] the procedures are seldom applicable to trifluoromethylation. For-

$$R^1R^2C{=}O + R_f\text{-}X \xrightarrow[\text{Solvent}]{M} R^2\text{—}C(R^1)(OH)\text{—}R_f \qquad (10.1)$$

R^1 = H, Alkyl, Aryl
R^2 = Alkyl, Aryl
R_f = C_2F_5, C_3F_7, etc
M = Zn, Mg, Mn, Ca, Ag, Li
Solvent = DMF, THF

$$CF_3M \longrightarrow MF + \ddot{C}F_2 \qquad (10.2)$$

mation of difluorocarbene by α elimination of metal fluoride is a serous side reaction (Eq. 10.2). Recently, electrochemical trifluoromethylation of carbonyl compounds was reported as a viable synthetic method, however, yields were poor with ketones and some aldehydes.[17]

Electrochemically reduced 2-pyrrolidone has been used by Shono et al.[18] to deprotonate trifluoromethane to generate the trifluoromethyl anion equivalent which reacts with aldehydes and ketones.

This chapter reports efficient nucleophilic perfluoroalkylation including trifluoromethylation using perfluoroalkyltrialkylsilanes.

R_fSiR_3

1a, $R_f = CF_3$, $R = CH_3$

1b, $R_f = CF_3$, $R = CH_2CH_3$

1c, $R_f = CF_3$, $R = CH_2CH_2CH_2CH_3$

1d, $R_f = C_2F_5$, $R = CH_3$

1e, $R_f = CF_2CF_2CF_3$

Over the years trimethylsilyl compounds substituted with electron-withdrawing substituents such as CN, I, Cl, Br, N_3, NCO, CNO, and so on, have been used as synthetic reagents to introduce these substituents to electron-deficient centers.[19] These reagents generally add substituents based on the hard–soft reactivity principle,[20] with the silicon acting as the hard acid and the electronegative substituent as the soft base. Accordingly, the bond between the pseudohalide trifluoromethyl, for example, and trimethylsilyl group in **1a** should be sufficiently polarized with the trifluoromethyl group bearing substantial negitive charge. If one considers the reaction of **1a** with a carbonyl group, the propensity of silicon to form strong bonds with the hard base oxygen can be a favorable thermodynamic process to drive the reaction (Eq. 10.3). The net result would be the addition of **1a** across the carbonyl. Thus Prakash, Krishnamurti, and Olah[21a] embarked on a study of trifluoromethyltrimethylsilane ($TMSCF_3$, **1a**) as a potential reagent for introducing the trifluoromethyl group into carbonyl compounds and found a long sought-after simple and efficient trifluoromethide equivalent reagent. Hartkopf and Meijere[21b] previously used trialkylsilyl(trifluoromethyl) diazenes as tailored reagents for nucleophilic trifluoromethylation. However, these reagents were not that easy to synthesize.

$$R^1R^2C{=}O \; + \; \mathbf{1a} \longrightarrow R^2{-}C(R^1)(OSiMe_3){-}CF_3 \qquad (10.3)$$

Recently, it was also found[22] that the nucleophilic trifluoromethylation works equally well with trifluoromethyltriethyl and trifluoromethyltri(*n*-butyl)silanes

1b and **1c** with benzoquinones. Even pentafluoroethyl and *n*-heptafluoropropyltrimethylsilanes **1d** and **1e** act as nucleophilic pentafluoroethylating and heptafluoropropylating agents, for carbonyl compounds, respectively. In addition to carbonyl groups, these reagents nucleophilically add to a variety of other electrophiles. Under certain conditions even singlet difluorocarbene can be generated from **1a**.

10.2. PREPARATION OF PERFLUOROALKYLTRIALKYLSILANES

Marchenko et al.[23] in 1980 and Ruppert et al. in 1984[24] independently reported the preparation of **1a** by the reaction shown in Eq. 10.4. The method has been further improved.[25]

$$(CH_3)_3SiCl + CF_3Br + (Et_2N)_3P \xrightarrow{\text{benzonitrile}} \underset{\textbf{1a}}{(CH_3)_3SiCF_3} \tag{10.4}$$

Pawelke[26] reported another route to **1a** using terakis(dimethylamino)ethylene and trifluoromethyl iodide (Eq. 10.5).

$$[(CH_3)_2N]_2C{=}C[N(CH_3)_2]_2 + (CH_3)_3SiCl + CF_3I \xrightarrow[\text{or } CH_2Cl_2]{\text{benzonitrile}} \underset{\textbf{1a}}{(CH_3)_3SiCF_3} \tag{10.5}$$

However, the latter method is not very convenient. Trifluoromethyltrimethylsilane is a colorless stable liquid with a bp 55–55.5 °C[25] (Ruppert reports a bp of 45 °C). Stahly and Bell[22] using Ruppert's procedure[24] have been able to prepare trifluoromethyltriethyl and trifluoromethyltri(*n*-butyl)silanes, **1b** and **1c** (in 69 and 64% yield, respectively). The boiling points of **1b** and **1c** are 52–54 °C/10 torr and 53–58 °C/0.5 torr, respectively. Attempts to prepare[25] perfluoroalkyltrimethylsilanes **1d** and **1e** by reacting the corresponding perfluoroalkyllithiums with chlorotrimethylsilane resulted only in very low yields ($\leqslant 5\%$ for **1d** and $\leqslant 15\%$ for **1e**). However, using a modified Ruppert's procedure[25] starting from pentafluoroethyl and heptafluoropropyl iodides, **1d** (bp 69–70°C) and **1e** (bp 87–89°C) were obtained (Eq. 10.6).

$$R_f\text{-}I + (Et_2N)_3P + Me_3SiCl \xrightarrow{\text{PhCN, -35 °C}} R_fSiMe_3 \tag{10.6}$$

$R_f = C_2F_5$: 50%
$R_f = C_3F_7$: 68%

In Chapter 11, Farnham reports[27] preparation of *n*-perfluorooctyltrimethylsilane, **1f**, by a modified Barbier-type reaction. Similarly, he has also prepared[26] 1,8-bistrimethylsilylperfluorooctane, **1g**.

$n\text{-}C_8F_{17}Si(CH_3)_3$ **1f** $(CH_3)_3Si(CF_2)_8Si(CH_3)_3$ **1g**

10.3. REACTIONS OF PERFLUOROALKYLTRIALKYLSILANES

10.3.1. Ketones and Aldehydes

Reactions of a carbonyl compound such as cyclopentanone with **1a** in the presence of a variety of Lewis acids such as $BF_3{:}Et_2O$, ZnI_2, $TiCl_4$, Et_2AlCl, and Et_2AlCl_2, and $SnCl_4$ were unsuccessful.[25] The ^{19}F NMR spectra of the reaction mixtures in all cases showed no trace of the trifluoromethylated product. Then the possibility of activating the silicon–carbon bond under nucleophilic conditions was considered. Silicon is known to form strong bonds with oxygen and fluorine. Many reactions of silicon compounds using fluoride ion as the mediator are well known.[19] Hiyama and his co-workers[28] reported a series of fluoride-initiated carbonyl addition reactions for a variety of α-halo carbanions derived from α-halo organosilicon compounds. From the vibrational spectra and force-field calculation data for **1a** and its perdeuterated analogue, Eujen[29] pointed out the apparent weakness of the Si—CF_3 bond in these and other silanes containing CF_3 groups. Thus it was envisaged that a similar carbonyl addition could occur with perfluoroalkyltrimethylsilanes (**1a, 1c,** and **1d**) under fluoride catalysis. Indeed, when cyclopentanone was treated with **1a** in THF in the presence of a catalytic quantity (2 mol%) of tetra-*n*-butylammonium fluoride trihydrate (TBAF)[30a] at 0 °C, ^{19}F and ^{13}C NMR spectra of the reaction mixture showed quantitative formation of the trifluoromethylated adduct (Eq. 10.7).

$$\text{cyclopentanone} \xrightarrow[\text{THF, 0 °C, 30 min}]{\textbf{1a}\text{, Cat. } (n\text{-Bu})_4N^+F^-} \text{1-(Me}_3\text{SiO)-1-(CF}_3\text{)cyclopentane} \tag{10.7}$$

Subsequently, it was found that **1a** adds equally well to a wide variety of aldehydes, ketones, and enones, and is generally unaffected by moisture.[25] The product trimethylsilyl ethers are hydrolyzed by aqueous acid to give trifluoromethylated carbinols in excellent overall yield. The results of the reaction of **1a** with aldehydes, ketones, and an enone are summarized in Table 10.1. In the case of cyclohexenone, the 1,2-adduct is obtained predominantly over the 1,4-Michael addition product (in the ratio 9:1). With hindered ketones such as 2-

TABLE 10.1. Fluride Ion Induced Trifluoromethylation of Carbonyl Compounds with Trifluoromethyltrimethylsilane (**1a**).

Carbonyl Compound	Product	Overall Isolated Yield (%)
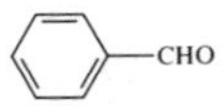 CHO	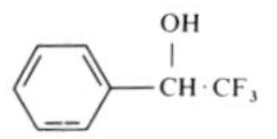 OH CH·CF_3	85
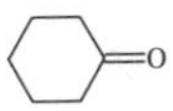 O	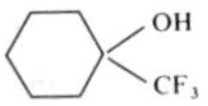 OH CF_3	77
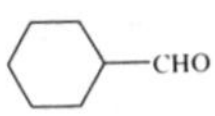 CHO	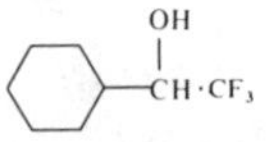 OH CH·CF_3	80
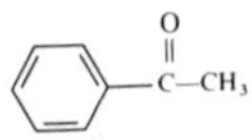 O C—CH_3	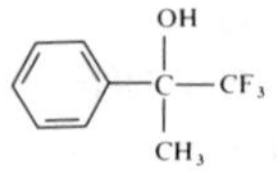 OH C—CF_3 CH_3	74
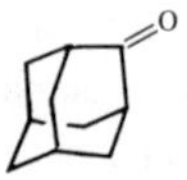 O	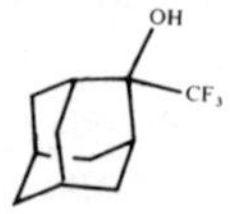 OH CF_3	72
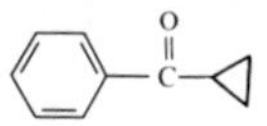 O C	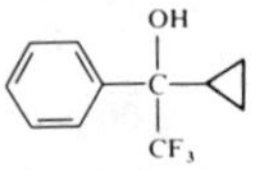 OH C CF_3	81
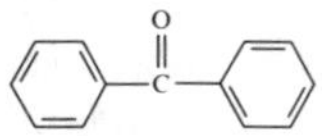 O C	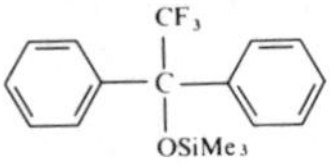 CF_3 C $OSiMe_3$	88
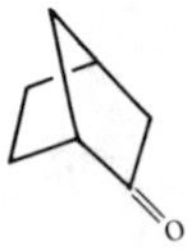 O	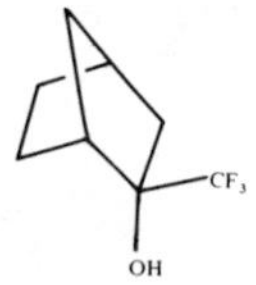 CF_3 OH	92
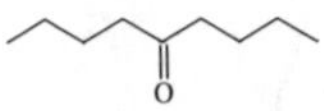 O	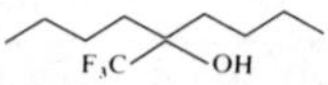 F_3C OH	87
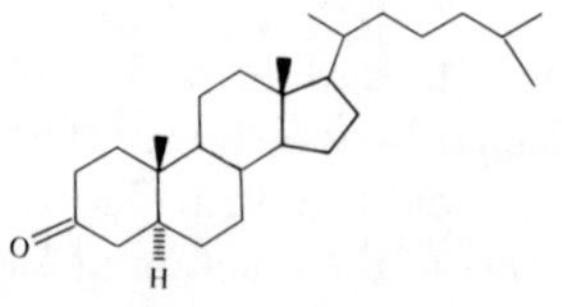 O H	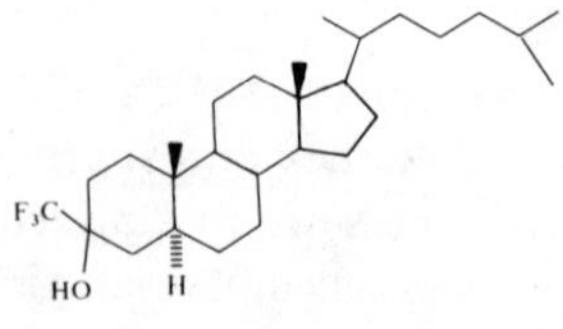 F_3C HO H	83[a]
O	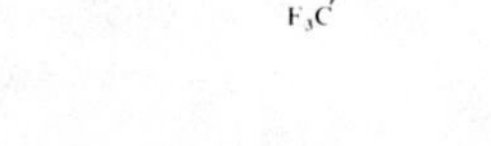 CF_3 OH + O F_3C	60

TABLE 10.1. Continued.

Carbonyl Compound	Product	Overall Isolated Yield (%)
CH_3O (estrone methyl ether)	HO, CF_3; CH_3O	62[b]

[a]Only one stereoisomer was obtained.
[b]Obtained as a mixture of 90% product (one isomer) and 10% starting ketone.

adamantanone and estrone methyl ether, the reactions were sluggish. Nevertheless, trifluoromethylated adducts were obtained on prolonged reaction. Extremely hindered ketones, such as 1,7,7-trimethylbicyclo[2.2.1]heptan-2-one, 3,3-dimethylbicyclo[2.2.1]heptan-2-one, and di(1-adamantyl) ketone, however, gave only traces ($<1\%$) of the respective trifluoromethylated adducts. An attempt was also made to ascertain if the carbonyl addition of **1a** proceeds diastereoselectively. Thus, 2-methylcyclohexanone was treated with **1a** in the presence of a catalytic amount of TBAF at 0°C as well as at −78°C. Surprisingly, however, the relative ratio of the two diastereomers remained unchanged (~60:40) at these two temperatures (Eq. 10.8).

O, CH_3 —**1a**, cat. TBAF, THF / 0 °C or -78 °C→ Me_3SiO, CF_3, CH_3

ca. 60/40 mixture of diastereomers

(10.8)

The proposed mechanism for the fluoride induced reaction depicted in Scheme I involves fluoride ion initation (indicated by the irreversible formation of fluorotrimethylsilane) to afford the trifluoromethylated oxyanion **2**, which then catalyzes the subsequent reaction. Support for this mechanism comes from the observation that other oxyanionic species, such as potassium *tert*-butoxide and sodium trimethylsilanolate (Me_3SiONa) (Ref-25) also are equally effective

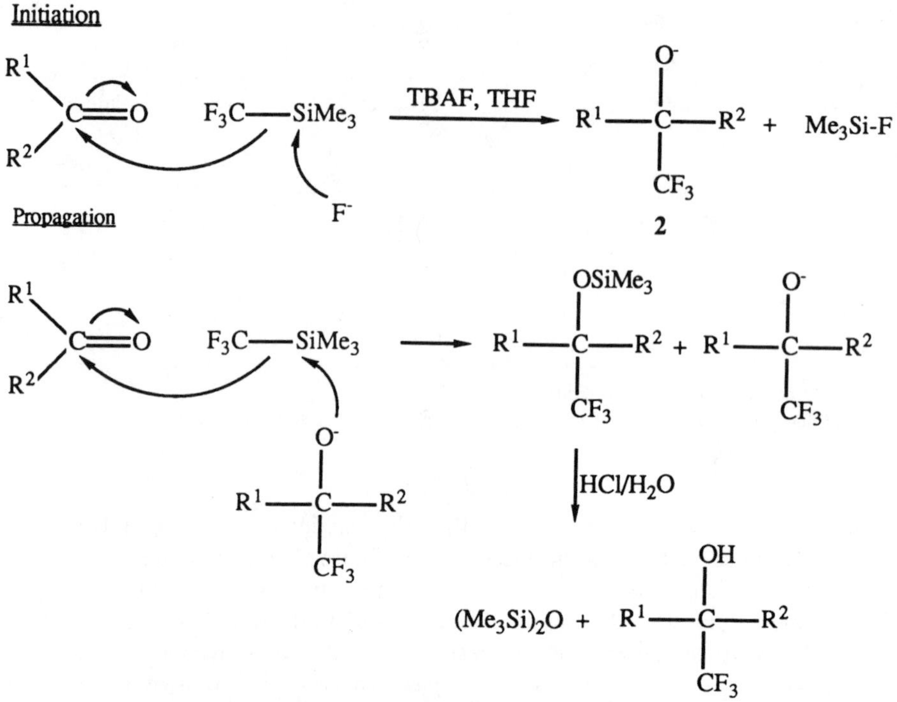

Initiation

$$R^1R^2C{=}O + F_3C{-}SiMe_3 \xrightarrow[F^-]{\text{TBAF, THF}} R^1{-}C(O^-)(CF_3){-}R^2 + Me_3Si{-}F$$

2

Propagation

$$R^1R^2C{=}O + F_3C{-}SiMe_3 \xrightarrow{R^1{-}C(O^-)(CF_3){-}R^2} R^1{-}C(OSiMe_3)(CF_3){-}R^2 + R^1{-}C(O^-)(CF_3){-}R^2$$

$$R^1{-}C(OSiMe_3)(CF_3){-}R^2 \xrightarrow{HCl/H_2O} (Me_3Si)_2O + R^1{-}C(OH)(CF_3){-}R^2$$

Scheme I

as catalysts. The mechanism, as depicted in Scheme I, should not imply a termolecular reaction. An intermediately formed pentavalent silicon adduct transfers the CF_3 group (a bimolecular reaction) to the carbonyl compound. Des Marteau and his co-workers[30b] reacted a series of polyfluorinated carbonyl compounds with trifluoromethyltrimethylsilane using potassium fluoride. Stahly and Bell[22] used trifluoromethyltrialkylsilanes **1b** and **1c** to introduce CF_3 groups into benzoquinone derivatives using excess of potassium fluoride as catalyst. The reaction is depicted in Eq. 10.9.

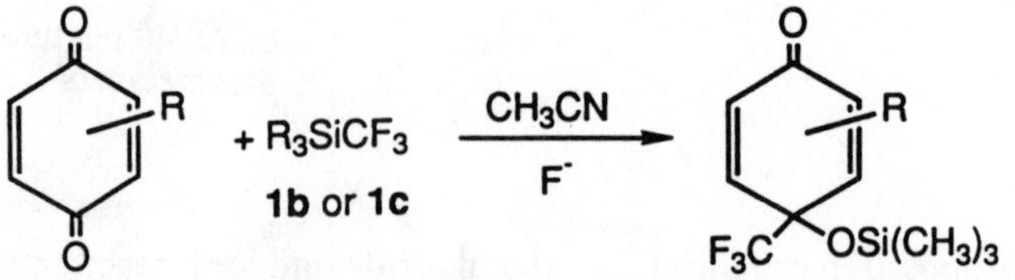

$$\text{R-benzoquinone} + R_3SiCF_3\ (\textbf{1b}\text{ or }\textbf{1c}) \xrightarrow[F^-]{CH_3CN} \text{4-}(F_3C)\text{-4-}(OSi(CH_3)_3)\text{-R-cyclohexa-2,5-dienone} \tag{10.9}$$

The monotrifluoromethylated adducts were further converted to industrially useful *p*-trifluoromethylated phenols and *p*-aminobenzotrifluoride derivatives (Eq. 10.10)

$$
\text{4-}CF_3\text{-4-}OSi(CH_3)_3\text{-cyclohexadienone (R)} \begin{cases} \xrightarrow{\text{Reduction}} \text{4-}CF_3C_6H_3(R)OH \\ \xrightarrow{\text{Reductive Amination}} \text{4-}CF_3C_6H_3(R)NH_2 \end{cases} \quad (10.10)
$$

Pentafluoroethyl and heptafluoropropyltrimethylsilanes **1d** and **1e** functioned as facile pentafluoroethylating and heptafluoropropylating agents, respectively, for carbonyl compounds under fluoride catalysis (Eq. 10.11). The products are best isolated as the carbinols.

$$
R^1R^2C{=}O \xrightarrow[\text{Cat. } Bu_4NF, \text{ THF}]{\textbf{1d} \text{ or } \textbf{1e}} R^1R^2C(R_f)OSiMe_3 \xrightarrow{Bu_4NF, \text{ THF}} R^1R^2C(R_f)OH \quad 66\text{-}86\% \quad (10.11)
$$

$R^1 = Ph, R^2 = Me$;
$R^1, R^2 = cyclo(C_6H_{10})$;
$R^1 = Ph, R^2 = H$

$R_f = C_2F_5, R^1 = Ph, R^2 = Me$
$R_f = C_2F_5, R^1, R^2 = cyclo(C_6H_{10})$
$R_f = C_2F_5, R^1 = Ph, R^2 = H$
$R_f = n\text{-}C_3F_7, R^1 = Ph, R^2 = Me$
$R_f = n\text{-}C_3F_7, R^1, R^2 = cyclo(C_6H_{10})$
$R_f = n\text{-}C_3F_7, R^1 = Ph, R^2 = H$

It must be emphasized that no enol silyl ether formation was observed with any of the enolizable aldehydes and ketones employed in the reactions with perfluoroalkyltrimethylsilanes. On the other hand, Hiyama and his co-workers[28] reported that (difluoromethyl)dimethylphenylsilane ($PhMe_2SiCHF_2$) reacts with carbonyl compounds under these conditions to give only enol silyl ethers. They suggested that destabilization of the negative charge at the α carbon in the difluoromethyl carbanion made it basic enough to induce enolization.[28,31]

10.3.2. α-Keto Esters

α-Keto esters react with trifluoromethyltrimethylsilane under fluoride initiation in THF,[32a] β,β,β-Trifluorolactic acid derivatives are produced in good yields in one step (Eq. 10.12). β,β,β-Trifluorolactic acids are important compounds in medicinal and biological chemistry.[33]

$$R\text{-}C(=O)\text{-}COOR' \xrightarrow[\text{THF, }F^-]{\textbf{1a}} R\text{-}C(OSi(CH_3)_3)(CF_3)\text{-}COOR' \xrightarrow{\text{aq. HCl}} R\text{-}C(OH)(CF_3)\text{-}COOR' \quad (10.12)$$

R= alkyl , aryl

R'= Alkyl

68-83%

10.3.3. Esters

Simple ester carbonyl was found unreactive towards trifluoromethyltrimethylsilane even with a stoichiometric amount of fluoride (TBAF) initiator. However, when the ester carbonyl is activated, as in a trifluoroacetate ester, the reaction does occur. For example, *n*-hexyl trifluoroacetate reacted with **1a** in the presence of a molar equivalent of TBAF to the extent of about 35% (^{19}F NMR) to give the silylated hemiketal (Eq. 10.13). Much of the silane **1a** was converted to the undesired trifluoromethane due to rapid quenching of the incipient trifluoromethide species by the protic impurities in the reaction mixture.[25]

$$n\text{-}C_6H_{13}O\text{—}C(=O)\text{—}CF_3 \xrightarrow{\textbf{1a}\text{, 1 eq., TBAF, THF}} n\text{-}C_6H_{13}O\text{—}C(OSiMe_3)(CF_3)\text{—}CF_3 \quad (10.13)$$

Other activated esters, such as oxalates are also reactive towards **1a** under fluoride catalysis as reported in the case of di-*tert*-butyl oxalate.[34] An efficient two-step synthesis of trifluoropyruric acid monohydrate has been carried out by

Broicher and Geffken[34] using trifluoromethyltrimethylsilane (Eq. 10.14).

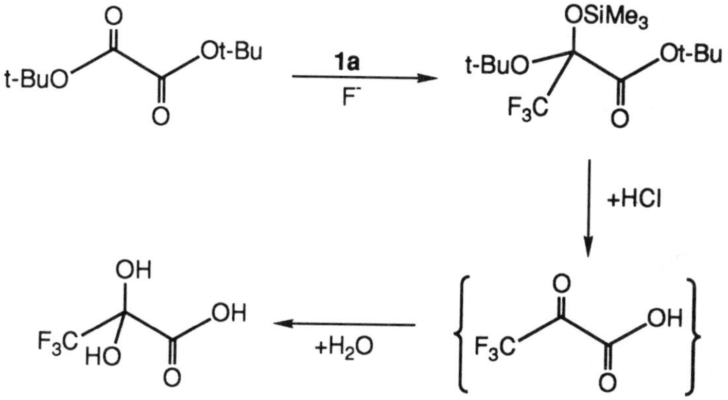

(10.14)

10.3.4. Lactones

Reactivity of lactones toward **1a** under fluoride catalysis in THF has been explored.[25] Indeed, **1a** adds smoothly to lactones under these conditions (Eq. 10.15), but only the silylated hemiketals in the case of five- and six-membered ring lactones could be isolated cleanly (in 70 and 75% yield, respectively). The four-membered ring adduct was formed clearly in solution, however, it decomposed extensively during distillation. In the case of the seven-membered ring lactone, a mixture of products resulted (Eq. 10.16).

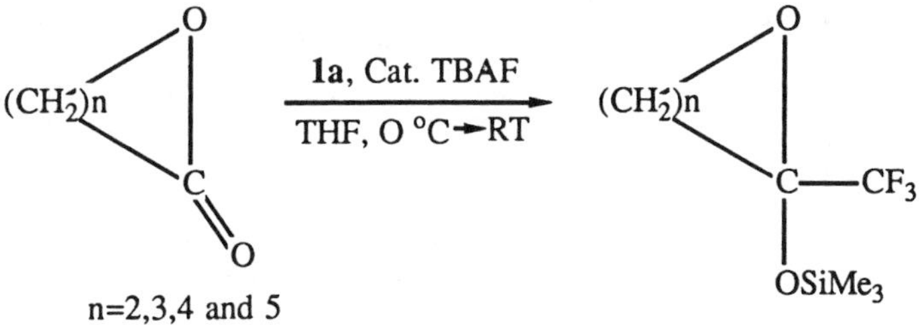

n=2,3,4 and 5

(10.15)

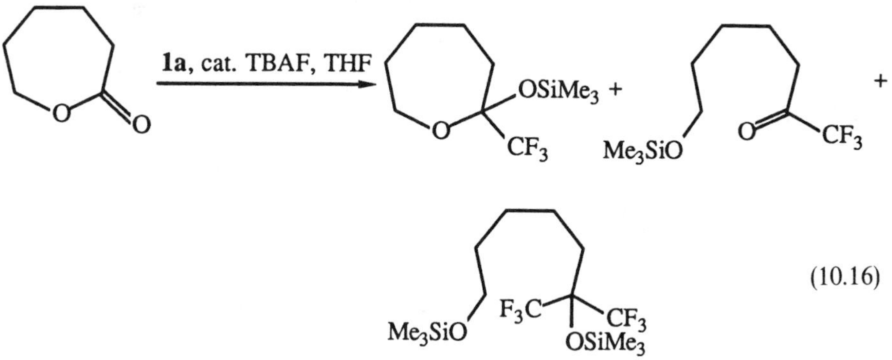

(10.16)

From NMR and IR data it appears that the mixture consists of the desired adduct as well as ring-opened products. The open-chain ketone perhaps arises from fluoride-mediated ring opening of the cyclic adduct. However, this ketone has a very reactive keto group. Thus, it reacts further with **1a** to give the bis(trifluoromethylated) product. These results show that a lactone carbonyl is sufficiently reactive (in contrast to simple ester carbonyls) to afford synthetically useful yields of interesting trifluoromethylated compounds, at least in the case of five- and six-membered ring systems.

10.3.5. Acid Halides

Acid halides such as benzoyl chloride also reacted with **1a**. However, a mixture of products were obtained[25] (Eq. 10.17). The reaction requires more than one molar equivalent of TBAF.

$$PhC(=O)Cl \xrightarrow{\textbf{1a}, \text{ TBAF, THF}} PhC(=O)CF_3 + PhC(OSiMe_3)(CF_3)CF_3 \qquad (10.17)$$

Farnham[27] has been successful in developing a general ketone synthesis by the reaction of *n*-perfluorooctyltrimethylsilane with acid fluorides (Eq. 10.18).

$$CF_3(CF_2)_3COF + C_8F_{17}Si(CH_3)_3 \xrightarrow[\text{glyme}]{\text{TASF}} \underset{88\%}{CF_3(CF_2)_3COC_8F_{17}} \qquad (10.18)$$

10.3.6. Cyclic Anhydrides

Cyclic anhydrides react quite well with $TMSCF_3$, **1a**, however, a stoichiometric amount of TBAF is required.[25] For example, succinic anhydride adds **1a** efficiently to initially form the adduct, which upon hydrolysis affords the trifluoromethyl-substituted keto carboxylic acid (Eq. 10.19).

$$\text{succinic anhydride} \xrightarrow{\textbf{1a}, \text{ TBAF, THF}} \text{2-CF}_3\text{-2-}(O^- \ N^+Bu_4)\text{-lactone} \xrightarrow{5\% \text{ aq. HCl}} HOOCCH_2CH_2C(=O)CF_3 \qquad (10.19)$$

Such compounds could act as versatile building blocks for synthesizing many other types of CF_3-substituted compounds. Acyclic anhydrides also react with **1a**, but less cleanly.

10.3.7. Amides and Imides

Simple amides, such as benzamide and acetamide do not react with **1a** even when a molar quantity of TBAF is used. Similarly, lactams, such as caprolactam do not react with **1a** under similar conditions. An activated amide carbonyl, such as that in *N*-methylsuccinimide, however, reacts smoothly to afford an interesting adduct,[25] which upon acid hydrolysis afforded the hemiaminal (Eq. 10.20). The trifluoromethylated hemiaminal could be useful in preparing heterocyclic derivatives containing the trifluoromethyl group.

1a, Cat. Bu_4NF, THF; 0 °C; quantitative → OSiMe$_3$ / CF$_3$ / N Me; δ ^{19}F NMR: -79.1 ppm; 1N HCl → OH / CF$_3$ / N Me

(10.20)

10.3.8. Nitroso Group

Nitroso compounds, such as nitrosobenzene, reacted in a facile manner to afford the *O*-silylated trifluoromethylated hydroxylamine quantitatively (Eq. 10.21).[35] Nonetheless, the adduct slowly decomposes to unidentified materials. This again shows that in order for **1a** to react successfully the generation of an oxyanionic intermediate species is highly desirable.

$$PhN{=}O \xrightarrow[\text{THF, 0 °C; quantitative}]{\textbf{1a}\text{, Cat. }Bu_4NF} (Ph)(F_3C)N{-}OSiMe_3$$

10.3.9. Alkyl, Allyl, Aromatic, and Vinylic Halides

Alkyl halides including benzylic and allylic (fluorides, chlorides, bromides, and iodides) do not react with **1a** even with molar amounts of TBAF.[35] Recently, Urata and Fuchikami[36] developed a convenient method for the trifluoromethylation of organic (aryl, vinyl, benzyl, and allyl) halides using trifluoromethyltrialkylsilane–KF–Cu^I in DMF solution (Eq. 10.22). Under these conditions CF_3Cu reagent is produced *in situ*

$$\text{R-X} + \text{CF}_3\text{SiR'}_3 \xrightarrow[\text{DMF, 60-80 °C}]{\text{KF/Cu}^{\text{I}}} \underset{\text{23-94\%}}{\text{R-CF}_3} \qquad (10.22)$$

The reaction also works with pentafluoroethyltrimethylsilane (Eq. 10.23)

$$\text{R-X} + \text{CF}_3\text{CF}_2\text{SiR'}_3 \xrightarrow[\text{DMF, 60 °C}]{\text{KF/Cu}^{\text{I}}} \underset{\text{41-92\%}}{\text{R-CF}_2\text{CF}_3} \qquad (10.23)$$

10.3.10. Aromatic and Olefinic Substrates

The scope of trifluoromethyltrimethylsilane as a nucleophilic trifluoromethylating agent for aromatic substrates has been explored.[35] Anhydrous TBAF was employed to minimize the hydrolysis of **1a** to CF_3H. Simple aromatics, such as bromo- and iodobenzene do not react with **1a**. Even hexafluorobenzene does not react. With a stoichiometric amount of TBAF, on the other hand, pentafluoronitrobenzene reacted readily at 0 °C to afford octafluorotoluene and bis(trifluoromethyl)tetrafluorobenzene in about a 90:10 ratio (Eq. 10.24). 2,4-Dinitrofluorobenzene, however, gave a mixture of trifluoromethylated products based on the ^{19}F NMR analysis of the reaction mixture (Eq. 10.25). The undesirable hydrolysis of **1a** to CF_3H also occurred during both reactions.

$$\text{C}_6\text{F}_5\text{NO}_2 \xrightarrow[\text{1 equiv. TBAF, 0 °C, 0.5 h, RT}]{\textbf{1a}\text{, THF}} \underset{90}{\text{C}_6\text{F}_5\text{CF}_3} + \underset{10\ (^{19}\text{F NMR})}{\text{C}_6\text{F}_4(\text{CF}_3)_2} \qquad (10.24)$$

$$\text{2,4-dinitrofluorobenzene} \xrightarrow[\text{1 equiv. TBAF, 0 °C}]{\textbf{1a}\text{, THF}} \text{Trifluoromethylated mixture of products} \tag{10.25}$$

Other aromatics, such as 4-fluoronitrobenzene do not react with **1a**. Obviously, the factors governing the reactivity of aromatics towards **1a** remain to be determined. All these reactions should involve the intermediacy of Meisenheimer complexes.

Farnham[27] reported nucleophilic addition of *n*-perfluorooctyl groups via *n*-perfluorooctyltrimethylsilane to perfluorocyclopentene involving an addition–elimination mechanism (Eq. 10.26).

$$C_8F_{17}Si(CH_3)_3 + \text{perfluorocyclopentene} \longrightarrow \text{1-}C_8F_{17}\text{-perfluorocyclopentene} \quad 95\% \tag{10.26}$$

Using 1,2-bis(trimethylsilyl)perfluorooctane, **1g**, as a synthon, perfluoropropene gives polymers of modest molecular weight.

10.3.11. Aryl Sulfonyl Fluorides

Aryl sulfonyl fluorides react with **1a** under fluoride initiation to give the corresponding aryl trifluoromethyl sulfones (Eq. 10.27). These reactions have been carried out by two groups independently.[32b,37] The same reaction also works with trifluoromethyltrimethyltin.[37]

$$ArSO_2F \xrightarrow[\text{THF, }F^-]{\textbf{1a}} ArSO_2CF_3 \tag{10.27}$$

10.3.12. Sulfur Dioxide

No reaction between trifluoromethyltrimethylsilane, and SO_2 occurred in the absence of an anionic initiator. Nevertheless, when one molar equivalent of TBAF was used, clean formation of tetra(*n*-butyl)ammonium trifluoromethanesulfinate was observed.[35] Although oxidation to the corresponding sulfonate occurred readily upon treatment with 30% hydrogen peroxide, attempts to liberate the free acid from the salt proved unsuccessful. When sodium trimethylsilanolate was used as the initiator, however, the reaction sequence (Eq. 10.28) was successful. The overall yield of CF_3SO_3H, however, was about 30%.

$$CF_3SiMe_3 + SO_2 \xrightarrow[-78\ ^\circ C]{1\ eq.\ Me_3SiONa,\ THF} CF_3SO_2^-\ Na^+ \xrightarrow[(ii)\ H_2SO_4]{(i)\ 30\%\ H_2O_2,\ reflux,\ 3\ h} \underset{30\%\ overall}{CF_3SO_3H} \quad (10.28)$$

10.3.13. Other Functional Groups

Trifluoromethyltrimethylsilane does not react with nitriles.[35] Furthermore, epoxides do not react either under nucleophilic or electrophilic conditions.[35] Acetals and ortho esters do not react with **1a**.[35] One can, therefore, selectively add **1a** to a carbonyl in the presence of an acetal or a ketal group as shown from the example in Eq. 10.29.[25]

$$p\text{-}(EtO)_2CH\text{-}C_6H_4\text{-}CHO \xrightarrow[0\ ^\circ C,\ quantitative]{\mathbf{1a},\ cat.\ TBAF,\ THF} p\text{-}(EtO)_2CH\text{-}C_6H_4\text{-}CH(CF_3)(OSiMe_3) \quad (10.29)$$

Attempts to add **1a** to imines under a variety of conditions also failed.[35] Perhaps the much weaker N—Si bond (as compared with the O—Si) does not provide sufficient driving force to push the reaction in the forward direction. Addition of **1a** to pyridine-*N*-oxide also failed under a variety of conditions.[35] Attempted reaction of conjugated nitro compounds, such as β-nitrostyrene, with **1a** and catalytic TBAF resulted in rapid polymerization of the nitro compound. The ^{19}F NMR spectrum of the reaction mixture showed that almost all of **1a** remains unreacted.

10.3.14. Generation of Difluorocarbene

Treatment of trifluoromethyltrimethylsilane with an anhydrous fluoride source such as TASF in THF does result in the generation of singlet difluorocarbene.[35,38] In the presence of an acceptor such as tetramethylethylene, the corresponding adduct can be isolated (Eq. 10.30).

$$(H_3C)_2C{=}C(CH_3)_2 + (CH_3)_3SiCF_3 \xrightarrow[\text{THF}]{\text{TASF}} (H_3C)_2C\overset{CF_2}{\triangle}C(CH_3)_2 \quad (10.30)$$

10.4. SUMMARY

The work described above enables one to utilize perfluoroalkyltrialkylsilanes to prepare a variety of perfluoroalkylated compounds. Particularly, trifluoromethyltrimethyltrimethylsilane, in the presence of a fluoride or oxyanion initiator, acts as a versatile and convenient equivalent for the exceedingly unstable trifluoromethyl carbanion. Perfluoroalkylation of aldehydes and ketones using the silanes **1** constitutes a general method for the preparation of a wide variety of perfluoroalkylated alcohols. Another attractive feature of this reaction is the absence of enolization when aldehydes and ketones with α hydrogens are used. Reactions of **1a** with heterocyclic substrates, such as lactones, cyclic imides, and cyclic anhydrides enables one to synthesize perfluoroalkylated heterocyclic compounds, which could be significant in the area of biologically active organofluorine compounds.

ACKNOWLEDGMENTS

The work reported in this chapter was supported by the Loker Hydrocarbon Research Institute and Office of the Naval Research. Professor G. A. Olah is thanked for his support and encouragement. The collaborative help of Dr. R. Krishnamurti, Dr. D. R. Bellew, and Dr. P. Ramaiah with the development of trifluoromethylation reaction is appreciated.

REFERENCES

1. (a) Liebman, J. F., Greenberg, A., and Dolbier, Jr., W. R., (Eds.), *Fluorine-containing Molecules, Structure, Reactivity, Synthesis*, VCH: New York, 1988. (b) Welch, J. T. and Eshwarakrishnan, S. (Eds.), *Fluorine in Bioorganic Chemistry*, Wiley: New York, 1991.

2. Bergstrom, D. E. and Awartling, D. J., *Fluorine-containing Molecules, Structure, Reactivity, Synthesis, and Applications*, Liebman, J. F. and Dolbier, Jr., W. R. (Eds.), VCH: New York, 1988, p. 259.
3. (a) Chambers, R. D., in *Fluorine in Organic Chemistry*, Wiley-Interscience: New York, 1973; (b) Hudlicky, M., in *Chemistry of Organic Fluorine Compounds*, 2nd ed., Ellis Horwood: Chichester, 1976.
4. (a) Gerstenberger, M. R. C. and Haas, A., *Angew. Chem. Int. Ed. Engl.* (1981), 20, 647; (b) Haas, A. and Lieb, M., *Chimia, 39*, 134 (1985).
5. For a review, see Hewitt, C. D. and Silvester, M. J., *Aldrichimica Acta, 21*, 3 (1988).
6. Banks, R. E. (Ed.), *Organofluorine Compounds and Their Industrial Applications*, Ellis Horwood: Chichester, 1979. Filler, R. and Kobayashi, Y. (Eds.), *Biomedicinal Aspects of Fluorine Chemistry*, Kodansha Elsevier: New York, 1982. Filler, R. (Ed.), *Biochemistry Involving Carbon–Fluorine Bonds*, American Chemical Society: Washington, DC, 1976; ACS Symp. Ser. No. 28.
7. Wiemers, D. M. and Burton, D. J., *J. Am. Chem. Soc. 108*, 832 (1986).
7. Wiemers, D. M. and Burton, D. J., *J. Am. Chem. Soc. 108*, 832 (1986).
8. (a) Kobayashi, Y., Nakazato, A., Kumadaki, I., and Filler, R., *J. Fluorine Chem. 32*, 467 (1986) and references cited therein; (b) Matsui, K., Tobita, E., Ando, M., and Kondo, K., *Chem. Lett.* 1719 (1981).
9. Olah, G. A. and Ohyama, T., *Synthesis*, 319 (1976).
10. Kondratendok, N. V., Vechirko, E. P., and Yaguploskii, L. M., *Synthesis*, 932 (1980).
11. Chen, Q.-Y. and Wu, S.-W., *J. Chem. Soc. Chem. Commun.* 705 (1989).
12. (a) Kitazume, T. and Ishikawa, N., *Chem. Lett.*, 1679 (1981); *137*, 1453 (1982); (b) O'Reilly, N. J., Maruta, M., and Ishikawa, N., *Chem. Lett.*, 517 (1984).
13. (a) Santini, G., Le Blanc, M., and Riess, J. G., *J. Organomet. Chem. 140*, 1 (1977).; (b) Santini, G., Le Blanc, M., and Riess, J. G., *J. Chem. Soc. Chem. Commun.* 678 (1975).
14. Kitazume, T. and Ishikawa, N., *Nippon Kagaku Kaishi*, 1725 (1984).
15. (a) Stacy, M., Tatlow, J. C., and Sharpe, A. B., *Adv. Fluorine Chem., 3*, 19 (1963); (b) Stone, F. G. A. and West, R., *Adv. Organomet. Chem., 1*, 143 (1964); (c) Denson, D. D., Smith, C. F., and Tamborski, C., *J. Fluorine Chem., 3*, 247 (1973/1974); (d) Smith, C. F., Soloski, E. J., and Tamborski, C., *J. Fluorine Chem., 4*, 35 (1974); (e) MeBee, E. T., Roberts, C. W., and Meiners, A. F., *J. Am. Chem. Soc., 79*, 335 (1957); (f) Thoai, N., *J. Fluorine Chem., 5*, 115 (1975); (g) Von Werner, K. and Gisser, A., *J. Fluorine Chem., 10*, 387 (1977).
16. (a) Haszeldine, R. N., *J. Chem. Soc.*, 2952 (1949); Pierce, O. R., McBee, E. T., and Judd, G. F., *J. Am. Chem. Soc., 76*, 474 (1954); (b) Johncock, P., *J. Organomet. Chem., 19*, 257 (1969); (c) Gassman, P. G. and O'Reilly, N. J., *Tetrahedron Lett.*, 5243 (1985). This is the most definitive paper to generate *in situ* pentafluoroethyllithium; (d) Also see, Grassman, P. G., O'Reilly, N. J. *J. Org. Chem., 52*, 2481 (1987).
17. Sibille, S., Mcharek, S., and Perichon, J., *Tetrahedron, 45*, 1423 (1989).
18. Shono, T., Ishifune, M., Okada, T., and Kashimura, S., *J. Org. Chem., 56*, 2 (1991).
19. (a) Weber, W. P., in *Silicon Reagents for Organic Synthesis*, Springer-Verlag: Berlin, 1983; (b) Colvin, E., in *Silicon in Organic Synthesis*, Butterworths: London, 1980.
20. Ho, T. L., in *Hard and Soft Acids and Bases Principles in Organic Chemistry*, Academic: New York, 1977.

21. (a) Prakash, G. K. S., Krishnamurti, R., and Olah, G. A., *J. Am. Chem. Soc.*, *111*, 393 (1989); (b) Hartkopf, V. and Meijere, A. de, *Angew. Chem. Int. Ed. Engl.*, *21*, 443 (1982).
22. Stahly, G. P. and Bell, D. R., *J. Org. Chem.*, *54*, 2873 (1989).
23. Marchenko, A. P., Miroshnichenko, V. V., Kodian, G. N., and Pinchuk, A. M., *Zh. Obsch. Khim.*, *50*, 1987 (1980); *Chem. Abstr.* *93*, 866 (1980).
24. (a) Ruppert, I., Schlich, K., and Volbach, W., *Tetrahedron Lett.*, 2195 (1984). (b) Kruse, A., Siegemund, G., Schumann, D. C. A., and Ruppert, I. (Hoechst, A.-G.) Ger.Offen. DE3805534, 1989; *Chem. Abstr.*, *112*, 56272p, 1990. (c) Kruse, A., Siegemund, G., and Schumann, D. C. A., (Hoechst A.-G.) U.S. Patent 4,968,848 (1990).
25. Krishnamurti, R., Bellew, D. R., and Prakash, G. K. S., *J. Org. Chem.*, *56*, 984 (1991).
26. Pawelke, G., *J. Fluorine Chem.*, *42*, 429 (1989).
27. Farnham, W. B., Chapter 11.
28. Fujita, M., Obayashi, M., and Hiyama, T., *Tetrahedron*, *44*, 4135 (1988).
29. Eujen, R. *Spectrochimica Acta*, *43A*, 1165 (1987).
30. (a) The water of hydration in TBAF does not pose any serious problem;; (b) Kotun, S. P., Anderson, J. D. O., and Des Marteau, D. D., *J. Org. Chem.*, **57**, 1124 (1992).
31. Nakamura, E., Murofushi, T., Shimizu, M., and Kuwajima, I., *J. Am. Chem. Soc.*, *98*, 2346 (1976).
32. (a) Prakash, G. K. S. and Ramaiah, P., unpublished results. (b) Ramaiah, P., Prakash, G. K. S., *Synlet*, *643*, (1991).
33. (a) Haydock, D. B., Mulholland, T. P. C., Telford, B., Thorp, J. M., and Wain, J. S., *Eur. J. Med. Chem.*, *19*, 205 (1984); (b) Pogolotti, Jr., A. and Rupley, J. A., *Biochem. Biopys. Res. Commun.*, *55*, 1214 (1973).
34. Broicher, V. and Geffken, D., *Tetrahedron Lett.*, *30*, 5243 (1989).
35. Prakash, G. K. S., Krishnamurti, R., and Bellew, D. R., unpublished results.
36. Urata, H. and Fuchikami, T., *Tetrahedron Lett.*, *32*, 91 (1991).
37. Kolomeitsev, A. A., Movchun, V. N., Kondratenko, N. V., and Yagupolski, Yu. L., *Synthesis*, 1151 (1991).
38. Gassman, P. G., unpublished results.

CHAPTER 11

Silicon Mediated Reactions in Organofluorine Chemistry

WILLIAM B. FARNHAM

Anionic activation of tetracoordinate organosilicon species has been utilized for a variety of C—C and C—heteroatom bond formation strategies.[1,2] With few exceptions, organosilicon reagents have not been employed in syntheses of highly fluorinated materials.[3–5] This chapter summarizes our work in this area on C—O and C—C bond formation.

11.1. CARBON–OXYGEN BOND FORMATION

Silyl ethers, used chiefly as protecting groups, are not particularly reactive species although carbon–oxygen bonds in certain aromatic systems have been prepared from silyl ethers.[6,7] We have found that under the influence of anionic catalysts, silyl ethers react with perfluorinated alkenes to produce exceptionally high yields of partially fluorinated vinyl ethers. The degree of substitution at the vinylic positions may be controlled using a variety of fluoroalkenes and silyl ethers. Specificities and yields of these reactions are sufficiently high that difunctional starting materials afford quantitative yields of condensation polymers of moderate molecular weight. Cyclic dimers may be obtained in high yields under modified conditions and are of interest as selective fluoride ion hosts.

The difunctional silyl ether starting materials are obtained conveniently in quantitative yield by controlled treatment of the corresponding diols with hexamethyldisilazane in the presence of catalytic trimethylsilyl chloride (Scheme

Synthetic Fluorine Chemistry,
Edited by George A. Olah, Richard D. Chambers, and G. K. Surya Prakash.
ISBN 0-471-54370-5

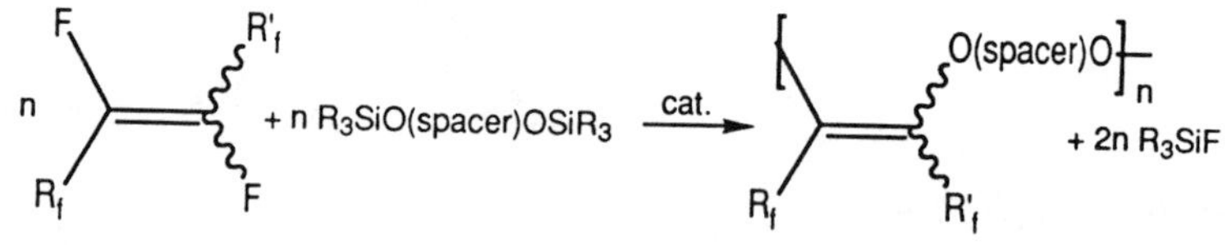

$$n\ R_fCF{=}CFR'_f + n\ R_3SiO(spacer)OSiR_3 \xrightarrow{cat.} [R_fC{=}C(R'_f)\text{-}O(spacer)O]_n + 2n\ R_3SiF$$

Scheme I

PREPARATION OF SILYL ETHERS

$$HO(CH_2)_nOH + (Me_3Si)_2NH \xrightarrow[\Delta]{Me_3SiCl\ (cat.)} Me_3SiO(CH_2)_nOSiMe_3 \quad \sim 100\%$$

$$HFPO \xrightarrow[steps]{several} HOCH_2CF_2CF_2OCF(CF_3)CF_2OCF(CF_3)CH_2OH$$

$$\downarrow (Me_3Si)_2NH$$

$$Me_3SiOCH_2CF_2CF_2OCF(CF_3)CF_2OCF(CF_3)CH_2OSiMe_3 \quad \sim 100\%$$

Scheme II

II. **Caution**: Can be exothermic). The reactivity of the silyl ethers (Scheme I) appears to depend on the silicon ligands, so a judicious choice of catalyst and fluoroalkene coreactant must be made to obtain optimum results in the substitution reaction. As shown in Scheme III, fluorosilicate catalysts activate silyl ethers containing simple hydrocarbon alkoxy ligands, and good control over the degree of substitution on the fluorinated double bond is realized. Partially fluorinated alkoxy silanes require milder conditions to achieve comparable selectivity. It should be noted that some of the catalysts which activate silyl ethers can cause side reactions with terminal fluoroalkenes. Nevertheless, the degree of fluorine substitution on hexafluoropropene can be controlled using sufficiently reactive silyl ethers and carefully controlled conditions.

A variety of anionic catalysts, including fluorosilicates, bifluoride, carboxylates, and other oxyanions are effective for the substitution reaction. Similar catalysts have been found useful for activating other silicon species.[8,9] Few limitations have been found in the choice of solvent, although dry, aprotic

SIMPLE VINYL ETHERS

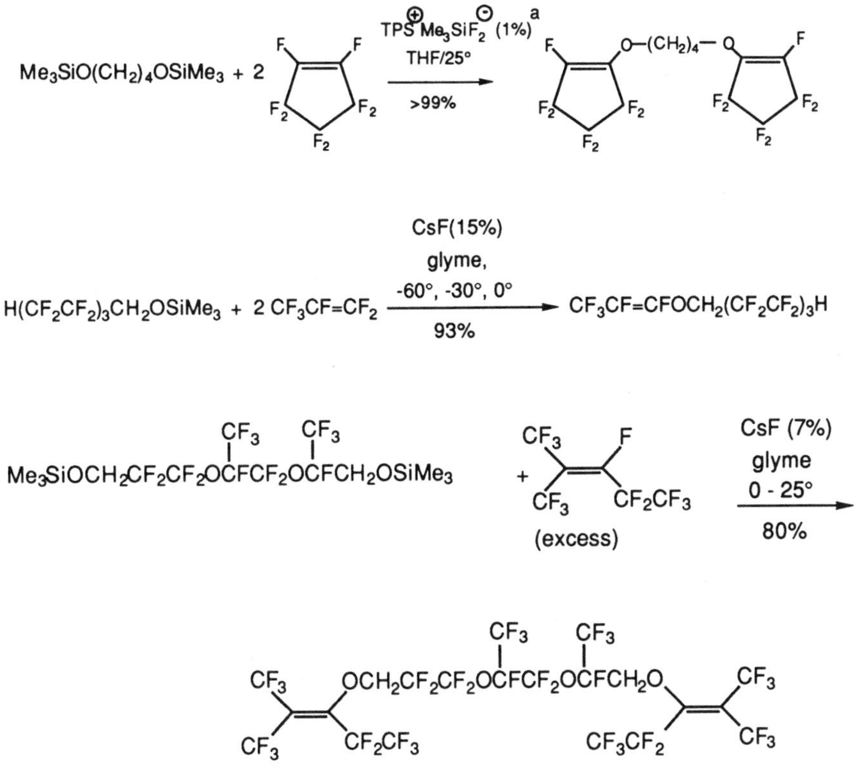

$$Me_3SiO(CH_2)_4OSiMe_3 + 2\ \text{(perfluorocyclopentene)} \xrightarrow[\text{THF/25°}\ >99\%]{TPS^{\oplus}Me_3SiF_2^{\ominus}\ (1\%)^{a}} \text{(perfluorocyclopentenyl)}\text{–}O\text{–}(CH_2)_4\text{–}O\text{–}\text{(perfluorocyclopentenyl)}$$

$$H(CF_2CF_2)_3CH_2OSiMe_3 + 2\ CF_3CF{=}CF_2 \xrightarrow[\text{-60°, -30°, 0°}\ 93\%]{CsF(15\%)\ \text{glyme,}} CF_3CF{=}CFOCH_2(CF_2CF_2)_3H$$

$$Me_3SiOCH_2CF_2CF_2OCF(CF_3)CF_2OCF(CF_3)CH_2OSiMe_3 + (CF_3)_2C{=}CF(CF_2CF_3)\ \text{(excess)} \xrightarrow[0\text{-}25°\ 80\%]{CsF\ (7\%)\ \text{glyme}}$$

$$(CF_3)_2C{=}C(CF_2CF_3)OCH_2CF_2CF_2OCF(CF_3)CF_2OCF(CF_3)CH_2OC(CF_2CF_3){=}C(CF_3)_2$$

a) TPS = tris(piperidino)sulfonium

Scheme III

media are preferred. Bulk polymerizations using difunctional silyl ethers and vinyl ethers (discussed later) can be carried out without solvent providing that very reactive catalysts are used.

Since the conversion of fluorinated alkenes to vinyl ethers is well known,[10] it should be emphasized that the principal benefit in our silyl ether chemistry is one of selectivity and yield. Whereas the base-catalyzed reaction of alcohols and fluoroalkenes almost always gives a mixture of substitution and addition products, clean substitution occurs with silyl ethers. Homogeneity, which persists during bulk polymerization, is an additional, attractive feature. Substitution of the first fluorine in a 1,2-disubstituted perfluorinated alkene proceeds more rapidly than subsequent fluorine substitution in the product vinyl ether. The 2:1 adducts shown in Scheme IV are therefore conveniently prepared in

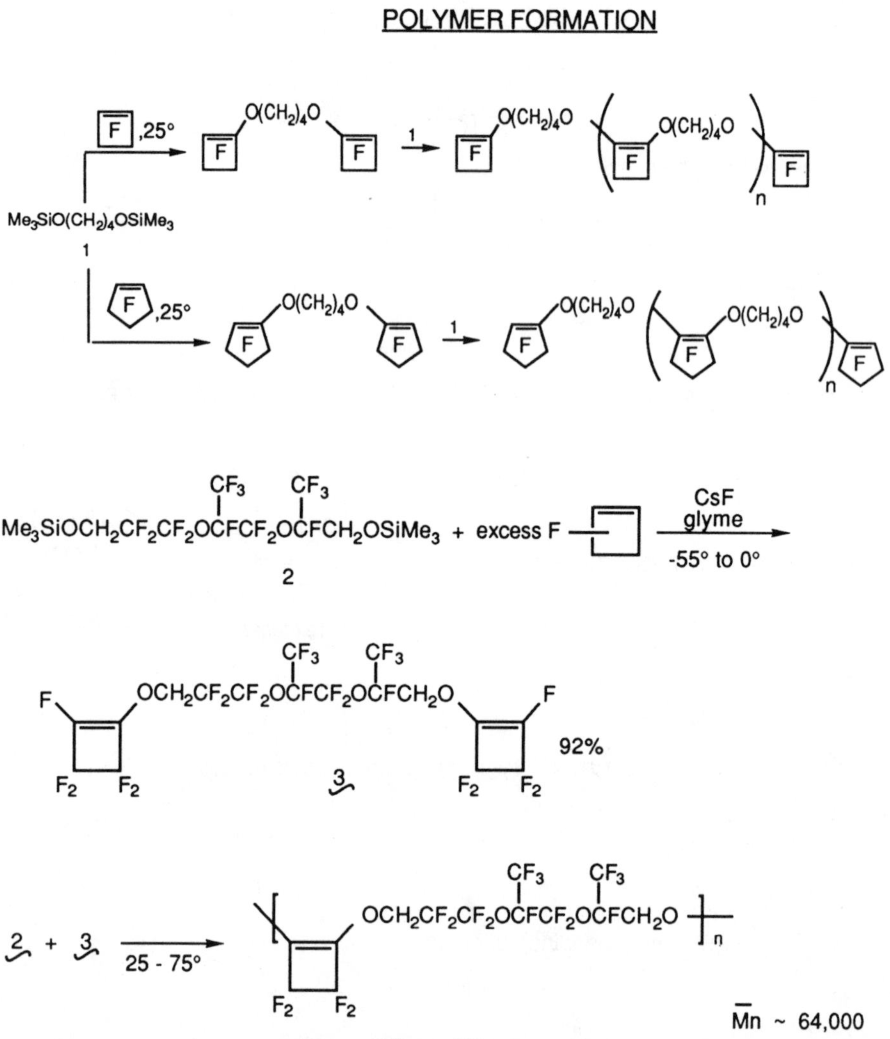

POLYMER FORMATION

$Me_3SiO(CH_2)_4OSiMe_3$ 1

$$\text{F-cyclobutene},\ 25^\circ \rightarrow \text{F}\!-\!O(CH_2)_4O\!-\!\text{F} \xrightarrow{1} \text{F}\!-\!O(CH_2)_4O\!\left(\!-\text{F}\!-\!O(CH_2)_4O-\right)_n\!\text{F}$$

$$\text{F-cyclopentene},\ 25^\circ \rightarrow \text{F}\!-\!O(CH_2)_4O\!-\!\text{F} \xrightarrow{1} \text{F}\!-\!O(CH_2)_4O\!\left(\!-\text{F}\!-\!O(CH_2)_4O-\right)_n\!\text{F}$$

$$\underset{2}{Me_3SiOCH_2CF_2CF_2OCF(CF_3)CF_2OCF(CF_3)CH_2OSiMe_3} + \text{excess F-cyclobutene} \xrightarrow[-55^\circ \text{ to } 0^\circ]{\text{CsF, glyme}}$$

$$\underset{3}{C_4F_5\text{-}OCH_2CF_2CF_2OCF(CF_3)CF_2OCF(CF_3)CH_2O\text{-}C_4F_5} \quad 92\%$$

$$2 + 3 \xrightarrow{25 - 75^\circ} \left[\text{-}C_4F_4\text{-}OCH_2CF_2CF_2OCF(CF_3)CF_2OCF(CF_3)CH_2O\text{-}\right]_n \qquad \overline{M}n \sim 64{,}000$$

Scheme IV

excellent yield. On a laboratory scale, these difunctional vinyl ethers permit good control of the stoichiometry required for polymer formation. Using model difunctional silyl ethers and vinyl ethers, number average molecular weights up to 64,000 have been achieved.

In the course of these polymerization studies, cyclic oligomers were found as significant by-products. Using moderate concentrations of difunctional silyl ethers and vinyl ethers, isolated yields of cyclic dimers approached 50%. Representative examples are shown in Schemes Va and b. Some of these cyclic dimers exhibit remarkable anion binding properties, and the structure and dynamics of a fluoride ion adduct (Scheme Vc) have been described in some

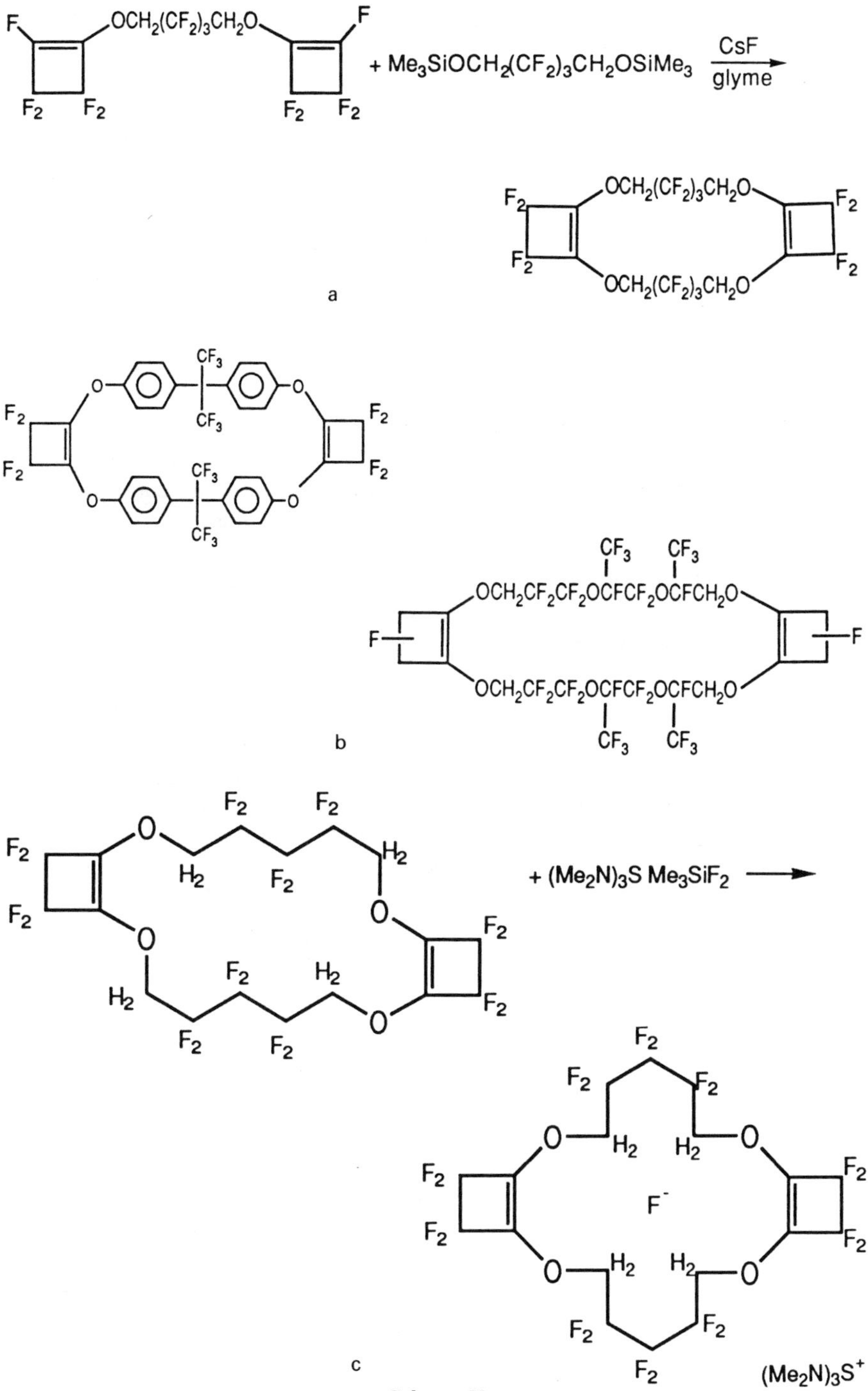

F OCH$_2$(CF$_2$)$_3$CH$_2$O F
F$_2$ F$_2$ F$_2$ F$_2$

$+ Me_3SiOCH_2(CF_2)_3CH_2OSiMe_3$ $\xrightarrow[\text{glyme}]{\text{CsF}}$

F$_2$ OCH$_2$(CF$_2$)$_3$CH$_2$O F$_2$
F$_2$ OCH$_2$(CF$_2$)$_3$CH$_2$O F$_2$

a

CF$_3$
F$_2$ O O F$_2$
CF$_3$
F$_2$ CF$_3$ F$_2$
O O
CF$_3$

CF$_3$ CF$_3$
OCH$_2$CF$_2$CF$_2$OCFCF$_2$OCFCH$_2$O
F F
OCH$_2$CF$_2$CF$_2$OCFCF$_2$OCFCH$_2$O
CF$_3$ CF$_3$

b

F$_2$ F$_2$
F$_2$ O H$_2$
H$_2$ F$_2$
F$_2$ O O F$_2$
F$_2$ H$_2$
H$_2$ F$_2$
F$_2$ F$_2$ O

$+ (Me_2N)_3S\ Me_3SiF_2 \longrightarrow$

F$_2$
F$_2$ F$_2$
O H$_2$ H$_2$ O
F$_2$ F$_2$
F$^-$
F$_2$ F$_2$
O H$_2$ H$_2$ O
F$_2$ F$_2$
F$_2$

c

$(Me_2N)_3S^+$

Scheme V

detail.[11] In this adduct, the fluoride ion is held within the host binding cavity by virtue of hydrogen bonding with the CH_2 groups.

One of the long-standing goals in fluorocarbon synthesis is the convenient preparation of high molecular weight perfluoropolyethers of well-defined structure.[12] Since the ring-opening polymerization of fluorocarbon epoxides is difficult to control because of chain-transfer processes, we adapted the polycondensation process of Scheme I for this application. Following the polymerization reaction, a subsequent fluorination step is required to saturate double bonds and replace hydrogens in the "spacer" groups with fluorine. We found that the partially fluorinated vinyl ethers produced in the polycondensation reaction undergo reaction with elemental fluorine to give the desired perfluorinated polyethers.[13] Ultraviolet irradiation is required to replace the least reactive hydrogens. Attainment of exceptionally high molecular weight material remains problematic. This is due, at least in part, to a minor cross-linking reaction involving polymeric vinyl ether chains and difunctional silyl ethers that takes place in the presence of the usual anionic catalysts under forcing conditions. Thus, treatment of vinyl ether-ended oligomer ($\bar{M}n \sim 12{,}000$) with excess difunctional silyl ether and catalyst at somewhat elevated temperatures (65–100 °C) resulted eventually in a product with altogether different physical properties (insoluble gel). This side reaction appears to be important especially when the concentration of terminal vinyl ether groups becomes quite small.

Our previous mechanistic studies of another silicon-mediated reaction, "group-transfer polymerization" of methacrylates,[8,14] indicated the involvement of pentacoordinate silicon species. Our working hypothesis for the conversion of silyl ether to fluorinated vinyl ether is that formation of pentacoordinate silicon species (Scheme VI) should increase reactivity and promote nucleophilic character in the alkoxy ligands. Other groups have described increased reactivity in pentacoordinate silicates[1,15] for other processes. Demonstrating that silicon remains bound to the oxygen during the product-forming step, however, is not a straightforward proposition. We prepared and isolated a pentacoordinate silicate that contains a partially fluorinated alkoxy ligand (Scheme VII). This complex ($M^+ = Na$ or TAS) undergoes the desired reaction with perfluorocyclobutene to give a vinyl ether product, but this process may involve predissociation (especially with

MECHANISTIC HYPOTHESIS

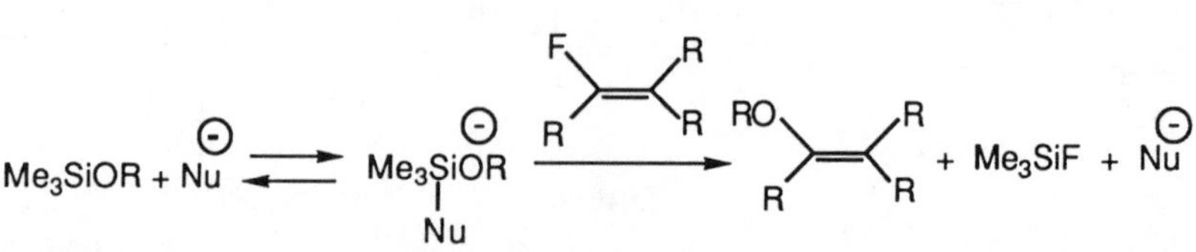

$$Me_3SiOR + Nu^{\ominus} \rightleftharpoons [Me_3Si(Nu)OR]^{\ominus} \xrightarrow{FRC=CR_2} ROCR=CR_2 + Me_3SiF + Nu^{\ominus}$$

Scheme VI

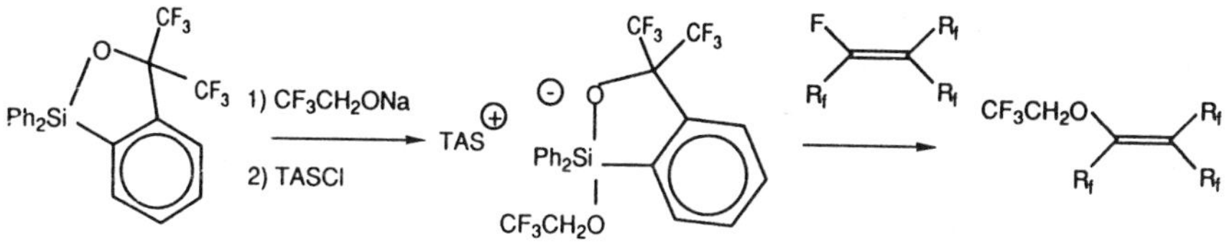

Scheme VII

M^+ = Na). Indirect evidence has been sought by examining the regiochemistry of substitution in perfluorocyclohexene. As shown in Scheme VIII, a counterion effect is certainly operative, but this does not directly address the question whether silicon is present at the product-determing step. It should be noted that the same isomer distribution results from either Me_3Si- or Et_3Si-substituted test reagents.

• Regioselectivity with 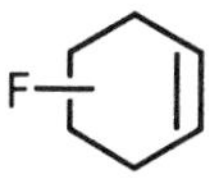 F–(cyclohexene)

reagent	products 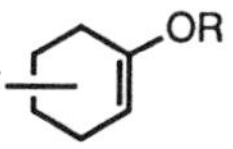	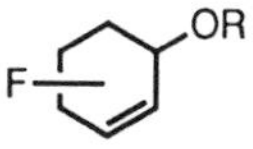
$H(CF_2)_6CH_2OSiMe_3$ + CsF	69.2%	30.8%
$CF_3CH_2OSiMe_3$ + CsF	70.0%	30.0%
$H(CF_2)_6CH_2OH$ + NaH	95.0%	5.0%
$H(CF_2)_6CH_2OH$ + Cs_2CO_3	70.5%	29.5%

Scheme VIII

11.2. CARBON–CARBON BOND FORMATION

In view of the beneficial influence of silicon for the reactions discussed above, we investigated other silicon reagents for utility in fluorocarbon synthetic schemes. Carbon–carbon bond formation seemed a particularly attractive goal, so we have examined perfluoroalkylsilanes (R_fSiMe_3) as reagents for adding perfluoroalkyl groups to perfluorinated systems. Prior to our work, perfluoroalkylsilanes had been prepared[16] but had not been utilized to synthetic advantage in this area. Prakash et al.[17] and Stahly and Bell[18] used the reaction of trifluoromethylsilanes with aldehydes to prepare trifluoromethyl-substituted compounds (see Prakash, Chapter 10).

We reinvestigated Gilman's preparation of perfluoroalkylsilanes and find somewhat better yields and more forgiving conditions than those reported previously (Scheme IX). Yields were consistently in the 80–85% range, and even those preparations that did not begin "normally" at low temperature (and resulted in exothermic excursions once the temperature reached ~25 °C) provided nearly equivalent yields. Production of R_fH was not a significant problem even with difunctional examples.

Synthesis of R_fSiMe_3

$$R_fBr + TMSCl \xrightarrow[-30° \text{ to } 25°]{Mg,\ THF} R_fSiMe_3$$

Yield ~ 80 to 85%

examples: $R_f = CF_3(CF_2)_n$- n = 5,7

$-(CF_2)_m$ m = 6, 8,10

Scheme IX

Activated by anionic catalysts (e.g., CsF, TAS fluorosilicates, carboxylates, fluoroalkoxides, fluorocarbanions, etc.), R_fSiMe_3 react readily with fluorinated acceptor molecules (A–F) to give perfluoroalkylated products and fluorotrimethylsilane. We have concentrated on acid fluorides and alkenes as acceptors, although certain reactive aromatics and some other carbonyl compounds function satisfactorily. Our heuristic mechanistic hypothesis for these reactions involves formation and trapping of pentacoordinate silicon intermediates.

Acid fluorides serve as precursors to ketones. As shown in Scheme X, product yield is somewhat higher for the perfluorinated acceptors. It should be noted

KETONE SYNTHESIS:

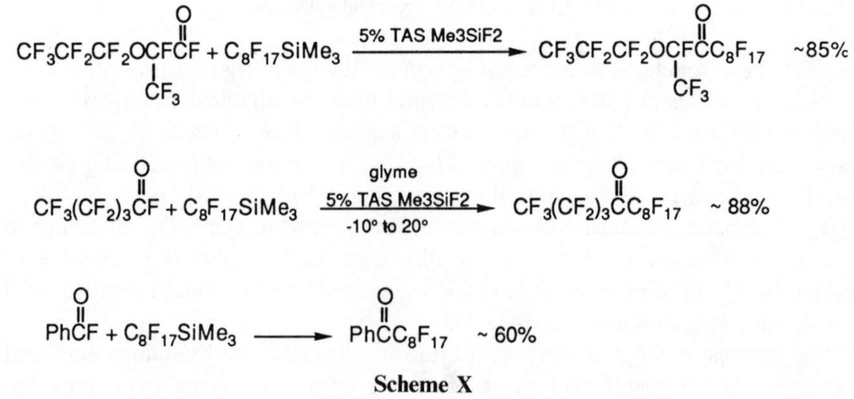

$$CF_3CF_2CF_2OCF(CF_3)C(=O)F + C_8F_{17}SiMe_3 \xrightarrow{5\% \text{ TAS Me3SiF2}} CF_3CF_2CF_2OCF(CF_3)C(=O)C_8F_{17} \quad \text{~85\%}$$

$$CF_3(CF_2)_3C(=O)F + C_8F_{17}SiMe_3 \xrightarrow[-10° \text{ to } 20°]{\text{glyme, } 5\% \text{ TAS Me3SiF2}} CF_3(CF_2)_3C(=O)C_8F_{17} \quad \text{~ 88\%}$$

$$PhC(=O)F + C_8F_{17}SiMe_3 \longrightarrow PhC(=O)C_8F_{17} \quad \text{~ 60\%}$$

Scheme X

that few general methods exist for preparation of perfluorinated ketones. Indeed, it is somewhat surprising that these materials do not react readily with perfluoroalkylsilane under these conditions to give tertiary alcohol derivatives. We observed this side reaction, but product yields are low and utilization of R_fSiMe_3 is relatively inefficient. A competitive reaction pathway involves anionic, catalytic destruction of R_fSiMe_3. Additional discussion of R_fSiMe_3 side reactions is warranted because of their impact on perfluoroalkylation reactions. Addition of trace amounts of anionic catalysts to solutions of R_fSiMe_3 rapidly leads to the condensation of R_f fragments, typically as a separate fluorocarbon layer. Thus, $C_8F_{17}SiMe_3$ provides a large number of $C_{16}F_{32}$ compounds presumed to be isomeric alkenes. Smaller proportions of trimers are also formed. In order to minimize this process during perfluoroalkylation reactions, it is advantageous to add the R_fSiMe_3 dropwise to a solution (or suspension) of acceptor and catalyst. It is often useful to employ the acceptor in modest excess since side reactions do not become serious until acceptor concentrations are low. An additional, yield-enhancing factor for the perfluorinated ketones is their low solubility in the reaction media employed. While the reaction of perfluorinated ketones with R_fSiMe_3 is inefficient, it is adequate to create cross-links between oligomers formed from terephthaloyl fluoride and α,ω-bis(trimethylsilyl) fluoroalkanes. Our attempts to prepare fluorinated polyketones using this C—C bond forming strategy have provided low molecular weight oligomers or in other cases, cross-linked and intractable material.

As discussed in Section 11.1, fluorinated alkenes react readily with many nucleophilic species. These acceptors react with R_fSiMe_3 to give either mono- or bis(perfluoroalkylated) products (Scheme XI). Product specificity is enhanced for cyclic alkenes whose higher reactivity and thermodynamically uphill

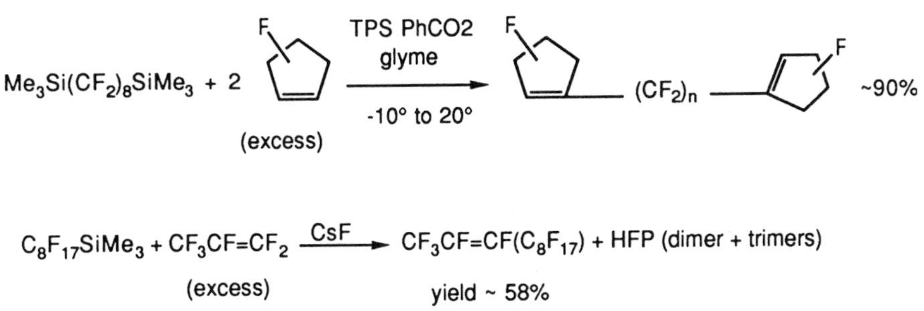

Alkene Synthesis

$C_8F_{17}SiMe_3$ + F [cyclopentene] ⟶ F [cyclopentene]—C_8F_{17} ~ 95%

$Me_3Si(CF_2)_8SiMe_3$ + 2 F [cyclopentene] (excess) —TPS PhCO2, glyme, -10° to 20°⟶ F [cyclopentene]—$(CF_2)_n$—[cyclopentene] F ~90%

$C_8F_{17}SiMe_3 + CF_3CF{=}CF_2$ (excess) $\xrightarrow{CsF}$ $CF_3CF{=}CF(C_8F_{17})$ + HFP (dimer + trimers) yield ~ 58%

Scheme XI

isomerization pathways account for this tendency. Perfluoroalkylation of hexafluoropropylene affords predominantly the *trans* 2-ene product, but isomerization occurs very easily in the presence of anionic cayalysts and isomeric composition is difficult to control. Moreover, hexafluoropropylene undergoes separate dimerization and trimerization processes that result in inefficient use of this component.

Because the yields of perfluoroalkylated alkenes in some cases are quite good and the principal side reactions involved coupling of the perfluoroalkylene fragments, we have attempted to use this chemistry for perfluorocarbon polymers. Using a model system (Scheme XII), preliminary reactions have provided low molecular weight oligomers (Mn ~3200). Improved selectivity is obviously required to produce high molecular weight polymer from these reactive species, and we are attempting to achieve this by judicious choice of metalloid ligands.

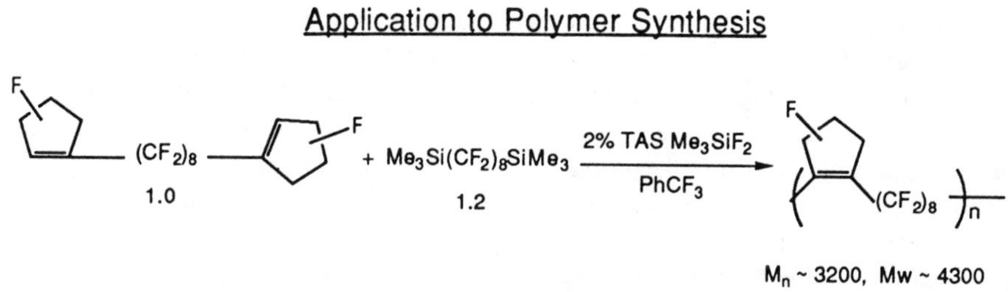

Application to Polymer Synthesis

$(CF_2)_8$ 1.0 + $Me_3Si(CF_2)_8SiMe_3$ 1.2 → (2% TAS Me_3SiF_2, $PhCF_3$)

M_n ~ 3200, Mw ~ 4300

Scheme XII

Silicon-based reagents are well suited to the construction of C—O and C—C bonds in highly fluorinated cystems. The ready availability of acceptors that contain fluorine as a potential leaving group makes such reagent–substrate combinations especially attractive. These and cognate systems should find wide applicability in fluorocarbon syntheses.

REFERENCES

1. Corriu, R. J. P. and Young, J. C., *The Chemistry of Organic Silicon Compounds*, S. Patai and Z. Rappoport (Eds.), Wiley, New York, 1989, p. 1241.
2. Larson, G. L., *J. Organometal. Chem., 360*, 39 (1989).
3. (a) Fujita, M. and Hiyama, T., *J. Am. Chem. Soc., 107*, 4085 (1985). (b) Fujita, M., Obayashi, J., and Hiyama, T., *Tetrahedron, 44*, 4135 (1988).
4. Yamazaki, T. and Ishikawa, N., *Chem. Lett.*, 521 (1984).
5. Boutevin, B. and Pietrasanta, Y., *Progress in Organic Coatings, 13*, 297 (1985).
6. Kricheldorf, H. R. and Bier, G., *Polym. Chem. Ed.*, 21, 2283 (1983).
7. Nambu, Y. and Endo, T. (1990), *Tetrahedron Lett., 31*, 1723 (1990).
8. Webster, O. W., Hertler, W. R., Sogah, D. Y., Farnham, W. B., and RajanBabu, T. V., *J. Am. Chem. Soc.*, 105 5706 (1983).

9. Dicker, I. B., Cohen, G. M., Farnham, W. B., Hertler, W. R., Laganis, E. D., and Sogah, D. Y., *Macromolecules*, *23*, 4034 (1990).
10. Chambers, R. D., *Fluorine in Organic Chemistry*, R. D. Chambers, Durham, UK, 1973, p. 138.
11. Farnham, W. B., Roe, D. C., Dixon, D. A., Calabrese, J. C., and Harlow, R. L., *J. Am. Chem. Soc.*, *112*, 7707 (1990).
12. Hill, J. T. and Erdman, J. P., *Ring-Opening Polymerization*, T. Saegusa and E. Goethals (Eds.), ACS Symposium Series 59, Washington DC, 1977, p. 269.
13. Farnham, W. B. and Nappa, M. J. (1990), unpublished results. For similar reactions, see Sievert, A. C., Tong, W. R., and Nappa, M. J., *J. Fluorine Chem.*, *53*, 397 (1991).
14. Sogah, D. Y. and Farnham, W. B., in *Organosilicon and Bioorganosilicon Chemistry*, H. Sakurai (Ed.), 1985, p. 219.
15. Sakurai, H., *Synlett*, *1*, 1 (1989).
16. Jukes, A. E. and Gilman, H. J., *Organometal. Chem.*, *18*, 33 (1969). See also Smith, M. R., Jr. and Gilman, H. J., *Organometal. Chem. 46*, 251 (1972).
17. Prakash, G. K. S., Krishnamurti, R., and Olah, G. A., *J. Am. Chem. Soc.*, *111*, 393 (1989).
18. Stahly, G. P. and Bell, D. R., *J. Org. Chem.*, *54*, 2873 (1989).

CHAPTER 12

Synthetic Aspects of Electrophilic ipso Reactions of Polyfluoroarenes

V. D. SHTEINGARTS

Polyfluoroarenes are an extraordinary type of chemical compound. At the beginning of the development of polyfluoroarene chemistry it was found that when fluorine atoms accumulate in the aromatic ring the arene chemistry becomes richer rather than poorer is electrons. In going from hydrocarbons to fluorocarbons the aromatic ring acquires an ability to undergo the transformations not characteristic of hydrocarbons. This number is constantly growing as a result of continuing efforts in a series of laboratories around the world. But it is fortunate that practically none of the hydrocarbon-type reactivity vanishes due to polyfluorination of the aromatic ring. Because of its wealth and variety of reactions, polyfluororene chemistry always has surprises for us.

For example, consider the electrophilic reactions most typical of arenes. It might be thought that progressive fluorine atom substitution into the aromatic ring will suppress this type of reactivity. However, it turns out that when the last hydrogen atom is present it can be replaced practically by any electrophile with which an aromatic hydrocarbon can react (Scheme I).[1]

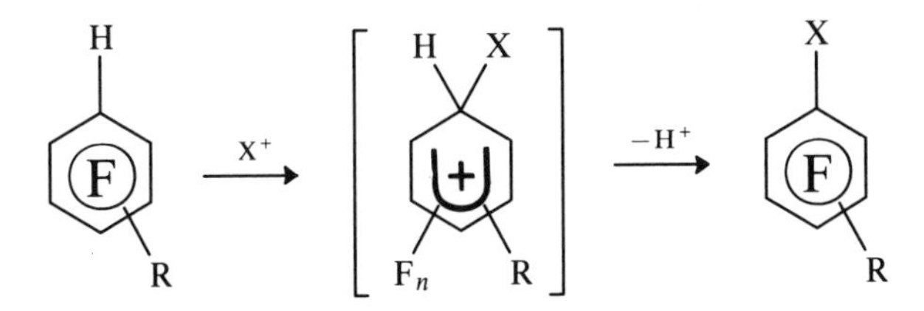

Scheme I

Synthetic Fluorine Chemistry,
Edited by George A. Olah, Richard D. Chambers, and G. K. Surya Prakash.
ISBN 0-471-54370-5

But what happens when the last hydrocarbon atom is replaced by fluorine? It was discovered[2,3] that despite the high electronegativity of fluorine, the electrophilic ipso attack[4] is possible at the position occupied by it. Since fluorine cannot be eliminated as a cation, the substitution does not occur, but the electrophilic addition is realized. This type of reactivity extends significantly the synthetic capabilities furnished by electrophilic reactions in polyfluoroarene chemistry (Scheme II).

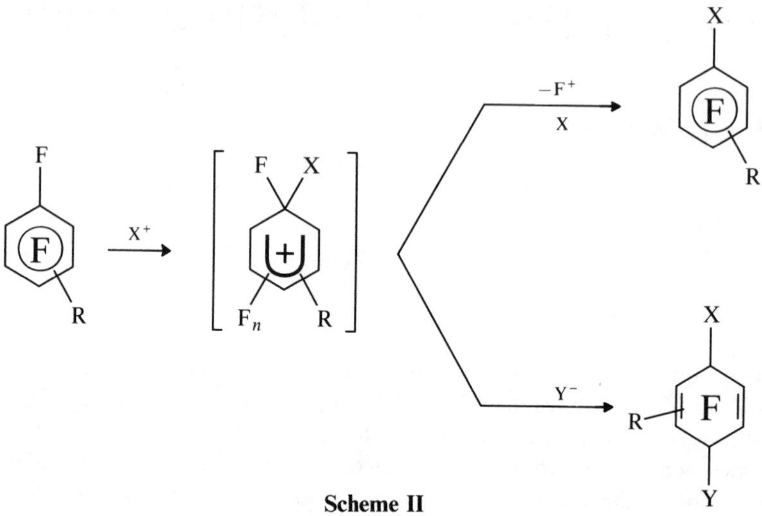

Scheme II

For the first time the possibility of electrophilic ipso reactions of polyfluorarenes has been demonstrated in the case of the addition of nitrating reagents. The electrophilic character of such reactions has been confirmed experimentally[3,5] and they have been shown to be typical for polyfluoroarenes occurring in two variantions: nitric acid addition yielding polyfluorinated nitrocyclohexadienones[2] and conjugated nitro fluorination with nitrocyclohexadiene formation (Scheme III).[3,6]

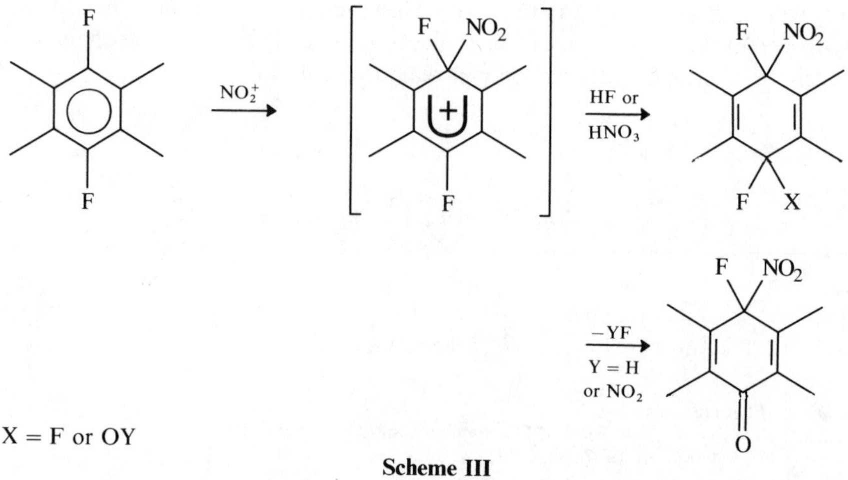

Scheme III

The regularities governing the regioselectivity due to the presence of substituents other than fluorine are important for the synthetic applications of these reactions. It was found that the total effect of all substituents results in the formally meta-orienting effect of the methyl group and the ortho–meta-orienting effect of halogens other than fluorine in the benzene ring.[6] The combined effect of the methyl group and halogen located para to each other leads to nitronium cation addition at the meta position with respect to the former and the ortho position with respect to the latter (NO_2^+ means HNO_3 solution in HF) (Scheme IV).[7]

X	Cl	Br
m/o	1.9	2.3

Scheme IV

The polyfluorinated polynuclear systems have the same most active positions as the parent hydrocarbons (Scheme V).[2,3,8]

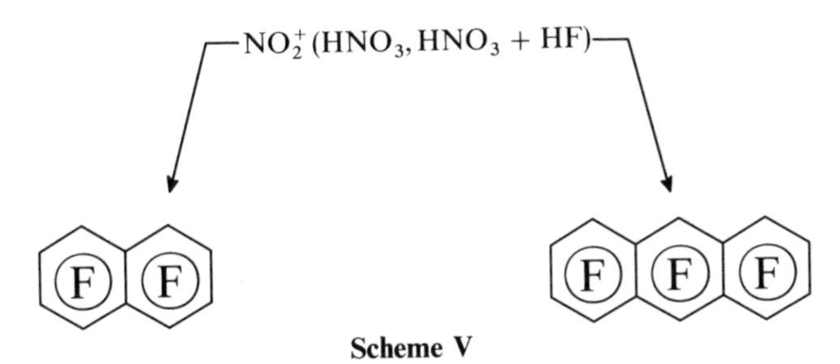

Scheme V

However, due to the presence of a substituent other than fluorine the equivalent positions of the polynuclear framework show different activities. The influence of β-substituents in polyfluorinated naphthalene on the relative reactivity of α positions is similar to their effects in the benzene ring.[9,10] When a nonfluorine substituent occupies the α position, there is a competition between the ipso and para-orienting effects of fluorine and nonfluorine substituents. In this situation the methyl group is a stronger substituent than fluorine in both effects,[11] whereas chlorine and bromine are weaker orienting groups (Scheme VI).[10]

It was shown that in general substituent effects on the regioselectivity of these reactions correspond to the relative stability of isomeric polyfluorinated arenium ions generated by various methods and modeled by their unsaturated part of the intermediates and, to some extent, the transition states of the reactions under consideration.[12]

The regioselectivity in the reactions shown above reveals what appears formally as the orienting effect of a nonfluorine substituent being virtually the result of the total effect of all substituents. For instance, the apparent meta-

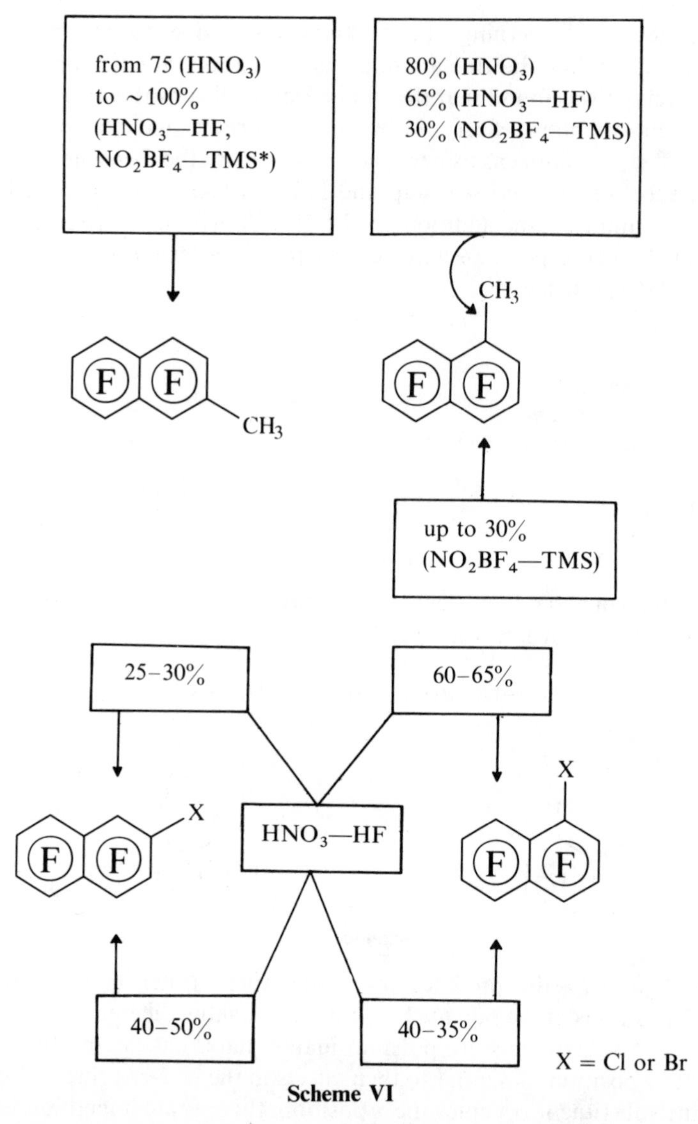

Scheme VI

orienting effect of the methyl group really reflects the tendency to avoid the meta-location of a fluorine atom in the arenium ion. The competition of the orienting effects of fluorine, on the one hand, and the chlorine or bromine atom, on the other hand, is more complicated and depends presumably on the electron demand of the electron-deficient framework as a function of its structure and reaction coordinate.[13] This dependence may account for the predominance of the ortho-location of chlorine in polyfluorinated benzenium and 1-naphthalenium ions contrary to the orienting effect of chlorine in the reactions of

*Tetramethylenesulfone = TMS.

polyfluoroarenes with the nitrating agents shown above. But the advantage of the para-location of fluorine in its competition with the heavier halogen atom is a general feature of the reactions of polyfluoroarenes with nitrating agents and formation of stable polyfluoroarenium ions.[12]

Unlike this, the effects of strong electron-donating substitutents, such as OH, OAlk, or OAr, dominate over the total fluorine effect leading to the ortho–para orientation with the former being predominant. Two types of transformations occur in these cases. Nitration of polyfluorinated hydroxy compounds proceeds as the phenol–dienone rearrangement yielding nitrocyclohexadienones.[2,5,14] When an ether group is present as a substituent, the electrophilic addition takes place with the formation of alkoxy- or aryloxycyclohexadienes (Scheme VII).[15]

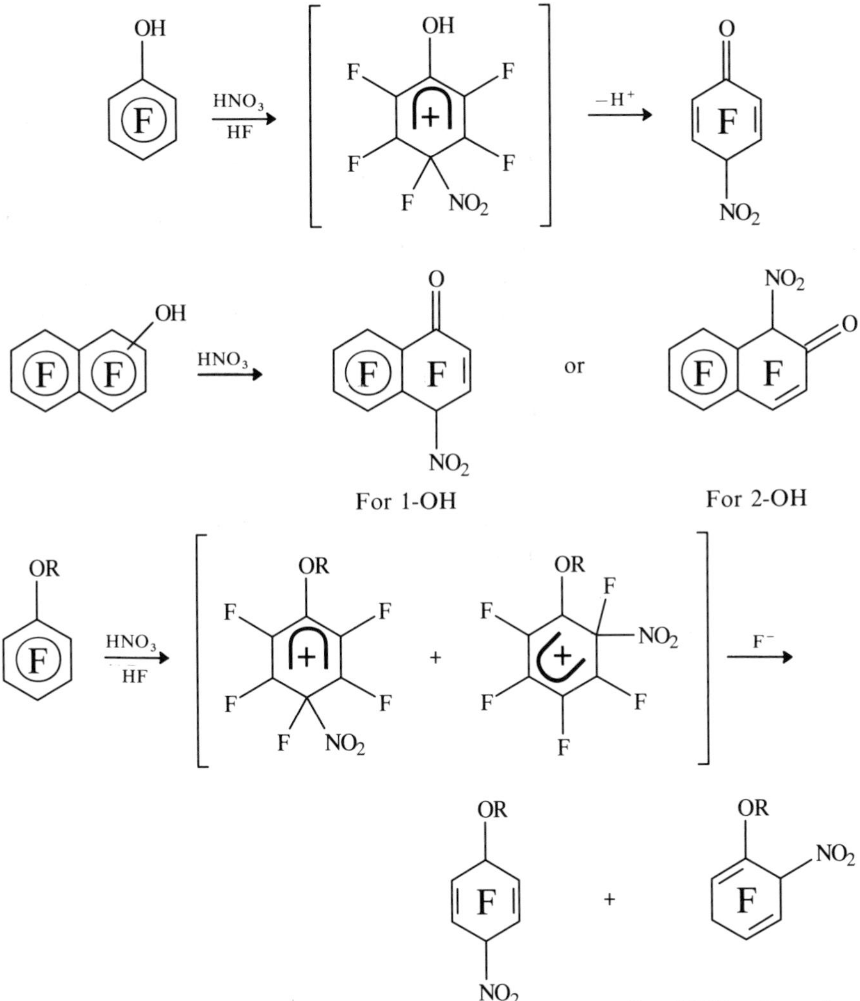

$R = CHF_2$ 9 : 1
$R = C_6F_5$ 3 : 1

Scheme VII

The reactions of partly fluorinated arenes with nitrating reagents, by contrast with other traditional electrophilic reactions, proceed not only as hydrogen substitution but also as ipso addition. As it was shown by *β-H*-heptafluoronaphthalene[16] and 1,2,3,4,9,10-hexafluoroanthracene[8] when the unsubstituted position is not related to the set of most active ones in this aromatic system ipso-addition occurs to the most reactive position despite its occupation by fluorine (Scheme VIII).

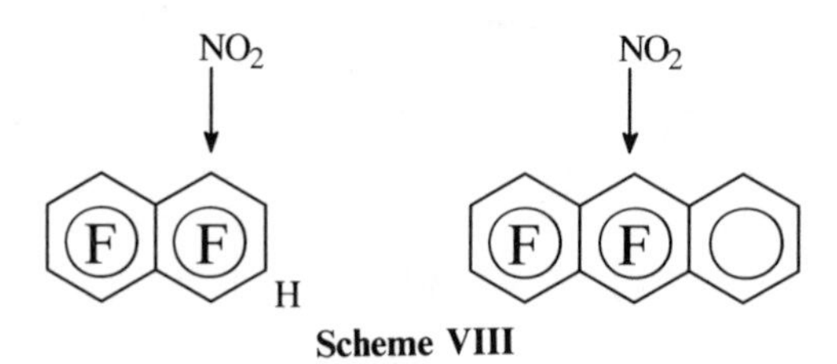

Scheme VIII

However, when positions that are either unsubstituted or occupied by fluorine are equivalent in the aromatic framework, the competition between the ipso and nonipso reaction takes place depending on the nature of the nitrating reagent. As a result, the share in the product mixture and the yield of pentafluoronitrobenzene (see Table 12.1) is the reactions of pentafluorobenzene with nitrating reagents were shown to depend on the nature of the nitrating system and to increase in going from the solution of nitric acid in sulfuric acid to the solution in hydrogen fluoride and further to nitronium tetrafluoroborate in TMS or as a result of antimony pentafluoride addition to the solution of nitric acid in hydrogen fluoride (Scheme IX).[5,6,17]

NO_2 (**1**) ← 1. NO_2^+ 2. $-H^+$ — H — 1. NO_2^+ 2. $+X^-$ → X, H, NO_2 (**2**)

Scheme IX

TABLE 12.1.

	Ratio (%)	
Nitrating System	**1**	**2**
HNO_3—H_2SO_4	~20	Not isolated
HNO_3—HF	70	30 (X = F)
NO_2BF_4—TMS HNO_3—HF—SbF_5	~100	~0

The ratio of the products of ipso-nitrofluorination and hydrogen substitution in the reaction of 1-X-2,3,5,6-tetrafluorobenzenes with HNO_3 in HF depends on the X identity with the share of the product of former reaction increasing in the order F < Cl < Br < CH_3 < H (see Table 12.2), which reflects the relative preference of the X substituent to be located in the ortho versus para position with respect to the reaction center (Scheme X). The case with X = CH_3 shows that in the situation with a lack of the competition of halogen, the methyl group reveals its orienting influence typical for electrophilic reactions.[7]

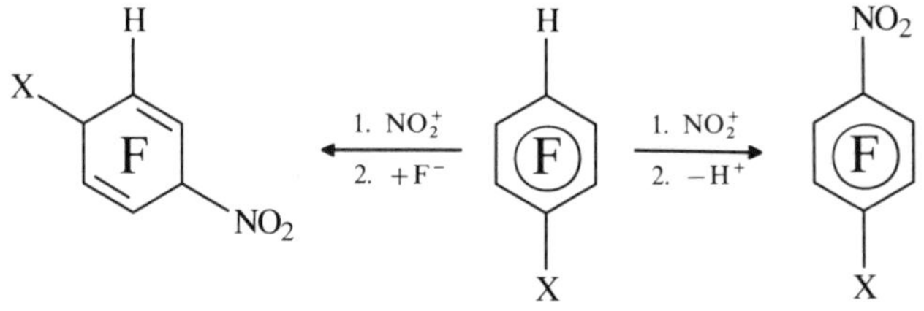

Scheme X

TABLE 12.2.

	Ratio (%)	
X	**3**	**4**
H	~100	0
CH_3	80	20
Br	75	25
Cl	66	34
F	30	70

The most illustrative reaction here is the interaction of α-*H*-heptafluoronaphthalene with nitrating agents. The reaction of this compound with HNO_3 gives the substitution product in yields up to 20% and ipso-addition at positions 5 and 8 as the main products. In hydrogen fluoride, the attack at the unsubstituted position increases to 50% resulting in this case in the formation of the addition product. No substitution product was obtained. Only treatment of this substrate with NO_2BF_4 in TMS leads mainly to the hydrogen substitution product—1-nitroheptafluoronaphthalene (Scheme XI).

Considering the origin of such regioselectivity variation one has to choose between two reasons: changing electrophilic properties of NO_2^+ with variation of solvation or ion association and changing the ratio between the rate of ipso-arenium ion capture by nucleophile, on the one hand, and the rate of this ion's rearrangement due to nitro group migration, on the other hand, as a result of variation of medium nucleophilicity. Currently, it is impossible to choose between these reasons.[18]

The transformations of principal ipso-nitration products, that is, nitrocyclohexadienones and nitrocyclohexadienes, offer many synthetic routes. The former

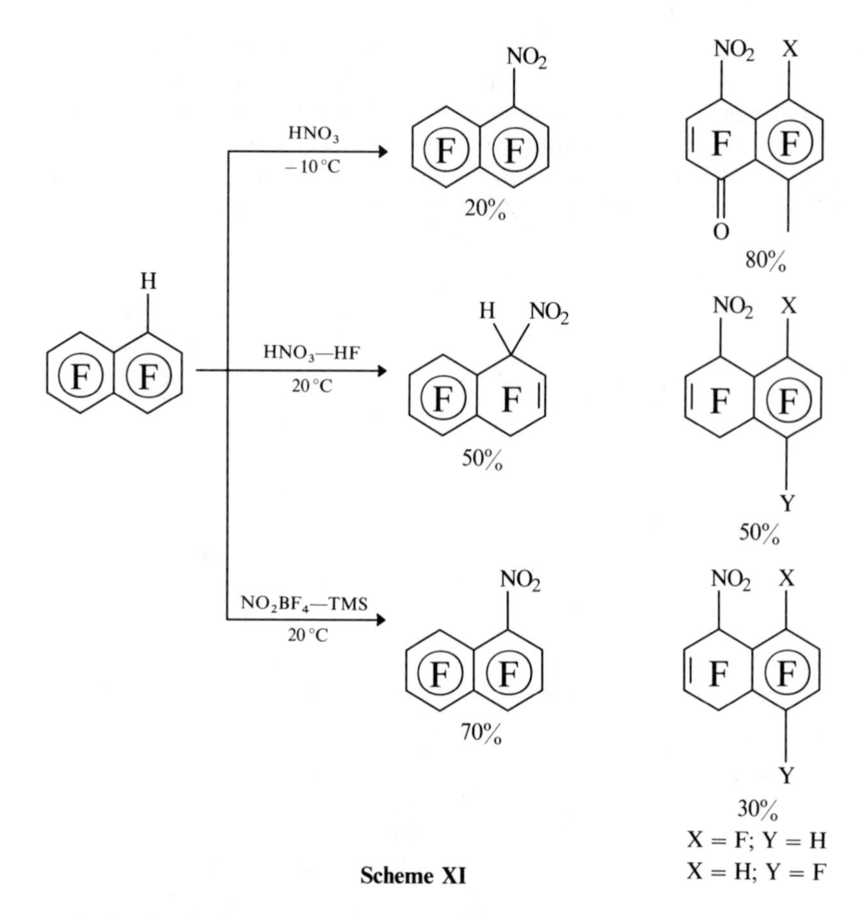

Scheme XI

compounds readily undergo reduction, which is a general way to polyfluorinated phenols[19] and thermolysis yielding isomeric quinones as a result of homolytic cleavage of C—N bonds followed by recombination of the aryloxy radical and nitrogen dioxide with C—O linking and nitrosyl fluoride elimination (Scheme XII).[2,3,20]

If nitrogen dioxide is trapped by a diene or converted to ions in a strong acid, the dimers of the corresponding aryloxy radicals are formed.[19,20] The decomposition of nitrocyclohexadienones may proceed in the presence of perhalogenated phenol with formation of a mixed dimer (Scheme XIII).[21]

Polyfluorinated nitrocyclohexadienes can be thermolyzed yielding cyclohexadienes and transforming into cyclohexadienes by heating with hydrogen fluoride (Scheme XIV).[3,6–11]

All the products of the reactions under consideration are extremely reactive compounds, which offer many possibilities for synthetic applications. One of these is connected with the possibility of generating stable polyfluorinated arenium ions via the interaction of polyfluorocyclohexadienes or -dihydroarenes with antimony pentafluoride (Scheme XV).[22]

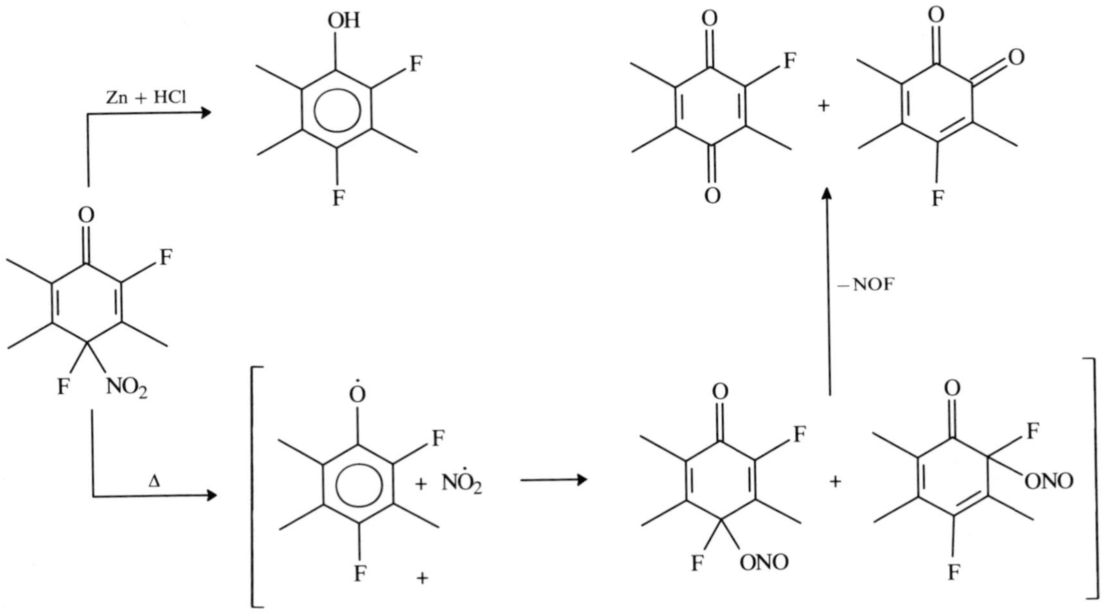

Scheme XII

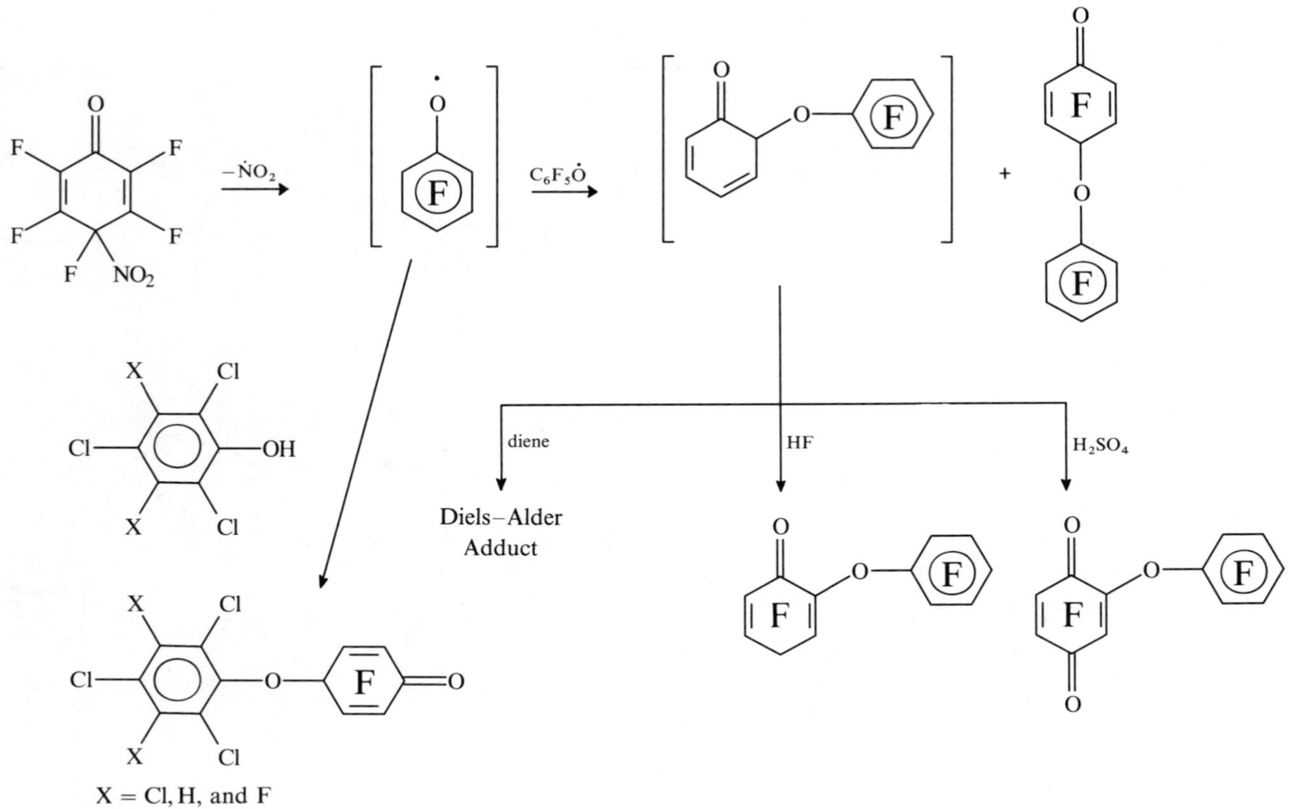

$-\dot{N}O_2$

$C_6F_5\dot{O}$

diene

HF

H_2SO_4

Diels–Alder Adduct

X = Cl, H, and F

Scheme XIII

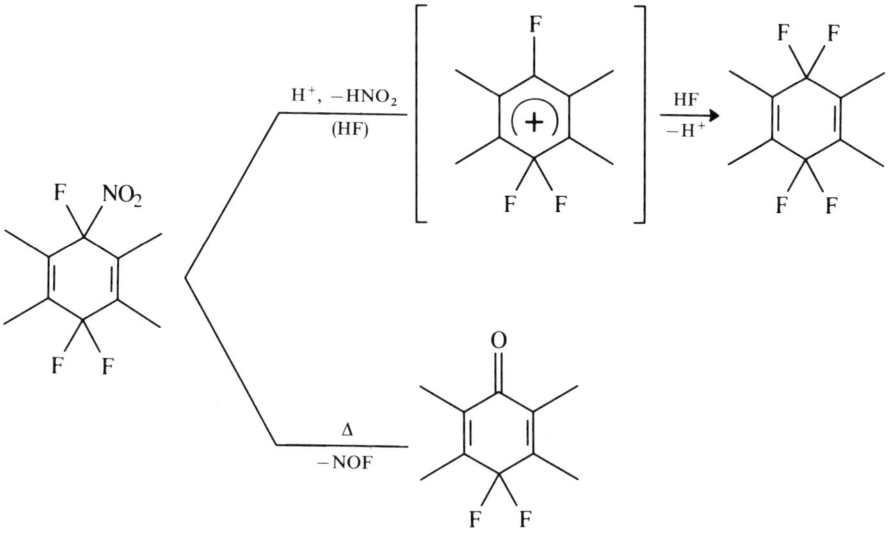

H^+, $-HNO_2$ (HF)

HF, $-H^+$

Δ, $-NOF$

Scheme XIV

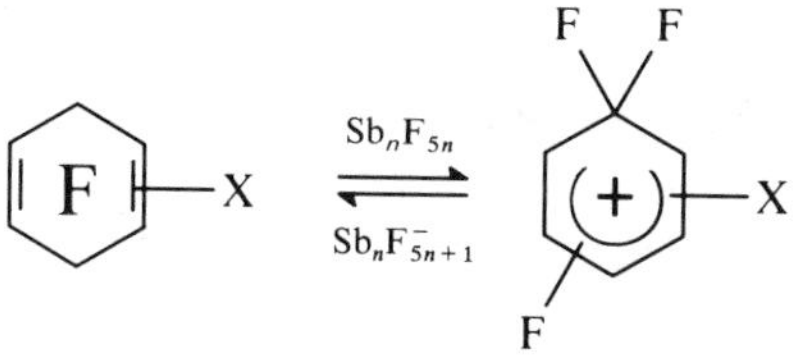

Sb_nF_{5n}

$Sb_nF_{5n+1}^-$

Scheme XV

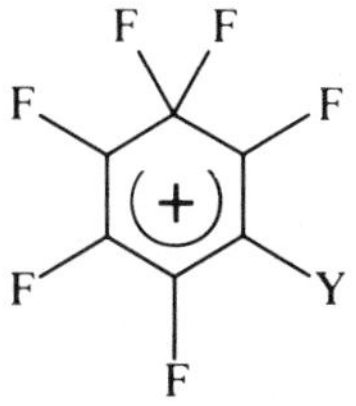

Y = F, H, CH_3, or C_6F_5

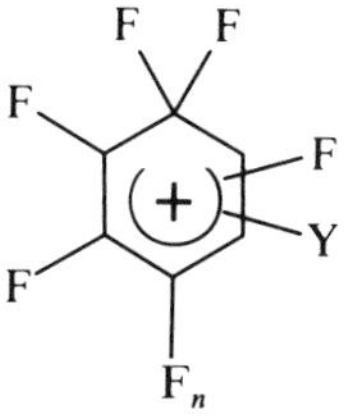

Y = Cl or Br

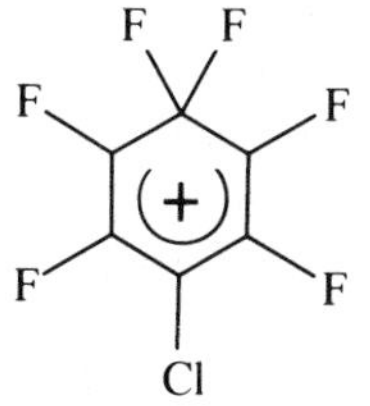

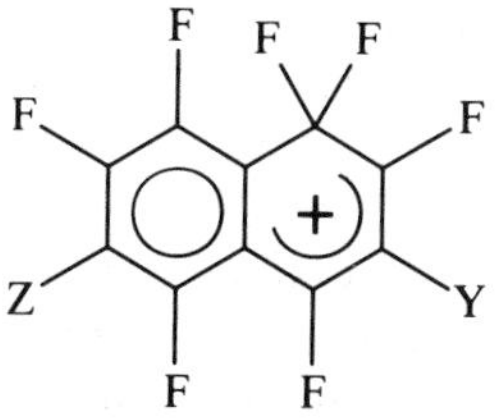

Y = F; Z = F, H, CH_3, or C_6F_5

Z = F; Y = CH_3 or C_6F_5

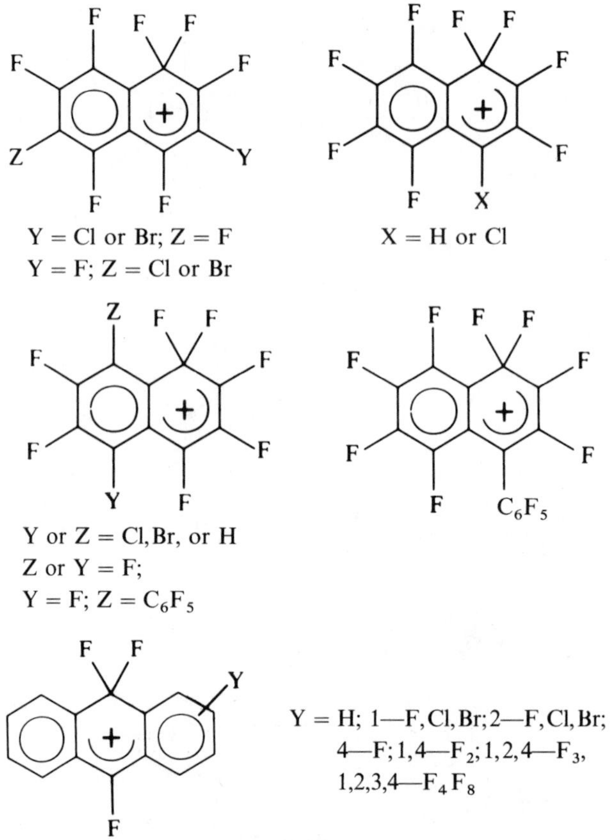

Y = Cl or Br; Z = F
Y = F; Z = Cl or Br

X = H or Cl

Y or Z = Cl, Br, or H
Z or Y = F;
Y = F; Z = C_6F_5

Y = H; 1—F, Cl, Br; 2—F, Cl, Br; 4—F; 1,4—F_2; 1,2,4—F_3, 1,2,3,4—F_4 F_8

Scheme XVI

A great number of polyfluorinated benzenium, naphthalenium, and anthracenium ions have been generated containing, along with fluorine atoms, other halogens, as well as methyl and pentafluorophenyl groups as substituents (Scheme XVI).[12]

At temperatures above $-20\,^{\circ}C$ the ratio of isomers differing in the location of substituent in the same ring is usually an equilibrium one, due to the reversibility of fluoride ion elimination from polyfluorinated dihydroarenes. When the isomers have substituents in different rings, as in the case of naphthalenium ions, the equilibration demands heating, because the isomerization involves the fluoride ion addition to the benzene ring with a lack of aromaticity. As pointed out above, the equilibrium ratios of isomeric ions correspond generally to regioselectivity of the reactions of polyfluoroarenes with nitrating agents, supporting an electrophilic mechanism for these reactions.[12] However, different isomer ratios may be obtained at low temperatures because of the disparity of the kinetic and thermodynamic control in elimination of the fluoride ion from cyclohexadiene for these arenium ions containing a chlorine or bromine nonfluorine substituent. These kinetically controlled ratios are in accordance with the regioselectivity of nitrofluorination of halopentafluorobenzenes and 2-

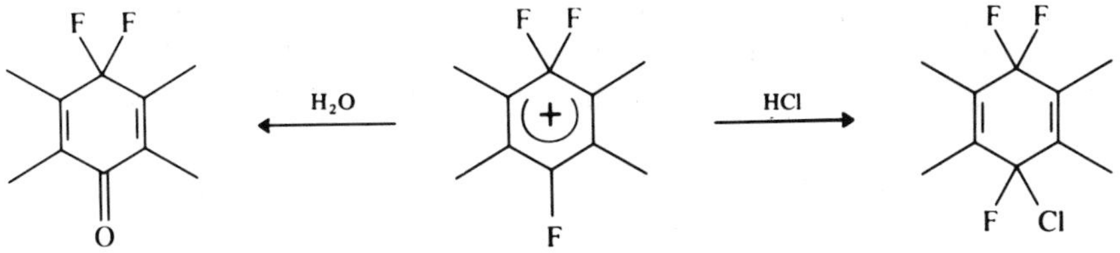

Scheme XVII

haloheptafluoronaphthalenes rather than the thermodynamically controlled ones.[13]

A principal approach to synthetic application of stable polyfluoroarenium ions involves the nucleophilic capture of the ion, for instance by water or hydrogen halide, leading to cyclohexadienones[19] and -dienes, respectively (Scheme XVII).[13,23]

As seen in the case of pentafluorotoluene, this method of cyclohexadienone preparation allows a different location of the carbonyl group with respect to the substituent than the locations provided by nitrofluorination with subsequent thermolysis of nitrocyclohexadiene. It means that these two routes are complementary to each other. The case is similar to the synthesis of polyfluorinated phenols by the reduction of cyclohexadienones (Scheme XVIII).[9,24]

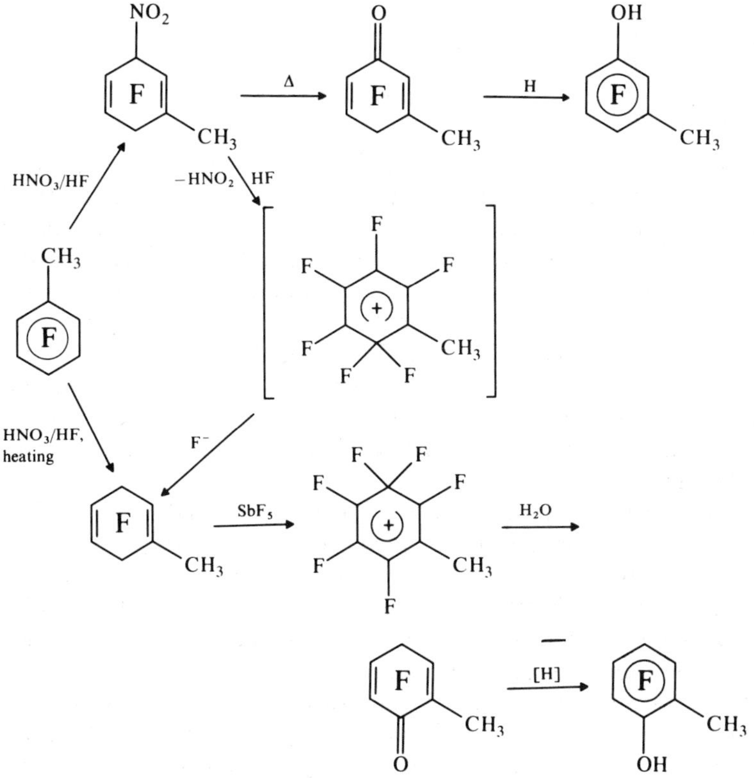

Scheme XVIII

Electrophilic ipso-reactions of polyfluoroaromatic compounds were successfully extended to other types of reagents. The system Cl_2—SbF_5 in SO_2 or SO_2ClF is as universal as are nitrating reagents. Formation of a variety of stable polyfluorinated arenium ions has been observed as a result of ipso-chlorination of polyfluoroarenes with regioselectivity similar to that in ipso nitration of polyfluoroarenes (Scheme XIX).[23,25]

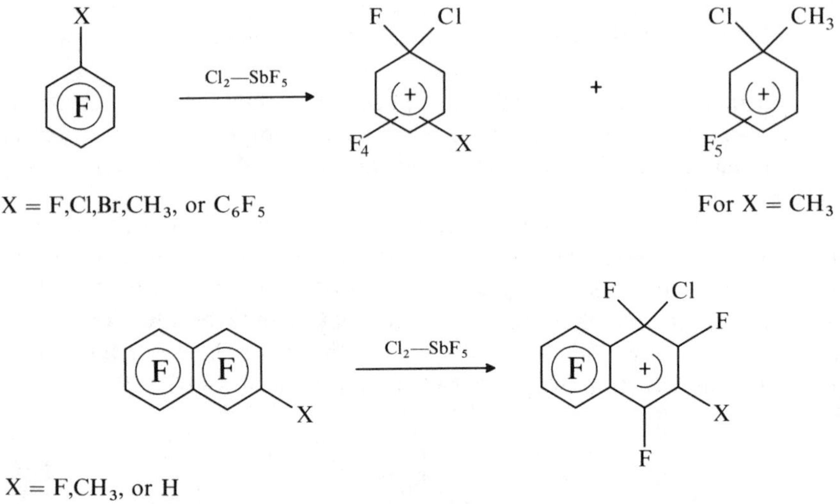

Cl_2—SbF_5

$X = F, Cl, Br, CH_3$, or C_6F_5

For $X = CH_3$

Cl_2—SbF_5

$X = F, CH_3$, or H

Scheme XIX

Polyfluoroarenium ions containing the chlorine atom at the sp^3-hybridized carbon turn into cyclo-1,4-dienes by halide ion capture. Interaction of these dienes with antimony pentafluoride at low temperature yields arenium ions with the chlorine atom in the para-position with respect to the sp^3-hybridized carbon. These ions are not the most stable isomers and undergo isomerization by two routes: alternating addition–elimination of the fluoride anion and intramolecular chlorine migration. This shows an opportunity to change the ion structure starting from the ion generation at low temperature and gradually freezing out the reaction mixture. Consequently, isomeric cyclohexadienes and -dienones may be synthesized by nucleophilic quenching of the corresponding ions (Scheme XX).[23]

Unlike the reactions with nitrating agents, chlorination of partially substituted polyfluoroarenes proceed only at the unsubstituted position if it is one of the preferable reaction centers of the parent aromatic system. In the chlorination of 1-*H*-heptafluoronaphthalene, a rare case of formation of a relatively stable arenium ion has been observed, which is the real intermediate of electrophilic hydrogen substitution (Scheme XXI).[26]

Easy ipso-halogenation with phenol–dienone rearrangement has been observed for the alkaline salts of polyfluorinated phenols and naphthols, as well as for naphthols themselves. Here the hydroxy function shows the pronounced

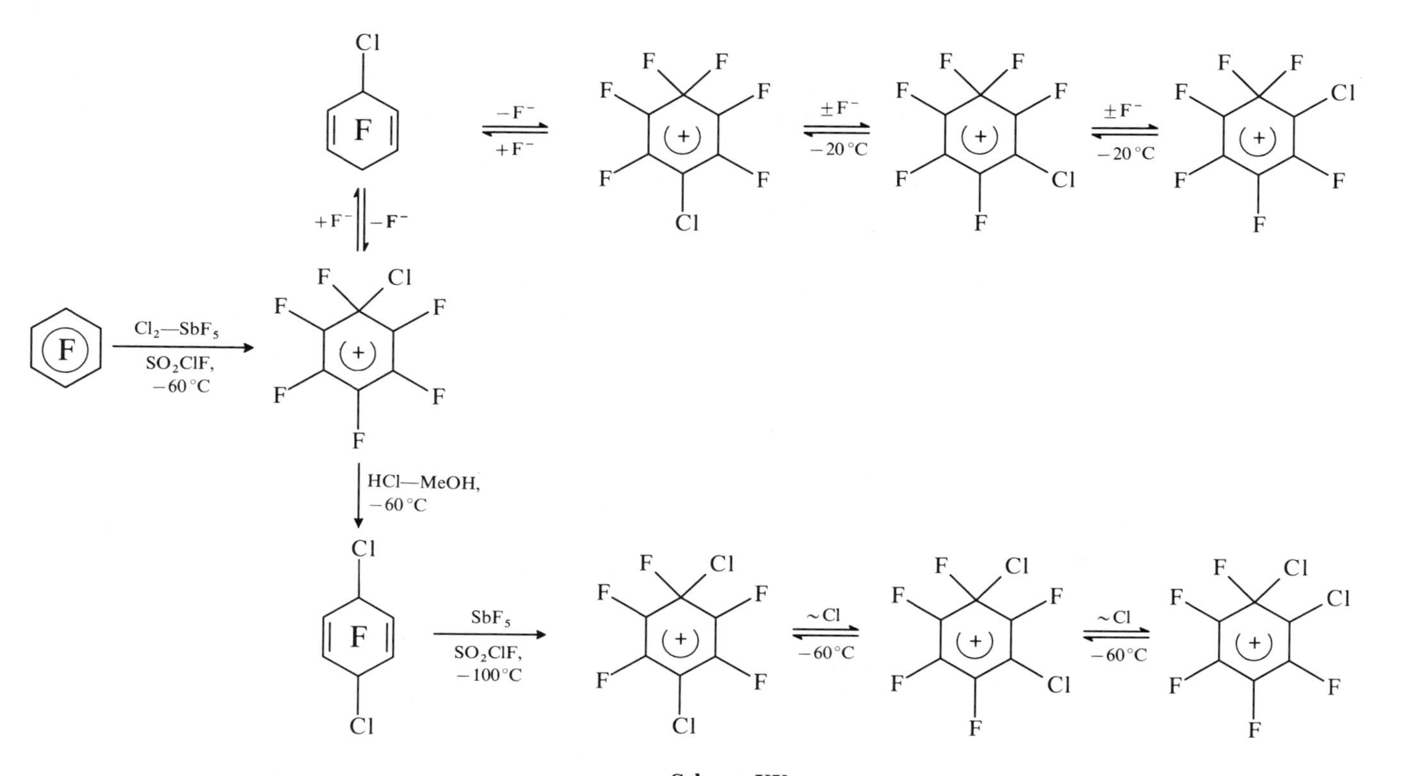
Cl₂—SbF₅
SO₂ClF,
−60 °C
−F⁻
+F⁻
±F⁻
−20 °C
±F⁻
−20 °C
HCl—MeOH,
−60 °C
SbF₅
SO₂ClF,
−100 °C
~Cl
−60 °C
~Cl
−60 °C

Scheme XX

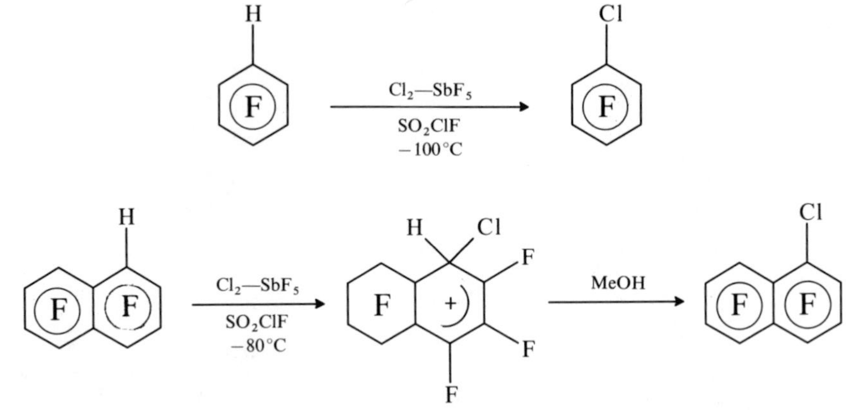

H F $\xrightarrow[\text{SO}_2\text{ClF},\ -100°\text{C}]{\text{Cl}_2\text{—SbF}_5}$ Cl F

H F F $\xrightarrow[\text{SO}_2\text{ClF},\ -80°\text{C}]{\text{Cl}_2\text{—SbF}_5}$ H Cl F + F F F $\xrightarrow{\text{MeOH}}$ Cl F F

NMR ^{19}F, ^{1}H, or ^{13}C

Scheme XXI

ortho–para orientation leading to the formation of polyfluorinated chloro- and bromocyclohexadienones. The ratio of isomeric cyclohexadienones in the reaction mixture depends on the nature of the halogen, the alkaline metal cation, and the degree of hydration of the salt. The latter dependence may result from stabilization of the chelate-like transition state in ortho-halogenation by metal cation coordination with water molecules (see Structure **5**) (Scheme XXII).[27]

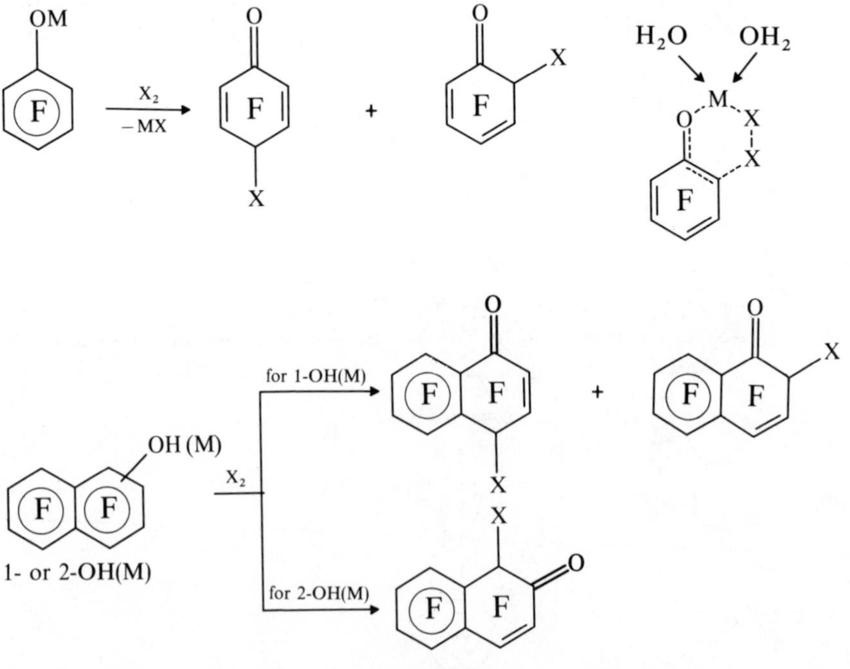

OM F $\xrightarrow[-\text{MX}]{X_2}$ O F X + O F X

H_2O OH_2 M O X X F

1- or 2-OH(M): OH (M) F F $\xrightarrow{X_2}$ for 1-OH(M) → O F F X + O F F X; for 2-OH(M) → X O F F

M = Na or K; X = Cl or Br

Scheme XXII

Various isomerizational transformations have been found for polyfluorinated cyclohexadienones occurring by fluorine migration. These are probably catalyzed by hydrogen fluoride, involving the cyclic transition state in the case of chlorocyclohexadienones, and by intermolecular bromine migration in the case of bromocyclohexadienones. These isomerizations substantially expand the range of synthetic results obtained in halogenation of polyfluorinated hydroxy-aromatic compounds (Scheme XXIII).[28]

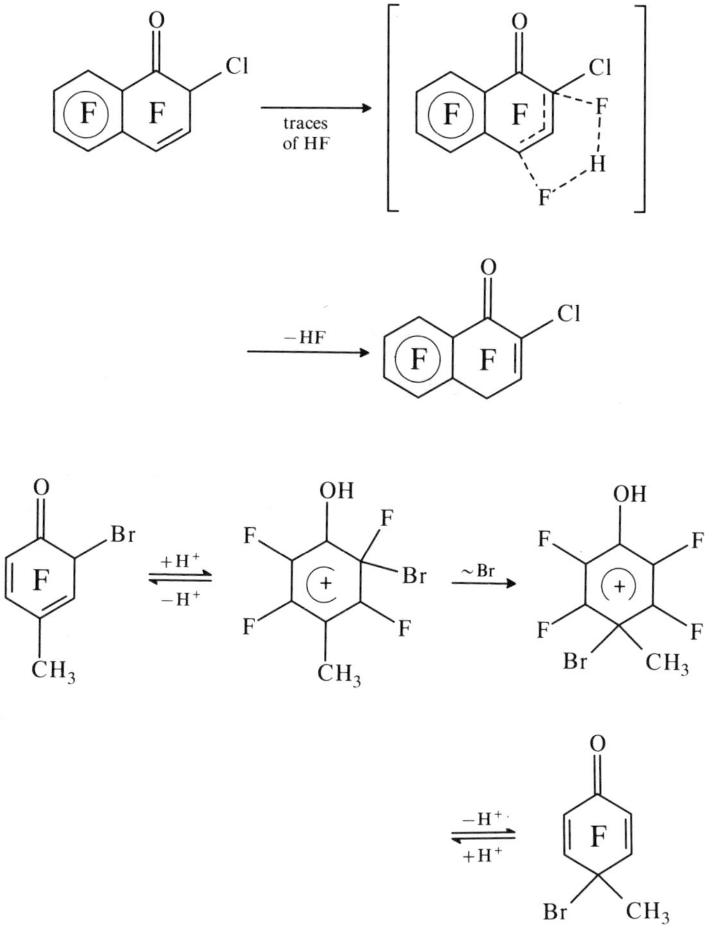

traces of HF

−HF

$+H^+$ / $-H^+$

~Br

$-H^+$ / $+H^+$

Scheme XXIII

Further extension of electrophilic ipso-reactions of polyfluorarenes led to methylation by MeF—SbF_5 to form stable arenium ions corresponding to methyl cation addition at the position of the aromatic ring occupied by fluorine or the methyl group. The kinetically controlled orientation in this reaction does not give the most stable isomers and the initially formed arenium ions are likely to undergo isomerization via migration of the methyl group. This allows one to vary the synthetic result of nucleophilic quenching of arenium ions produced by electrophilic methylation of polyfluoroarenes (Scheme XXIV).[29]

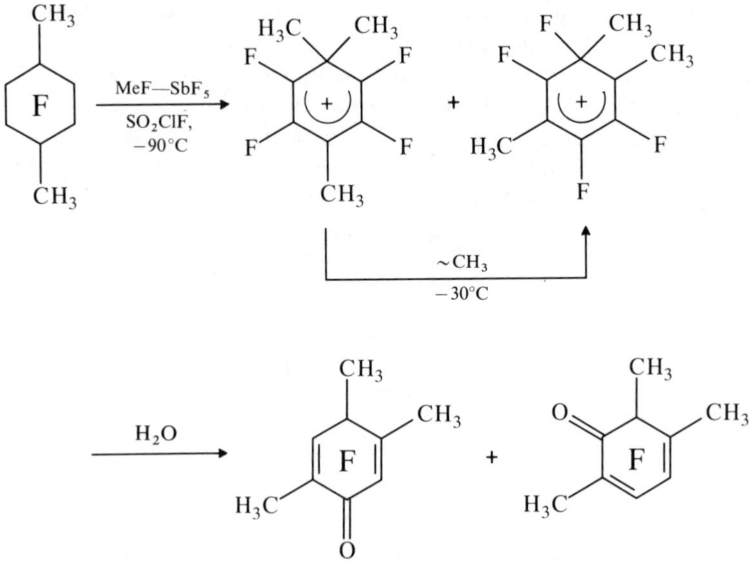

$$\xrightarrow[\text{SO}_2\text{ClF},\ -90^\circ\text{C}]{\text{MeF—SbF}_5}$$

$\sim CH_3$, $-30°C$

$\xrightarrow{H_2O}$

Scheme XXIV

It is quite surprising that upon treatment of octafluoronaphthalene with the MeF—SbF_5 complex, the fluorine atom is substituted by the methyl group.[30] This result is believed to be due to aromatization of intermediate naphthalenium ions formally equivalent to elimination of the fluorine cation. Because the latter route is unreal, the actual mechanism is not clear now (Scheme XXV).

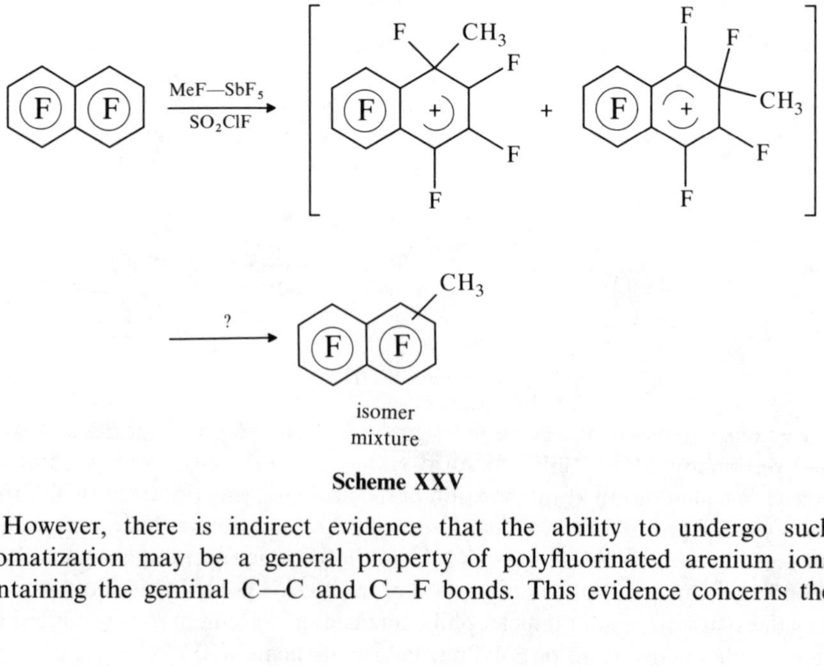

$$\xrightarrow[\text{SO}_2\text{ClF}]{\text{MeF—SbF}_5}$$

isomer mixture

Scheme XXV

However, there is indirect evidence that the ability to undergo such aromatization may be a general property of polyfluorinated arenium ions containing the geminal C—C and C—F bonds. This evidence concerns the reactions of pentafluorobenzene with nontraditional electrophiles—polyfluorinated arenium ions and radical cations; the arenium ions of the type mentioned above probably being intermediates in these reactions.

The interaction of the perfluorobenzenium ion with one equivalent of pentafluorobenzene in the SbF_5 medium leads to the stable ion **8**, which is believed to be a result of the isomerization of the initially formed ions **6** and **7**.[31] A similar result involves the interaction of a second equivalent of pentafluorobenzene. In both cases the corresponding products of nucleophilic quenching of the final ions have been obtained (Schemes XXVI and XXVII).[32]

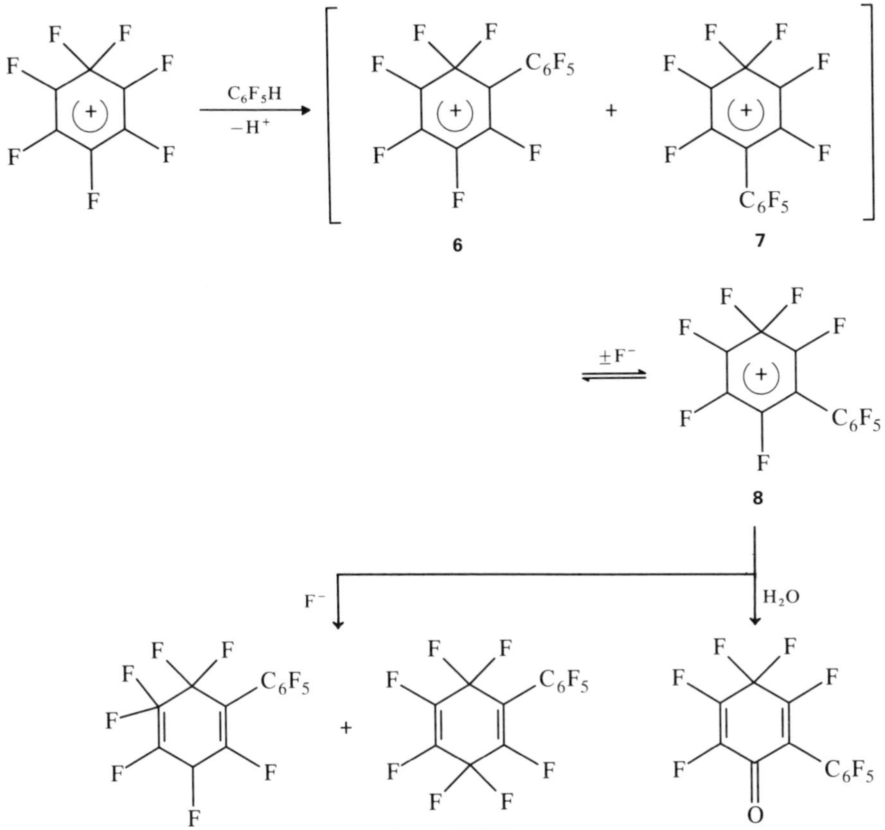

C_6F_5H, $-H^+$; **6**; **7**; $\pm F^-$; **8**; F^-; H_2O

Scheme XXVI

In contrast, no formation of stable arenium ions was observed in the reaction of the perfluorobenzenium ion with three equivalents of pentafluorobenzene leading to perfluoro-1,3,5-triphenylbenzene (yield ~70%). This fact reflects the fast aromatization of the cationoid cyclohexadienyl system after the addition of the third pentafluorophenyl group. Taking into account the tendency of the meta-position to be occupied by the pentafluorophenyl group and the lack of such easy aromatization in the reactions with one or two equivalents of

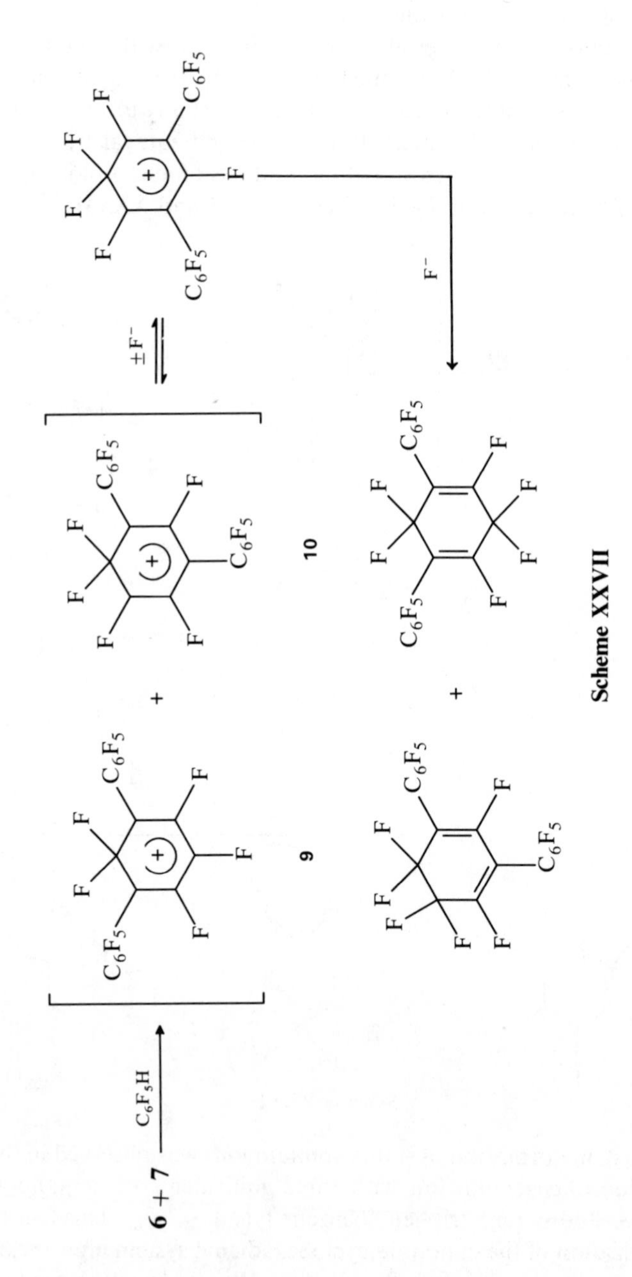

Scheme XXVII

pentafluorobenzene, the ion **11** can be considered a most probable precursor of the final product (Scheme XXVIII).[32]

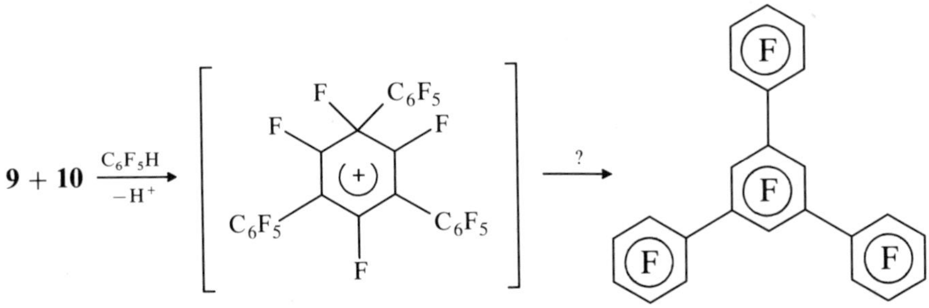

$$9 + 10 \xrightarrow[-H^+]{C_6F_5H} [\mathbf{11}] \xrightarrow{?}$$

11

Scheme XXVIII

Reactions of this type are very promising with respect to the construction of polyfluorinated polynuclear systems (Schemes XXIX and XXX).[12,33,34]

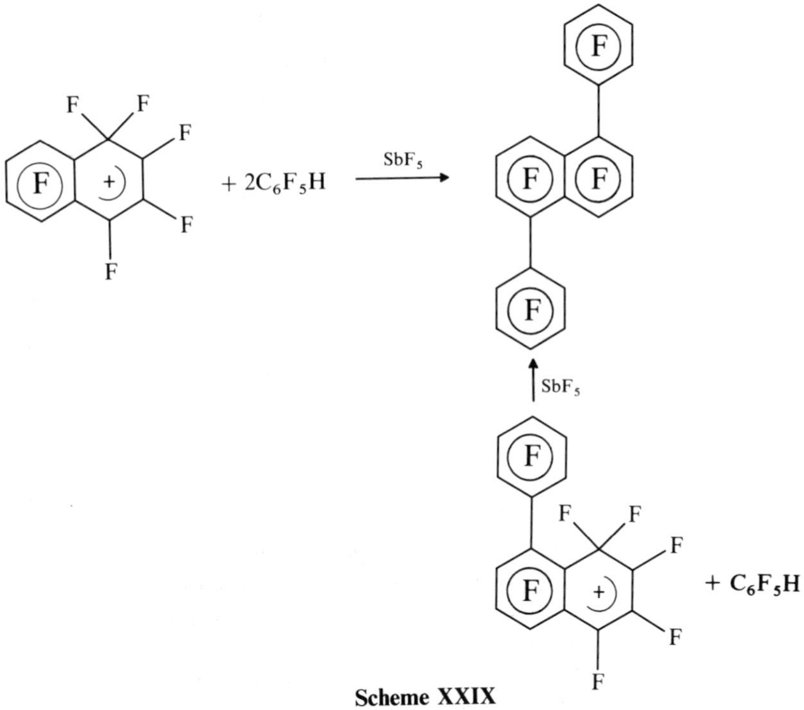

$$+ 2C_6F_5H \xrightarrow{SbF_5}$$

$$\uparrow SbF_5$$

$$+ C_6F_5H$$

Scheme XXIX

The formation of stable radical cations has been shown to be the general property of polyfluoroarenes as a result of their interaction with SO_3 or SbF_5.[35,36] Polyfluoronaphthalenes undergo this transformation nearly quantitatively when dissolved in antimony pentafluoride in rather high concentrations,[36] which makes these solutions applicable for synthetic use. The

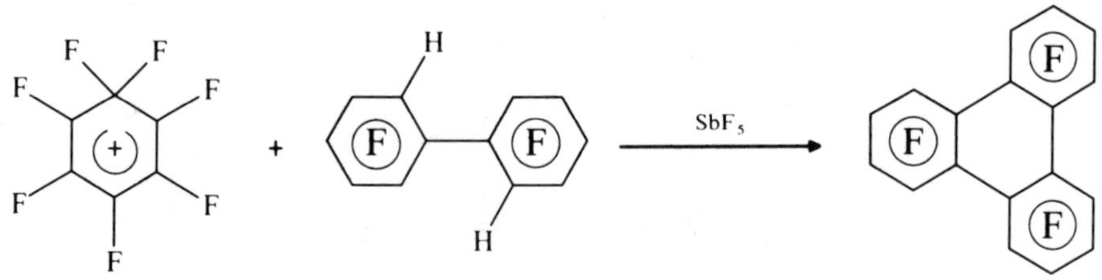

Scheme XXX

interaction of the radical cations of 2-X-heptafluoronaphthalenes (X = F, H, or CH_3) with penta- or tetrafluorobenzenes in SbF_5 leads to polyfluorinated 3-X-1-phenylnaphthalenes, which allows one to suggest the intermediate formation of arenium ions, **12** (Scheme XXXI).[33,37]

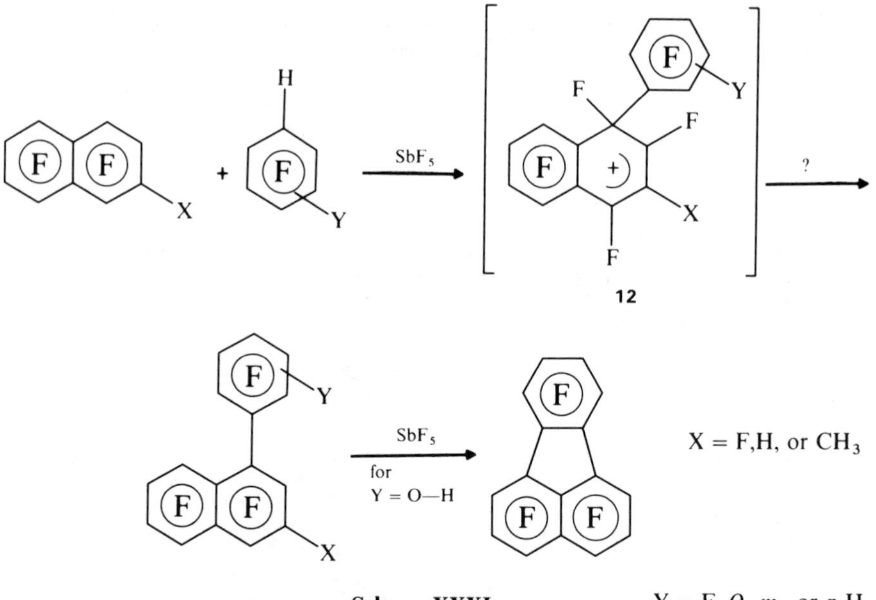

X = F,H, or CH_3

Scheme XXXI

Y = F, *O*-, *m*-, or *p*-H

As mentioned above, one of the principal types of compounds obtained in the electrophilic ipso-reactions of polyfluoroarenes are cyclohexadienones, which are the precursors of polyfluoroarenes. Fortunately, polyfluorinated cyclohexadienones may be easily modified by fluorine nucleophilic substitution in the β position with respect to the carbonyl group. The isomerization of ortho- to para-cyclohexadienones with fluorine migration shown above is also a promising method for modification, allowing the chlorine atom to be fixed in a position in which it can be retained upon the reduction of cyclohexadienone into the respective phenol. The reduction of cyclohexadienones modified by these methods is a basic route to a great variety of polyfluorinated hydroxyarenes containing diverse nonfluorine substituents (Scheme XXXII).[38]

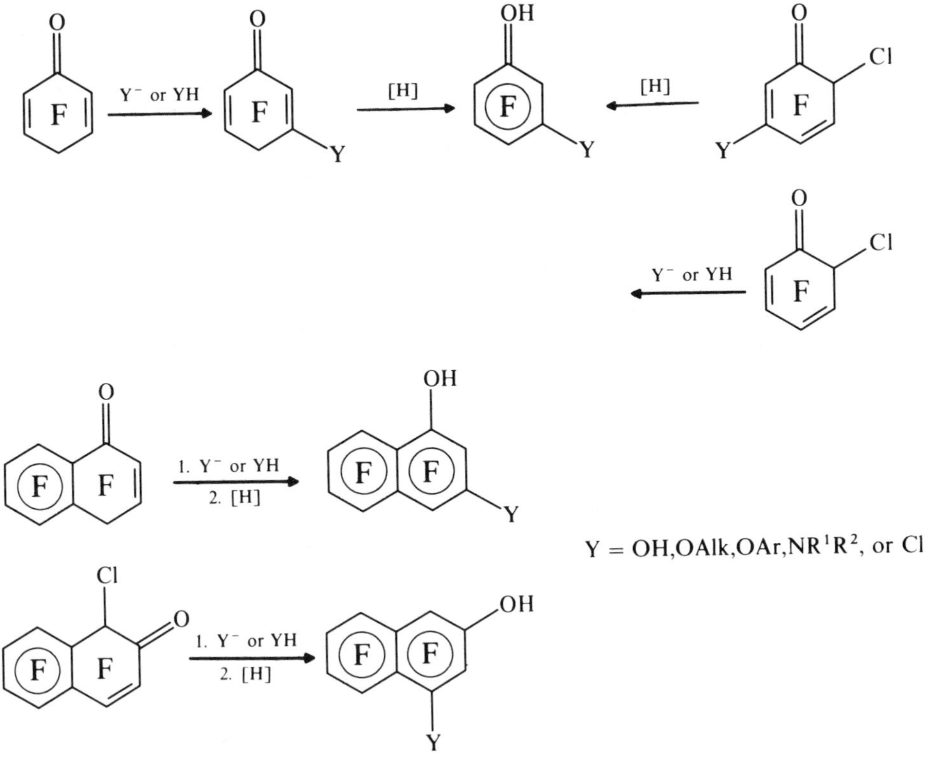

$Y = OH, OAlk, OAr, NR^1R^2$, or Cl

Scheme XXXII

Similarly, the polyfluorodihydroarenes obtained by nitrofluorination of polyfluoroarenes and the subsequent nitro group replacement by fluorine in nitrodihydroarene (see above) after modification can be transformed into substituted polyfluoroarenes by reductive defluorination (Schemes XXXIII and XXXIV).[32,39]

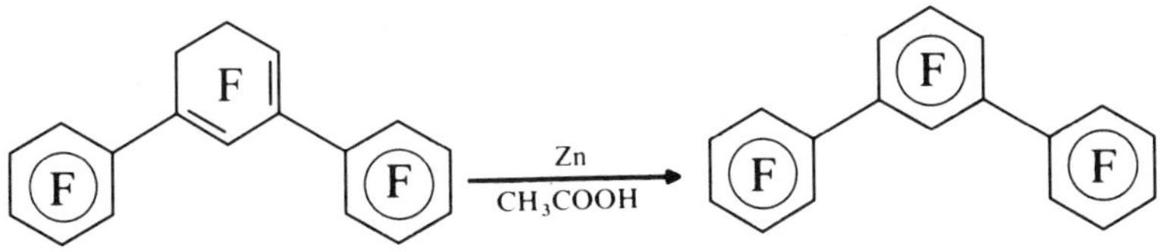

Scheme XXXIII

Recently, it has been demonstrated that there is another route to transform polyfluoro-ortho-cyclohexadienone into arene—the formation of the Diels–Alder adduct by reaction with an alkyne. This adduct readily undergoes the haloform-type decomposition with aromatization yielding polyfluoroarylacetic acid derivatives with the adjacent positions occupied by fluorines and nonfluorine substituents, respectively.[40] In view of the wide possibility of structure modification of cyclohexadienone, dienophile, and Diels–Alder adduct, it is not

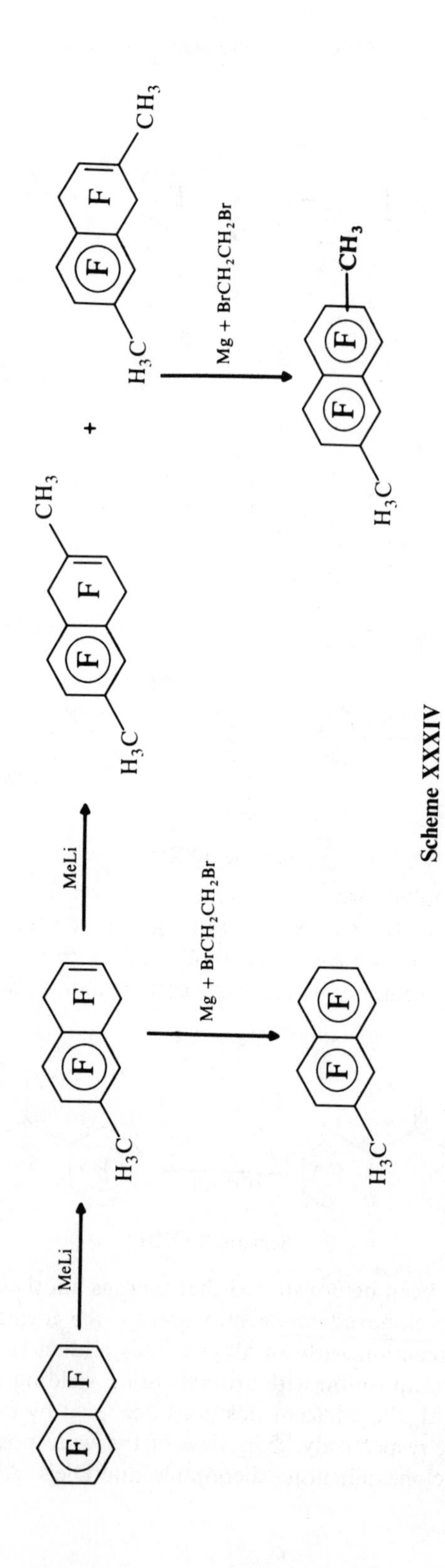

Scheme XXXIV

difficult to imagine the lavishness of a new synthetic field that may be opened by this type of chemistry (Scheme XXXV).

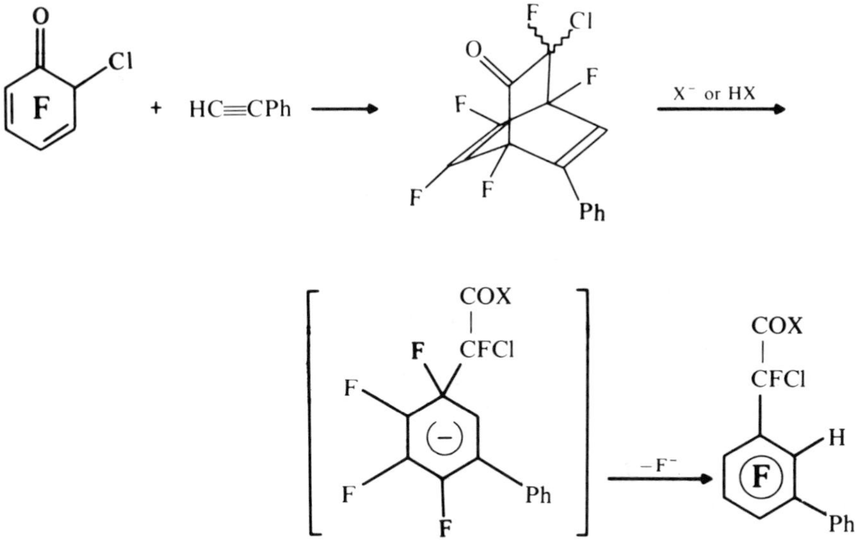

$X = OH, OC_2H_5$, or $N(C_2H_5)_2$

Scheme XXXV

Summing up, it is possible to present the general strategy of the synthetic application of polyfluoroarene electrophilic ipso-reactions. They yield structures that can be qualified as "proarenes" because their typical reactions allow us to restore the aromatic system. At the same time they can be readily modified beforehand to give "proarenes" with changed structures followed by the creation of modified aromatic systems. Subsequently, it is possible to return to the starting point for a new circle to realize the synthetic "perpetuum mobile."

REFERENCES

1. V. D. Shteingarts and G. G. Yakobson, *Zh. Vses. Khim. Ova.*, *15*, 72–80 (1970).
2. G. G. Yakobson, V. D. Shteingarts, and N. N. Vorozhtsov, *Zh. Vses. Khim. Ova.*, *9*, 702–704 (1964).
3. V. D. Shteingarts, G. G. Yakobson, and N. N. Vorozhtsov, *Dokl. Akad. Nauk. SSSR*, *170*, 1348–1351 (1966).
4. C. L. Perrin and G. A. Skinner, *J. Am. Chem. Soc.*, *93*, 3389–3394 (1973).
5. V. D. Shteingarts, O. I. Osina, G. G. Yakobson, and N. N. Vorozhtsov, *Zh. Vses. Khim. Ova.*, *11*, 115–116 (1966).
6. A. A. Shtark and V. D. Shteingarts, *Zh. Org. Khim.*, *12*, 1499–1508 (1976).

7. A. A. Shtark and V. D. Shteingarts, *Izv. Sib. Otd. Akad. Nauk SSSR Ser. Khim. Nauk*, *4*, 123–128 (1976).
8. B. G. Oksenenko, V. D. Shteingarts, and G. G. Yakobson, *Zh. Org. Khim.*, *7*, 745–751 (1971).
9. A. A. Shtark, V. D. Shteingarts, and A. G. Majdanyuk, *Izv. Sib. Otd. Akad. Nauk SSSR Ser. Khim. Nauk*, *6*, 117–124 (1974).
10. O. I. Osina, T. V. Tschujkova, and V. D. Shteingarts, *Zh. Org. Khim.*, *16*, 805–817 (1980).
11. O. I. Osina and V. D. Shteingarts, *Zh. Org. Khim.*, *19*, 1053–1061 (1983).
12. V. D. Shteingarts, *Izv. Sib. Otd. Akad. Nauk SSSR Ser. Khim. Nauk*, *3*, 53–63 (1980); V. D. Shteingarts, *Usp. Khim.*, *50*, 1407–1436 (1981).
13. P. N. Dobronravov, T. V. Tschujkova, Yu. V. Pozdnyakovich, and V. D. Shteingarts, *Zh. Org. Khim.*, *16*, 796–805 (1980).
14. V. D. Shteingarts, A. G. Budnik, G. G. Yakobson, and N. N. Vorozhtsov, *Zh. Obshch. Khim.*, *37*, 1537–1542 (1967).
15. A. A. Shtark, Ph.D Thesis, *Kand. Khim. Nauk Novosibirsk*, 1–44 (1977).
16. V. D. Shteingarts, O. I. Osina, N. G. Kostina, and G. G. Yakobson, *Zh. Org. Khim.*, *6*, 833–840 (1970).
17. A. A. Shtark and V. D. Shteingarts, *Zh. Org. Khim.*, *22*, 831–837 (1986).
18. O. I. Osina and V. D. Shteingarts, *Zh. Org. Khim.*, *10*, 335–343 (1974).
19. L. S. Kobrina and V. D. Shteingarts, *J. Fluorine Chem.*, *41*, 111–162 (1988).
20. A. G. Budnik, V. D. Shteingarts, and G. G. Yakobson, *Izv. Akad. Nauk SSSR Ser. Khim.*, 2485–2492 (1969); A. G. Budnik, V. D. Shteingarts, and G. G. Yakobson, *Izv. Akad. Nauk SSSR Ser. Khim.*, 1594–1602 (1970).
21. A. G. Budnik, V. D. Shteingarts, and G. G. Yakobson, *Zh. Org. Khim.*, *6*, 1198–1207 (1970).
22. V. D. Shteingarts, Yu. V. Pozdnyakovich, and G. G. Yakobson, *Chem. Commun.* 1264–1265 (1969). V. D. Shteingarts and Yu. V. Pozdyakovich, *Zh. Org. Khim.*, *7*, 734–744 (1970).
23. V. D. Shteingarts and P. N. Dobronravov, *Zh. Org. Khim.*, *12*, 2005–2012 (1976); P. N. Dobronravov, and V. D. Shteingarts, *Zh. Org. Khim.*, *13*, 461 (1977).
24. T. V. Tschujkova, A. A. Shtark, and V. D. Shteingarts, *Zh. Org. Khim.*, *10*, 1712–1721 (1974).
25. V. D. Shteingarts, *Zh. Org. Khim.*, *11*, 461 (1975).
26. P. N. Dobronravov and V. D. Shteingarts, *Zh. Org. Khim.*, *13*, 1679–1684 (1977).
27. N. E. Akhmetova, A. A. Shtark, and V. D. Shteingarts, *Zh. Org. Khim.*, *9*, 1218–1227 (1973).
28. N. E. Akhmetova and V. D. Shteingarts, *Zh. Org. Khim.*, *13*, 1269–1276 (1977).
29. P. N. Dobronravov and V. D. Shteingarts, *Zh. Org. Khim.*, *19*, 995–1004 (1983).
30. P. N. Dobronravov and V. D. Shteingarts, *Zh. Org. Khim.*, *17*, 2245–2246 (1981).
31. Yu. V. Pozdnyakovich, T. V. Tschujkova, and V. D. Shteingarts, *Zh. Org. Khim.*, *11*, 1689–1698 (1975).
32. Yu. V. Pozdnyakovich and V. D. Shteingarts, *Zh. Org. Khim.*, *13*, 1911–1917 (1977).
33. B. A. Selivanov, Yu. V. Pozdnyakovich, T. V. Tschujkova, O. I. Osina, and V. D. Shteingarts, *Zh. Org. Khim.*, *16*, 1910–1924 (1980).

34. Yu. V. Pozdnyakovich, T. V. Tschujkova, V. V. Bardin, and V. D. Shteingarts, *Zh. Org. Khim.*, *12*, 690–691 (1976).
35. N. M. Bazhin, N. E. Akhmetova, L. V. Orlova, V. D. Shteingarts, L. N. Shchegoleva, and G. G. Yakobson, *Tetrahedron Lett.*, 4449–4452 (1968).; C. Thomson and W. J. MacCulloch, *Tetrahedron Lett.*, 5899–5900 (1968): N. M. Bazhin, Yu. V. Pozdnyakovich, V. D. Shteingarts, and G. G. Yakobson, *Izv. Akad. Nauk SSSR Ser. Khim.*, 2300–2302 (1969); C. Thomson, W. J. MacCulloch, *Mol. Phys.*, *19*, 817–832 (1970).
36. N. E. Akhmetova, N. M. Bazhin, Yu. V. Pozdnyakovich, V. D. Shteingarts, and L. N. Shchegoleva, *Teor. Eksp. Khim.*, *10*, 613–623 (1974).
37. B. A. Selivanov and V. D. Shteingarts, *Zh. Org. Khim.*, *16*, 2108–2112 (1980).
38. N. E. Akhmetova and V. D. Shteingarts, *Zh. Org. Khim.*, *13*, 1277–1285 (1977); N. E. Akhmetova, N. G. Kostina, and V. D. Shteingarts, *Zh. Org. Khim.*, *15*, 2137–2147 (1979).
39. T. V. Tschujkova and V. D. Shteingarts, *Zh. Org. Khim.*, *9*, 1733–1740 (1973).
40. L. S. Kobrina and V. D. Shteingarts, *Zh. Org. Khim.*, *24*, 1344–1345 (1988).

CHAPTER 13

The Perfluorobenzene Oxide–Perfluorooxepin System

NATALIE E. TAKENAKA and DAVID M. LEMAL[1]

13.1. INTRODUCTION

Benzene oxide (**1**) and oxepin (**2**) comprise a pair of fluxional valence isomers with interconversion barriers below 10 kcal mol^{-1}.[2–5] Their equilibrium is balanced very delicately. By virtue of its substantially greater entropy content, the boat-shaped oxepin predominates over benzene oxide, the isomer with lower enthalpy, at room temperature in nonpolar media. In highly polar solvents, however, the latter isomer is dominant.

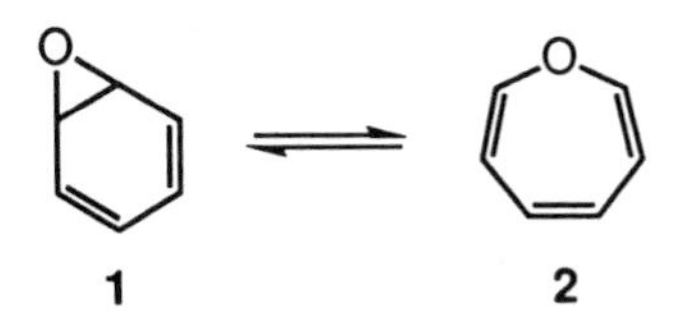

Synthetic Fluorine Chemistry,
Edited by George A. Olah, Richard D. Chambers, and G. K. Surya Prakash.
ISBN 0-471-54370-5

As part of a program to measure and understand fluorine substituent effects, we wished to synthesize perfluorobenzene oxide (**3**) and perfluorooxepin (**4**) and to compare their properties and chemistry with those of the parent compounds.

The parent molecules were prepared by dehydrobromination of oxabicycloheptane **5** with sodium methoxide in refluxing ether.[2] Early attempts to synthesize the fluorinated analogues were unsuccessful, however. None of the desired oxepin (or benzene oxide) was obtained from efforts in England to dechlorinate oxacycloheptadiene **6**.[6] Oxabicycloheptene **7** was synthesized in our laboratory and subjected to a variety of reducing agents. Pentafluorophenol (**8**) was obtained with zinc dust in acetic acid or sodium iodide in acetone, but no **3** or **4** was detected under any conditions.[7] The English group also synthesized a potential precursor to **3**, oxatricycloheptene **9**. Even under gentle conditions, pyrolysis of **9** gave cyclohexadienone **10**, among other things, but no **3** or **4**.[6]

We concluded that the **3**–**4** system is very sensitive to reducing conditions and perhaps thermally labile as well. A new synthetic approach that does not have a reductive or thermal final step was therefore designed (Scheme I). This approach is based on a somewhat unusual tactic for protection of a C—C double bond, namely, the use of a two-carbon bridge. The photocycloaddition of *trans*-1,2-dichloroethylene to perfluorobenzene, and the subsequent thermal ring opening to dichlorobicyclooctadiene **12** had been carried out previously in this laboratory.[8] Oxidation of **12** with peroxytrifluoroacetic acid had a close precedent in the synthesis of **7**.[7] We anticipated dechlorinating dichloroepoxide **13** to diene epoxide **14** using zinc in dimethyl sulfoxide (DMSO) with the assistance of ultrasound. Ozonation of the unsubstituted double bond followed by photolysis of the ozonide[9] offered promise of giving the desired **3**–**4**, with the added benefit of great flexibility in the choice of reaction conditions for the crucial final step.

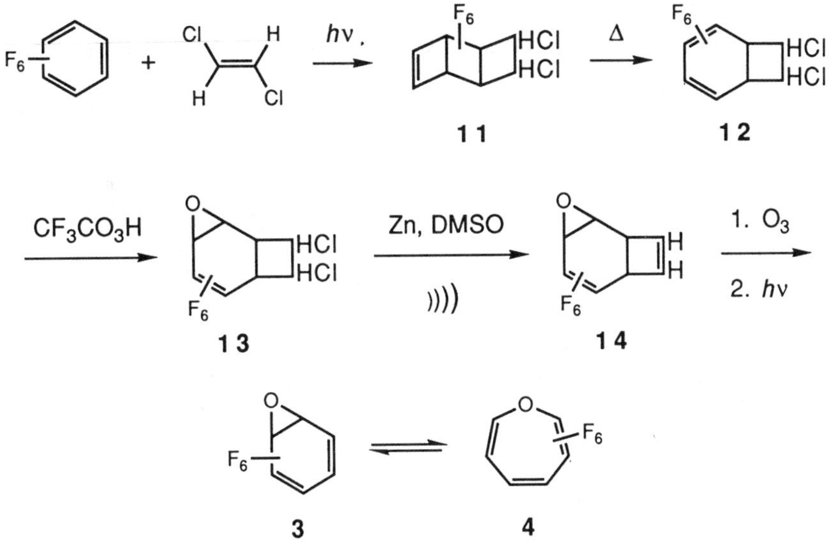

Scheme I

13.2. SYNTHETIC STUDIES

13.2.1. A Model Reaction

Ozonation of an alkene followed by photolysis is a powerful method for excising an ethylene or substituted ethylene unit from a molecule. In the first example of this reaction sequence, cyclopentene ozonide (**15**) was irradiated at wavelengths >280 nm to give cyclopropane (**16**) via double β scission of the initially formed

15 $\xrightarrow{h\nu}$ [ ] $\longrightarrow$ + H(C=O)O(C=O)H

$\longrightarrow$ △ + HCO_2H + CO

16

biradical.[10] When the substrate is a cyclobutene, a new π bond is created,[9] as proposed in Scheme I.

Since we regarded this sequence as the most critical part of our synthetic plan, we tested it at the outset on a model compound, tricyclooctadiene **17**. Ozonation of **17** at 0°C in methylene chloride gave **18**, as revealed by ^{19}F NMR signals at 121.2 (vinyl F group), 190.5 and 192.6 ppm (bridgehead F groups) upfield from internal trichlorofluoromethane. Attack had occurred selectively, as expected, at the more electron-rich double bond. Only one stereoisomer was formed, and it was assigned the configuration shown on steric grounds.

17 $\xrightarrow[0^\circ\,C]{O_3}$ 18 $\xrightarrow{h\nu}$ 19 + C_6F_6

^{19}F NMR played a central role in the determination of structure and configuration throughout the work described here. Only a limited amount of NMR data will be presented in this chapter, but a few generalizations regarding the ^{19}F chemical shifts of the molecules to be discussed will be useful. All chemical shifts are given in parts per million (ppm) *upfield* from trichlorofluoromethane. In systems containing fused four-membered rings, bridgehead fluorines typically appear at high fields (150–200 ppm), while vinyl fluorines are generally found near 120 ppm (cyclobutenes) and 150 ppm (larger rings, conjugated systems).[11] A bridgehead fluorine vicinal and cis to a bromine, chlorine, or hydrogen is deshielded relative to one that has another fluorine in this position.

When ozonide **18** was irradiated at room temperature with a medium pressure mercury lamp through a Vycor filter, hexafluoro Dewar benzene (**19**) and hexafluorobenzene were formed cleanly in the ratio 7:1. The same products in the same ratio were obtained when the photolysis of **18** was carried out with a 340-nm cutoff filter, though the reaction was far slower. Benzophenone was added as a triplet sensitizer and this filter was used to assure that the sensitizer absorbed essentially all of the light. Reaction was again rapid and the benzene was the exclusive end product. A control experiment showed that benzophenone efficiently sensitized the isomerization of **19** to the benzene, so the ratio of these compounds formed initially in the sensitized photolysis is not known.

How the small amount of benzene is formed in the direct photolysis of **18** is an interesting question. It might arise from ring opening of Dewar born vibrationally hot, but since excess vibrational energy would be lost very fast in the condensed phase, this argument is not compelling. The alternative possibility that the benzene arises from subsequent photolysis of the Dewar isomer can be ruled out on the basis that **19** absorbs only at very short wavelengths, yet the Dewar benzene:benzene ratio was the same whether or not the 340-nm cutoff filter was employed.

The observation that benzophenone sensitizes the Dewar benzene–benzene isomerization implies the intermediacy of triplet **19**, which in turn raises the interesting possibility that the benzene formed in the ozonide photolysis arises from **19** born in the T_1 state. If intersystem crossing occurs in the biradical formed when ozonide is irradiated, adiabatic fragmentation of the resulting triplet biradical would indeed yield triplet Dewar benzene. This fragmentation is sufficiently exothermic to make the triplet state of **19** accessible in a thermodynamic sense. The question whether electronically excited products are produced in ozonide photolyses deserves further investigation.

Mechanistic questions aside, the clean formation of Dewar benzene and benzene in the photolysis of ozonide **18** encouraged us to undertake the chemistry outlined in Scheme I.

13.2.2. Dichlorobicyclooctadiene 12

Photocycloaddition of hexafluorobenzene to *trans*-1,2-dichloroethylene gave *trans*- and *exo, cis*-dichlorotricyclooctene **11** in yields up to 45%.[8] Fractional distillation separated the trans and cis isomers, giving crystalline solids in the ratio of 9:1, respectively. The formation of a cis isomer indicates that the photocycloaddition is a stepwise process; the absence of endo, cis isomer is probably traceable to steric hindrance.

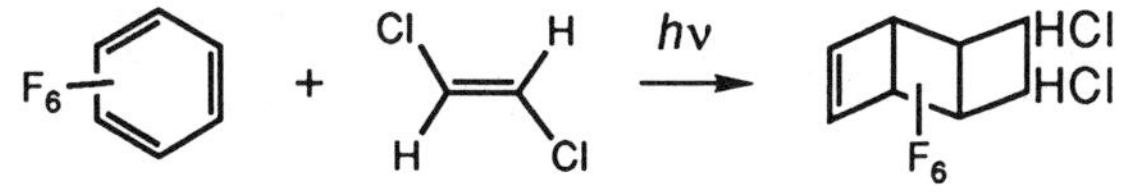

Ring opening of both isomers of **11** to yield diene **12** was accomplished in >90% yield by heating at ~160°C.[8] Only a single cis isomer was obtained; the chlorines were assigned the endo configuration on the basis of NMR chemical shift comparisons with model compounds and analogy to the close relative *cis*-**21**, whose stereochemistry was proven by nuclear Overhauser enhancement (NOE) measurements.[12] The high temperatures at which both *cis*-**12** and *cis*-**21**

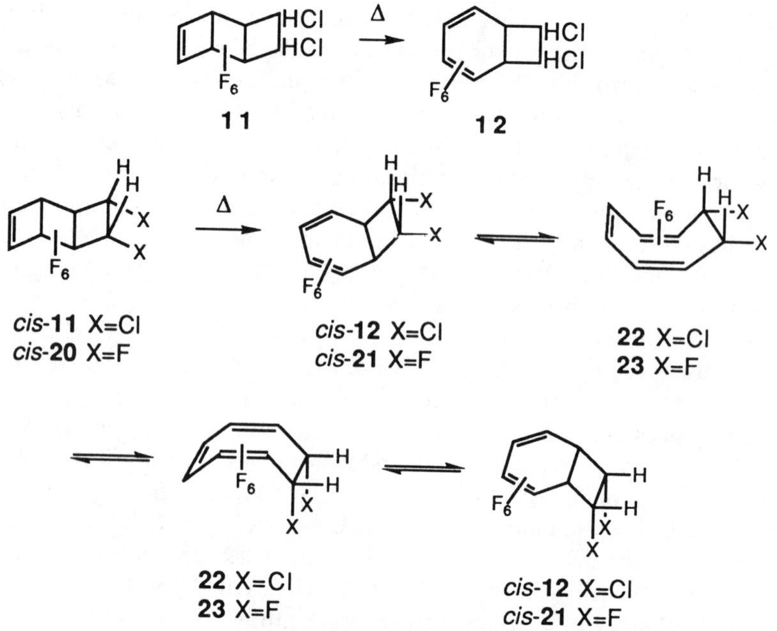

Scheme II

are formed assures thermodynamic control of the endo–exo stereochemistry, because their rather facile, reversible electrocyclic ring opening to cyclooctatrienes equilibrates endo and exo isomers (Scheme II). The greater stability of the endo compounds can be understood in terms of smaller dipole–dipole repulsion relative to their exo isomers, Fig. 13.1.[12]

Overheating during the ring opening of *cis*-**11** to *cis*-**12** gave an unexpected

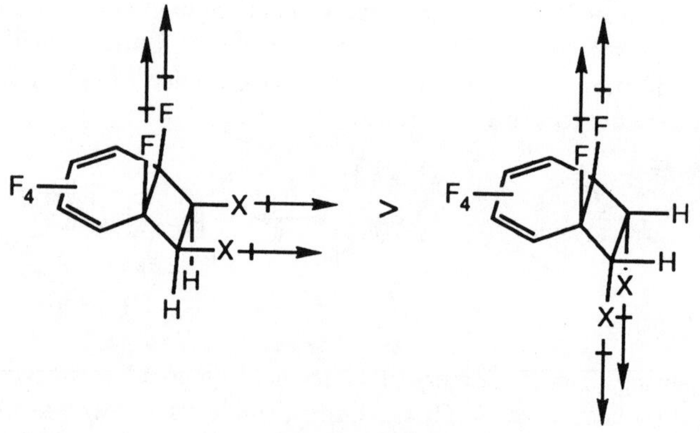

Figure 13.1.

product, *cis*-3,4-dichlorobicyclooctadiene **24**, formed by rearrangement of *cis*-**12**. When the ring opening was carried out at 180°C for 24 h, for example, the product mixture was 13% *cis*-**12** and 87% *cis*-**24**. Several lines of evidence established the structure and configuration of *cis*-**24**, but the doublet splitting of 58.1 Hz characteristic of fluorine geminal to hydrogen in its highest field ^{19}F NMR signal was particularly revealing.

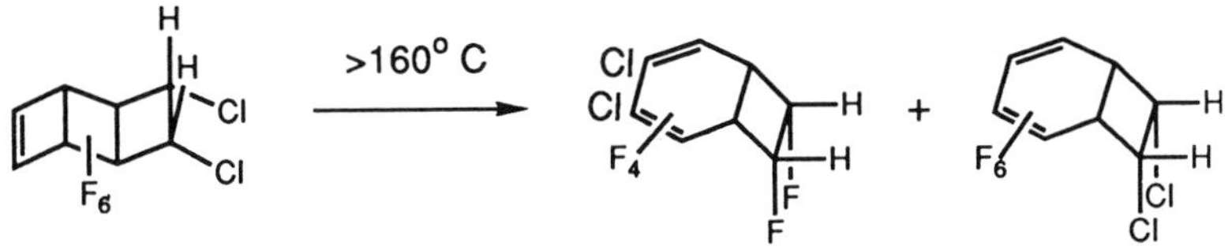

>160° C

cis-**11** *cis*-**24** *cis*-**12**

Similarly, heating *trans*-**12** with calcium carbonate as an acid scavenger gave rise to *trans*-**24**, but after 50 h at 250°C the reaction had proceeded only about halfway. After purification by gas chromatography, the *trans*-**24** still contained about 13% of another compound, tentatively identified by its ^{19}F NMR and IR spectra as the monocyclic valence isomer *trans*-**25**. The presence of this isomer is not a great surprise, for model compound **21** exists in equilibrium with a large amount of the monocyclic form.[12] For carrying forward our synthesis, it is fortunate that *cis*- and *trans*-**12** are not contaminated with appreciable amounts of their monocyclic counterparts.

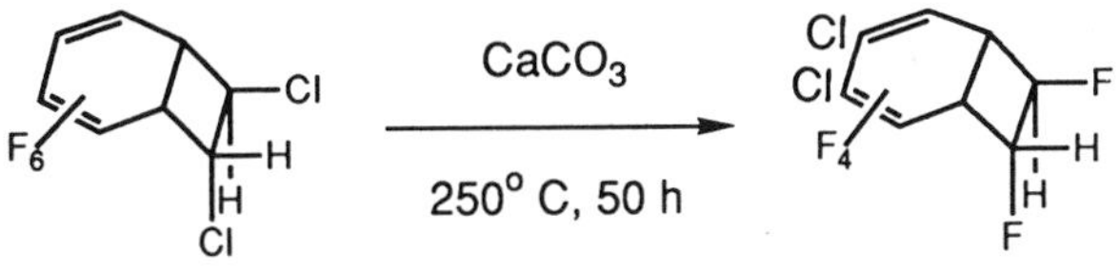

$CaCO_3$

250° C, 50 h

trans-**12** *trans*-**24**

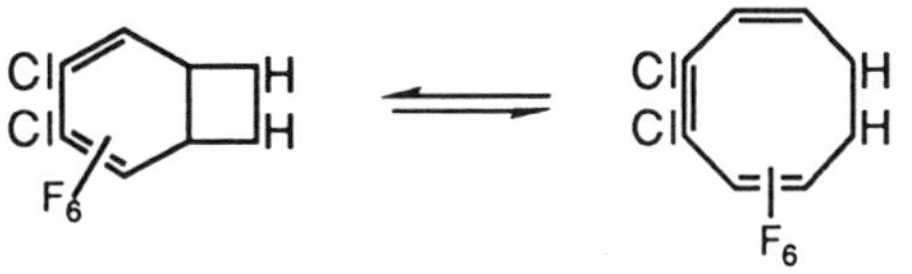

trans-**24** *trans*-**25**

Two possible mechanisms for the rearrangement of *cis*-**12** to *cis*-**24** are presented in Scheme III. The first involves ring opening followed by two sequential, thermally allowed [1,5] H-shifts; the second entails ring opening followed by a type II dyotropic rearrangement,[13] a process akin to diimide reduction of double bonds.[14] The hydrocarbon parent of *cis*-**12** rearranges analogously, and the [1,5] H-shift mechanism has been shown to operate in this case.[15] *trans*-Dichlorocyclooctadiene **12** cannot rearrange via the dyotropic

Sequential [1,5] H-Shift Pathway

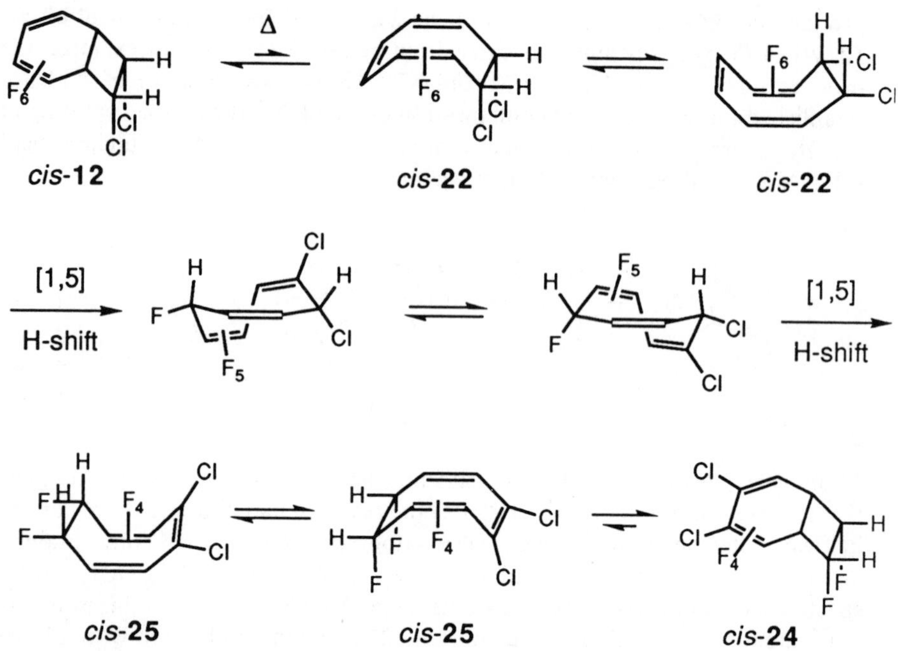

cis-**12** $\xrightarrow{\Delta}$ *cis*-**22** ⇌ *cis*-**22**

[1,5] H-shift ⇌ [1,5] H-shift

cis-**25** ⇌ *cis*-**25** ⇌ *cis*-**24**

Type II Dyotropic Pathway

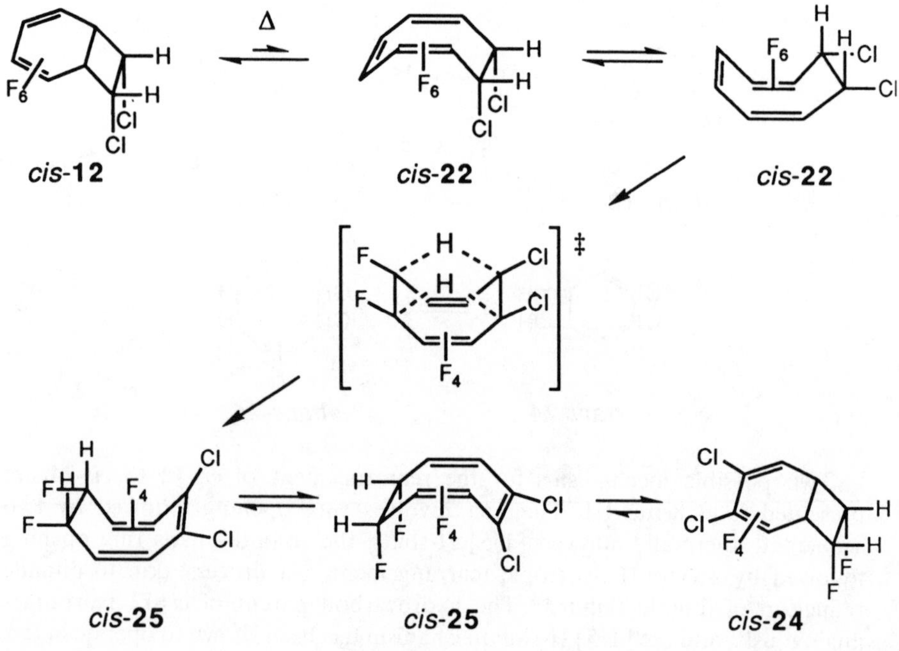

cis-**12** $\xrightarrow{\Delta}$ *cis*-**22** ⇌ *cis*-**22**

$[\ ]^{\ddagger}$

cis-**25** ⇌ *cis*-**25** ⇌ *cis*-**24**

Scheme III

pathway, so the [1, 5] H-shift mechanism presumably operates here as well. However, the fact that the trans isomer reacts considerably slower than the cis even at a temperature 70°C higher argues for two separate mechanisms of reaction, for the [1, 5] H-shift pathway should not be influenced strongly by the configuration of the starting material. Thus, we suggest that *cis*-**12** may undergo the interesting simultaneous transfer of two hydrogens.

13.2.3. Dichlorooxirane 13

Dry peroxytrifluoroacetic acid was needed for the transformation of **12** into the oxirane **13**. Since the 90% hydrogen peroxide used to prepare this reagent is no longer commercially available, we developed a procedure that utilizes the readily obtainable 30% hydrogen peroxide. Two equivalents of trifluoroacetic anhydride were added slowly to the hydrogen peroxide, then concentrated sulfuric acid was introduced as a drying agent. Vacuum transfer to a trap containing methylene chloride followed by drying with sodium sulfate gave a methylene chloride solution that was about 2.0 *M* in peroxytrifluoroacetic acid (90% yield from hydrogen peroxide).

cis-Dichlorobicyclooctadiene (**12**) was epoxidized at room temperature with this reagent in the presence of disodium hydrogen phosphate as a buffer. A single stereoisomer was obtained in 54% yield, and it was assigned the anti configuration on steric grounds. A characteristic IR band at 1500 cm^{-1} revealed the presence of the fluorinated oxirane moiety.[16] Epoxidation of *trans*-**12** was carried out as with the cis isomer, albeit in lower yield.

$$cis\text{-}\mathbf{12} \xrightarrow[Na_2HPO_4]{CF_3CO_3H} cis\text{-}\mathbf{13}$$

We found both the cis and trans oxiranes to be very sensitive compounds. They could not withstand the conditions that have become standard in our laboratory for vicinal dechlorination of chlorofluorocarbons, namely, zinc dust in DMSO with ultrasound.[17] The solvent alone destroyed these compounds at room temperature, probably by attack on the electrophilic vinyloxirane moiety. Some success was eventually achieved with another dipolar aprotic, but less nucleophilic solvent, benzonitrile. Reaction of *cis*-**13** with zinc in benzonitrile at 45°C in an ultrasonic bath proceeded very slowly and resulted in extensive degradation, as revealed by the presence of a broad and intense "HF/F^-" signal in the ^{19}F NMR spectrum. The desired diene epoxide **14** was obtained by preparative gas chromatography, but in only 7% yield. The IR spectrum of this key compound displayed bands at 1745, 1575, and 1490 cm^{-1} for the fluorinated double bond, cyclobutene double bond and fluorinated oxirane, respectively. Its

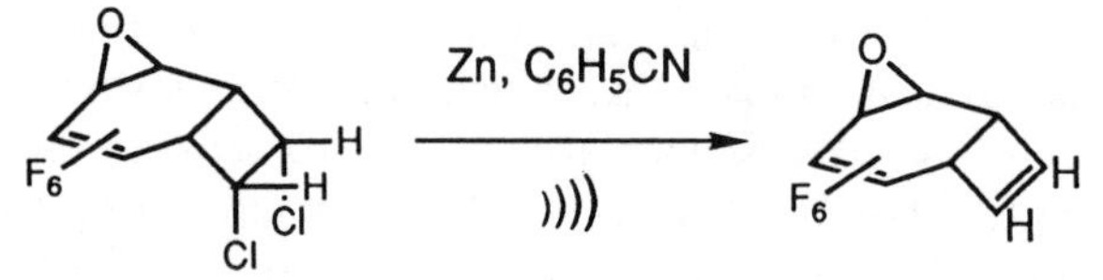

^{19}F NMR spectrum displayed six signals of equal intensity, which were assigned as shown with the help of spin decoupling experiments.

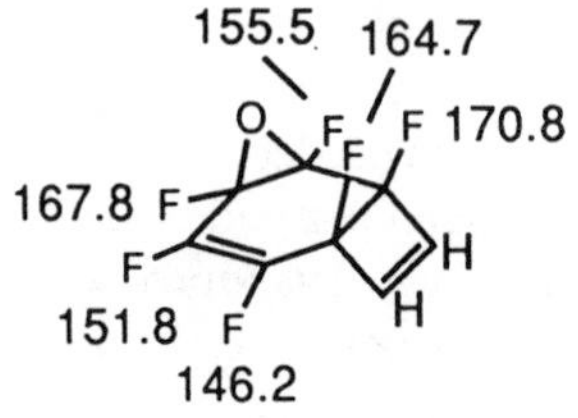

Attempts to improve the yield by predrying the solvent, adding calcium carbonate as an acid scavenger, adding zinc chloride as a catalyst, and using zinc amalgam all gave similar results. Probably attack at the sensitive electrophilic vinyloxirane moiety is competing effectively with dechlorination. Other reagents for vicinal dehalogenation, namely, chromous ion, magnesium–iodine,[18] and bis(tributyltin)/$h\nu$,[19] were also tried, but to no avail.

We speculated that thermolysis of the sensitive vinyloxirane would yield the stabler isomer **26** and/or **27** by analogy to the vinylcyclopropane rearrangements suffered by cyclopropanes similar in structure to **13**.[7] Since we also envisioned schemes for transforming either **26** or **27** into the desired oxepin, the idea was pursued. Pyrolysis in a silylated Pyrex ampoule, with calcium carbonate as an acid scavenger, proceeded slowly and messily at 200 °C. Flash pyrolysis at 350 °C and 500 mtorr, however, transformed cis-**13** into bicyclooctenone **28**. The presence of the carbonyl function was apparent in the IR

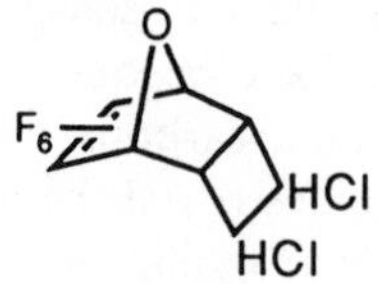

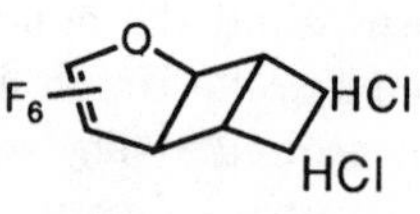

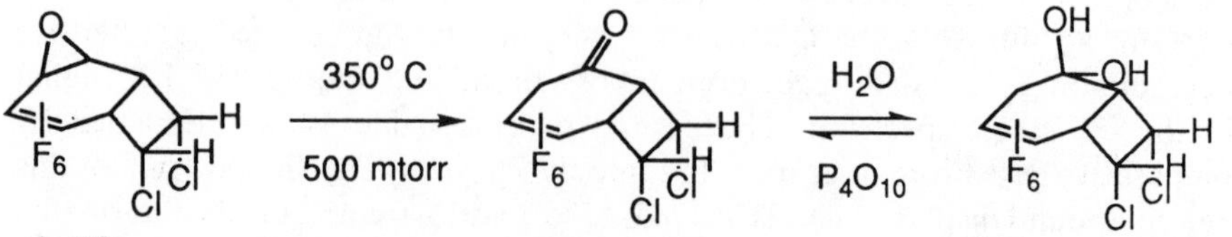

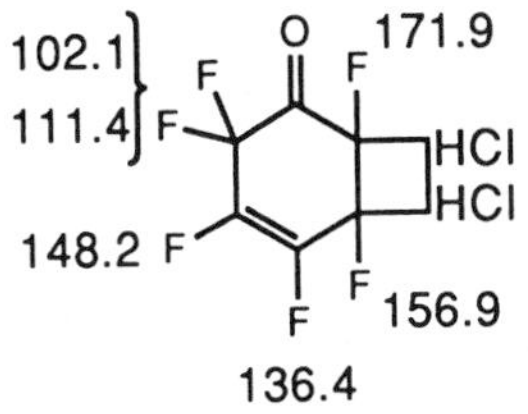

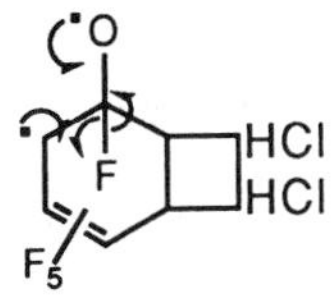

Figure 13.2.

spectrum, which revealed bands at 1645 ($\nu_{C=C}$) and 1780 cm^{-1} ($\nu_{C=O}$), and was confirmed by the facile formation of a hydrate (**29**). Hydrate formation could be reversed by treatment with phosphorus pentoxide.

The low intensity of the ketone's near UV maximum (218 nm, ε1300, cyclohexane) is consistent with the β,γ-unsaturated structure shown,[20] and convincing evidence that the carbonyl is not α,β-unsaturated was provided by analysis of its ^{19}F NMR spectrum. Strong coupling ($J = 21$ Hz) between the bridgehead fluorine at 156.9 and the vinyl fluorine at 136.4 ppm ruled out the alternative structure (Fig. 13.2).

Formation of a ketone instead of an oxanorbornene following cleavage (presumably homolytic) of the allylic C—O bond in *cis*-**13** may be ascribable to the great strength of the C=O linkage (Fig. 13.2), since as noted above all-carbon counterparts yield vinylcyclopropane rearrangement products.

Crawford, Lutener, and Cockcroft[21] found that gas-phase pyrolysis of the parent vinyloxirane (**30**) gave both dihydrofuran (**32**) and butenals (**33** and **34**) (Scheme IV). They proposed different mechanisms for the formation of the two types of product: conrotatory ring opening leading to **32** via the dipolar intermediate **31**, and C—O homolysis followed by H migration to give **33** and **34**. In *cis*-**13** conrotatory opening of the epoxide is geometrically inhibited.

13.2.4. Dibromobicyclooctadiene 35

Interesting as it was, the pyrolysis of **13** did not provide a way to circumvent the troublesome dechlorination en route to the sought-after benzene oxide–oxepin.

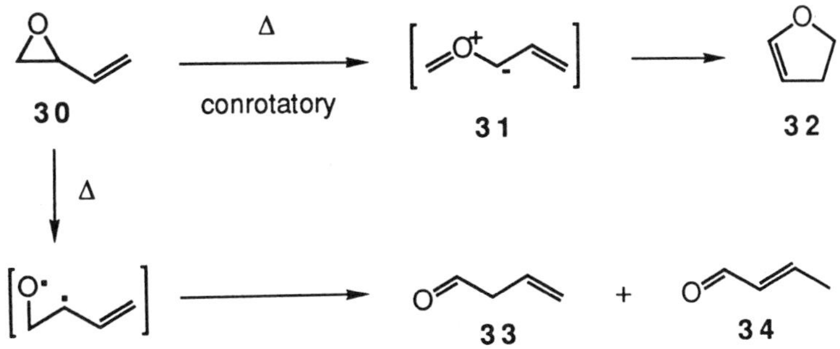

Scheme IV

Because vicinal bromines are much more readily eliminated reductively than vicinal chlorines, we decided to prepare the bromine analogue (**35**) of **13**. It was not possible to parallel exactly the synthetic route leading to **13**, for attempts to add 1,2-dibromoethylene photochemically to perfluorobenzene result in extensive carbon–bromine bond cleavage.[8] The desired adduct **36** could be prepared less directly, however, by selective addition of bromine to the unsubstituted double bond of tricyclooctadiene **17**. Ring opening to **37** would follow, then epoxidation to yield **35**.

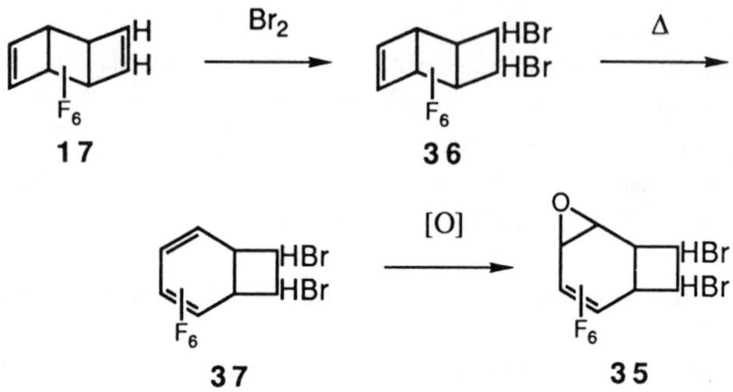

In principle, diene **17** could be synthesized by photocycloaddition of acetylene to perfluorobenzene (Scheme V). The reaction must be conducted at low temperatures to preclude thermal ring opening of the initially formed bicyclic adduct **38**. Bicyclooctatriene **38** was independently synthesized and found to ring open spontaneously to cyclooctatetraene **39**, which underwent bond-shift isomerization to **40**.[22] Acetylene did indeed add to perfluorobenzene upon irradiation, giving diene **17**, but the reaction was prohibitively slow and even at −30°C an equal quantity of **40** was formed.

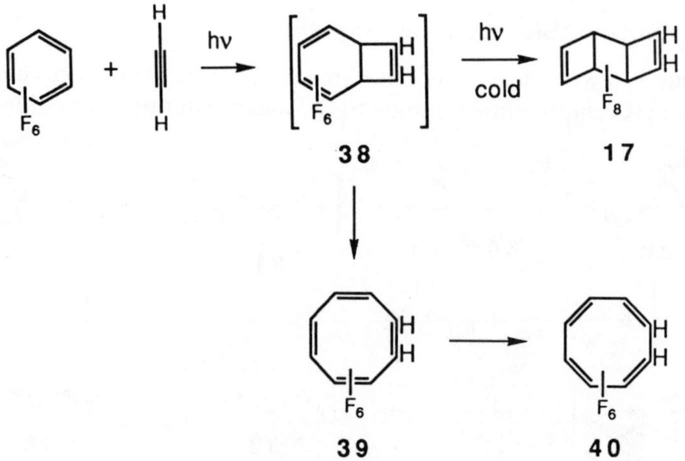

Scheme V

Diene **17** was obtained in 70% yield by reduction of photoadduct **11** with chromous chloride at room temperature.[8] Its ^{19}F NMR spectrum had signals at 120.4 (F_3, F_4), 177.8 (F_2, F_5), and 189.1 (F_1, F_6) ppm.

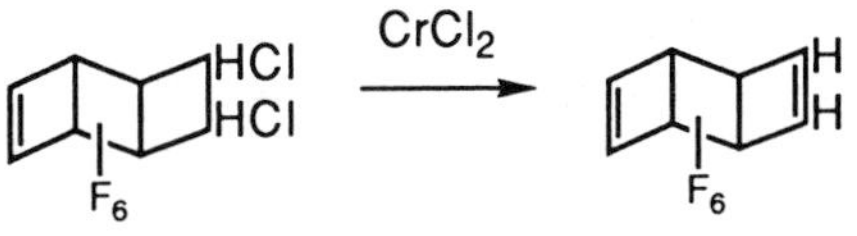

$CrCl_2$

11 **17**

For steric reasons, bromine reacts with the fully fluorinated tricyclooctadiene **41** to give exclusively *exo, cis*-dibromotricyclooctene **42**.[11] Thus, addition of bromine to the electron-rich double bond of **17** was expected to give *exo, cis*-**36**.

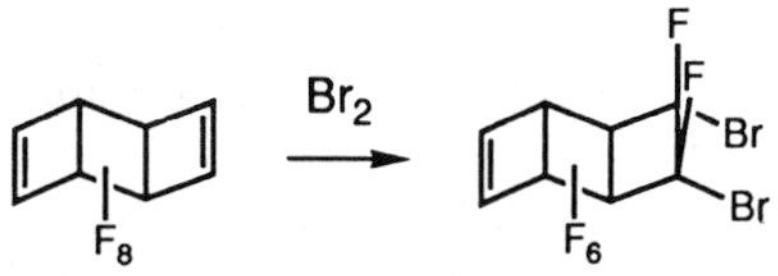

Br_2

41 **42**

In the event, photostimulated bromination of 0°C in trichlorofluoromethane yielded a mixture of *exo, cis* and *trans*-**36** in a 2:1 ratio. The stereoisomers were separated by flash chromatography.

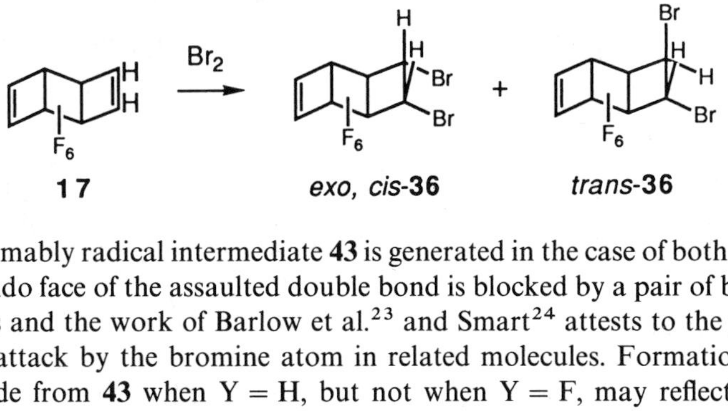

Br_2

17 *exo, cis*-**36** *trans*-**36**

Presumably radical intermediate **43** is generated in the case of both **41** and **17**, as the endo face of the assaulted double bond is blocked by a pair of bridgehead fluorines and the work of Barlow et al.[23] and Smart[24] attests to the preference for exo attack by the bromine atom in related molecules. Formation of trans dibromide from **43** when Y = H, but not when Y = F, may reflect a greater tendency for bridging (**43b**) in the former case.

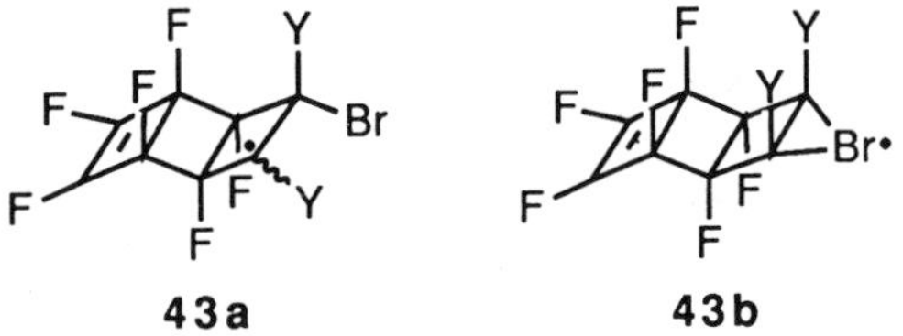

43a **43b**

Y = F, H

Heating *cis*-**36** in a sealed ampoule at 140°C with calcium carbonate as an acid scavenger and 2,5-di-*t*-butylhydroquinone as a free radical scavenger effected ring opening to give *cis*-**37** in 76% yield. The ^{19}F NMR spectrum displayed three signals of equal intensity at 152.8 (F_3, F_4), 155.0 (F_2, F_5), and 160.7 (F_1, F_6) ppm. Similarity between the spectra of *cis*-**37** and its chlorine analogue **12** prompted us to assign the bromines in *cis*-**37** as endo also.

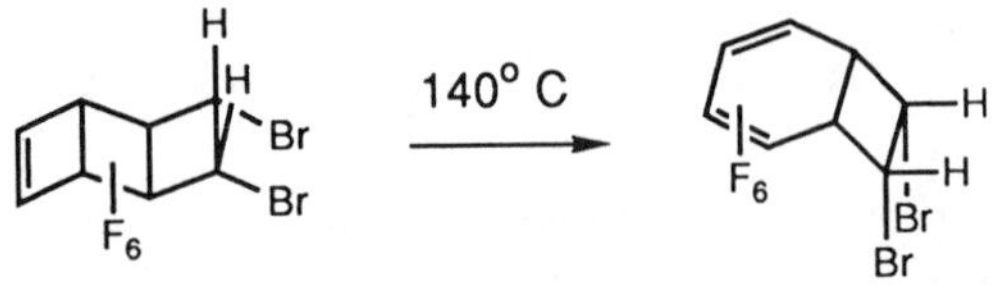

H H Br Br F_6 140° C F_6 H H Br Br

cis-**36** *cis*-**37**

Thermolysis of *trans*-**36** under the conditions described for the cis isomer resulted in the formation of *trans*-**37** and 1*H*,8*H*-perfluorocyclooctatetraene (**40**) in a 2:3 ratio. That the latter was produced from the former was revealed by a decrease with time of the ^{19}F NMR signals for *trans*-**37**, while those of **40** continued to grow. The surprising difference in stability of *cis*- and *trans*-**37** may be another consequence of the importance of bridging in bromine radicals. We suggest that debromination of the trans compound most likely occurs in the monocyclic valence isomer **44** in which the bromines are allylic. Though this isomer has not been observed, it should be present in small amount; it is related to **37** by a thermally allowed electrocyclic process, and it should not differ greatly in energy from the preferred form. Loss of a bromine atom to give an intermediate radical could be assisted by bridging (**45**) only in the case of the trans isomer.

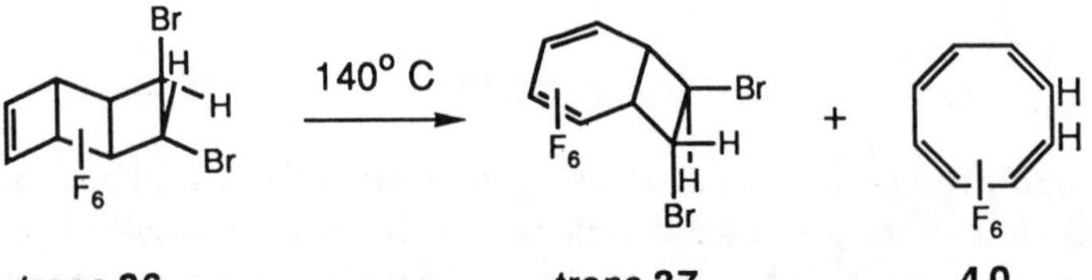

Br H H Br F_6 140° C F_6 Br H H Br + H H F_6

trans-**36** *trans*-**37** **40**

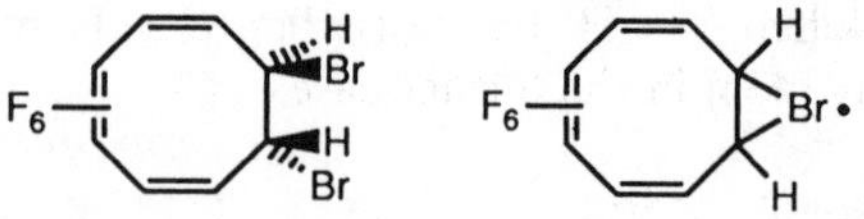

F_6 H Br H Br F_6 H Br• H

44 **45**

After cyclooctatetraene **40** had been carefully purified by gas chromatography, there remained many small signals in its ^{19}F NMR spectrum. A combination of spin-saturation transfer experiments,[25] spin decoupling experiments, and comparisons with model compounds[22] revealed that these signals arose from three valence isomers of **40**, and that all four isomers exist in dynamic equilibrium at ordinary temperatures (Scheme VI).[26] In principle, tetraenes **40** and **39** can exist in equilibrium with five bicyclo[4.2.0] valence isomers, but only two (**46** and **47**) were present in large enough amount to be detected in our experiments.

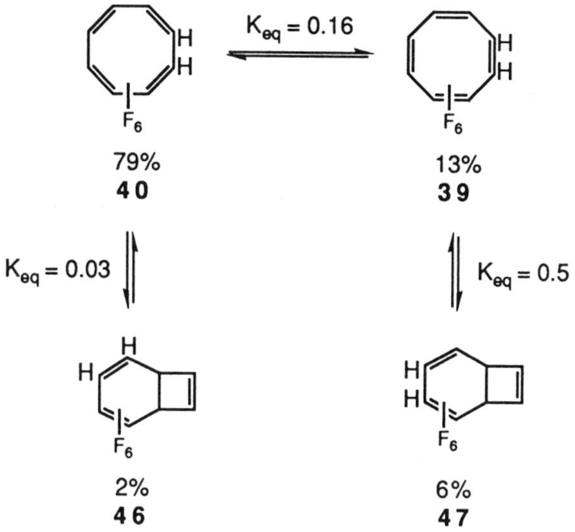

$K_{eq} = 0.16$

79% **40** 13% **39**

$K_{eq} = 0.03$ $K_{eq} = 0.5$

2% **46** 6% **47**

[a]Data are for 98°C in *o*-dichlorobenzene.

Scheme VI

These results are particularly interesting, we believe, because of the light they shed on the question of destabilization of C═C bonds by fluorine substitution Analysis of the data in Scheme VI leads to the conclusion that incorporation of a second cis fluorine into a vinyl fluoride is only slightly destabilizing (~1.4 kcal mol^{-1}), in contrast to a literature estimate[27] of about 8 kcal mol^{-1} based on differences in heats of formation.

13.2.5. Synthesis to Perfluorobenzene Oxide (3)–Perfluorooxepin (4)

Because so much debromination to **40** attended ring opening of *trans*-**36**, the synthesis was continued with *cis*-**36** only. The trans isomer was recycled, as it could be debrominated to **17** in high yield. *cis*-Dibromobicyclooctadiene **37** was epoxidized at room temperature with peroxytrifluoroacetic acid and disodium hydrogen phosphate as a buffer. A single stereoisomer, assigned the anti configuration on steric grounds (**35**), was obtained in moderate yield. Treatment of **35** with zinc and ultrasound in tetraglyme at room temperature gave epoxydiene **14** in 44% yield after purification. Though modest, this yield

CF_3CO_3H, Na_2HPO_4

cis-**37** → *cis*-**35**

represented a dramatic improvement over the 7% obtained from the chlorinated precursor **13**, and it was now practical to proceed with the final reactions in the synthetic scheme.

Zn, Tetraglyme,))))

cis-**35** → **14**

Ozonation of epoxydiene **14** in methylene chloride at 0°C yielded two stereoisomeric ozonides (**48**) in a 2:1 ratio, as shown by ^{19}F NMR. It is not known whether **48a** or **48b** is the major isomer. The ozonide mixture was used without purification for the last step.

O_3

14 → **48a** + **48b**

Irradiation of the ozonides in methylene chloride with a Pyrex-filtered medium pressure mercury arc produced a rather complex mixture as judged by ^{19}F NMR, but vacuum transfer of the mixture gave a single volatile product whose spectrum showed three signals of equal intensity at 148.6, 152.6, and 154.5 ppm (methylene chloride). The mass spectrum displayed a parent peak for C_6F_6O at m/z 202 and fragment peaks at m/z 186, 174, 155, and 124 (base) for the loss of O, CO, CO plus F, and CO plus CF_2, respectively. Bands at 1725 and 1445 cm^{-1} in the IR spectrum (carbon tetrachloride) were assigned to fluorinated double bonds and fluorinated oxirane, respectively. Diene **14** had a

hν, ?

48 → **3** ⇌ **4**

band at 1495 cm^{-1}, typical of other fluorinated oxiranes;[16] the band at 1445 cm^{-1} thus had a somewhat lower frequency than expected for this structural feature. A maximum at 246 nm ($\varepsilon = 7400$, cyclohexane) in the UV spectrum was consistent with the presence of a cyclohexadiene chromophore. For comparison, a model benzene oxide bridged to prevent ring opening, 8,9-indane oxide, has $\lambda_{max} = 258$ nm ($\varepsilon = 1800$). In contrast, a model oxepin, the 2,7-dimethyl derivative, has $\lambda_{max} = 297$ nm ($\varepsilon = 4900$) with absorption tailing far enough into the visible region to make the compound yellow-orange.[4]

The ^{19}F chemical shifts of model compounds (Fig. 13.3) confirmed that the volatile new compound was perfluorobenzene oxide (**3**), not perfluorooxepin (**4**). Trienes **49**[39] and **50**[28] imply that the fluorines geminal to oxygen in **4** should appear between 90 and 100 ppm, very far from any of the signals of the new compound. The fluorines on the oxirane ring in models **7**[7] and **14**, on the other hand, are in the same spectral region as the observed signals. Variable temperature ^{19}F NMR, to be discussed, revealed that the oxirane fluorines in **3** give rise to the 154.2-ppm resonance. When the spectrum is measured at $-38\,^{\circ}C$, the signals are sharp, with well resolved spin–spin splitting. The lowest field signal is substantially broader than the middle one. Resonances of terminal fluorines on the diene moiety in perfluorocyclohexadienes are typically significantly broader than those of internal ones. Together with the similarities in chemical shifts between these fluorines and their counterparts in cyclohexadiene *cis*-**51**[29] (Fig. 13.3), this observation permitted assignment of the two vinyl fluorine signals as shown.

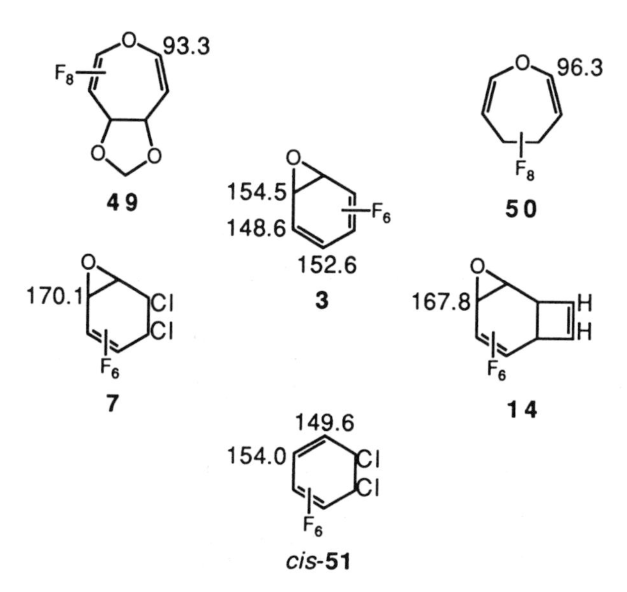

Figure 13.3. The ^{19}F NMR chemical shifts of perfluorobenzene oxide and model compounds.

13.3. Dynamic NMR Behavior of Perfluorobenzene Oxide (3) and Perfluorooxepin (4)

It rapidly became apparent that benzene oxide **3** exists in dynamic equilibrium with a small amount of oxepin **4**. The behavior of the ^{19}F NMR spectrum of the oxide in chlorobenzene (Fig. 13.4) as a function of temperature is characteristic of chemical exchange. As the temperature is increased each signal broadens and then resharpens (Table 13.1). In an exchanging system, the width at half-height for each signal is influenced by the chemical shift difference between exchanging sites as well as the fraction of each component involved in the dynamic process.[30] Therefore, each signal should have a different maximum width at half-height, and each should reach it at a different temperature. Based on the model compounds discussed above, the signal for the fluorines geminal to oxygen in **3** exchanges with a signal in **4** approximately 60-ppm downfield from it. In contrast, the other resonances for **3** exchange with resonances in **4** which should lie nearby. In the model compound perfluorotropilidene, for example, the corresponding fluorines appear at 148.4 and 150.4 ppm (acetonitrile). It follows

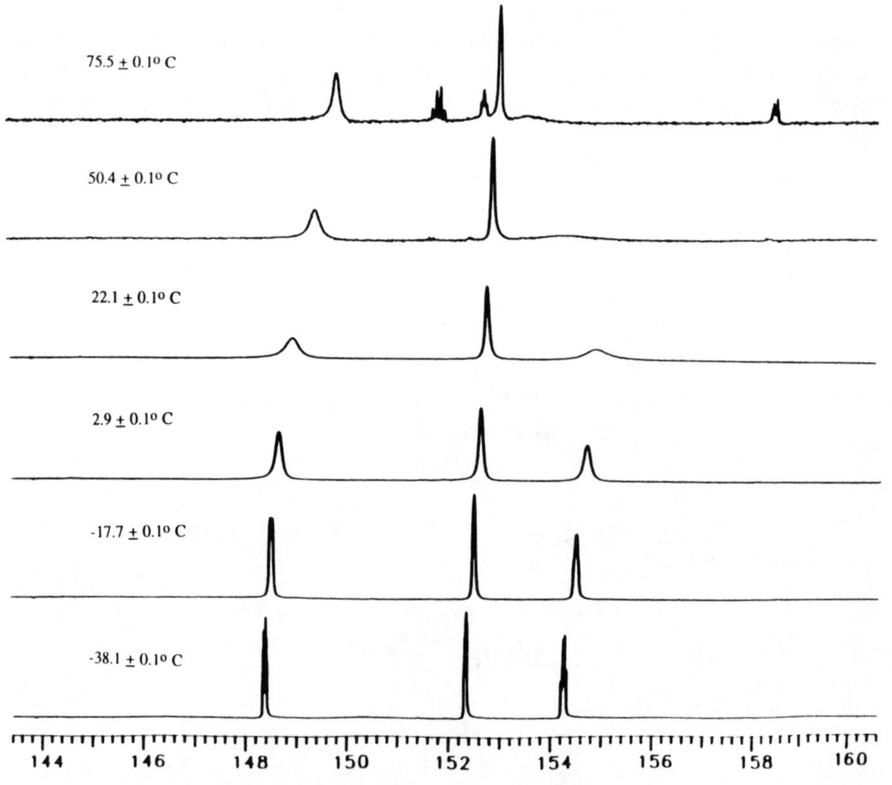

Figure 13.4. 282.2 MHz ^{19}F NMR spectrum (C_6H_5Cl + *o*-$C_6D_4Cl_2$, 10:1) of perfluorobenzene oxide at various temperatures.

TABLE 13.1. Temperature Dependence of Line Widths for Benzene Oxide 3.[a]

Temperature	Half-Height Width (Hz)		
(±0.1 °C)	149 ppm	152 ppm	154 ppm
−17.7	26	19	29
2.9	41	28	45
21.4	90	27	143
22.1	86	26	144
50.4	65	20	323
55.1	61	20	349
60.1	54	19	323
75.5	38	18	212
95.5	28	17	

[a]Spectra recorded in chlorobenzene–1,2-dichlorobenzene-d_4 (10:1).

that the signal for the fluorines geminal to oxygen should become much broader than the other two, reach maximum breadth at a higher temperature, and move downfield as the temperature is raised. This is what happens with the highest field signal (154.2), which reaches its maximum breadth at ~55 °C, and shifts downfield rapidly as it resharpens. All these observations are consistent with an equilibrating system in which benzene oxide **3** is directly observed and oxepin **4** is evident only by virtue of its influence on the spectrum of the major component. The upper limit of this variable temperature NMR study was set by the thermal lability of the benzene oxide, described more fully later.

When a careful search for oxepin signals in equilibrated samples had failed, we examined at low temperatures a solution which we had tried to enrich in that isomer. Vogel found $\Delta H^\circ = 1.7 \pm 0.4\,\text{kcal mol}^{-1}$ and $\Delta S^\circ = 11 \pm 5\,\text{eu}$ for the ring opening of the parent benzene oxide.[4] Based on the parent system as a model, **4** has a larger entropy content than **3**. If so, the predominance of **3** requires that ΔH° for the **3**–**4** equilibrium must be quite large, and K_{eq} must therefore vary considerably with temperature. As the temperature is increased from low values, the concentration of **4** must increase, but the signals become broadened by fast exchange with **3**. In order to observe **4** directly, an adequate concentration must be achieved at a temperature where its signals are narrow. Since this would require a nonequilibrium concentration, the temperature chosen must be low enough to allow time for making measurements before equilibrium is reestablished. If the **3**–**4** equilibrium close to room temperature could be frozen out by rapid quenching in liquid nitrogen, examination at low temperature might reveal the signals for **4**. An NMR sample of **3** was lowered into liquid nitrogen and then placed quickly in the spectrometer probe, which had previously been cooled to −120 °C. The resulting spectrum, obtained with a high signal-to-noise ratio ($\sim$4500:1), did not reveal any signals other than those previously assigned to **3**. If the isomer interconversion barrier is low, equilibrium may not be frozen out until far below the starting temperature, even under rapid quenching conditions.[31] Our failure to observe the oxepin directly in this experiment may reflect an effective "freezing-out temperature" far below room temperature.

Computer simulations[32] of the exchange between the fluorines geminal to oxygen in benzene oxide **3** and oxepin **4** permitted us to estimate the percent of **4** in the intermediate exchange rate regime. We assumed, as mentioned above, that the difference between the two signals of interest is 60 ppm ($\sim$17,000 Hz); this is probably correct to within several parts per million. The maximum signal width (half-height width) was then calculated for various ratios of **3**:**4**, and the experimental value of 350 Hz was matched with the ratio 98:2.

An alternative way to calculate the fraction of **4** present would be based on the change in chemical shift of the same signal from the slow to the fast exchange limit. It was not possible to measure this directly because the thermal lability of the benzene oxide prevented observation of the fast exchange limit. That limit was estimated as follows. In the above computer simulation, the drift in chemical shift reached the halfway point where the signal breadth was maximal. Thus, measurement of that drift from the slow exchange limit to 55 °C, where the line width was greatest, and doubling the result gave a estimate of the total chemical shift change. In practice, this measurement required correction for temperature dependence of the chemical shift in the absence of exchange, which was determined in the slow exchange region and extrapolated to 55 °C. The calculated drift due to exchange was about 2.3 ppm, corresponding to a ratio of **3**:**4** of 96:4. The percentage of **4** at 55 °C is thus estimated by our two methods to fall in the range 2–4%. If indeed the percentage is that high, it seems surprising that we were unable to detect the oxepin directly. However, the temperature dependence of the **3**–**4** equilibrium is probably rather steep, as noted above, making the percentage of **4** present much smaller at temperatures sufficiently low for its signals to be narrow enough to observe.

With the amount of oxepin estimated as 3% at 55 °C, simulation yields a rate constant of 3200 s^{-1} for **4** → **3**. While this value is very approximate, it is clear that the fluorinated valence isomers interconvert far slower than the parent heterocycles.[4] The extrapolated rate constant at 55 °C for cyclization of oxepin is $2 \times 10^7 s^{-1}$, >6000 times the estimated rate for **4**.

13.4. CHEMISTRY OF THE BENZENE OXIDE

The highest temperature spectrum (75.5 °C) in Fig. 13.4 reveals the presence of perfluorocyclohexa-2,4-dienone (**10**),[6,33] formed by thermal rearrangement of **3** in chlorobenzene. The beginning of this rearrangement is perceptible even in the 50.4 °C spectrum. Dienone formation is much faster in polar than in nonpolar solvents; it occurs spontaneously and cleanly at room temperature in acetone and acetonitrile ($t_{1/2} \sim 1.5$ h in the latter solvent). The presence of a Lewis acid such as boron trifluoride was found to accelerate the reaction in methylene chloride (Scheme VII). These results are consistent with the hypothesis of spontaneous (or catalyzed in the last case) opening of the oxirane ring to give zwitterion **52**, followed by a 1,2-shift of fluoride.

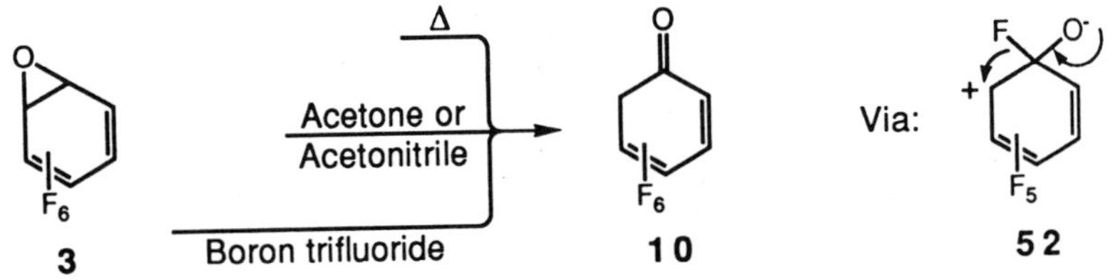

Δ; Acetone or Acetonitrile; Boron trifluoride; 3; 10; Via: 52; F_6; F_5

Scheme VII

Barlow found that thermolysis of the tricyclic epoxide **9** in the liquid or gas phase gives the same cyclohexadienone (**10**) as the primary product.[6] Our findings strongly support his suggestion that **3** is formed as an intermediate and rearranges by the pathway just described. Perhaps the reaction proceeded via the biradical counterpart of **52** in his high temperature gas-phase pyrolysis.

Benzene oxide itself also suffers facile thermal ring-opening rearrangement to yield phenol.[4] The reaction is catalyzed by acids and proceeds via the NIH shift.[34,35] The mechanism thus parallels the polar pathway proposed for the perfluoro system, with the exception that aromatization of the intermediate dienone is not possible in the latter case.

Irradiation of **3** with benzophenone as a triplet sensitizer, using a 340-nm cutoff filter to assure exclusive excitation of the sensitizer, again produced the dienone **10**, Scheme VIII. Direct photolysis of **3** through a Vycor filter gave instead the unsaturated acid **53**. When a sample of **10** was irradiated under these reaction conditions, the same acid was formed. The photolysis of **3** presumably proceeds via **10**, which undergoes electrocyclic ring opening to yield the ketene

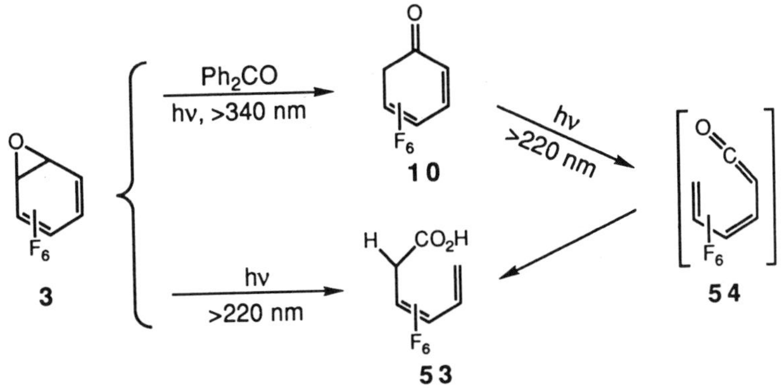

Ph_2CO, hν, >340 nm; hν, >220 nm; hν >220 nm; 3; 10; 53; 54; F_6; CO_2H

Scheme VIII

54. Hydrolysis of **54** by traces of water in the solvent then gives the acid.[36] Thus, the benzene oxide rearranges cleanly to dienone **10** in the ground state or upon excitation into either its S_1 or T_1 state.

The photochemistry of the parent benzene oxide–oxepin system exhibits both differences from and parallels with its perfluoro analog. Irradiated at long wavelengths where oxepin absorbs, the mixture yielded 2-oxabicyclo[3.2.0]hepta-3,6-diene, the product of a four-electron cyclization of the oxepin.[4,37] At wavelengths where benzene oxide absorbs intensely, however, phenol was the principal product, accompanied by some benzene. Direct photolysis at 77 K revealed that the putative intermediate 2,4-cyclohexadienone en route to phenol undergoes reversible ring opening to a ketene analogous to **54**.[37] Triplet sensitization of either isomer gives phenol,[38] and sensitized photolysis of deuterium-labeled benzene oxide revealed the presence of a degenerate "oxygen walk" process as well.[37]

Earlier attempts to synthesize **3** in our laboratory had entailed the reduction of oxabicycloheptene **7**, Scheme IX. Dailey had found perfluorophenol (**8**) as the major product when either sodium iodide in acetone or zinc in acetic acid was used as reducing agents.[7] We have now found that sodium iodide in acetone reduces **3** to **8** at room temperature (Scheme IX), consistent with the intermediacy of **3** in Dailey's reduction. Reduction of **3** occurs faster than its spontaneous rearrangement to the dienone **10** in acetone. Dienone **10** was ruled out as an intermediate by the finding that it reacted differently under the same reaction conditions. We do not know whether reduction of **3** occurs via electron transfer from or nucleophilic attack by iodide ion.

Nucleophilic attack on benzene oxide **3** might proceed via any of three mechanisms, S_N2, S_N2', and S_N2''. Depending on the reagent, nucleophilic opening of the oxirane ring in benzene oxide itself has been found to occur by the S_N2 and/or S_N2'' mechanisms, the latter in either suprafacial or antarafacial fashion.[39–41] To differentiate among the possible pathways, **3** was allowed to

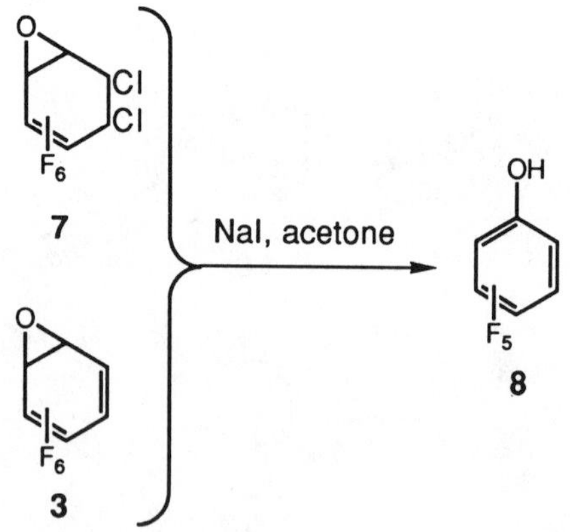

Scheme IX

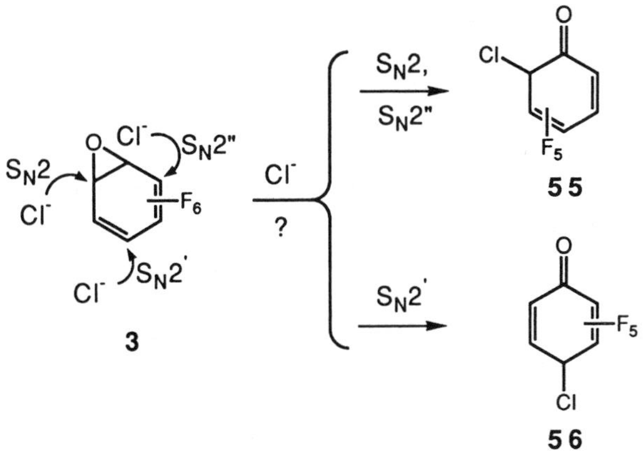

Scheme X

react with chloride ion (Scheme X). If chloride attacked in S_N2 or S_N2'' fashion, the resulting product should be chlorocyclohexadienone **55**, but if attack occurred via the S_N2' mechanism the product should be chlorocyclohexadienone **56**. Both dienones are known. Unfortunately, both tetrabutylammonium chloride and lithium chloride in acetonitrile destroyed **3** and gave only unidentified products.

All of the chemistry of the **3**/**4** system which we have worked out is apparently derived from **3**. Wishing to trap the small amount of **4** present in the equilibrium mixture, we tried brominating the mixture in the dark. Since the enol ether double bonds of the oxepin should be more susceptible to electrophilic attack than the π bonds of the benzene oxide, as dipolar resonance forms suggest, there was reason to hope for selective addition of bromine to **4**, giving **57**. However, ^{19}F NMR monitoring of this reaction revealed only the spontaneous formation of dienone **10**.

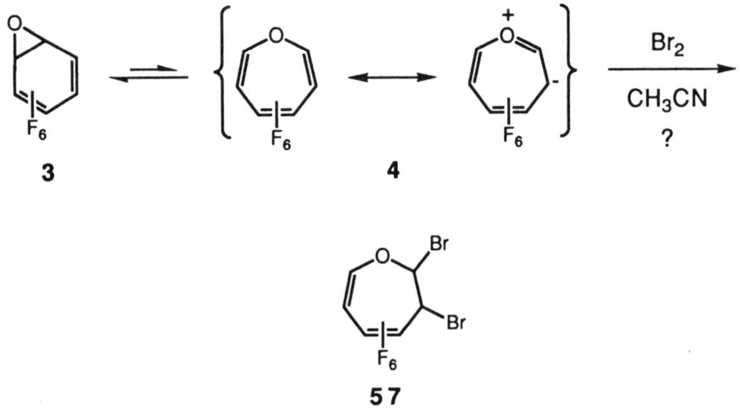

13.5. CONCLUSION

Whereas earlier attempts to synthesize perfluorobenzene oxide and perfluoro-oxepin based on ground state chemistry had failed, we finally succeeded by adopting an approach that culminated in a photochemical step. The synthetic plan outlined in Scheme I was successfully executed, but because of the low yield in the dechlorination step a less direct but quite parallel route employing bromines for double-bond protection proved to be more efficient. The high reactivity displayed by the benzene oxide explains the disappointment attending earlier synthetic attempts. Opening of the oxirane ring in this compound is extremely facile; it occurs spontaneously in polar media, photochemically, under mild reducing conditions, and with acid catalysis.

The parent benzene oxide–oxepin equilibrium is quite evenly balanced. At the outset of our work, we predicted that destabilization of benzene oxide **3** caused by fluorine substitution on a three-membered ring[42] would outweigh destabilization of the oxepin by its additional fluorinated double bond. An equilibrium dominated by the oxepin was thus anticipated. In a byway along our path, we found evidence described above that destabilization of a double bond by cis, vicinal fluorines is not significant. It was therefore a real surprise to discover that the fluorinated benzene oxide–oxepin equilibrium lies far on the benzene oxide side.[43] Our finding suggests that the marked destabilization of cyclopropane rings characteristic of fluorine substitution does not carry over to oxirane rings, but this idea can be taken seriously only after further testing. The barrier to interconversion of the valence isomers is substantially higher in the fluorinated than in the parent system.

The properties and chemical behavior of the perfluorobenzene oxide–perfluorooxepin system remind one of a truism in organofluorine chemistry, namely, that whatever a hydrocarbon derivative is like, its perfluorinated counterpart is expected to be refreshingly different.

ACKNOWLEDGMENTS

The authors wish to thank Robert Hamlin for laying the groundwork for our synthesis of perfluorobenzene oxide–perfluorooxepin and Wayne P. Casey, who provided valuable guidance in the NMR work. Support of this research by the Air Force Office of Scientific Research is gratefully acknowledged.

REFERENCES

1. This chapter is based on the Ph.D. Thesis of N. E. Takenaka, Dartmouth College, 1990. A preliminary account of some of the work has appeared in *J. Am. Chem. Soc.*, *112*, 6715 (1990).
2. Vogel, E., Böll, W. A., and Günther, H., *Tetrahedron Lett.*, 609 (1965).

3. Günther, H. *Tetrahedron Lett.*, 4085 (1965).
4. For a review see Vogel, E. and Günther, H., *Angew. Chem. Int. Ed. Engl.*, *6*, 385 (1967).
5. Theoretical calculations on this pair of valence isomers are presented in the following papers. Bock, C. W., George, P., Stezowski, J. J., and Glusker, J. P., *Struct. Chem.*, *1*, 33 (1990). Cremer, D., Dick, B., and Christeu, D., *Theochem.*, *19*, 277 (1984). Schulman, J. M., Disch, R. L., and Sabio, M. L., *J. Am. Chem. Soc.*, *106*, 7696 (1984). Thieme, R. and Weiss, C. *Stud. Biophys.*, *93*, 273 (1983). Boyd, D. R. and Stubbs, M. E., *J. Am. Chem. Soc.*, *105*, 2554 (1983).
6. Barlow, M. G., Haszeldine, R. N., and Peck, C. J., *J. Chem. Soc. Chem. Commun.*, 158 (1980).
7. Dailey, W. P., Ralli, P., Wasserman, E., and Lemal, D. M., *J. Org. Chem.*, *54*, 5516 (1989).
8. Hamlin, R. and Lemal, D. M., unpublished results.
9. Adam, W., *Angew. Chem. Int. Ed. Engel.*, *13*, 619 (1974).
10. Story, P. R., Morrison, W. H., III, T. K., Farine, J.-C., and Bishop, C. E., *Tetrahedron Lett.*, 3291 (1968).
11. Barefoot, A. C., III, Saunders, W. D., Buzby, J. M., Grayston, M. W., and Lemal, D. M., *J. Org. Chem.*, *45*, 4292 (1980).
12. Rahman, M. M., Secor, B. A., Morgan, K. M., Shafer, P. R., and Lemal, D. M., *J. Am. Chem. Soc.*, *112*, 5986 (1990).
13. Reetz, M. T., *Angew. Chem. Int. Ed. Engl.*, *11*, 129, 130 (1972).
14. For reviews on hydrogenation with diimide see House, H. O., *Modern Synthetic Reactions*, 2nd ed., W. A. Benjamin: New York, 1972, pp. 248–256; Miller, C. E., *J. Chem. Educ.*, *42*, 254 (1965).
15. Roth, W. R., *Liebigs Ann. Chem.*, *671*, 25 (1964).
16. Tarrant, P., Allison, C. G., Barthold, K. P., and Stump, E. C., in *Fluorine Chemistry Reviews*; Dekker: New York, 1971; Vol. 5.
17. See, for example, Waldron, R. F., Barefoot, A. C., III; and Lemal, D. M., *J. Am. Chem. Soc.*, *106*, 8301 (1984).
18. March, J., *Advanced Organic Chemistry*, Wiley: New York, 1985, p. 924.
19. Kuivila, H. G. and Pian, C. H.-C., *Tetrahedron Lett.* 2561 (1973).
20. Model β,γ-unsaturated ketones include 4-methylcyclohex-3-enone ($\lambda_{max} = 226$ nm, $\varepsilon = 182$), 3-methylenecyclobutanone ($\lambda_{max} = 214$ nm, $\varepsilon = 1500$) and bicyclo-[2.2.2]oct-5-en-2-one ($\lambda_{max} = 202$ nm, $\varepsilon = 3000$). Grasselli, J. G. (Ed.), *CRC Atlas of Spectral Data and Physical Constants of Organic Compounds*, CRC Press: Cleveland, 1973, p. B-459. Ferguson, L. N. and Nnadi, J. C., *J. Chem. Educ.*, *42*, 529 (1965).
21. Crawford, R. J., Lutener, S. B. and Cockcroft, R. D., *Can. J. Chem.*, *54*, 3364 (1976).
22. Spector, T., Ph.D. Thesis, Dartmouth College, 1987.
23. Barlow, M. G., Haszeldine, R. N., Morton, W. D., and Woodward, D. R., *J. Chem. Soc. Perkin Trans. I*, 2170 (1972).
24. Smart, B. E., *J. Org. Chem.*, *38*, 2027 (1973).
25. Forsen, D. and Hoffman, R. A., *J. Chem. Phys.*, *39*, 2892 (1963). Sandstrøm, J., *Dynamic NMR Spectroscopy*, Academic: London, 1982, pp. 53–58.
26. Takenaka, N. E., Hamlin, R., and Lemal, D. M., manuscript in preparation.

27. Smart B. E., *Mol. Struct. Energ.*, *3*, 141 (1986).
28. Toy, M. S. and Stringham, R. S., *J. Org. Chem.*, *44*, 2813 (1979). Toy, M. S. and Stringham, R. S., *J. Fluo. Chem.*, *12*, 23 (1978).
29. Dailey, W. P., Correa, R. A., Harrison, E., III; and Lemal, D. M., *J. Org. Chem.*, *54*, 5511 (1989).
30. See, for example, Binsch, G., in *Dynamic Nuclear Magnetic Resonance Spectroscopy*, L. M. Jackman, and F. A. Cotton (Eds.), Academic: New York, 1974, Chap. 3.
31. Faster quenching could be achieved from the gas phase. See, for example, Squillicote, M. E. and Neth, J. M., *J. Am. Chem. Soc.*, *109*, 198 (1987).
32. The DNMR3 program of D. A. Kleier and G. Binsch was used as modified by Bushweller's group (Bushweller, C. H., Bhat, G., Letendre, L. J., Burnelle, J. A., Bilofsky, H. S., Ruben, H., Templeton, D. H., and Zalkin, A., *J. Am. Chem. Soc.*, *97*, 65 (1975)).
33. Soelch, R. S., Mauer, G. M., and Lemal, D. M., *J. Org. Chem.*, *50*, 5845 (1985).
34. Kasperek, G. J., Bruice, T. C., Yagi, H., and Jerina, D. M., *J. Chem. Soc. Chem. Commun.* 784 (1972). Kasperek, G. J. and Bruice, T. C., *J. Am. Chem. Soc.*, *94*, 198 (1972).
35. George, P., Bock, C. W., and Glusker, J. P., *J. Phys. Chem.*, *94*, 8161 (1990). Ferrell, J. E., Jr. and Loew, G. H., *J. Am. Chem. Soc.*, *101*, 1385 (1979).
36. See, for example, Cowan, D. O. and Drisko, R. L., *Elements of Organic Photochemistry*; Plenum: New York, 1976, pp. 321–323.
37. Jerina, D. M., Witkop, B., McIntosh, C. L., and Chapman, O. L., *J. Am. Chem. Soc.*, *96*, 5578 (1974).
38. Holovka, J. M. and Gardner, P. D., *J. Am. Chem. Soc.*, *89*, 6390 (1967).
39. Rastetter, W. H., Chancellor, T., and Richard, T. J., *J. Org. Chem.*, *47*, 1509 (1982).
40. Reuben, D. M. E. and Bruice, T. C., *J. Am. Chem. Soc.*, *98*, 114 (1976).
41. Jeffrey, A. M., Yeh, H. J. C., Jerina, D. M., DeMarinis, R. M., Foster, C. H., Piccolo, D. E., and Berchtold, G. A., *J. Am. Chem. Soc.*, *96*, 6929 (1974). Berchtold, G. A. and Foster, C. H., *J. Am. Chem. Soc.*, *93*, 3831 (1971). DeMarinis, R. M. and Berchtold, G. A., *J. Am. Chem. Soc.*, *91*, 6525 (1969).
42. Dolbier, W. R., Jr., *Acc. Chem. Res.*, *14*, 195 (1981). O'Neal, H. E. and Benson, S. W., *J. Phys. Chem.*, *72*, 1833 (1969).
43. For molecular orbital calculations on the effect of *mono*fluorination at various sites on the energy of oxepin, see Hayes, D. M., Nelson, S. D., Garland, W. A., and Kollman, P. A., *J. Am. Chem. Soc.*, *102*, 1255 (1980).

CHAPTER 14

Perhalodioxins and Perhalodihydrodioxins

CARL G. KRESPAN AND DAVID A. DIXON

The recent report[1] of reactive *F*-dioxoles such as **1** giving homopolymers of extraordinarily high T_g led us to consider six-membered cyclic difluorovinylene diethers **2**–**4** as candidate monomers leading to related amorphous polymers of high T_g. The level of reactivity of **2**–**4** toward free radicals was also of interest, since **1** has been shown to polymerize with ease while the acyclic relatives, trifluorovinyl ethers, do not homopolymerize under normal conditions of temperature and pressure.

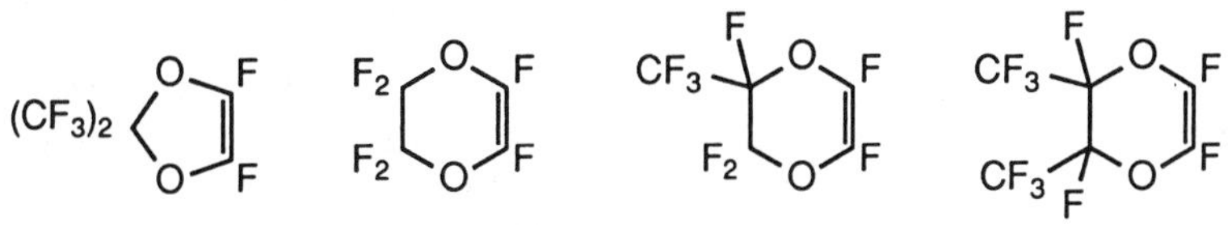

Synthetic Fluorine Chemistry,
Edited by George A. Olah, Richard D. Chambers, and G. K. Surya Prakash.
ISBN 0-471-54370-5

14.1. SYNTHESES AND REACTIONS

F-2,3-Dihydro-1,4-dioxin (**2**) is a known compound,[2] but the synthetic route involved steps with very low yields. Our route, shown in Scheme I, allowed selective perfluorination of one side of the ring followed by introduction of the difluorinated double bond. Note that hexachloro-2,3-dihydro-1,4-dioxane (**5**), a new compound, is easily obtained in quantity by this method.

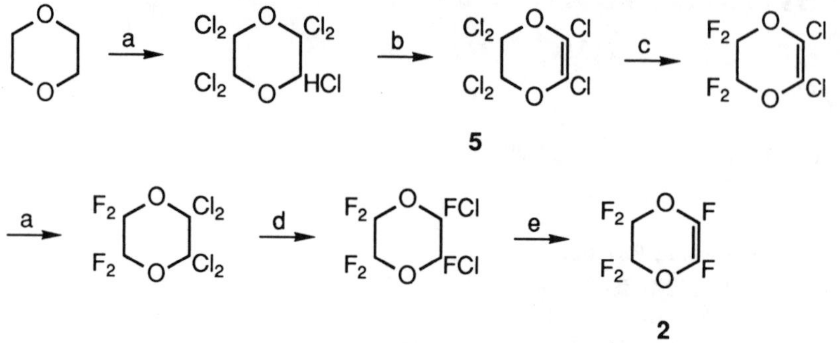

Scheme I. (*a*) Cl_2; (*b*) $NaOH/CH_3OH$; (*c*) $SbF_3/SbCl_5$ at 100 °C; (*d*) $SbF_3/SbCl_5$ at reflux; (*e*) *Zn* + *N*,*N*,-dimethylformamide (DMF).

F-2-Methyl-2,3-dihydro-1,4-dioxin (**3**) was also obtained by a synthesis that allowed formation of the perfluorinated dihydro side of the ring first, then introduction of the difluorovinylene unit in a second stage. As indicated in Scheme II, **6**, the product from the first stage of reaction, can be obtained by either of two pathways. Schwertfeger and Siegemund[3] prepared dioxane (**6**) earlier by a low yield route.

Photochemical chlorination of **6** cleanly gave the tetrachloro derivative, which was converted to **3** by the usual fluorination–dechlorination reactions (Scheme III).

Synthesis of **4** started from *F*-2,3-epoxybutane, which was itself readily obtained by reaction of *F*-butene-2 with aqueous NaOBr at 0 °C. Scheme IV shows the steps involved.

Radical-catalyzed polymerizations of **2** and **3** were carried out under conditions similar to those under which dioxole (**1**) rapidly forms amorphous homopolymers with T_gs in the range 173–330 °C. The neat monomers held at 40–45 °C under nitrogen were treated intermittently with small amounts of *F*-propionyl peroxide catalyst while viscosity slowly increased over a period of days. Workup afforded in both cases about 30% conversions to amorphous solid homopolymer along with recovered monomer. Differential scanning calorimetry (DSC) measurements indicated glass transition temperatures of 110 °C (polymer of **2**) and 150 °C (polymer of **3**). Absence of carbonyl absorption in the IR spectra is good evidence that ring opening of intermediate radical chain ends does not occur. The level of reactivity exhibited by these monomers is clearly

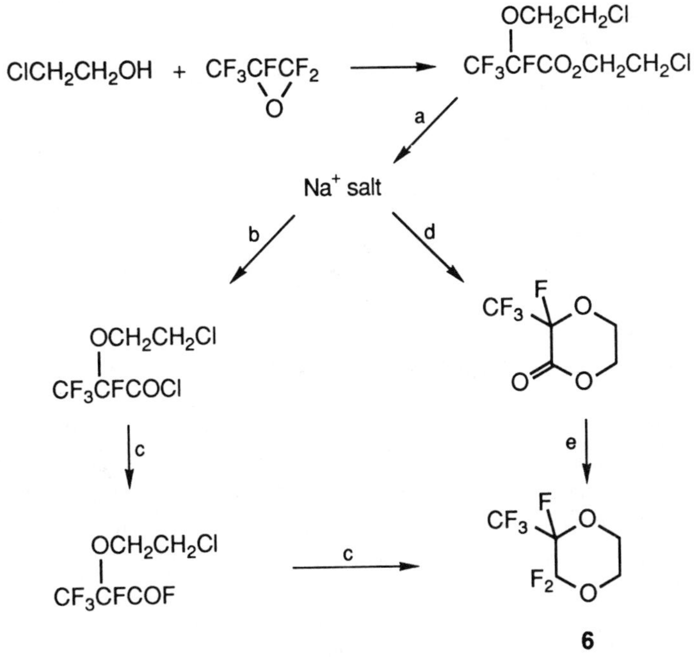

$ClCH_2CH_2OH$ + CF_3CFCF_2 (epoxide, O) → $CF_3CF(OCH_2CH_2Cl)CO_2CH_2CH_2Cl$

a → Na^+ salt

b → $CF_3CF(OCH_2CH_2Cl)COCl$ —c→ $CF_3CF(OCH_2CH_2Cl)COF$ —c→ 6

d → CF_3, F, O, O, O (lactone) —e→ CF_3, F, O, F_2, O

6

Scheme II. (*a*) NaOH; (*b*) $SOCl_2$/DMF; (*c*) KF; (*d*) 185°C; (*e*) SF_4/HF at 160°C.

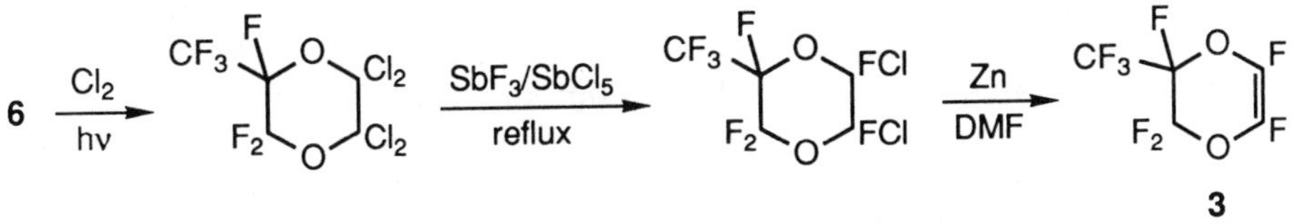

6 —Cl_2, hν→ CF_3, F, O, Cl_2, F_2, O, Cl_2 —$SbF_3/SbCl_5$, reflux→ CF_3, F, O, FCl, F_2, O, FCl —Zn, DMF→ CF_3, F, O, F, F_2, O, F

3

Scheme III

CF_3—CFCF—CF_3 (epoxide, O) + $HOCH_2CH_2OH$ —a, b→ $CF_3CF(OCH_2CH_2OH)C(OH)_2CF_3$

c → $CF_3CF(OCH_2CH_2Cl)COCF_3$

d → CF_3, F, O, CF_3, F, O —e, f, g→ CF_3, F, O, F, CF_3, F, O, F

4

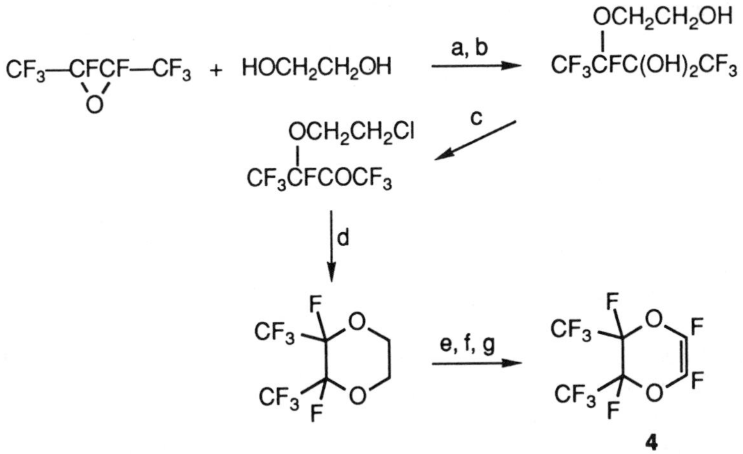

Scheme IV. (*a*) $KOC(CH_3)_3$; (*b*) $H^+(H_2O)$; (*c*) $SOCl_2$; (*d*) KF; (*e*) $Cl_2 + h\nu$; (*f*) $SbF_3/SbCl_5$; (*g*) *Zn*, DMF.

much greater than that of the acyclic trifluorovinyl ethers, which do not homopolymerize, but lower than that of dioxole (**1**), which polymerizes much more rapidly. Furthermore, the observed glass transitions are well above those of many fluoropolymers, yet not as high as those from the *F*-dioxole class of monomers. Monomer **4** gave oils with only small amounts of solid polymer under similar polymerization conditions.

Copolymerizations of **3** with three monomers of varying electron demand (tetrafluoroethylene, ethylene, and vinyl acetate) proceeded readily at 25–55 °C. The easy incorporation of 5–20 wt% of **3** into the solid copolymers at such low temperatures indicates receptivity to attack by a spectrum of radical types.

Addition of carbon-based free radicals to the boat-shaped 2,3-dihydro-1,4-dioxin system is clearly a favorable reaction. Subsequent addition of the resulting radical to another dihydrodioxin molecule proceeds, but the rate appears to be governed by steric constraints. Dihydrodioxins **2** and **3** give solid homopolymers even at 40 °C, but the reaction is slow. Addition of a second trifluoromethyl group to the ring results in a dihydrodioxin (**4**), which polymerizes even more slowly at 40 °C to give viscous oils rather than solid polymer. The radical adduct contains a puckered ring which, in the latter case, can adopt few conformations capable of propagating the chain. The preferred conformational structures for the homopolymer of **2** have been determined (see below).

The foregoing results, while encouraging, were not as striking as the reactivity exhibited by the perhalodioxins. The first perhalodioxin synthesized was tetrachloro-1,4-dioxin (**7**), formed as a coproduct with **5** during dehydrochlorination of an impure sample of heptachloro-1,4-dioxane. The crude product became exceedingly hot when filtered to remove drying agent, a phenomenon shown to be due to the reaction of oxygen with the unexpected by-product **7**. To the extent that underchlorination produced hexachloro-1,4-dioxane isomers with hydrogen atoms distributed one on each side of the ring, dehydrochlorination led to **7**. Fractionation of the mixture under a nitrogen atmosphere easily afforded **7** of 90–95% purity.

$$\text{Cl}_2,\ \text{HCl},\ \text{HCl}_3\text{-1,4-dioxane} \xrightarrow{\text{NaOH}} \text{Cl}_4\text{-1,4-dioxin}\ (\mathbf{7}) \qquad (14.1)$$

Related dioxin (**8**) was also prepared, starting from dihydrodioxin (**5**). Partial fluorination at 55 °C gave 2,3,5,6-tetrachloro-2,3-difluoro-2,3-dihydro-1,4-dioxin in high yield. Then dechlorination with zinc in DMF gave 2,3-dichloro-5,6-difluoro-1,4-dioxin (**8**) as a distillate contaminated with DMF. This mixture could be stirred in air without noticeable change, but reaction of **8** with oxygen proved to be even more vigorous than that of **7** after the DMF had been removed by a water wash.

$$\mathbf{5} \xrightarrow[55^\circ C]{SbF_3/SbCl_5} \text{(FCl, FCl, Cl, Cl dioxene)} \xrightarrow[DMF]{Zn} \mathbf{8} \qquad (14.2)$$

Further investigation of the reaction between **7** and oxygen showed it to be spontaneous and reproducible, as well as exothermic. A sample of **7** and excess oxygen gave crude product, which slowly hydrolyzed in air, leaving a residue of oxalic acid hydrate. When the oxidation was conducted under anhydrous conditions, nearly 0.5 mol O_2/mol **7** was absorbed to produce one major product, epoxide **9**, in over 60% yield. Minor products included phosgene and oxalyl chloride.

$$\mathbf{7} \xrightarrow{O_2} \mathbf{9} + COCl_2 + ClCOCOCl \xrightarrow{H_2O} HCl + CO_2 + HO_2CCO_2H \qquad (14.3)$$

The formation of epoxide as an isolable product from direct, uncatalyzed reaction of oxygen with alkene is similar to reactions known for many fluoroalkenes. Tetrafluoroethylene and other perhaloalkenes will interact with oxygen at elevated temperatures by a multistep mechanism to form epoxides as major products and acid halides from carbon–carbon bond cleavage pathways as minor products.[4] A free radical mechanism is generally accepted for such reactions, and indeed, peroxidic products formed from oxygen and either tetrafluoroethylene or hexafluoropropylene at low temperature could be decomposed at 100–250 °C to give carbon-based free radicals observable by ESR.[5] A similar mechanism involving both oxygen- and carbon-based free radical intermediates would explain the products derived from dioxin (**7**) and oxygen, but some major differences should be noted. Oxidation of **7** initiates readily at only 25 °C, the products are derived mainly or entirely from reaction at only one of the two double bonds available, and no polyethers or other products of oligomerization seem to form despite the low temperatures involved.

Attempts to capitalize on the high energy content of dioxin (**7**) indicated it to have low reactivity in nonradical reactions. Little interaction occurred between **7** and either maleic anhydride or tetrafluoroethylene at 150 °C; no cyclo adducts were detected and **7** was largely recovered unchanged. Cycloaddition of 2,3-dimethylbutadiene to **7** did occur slowly at 150 °C to produce a low yield of **10** along with considerable diene polymer. Iodine did not add to **7** at 25 °C, and exposure to excess methanolic sodium hydroxide during synthesis seemed to have little effect.

$$7 + CH_2{=}C(CH_3){-}C(CH_3){=}CH_2 \longrightarrow \mathbf{10} \qquad (14.4)$$

As was mentioned earlier, purified dioxin (**8**) was even more reactive toward oxygen than **7**, as might be expected from replacement of vinylic chlorine with vinylic fluorine. The spectrum and amounts of products obtained showed that attack occurred at both double bonds, with some selectivity for oxidation of the fluorinated double bond. The formation of low molecular weight oligomers may occur via radical intermediates during the oxidation or, less likely, by polymerization of fluoroepoxide (**11**). Tarrant et al.,[4] discussed polymerization of preformed fluoroepoxides using anionic catalysts. This source of the oligomeric polyethers from oxidation of **8** would seem to require conditions different from the acidic environment actually present. Our interpretation is that the fluorinated double bond in **8** is susceptible to attack by fluoroalkyloxy radical intermediates, leading to polyethers in a fashion similar to that in which other fluoroalkenes and oxygen form polyethers at low temperature.

$$8 \xrightarrow{O_2} COF_2 + COCl_2 + FCOCOF + ClCOCOCl + \mathbf{11} + FCO_2CCl_2COF + ClCO_2CFClCOF + \text{low oligomers} \qquad (14.5)$$

The apparent reactivity of the difluorovinylene moiety in **8** toward oxy radicals raises the question of reactivity toward carbon-based radicals, especially those derived from **8**. In other words, will the high energy content of this dioxin system (see below) be a sufficient condition for facile homopolymerization to occur as it does with dioxole (**1**) and more readily than with dihydrodioxins (**2**) and (**3**)? In fact, polymerization of **8** proceeded only slowly under conditions similar to those for polymerization of **3**. The polymerization appeared to proceed selectively through the fluorinated double bond, as expected, to give the polymer structure shown, although the level of selectivity is unknown due to the very low intensity of the IR absorption by the cyclic difluorovinylene unit.

$$\mathbf{8} \xrightarrow{R\cdot} [\text{poly}]_n \qquad (14.6)$$

The high reactivity of dioxin (**8**) with oxygen and the tendency to form some polyether makes the response of an active *F*-dihydrodioxin such as **3** to oxygen at low temperature of interest. Indeed, a slow but steady reaction of **3** with 1 atm O_2 did occur at 25 °C, and with continuously increasing viscosity. Nearly 0.5 mol O_2/mol **3** had been absorbed before reaction slowed after 8 days. Further exposure to oxygen caused a small general increase in degree of oligomerization. Gas chromatography–mass spectrometry analysis supported the structures shown below.

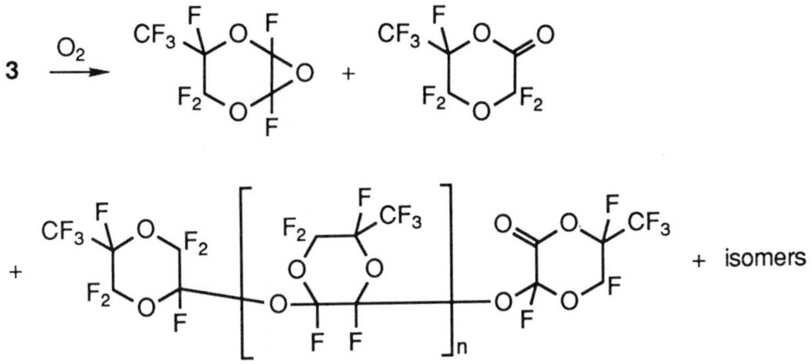

3 $\xrightarrow{O_2}$ + + isomers (14.7)

As we have seen with dioxin (**8**) then, but to a markedly greater extent, dihydrodiozin (**3**) also oxidizes at 25 °C via free radical intermediates to form polyethers.

14.2. THEORETICAL MODELS

Due to the novel chemistry of these materials, we decided to investigate their electronic and molecular structure[6] by using ab initio molecular orbital theory with extended basis sets[7] and MP-2 correlation corrections[8] for the energies. The calculations were done with the program system GRADSCF, which is an ab initio program system designed and written by A. Komornicki at Polyatomics Research.

14.2.1. Dioxins

Halogenation in the dioxins (see Table 14.1) leads to a decrease in the C═C bond length consistent with previous calculations on the fluoroethylenes.[9] The C—O bond adjacent to the C—F bond also shortens by as much as 0.035 Å, whereas the C—O bond adjacent to a C—H bond lengthens by as much as 0.019 Å for the difluoro-substituted compounds. The shortening of the C—O bond adjacent to the C—F bond is consistent with the result found in the fluoroethylenes where addition of fluorines leads to a shortening of the C—F bonds. Since O is also quite electronegative, it follows the same trend of a

TABLE 14.1. Bond Distances (Å).

Molecule[a]	r(C=C)[b]	r(C–O)[c,d]	r(C–O)[c,e]
	Dioxins		
$C_4O_2H_4$	1.318	1.369	1.369
$C_4O_2F_4$	1.306	1.356	1.356
$C_4O_2Cl_4$	1.312	1.364	1.364
$C_4O_2H_2F_2$(2,5)	1.313	1.334 (F)	1.388 (H)
$C_4O_2H_2F_2$(2,6)	1.311	1.353 (F)	1.371 (H)
$C_4O_2H_2F_2$(2,3)	1.307(F), 1.317(H)	1.351 (F)	1.374 (H)
$C_4O_2Cl_2F_2$(2,5)	1.309	1.344 (F)	1.374 (Cl)
$C_4O_2Cl_2F_2$(2,6)	1.309	1.353 (F)	1.367 (Cl)
$C_4O_2Cl_2F_2$(2,3)	1.309(F), 1.315(Cl)	1.344 (F)	1.374 (Cl)
	Dihydrodioxins		
$C_4O_2H_6$	1.324	1.360	1.407
$C_4O_2F_6$	1.303	1.353	1.364
	Dioxoles		
$C_3O_2H_4$	1.316	1.372	1.406
$C_3O_2H_2F_2$ (4,5)	1.303	1.356	1.413
$C_3O_2F_4$	1.299	1.363	1.367

[a]Fluorine positions in parentheses. [b]Labels in parentheses for unique double bonds. [c]Labels in parentheses denote other substituent attached to carbon. [d]Bond distance between the O and the ethylenic carbon for the diydrodioxins and the dioxoles. [e]Bond distance between the O and the CR_2 for the dihydrodioxins and the dioxoles.

decrease in bond distance with increasing fluorination. The effect of chlorine substitution on the shortening of the C=C and C—O bonds is not as pronounced as for fluorination.

Although the parent compound and the difluoro isomers are planar, the perhalogenated compounds are nonplanar (see Table 14.2). The perfluorinated compound has torsions about the C—O bonds of 12° and an inversion barrier of 0.41 kcal mol^{-1} at the MP-2 level. Substitution of chlorine for fluorine leads to a significant increase in nonplanarity. Substitution of two chlorines in tetrafluorodioxin leads to three isomers and the torsions about the C—O bonds range from 20–24°.

The inversion barriers are higher with values between 1.4 and 1.9 kcal mol^{-1} at the MP-2 level. The largest barrier is found for the tetrachloro compound with a barrier of 3.25 kcal mol^{-1} at the MP-2 level.

TABLE 14.2. Inversion Barriers (kcal mol^{-1}).

Compound	SCF	MP-2	iv[a]	τ(C–O)[b]
	Dioxins			
$C_4O_2F_4$	0.04	0.41	17	12
$C_4O_2Cl_2F_2$ (2,3)	0.41	1.92	51	24, 24
$C_4O_2Cl_2F_2$ (2,5)	0.23	1.41	38	20, 21
$C_4O_2Cl_2F_2$ (2,6)	0.26	1.52	44	20, 23
$C_4O_2Cl_4$	1.01	3.25	52	30
	Dioxoles			
$C_3O_2H_4$	0.21	0.59	112	11, 18
$C_3O_2F_4$	0.48	1.19	138	14, 22

[a]Imaginary frequency in cm^{-1} for the planar structure. [b]Torsion angle in degrees about the C–O bond. If more than one value is given, the first corresponds to fluorine substitution on the carbon for the dioxins. For the dioxoles, the first corresponds to torsion about the CR–O bond and the second about the CR_2–O bond.

Given the heat of formation of the hydrocarbon dioxin, it is possible to derive the heats of formation of the substituted dioxins from a series of isodesmic reactions. The heat of formation of the parent can be calculated from the following reaction ΔH of $= -9.8$ kcal mol^{-1} yielding ΔH_f(dioxin) $= -19.8$ kcal mol^{-1}.

$$\text{cyclo-1,4-}C_4O_2H_8 + 2\,C_2H_4 \longrightarrow \text{cyclo-1,4-}C_4O_2H_4 + 2\,C_2H_6 \tag{14.8}$$

This can be used to derive the heats of formation in Table 14.3.

There are a number of ways of characterizing the reactivity of a molecular system. We first focus on the stability of the double bond in these systems. The π bond strengths of some halogenated ethylenes are given in Table 14.4. After accounting for the difference in the π bond strengths of the two ethylene fragments, formation of the FC═CF unit leads to a destabilization of the dioxin by 2 kcal mol^{-1}. Substitution of a second FC═CF unit leads to a destabilization of the ring π bonds by 7 kcal mol^{-1}. Substitution of a ClC═CCl unit to 2,3-difluorodioxin leads to similar destabilization of the dioxin by 10 kcal mol^{-1}. We can also compare the direct addition of the tetrahalogenated alkenes to the dioxin. This leads to a destabilization of the dioxin by 21 kcal mol^{-1} for the tetrafluoro-, by 17 kcal mol^{-1} for the difluorodichloro-, and 13 kcal mol^{-1} for the tetrachlorohalogenated. Clearly, perhalogenation leads to destabilization of the dioxin in terms of the π bond strengths.

TABLE 14.3. Heats of Formation (kcal mol^{-1}).

Compound	ΔHf
$C_4O_2H_4$	-19.8
$C_4O_2H_2F_2$ (2,3)	-99.2
$C_4O_2H_2F_2$ (2,5)	-103.2
$C_4O_2H_2F_2$ (2,6)	-102.2
$C_4O_2F_4$	-177.3
$C_4O_2Cl_2F_2$ (2,3)	-104.8
$C_4O_2Cl_2F_2$ (2,5)	-106.8
$C_4O_2Cl_2F_2$ (2,6)	-106.6
$C_4O_2Cl_4$	-33.2
$C_4O_2H_6$	-51.8
$C_4O_2F_6$	-324.1
$C_3O_2H_4$	-43.4
$C_3O_2F_2H_2$ (2,2)	-119.2
$C_3O_2F_4$	-234.5

TABLE 14.4. Pi Bond Strengths (kcal mol^{-1}).[10-14]

$CH_2{=}CH_2$	64
$CH_2{=}CF_2$	65.5
cis-CHF=CHF	55
$CF_2{=}CF_2$	52
$CCl_2{=}CCl_2$	53

Another way to predict reactivity is to look at the frontier molecular orbitals. The highest occupied molecular orbital (HOMO) and lowest unoccupied molecular orbital (LUMO) energies are given in Table 14.5. Halogenation stabilizes the HOMO even though the π bond is destabilized in these compounds. The HOMO of the parent hydrocarbon is derived from the π orbitals of benzene and the LUMO is derived from an antibonding π of benzene.

```
    −              •
 +     +        +     −
 +     +        −     +
    −              •
  HOMO           LUMO
```

TABLE 14.5. Orbital Energies (eV).

Molecule	HOMO	LUMO
	Dioxins	
$C_4O_2H_4$	8.24	-4.03
$C_4O_2H_2F_2$ (2,3)	9.12	-3.67
$C_4O_2H_2F_2$ (2,5)	9.15	-3.89
$C_4O_2H_2F_2$ (2,6)	9.19	-3.93
$C_4O_2F_4$	10.05	-3.87
$C_4O_2Cl_2F_2$ (2,3)	9.98	-3.43
$C_4O_2Cl_2F_2$ (2,5)	9.92	-3.57
$C_4O_2Cl_2F_2$ (2,6)	9.94	-3.57
$C_4O_2Cl_4$	9.88	-3.22
	Dihydrodioxins	
$C_4O_2H_6$	8.81	-5.29
$C_4O_2F_6$	11.55	-4.16
	Dioxoles	
$C_3O_2H_4$	8.75	-4.86
$C_3O_2H_2F_2$ (2,2)	9.55	-5.64
$C_3O_2F_4$	10.95	-4.95

The basic form of the orbital is independent of halogenation. The fluorine and chlorine π lone pairs participate in the HOMO and LUMO in an out-of-phase interaction, There are small perturbations due to the halogen substitution such as which π bond has more density but the differences are not large.

14.2.2 Dihydrodioxins

We also calculated the structure of the parent dihydrodioxin and the perfluoro derivative. The structures (see Table 14.1) are shown in Fig. 14.1 and it is clear that both are folded. However, the parent shows twisting about the single bond to give a structure with a C_2-like symmetry similar to a twist-boat configuration. The fluorocarbon assumes a boatlike structure with a C_s-like configuration with eclipsed CF_2 groups. The C═C and C—O bond distances show the same decrease on fluorination as found in the dioxins.

The heat of formation of the parent can be calculated from a reaction analogous to that used for the dioxins yielding a value for ΔH_f of $-51.8\,\mathrm{kcal\,mol^{-1}}$. This can be used to derive a heat of formation for the perfluoro derivative (see Table 14.3). The following reaction can be used to gain insight into the π bond energies of the perfluoro derivative of the dioxin.

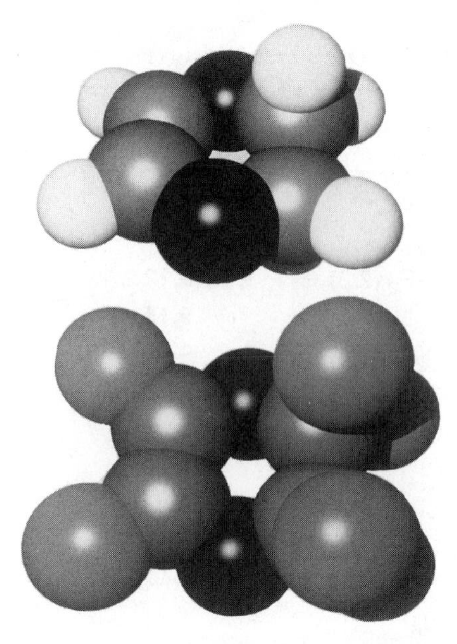

Figure 14.1. Geometries of the parent dihydrodioxin and the perfluoro derivative.

$$\text{cyclo-1,4-}C_4O_2F_6 + \textit{cis}\text{-CHF=CHF} \longrightarrow \text{cyclo-1,4-}C_4O_2F_4 + CF_2HCF_2H \quad (14.9)$$

The value for ΔE for this reaction is $13.4\,\text{kcal}\,\text{mol}^{-1}$, whereas for the pure hydrocarbon reaction, the reaction is thermoneutral. The large positive value for Reaction 14.9 suggests that the π bond in the perfluorodihydro derivative is not that destabilized as compared to the perfluorodioxin.

The HOMO of the dihydrodioxin is localized in the π bond (see Table 14.5). Clearly, fluorination stabilizes the π bond. The LUMO is somewhat lower in energy at 4.16 eV. Comparison to the dioxins shows that the HOMO in the perfluorodihydrodioxin is stabilized with respect to the dioxin values and the LUMO is also stabilized.

14.2.3. Dioxoles

The geometries of the dioxoles (see Table 19.1) show a similar behavior to the dioxins if the CR_2 group has R=H. The C=C bond shortens as does the CH—O bond on fluorination. The CH_2—O bond lengthens when F is substituted on the double bond. The effect of perfluorination leads to only a small change in C=C but to a significantly larger change in the C—O bonds with the bonds becoming essentially equal. The largest change is in the CF_2—O bond as compared to the CH_2—O bond.

The effect on halogenation on the planarity of the dioxoles is exactly opposite to that found in the dioxins (see Table 14.2). The parent compound is not planar but has a C_{2v} structure with a small inversion barrier. Substitution of F for H at the double bond yields a nonplanar structure with a higher inversion barrier and perfluorination yields a planar structure.

The heat of formation of the parent dioxole can be derived from the following reaction (14.10) whose ΔH is 4.9 kcal mol^{-1}. The heat of formation of the fluorinated compounds can then be derived similarly from the following reactions whose respective ΔH values are 7.2 and -0.5 kcal mol^{-1} (see Table 14.3).

$$\text{1,3-dioxole} + C_2H_6 \longrightarrow \text{1,3-dioxolane} + C_2H_4 \quad (14.10)$$

$$\text{1,3-dioxole} + cis\text{-CHF=CHF} \longrightarrow \text{1,3-dioxole-}F_2\,(4,5) + C_2H_4 \quad (14.11)$$

$$\text{1,3-dioxole-}F_2\,(4,5) + CF_4 \longrightarrow \text{1,3-dioxole-F4} + CF_2H_2 \quad (14.12)$$

Fluorination of the double bond as shown by Reaction 14.11 leads to a slight stabilization of the π bond of 2 kcal mol^{-1} after the effect of the difference in the π bond strengths of the ethylenes (9 kcal mol^{-1}) is accounted for. The calculation of the π bond strengths in the perfluordioxole is complicated by the difference in the CF_2 group and its effect on the C—O bond strengths. Reaction 14.12 is essentially thermoneutral, suggesting that the π bond strengths in the two dioxoles are very similar if the CF_2 bond strengths are similar. In order to provide more insight, we calculated the structure of 2,2-difluoro-1,3-dioxole (**12**). Reaction 14.13 suggests that the π bond strength in the perfluorodioxole is less than that of **12** as the ΔH is 12.3 kcal mol^{-1}. Accounting for the π bond strength in the ethylenes leaves a value of 3 kcal mol^{-1} destabilization of the π bond in $C_3O_2F_4$. The effect of fluorination on the CR_2 group can be estimated from Reaction 14.14 for which ΔH is -5.5 kcal mol^{-1}.

$$\mathbf{12} + cis\text{-CHF=CHF} \longrightarrow C_3O_2F_4 + C_2H_4 \quad (14.13)$$

$$C_3O_2H_4 + CF_4 \longrightarrow \mathbf{12} + CF_2H_2 \quad (14.14)$$

The behavior of the HOMO and LUMO (see Table 14.5) dioxoles is similar with halogenation stabilizing the orbital energy of the HOMO. The HOMO is derived from a butadiene-like orbital with the amount of oxygen participation decreasing with increasing fluorination. The LUMO is predominantly the C═C π^* orbital. The LUMO is destabilized by halogenation of the double bond and substitution of the CR_2 group leads to a LUMO energy the same as predicted for the parent.

– –
+ + HOMO

14.3. CONCLUSIONS

The prediction of reactivity in these systems is not simple. If frontier molecular orbital theory is important, then the reactivity is governed by the behavior of the HOMO and LUMO. The form of the HOMO and LUMO is not strongly dependent on the substituents. The orbital energies for the HOMO actually become more stable on halogenation and are less accessible. The LUMO energies show some dependency on halogenation but in the very reactive fluorinated dioxins and dioxoles, the effect of fluorination on the LUMO is very small, which is not consistent with the differences in reactivity expected for the parent and the halogenated species. Thus there is no strong correlation between the properties of the orbitals and the observed reactivity.

The above results show that the prediction of π bond strengths is complicated. Our estimates suggest that the π bond strengths in the halogenated dioxins are significantly weaker than those in the dihydrodioxins or in the dioxoles. Thus the reactivity of the dioxins should be higher than those of the other two series if only the electronic effects that we have considered are included. However, steric effects may also play a role.

$$\text{1,3-}C_3O_2F_6 + \textit{cis}\text{-CFH=CFH} \longrightarrow \text{1,3-}C_3O_2F_4 + CF_2HCF_2H \qquad (14.15)$$

$$\text{1,4-}C_4O_2F_8 + \textit{cis}\text{-CHF=CHF} \longrightarrow \text{1,4-}C_4O_2F_6 + CF_2HCF_2H \qquad (14.16)$$

$$\text{1,4-}C_4O_2F_8 + 2\ \textit{cis}\text{-CHF=CHF} \longrightarrow \text{1,4-}C_4O_2F_4 + 2\ CF_2HCF_2H \qquad (14.17)$$

The most important effect seems to be the change in energy content brought about by introducing a double bond into the system. Our calculations establish that introduction of a double bond into the perfluoro-1,3-dioxolane ring is accompanied by a large increase in energy content (Reaction 14.15) (18.3 kcal mol^{-1}) and that the resulting perfluoro-1,3-dioxole is essentially planar. The high reactivity observed for such systems in radical polymerizations is thus explicable on the basis of very favorable energetics in converting the difluorovinylene moiety into a saturated system and to the lack of steric complications in the chain propagation process. The formation of perfluoro-2,3-dihydro-1,4-dioxin from perfluoro-1,4-dioxane is accompanied by an increase in energy of only 12.0 kcal mol^{-1} (as shown by Reaction 14.16). This monomer is also boat shaped (see Fig. 14.1) and participation in a polymerization converts it into an even more sterically congested system. Therefore this case, in which radical polymerization is found to proceed, but relatively slowly, seems to be one in which somewhat less favorable energy changes are also countered by steric crowding in the propagation step. The very large increase in energy involved in formation of perfluoro-1,4-dioxin from perfluoro-1,4-dioxane (25.4 kcal mol^{-1}, as shown by Reaction 14.17) is reflected in the vigorous reactions of the perhalodioxins with oxygen, a small molecule, but not in the homopolymerization of one representative (dioxin, **8**). We propose that in this case, also, the relatively slow polymerization to form a polymer composed of heavily substituted boat-shaped structures (Eq. 14.6) is a consequence of steric retardation.

ACKNOWLEDGMENT

We thank Dr. Bruce E. Smart of these laboratories for most helpful discussions and continued support.

REFERENCES

1. Resnick, P. R., *Polymer Preprints, ACS Division of Polymer Chem., 31*, 1, 199th National ACS Meeting, Boston, MA, 1990, 312–313.
2. Coe, P. L., Dodman, P., and Tatlow, J. C., *J. Fluorine Chem., 6*, 115–128 (1975).
3. Schwertfeger, W. and Siegemund, G., *Angew. Chem. Int. Ed. Engl., 19*, 126 (1980).
4. Tarrant, P., Allison, C. G., and Barthold, K. P., in *Fluorine Chemistry Reviews* (P. Tarrant, Ed.), Vol. 5, Marcel Dekker, New York, NY, 1971, pp. 77–113.
5. Faucitano, A., Buttafava, A., Caporiccio, G., and Viola, C. T., *J. Am. Chem. Soc., 106*, 4172–4174 (1984).
6. Krespan, C. G. and Dixon, D. A., *J. Org. Chem., 56*, 3915–3923 (1991).
7. Dunning, T. H., Jr. and Hay, P. J., in *Methods of Electronic Structure Theory*, (Schaefer, H. F., III, Ed.), Plenum, New York, 1977, pp. 1–27.
8. Pople, J. A., Binkley, J. S., and Seeger, R., *Int. J. Quantum Chem. Symp., 10*, 1–19 (1976).
9. Dixon, D. A., Fukunga, T., and Smart, B. E., *J. Am. Chem. Soc., 108*, 1585–1588 (1986).
10. Smart, B. E., in *Molecular Structure and Energetics. Studies of Organic Molecules.* (J. F. Liebman and A. Greenberg, Eds.), Vol. 3, VCH, Deerfield Beach, FL, 1986, pp. 141–191.
11. Pickard, J. M. and Rodgers, A. S., *J. Phys. Chem., 98*, 6115–6118 (1976).
12. Wu, E.-C. and Rodgers, A. S., *J. Phys. Chem., 98*, 6112–6115 (1976).
13. McMillen, D. F. and Golden, D. M., *Ann. Rev. Phys. Chem., 33*, 493–532 (1982).
14. Pedley, J. B., Naylor, R. D., and Kirby, S. P., in *Thermochemical Data of Organic Compounds*, 2nd ed., Chapman and Hall, London, 1986.

CHAPTER 15

The Fluoroacetamide Acetal Claisen Rearrangement as a Tool for Asymmetric Synthesis

JOHN T. WELCH, TAKASHI YAMAZAKI, AND RAYOMAND H. GIMI

Synthetic Fluorine Chemistry,
Edited by George A. Olah, Richard D. Chambers, and G. K. Surya Prakash.
ISBN 0-471-54370-5

The current level of interest in the preparation of regio- as well as stereoselectively fluorinated compounds is indicated by the increasing number of publications and presentations in this area.[1] It has been known for some time that fluorine can have profound and often unexpected effects on activity. Selective fluorination has been an extremely effective synthetic tool for modifying and probing reactivity. Replacement of hydrogen or hydroxyl by fluorine in a biologically important molecule often yields an analogue of that substance with improved selectivity or a modified spectrum of activity.[2] A number of very valuable monographs and other general reviews are available for guiding chemists in the organic chemistry of fluorine.[3]

The asymmetric synthesis of fluorinated materials is important not only for the preparation of biologically active substances[4] or for the construction of novel ferroelectric devices[5] but also as a test of synthetic strategy. However, optically active fluorinated compounds where at least one of the asymmetric carbon atoms bears a fluorine or a fluoroalkyl group are extremely difficult to prepare. Important differences in the chemical reactivity of fluorinated compounds are based upon the difference in electronegativity between fluorine and hydrogen, on the higher carbon–fluorine bond strength versus the strength of the carbon–hydrogen bond, and on the ability of fluorine to participate in hydrogen bonding as an electron pair donor. The effects of fluorination have been thoroughly summarized by Chambers,[3c] Smart,[3a] and others.[6]

Fluorine has a pronounced electron-withdrawing effect by relay of an induced dipole along the chain of bonded atoms, a sigma-withdrawing effect, (I_σ), or this trend may also result from a through space electrostatic interaction also known as a field effect.[7,8] These effects are apparent when the acidity of trifluoroacetic acid (pK_a0.3) is compared with that of acetic acid (pK_a2.24). However, in the gas phase, although fluoroacetic acid is more acidic than acetic acid, it is less acidic than chloroacetic acid, presumably as a result of the lower polarizability of the carbon–fluorine bond, 0.53 Å^3, relative to that for the carbon–chlorine bond, 2.61 Å^3.[9] This diminished charge induced dipole effect of the carbon–fluorine bond, relative to the carbon–chlorine bond, is much closer to the polarizability of the carbon–hydrogen bond. The electronic effect of fluorine directly attached to a π system can be especially complex, as electrons from fluorine may be donated back to the π system in an I_π repulsive interaction along with the opposite I_σ effect. These interactions will be shown to be important in the amide acetal Claisen rearrangement and in the formation of the orthoamide and imidate intermediates that are involved.

15.1. INTRODUCTION

15.1.1. [3, 3]-Sigmatropic Rearrangements

A sigmatropic rearrangement is a pericyclic reaction[10a,b] that involves the migration of a σ bond within a conjugated π-electron framework to a new position. If the migrating group is chiral, retention or inversion of configuration can occur. The antara- or suprafacial stereochemistry of the π system is dependent on the number of electrons involved in the process, as well as the experimental conditions. [3, 3]-Sigmatropic rearrangements include the Claisen,[10c,d] Cope, and related rearrangements, which occur via highly ordered, six-membered cyclic transition states, and therefore are useful in the creation of new asymmetric centers in a predictable fashion.

15.1.1.A. The Claisen and Related Rearrangements

Formally, the term Claisen rearrangement as it is generally used today, implies the sigmatropic thermo reorganization (150–200°C) of an allyl vinyl ether (**1**) to an isomeric γ,δ-unsaturated carbonyl system (**2**).[10c–e] The process is considered to be a concerted, intramolecular S_N2' addition of a carbonyl enol to an allylic alcohol to form a carbon–carbon σ bond ([3, 3]-sigmatropic rearrangement) with concomitant double-bond migration. The all-carbon analogue of the Claisen rearrangement is the Cope rearrangement.

2 3 O 1 3' 1' 2' 150 - 200 °C O

1 2

If a chairlike transition state (**3**), which minimizes 1,3-diaxial interactions, is assumed, then the stereochemistry of the rearrangement (**4**) can be predicted. Thus, the Claisen rearrangement has been a topical area of research in recent years since it: (1) provides a method of introducing functionality stereo- and regiospecifically, (2) generates two of the most useful functional groups in preparative organic chemistry (the carbonyl group and the alkene), and (3) fixes the geometry of the newly formed carbon–carbon double bond.

3 2 O 1 3' 2' 1' O

3 4

Acyclic substrates that are unsubstituted or alkyl substituted undergo the rearrangement at 150–200°C. However, it has been shown that the presence of electron-donating substituents like dialkylamino, alkoxy, and trimethylsilyloxy at C-2 could lower the rearrangement temperature to ambient. These variants of the Claisen rearrangement make use of the ketene acetal tautomers of carboxylic acid derivatives to give amides, esters, or acids upon transformation while facilitating the reaction pathway.

15.1.1.B. Variants of the Claisen Rearrangement

15.1.1.B.1. Carroll Rearrangement. A β-ketoester (**5**)[11a–c] whose enolic form (**6**) generates the double bond required for rearrangement with the allylic fragment of the ester will rearrange, on pyrolysis, to a β'-keto-γ, δ-unsaturated acid (**7**) that loses carbon dioxide to give γ, δ-unsaturated ketones (**8**).

5 6 7 8

15.1.1.B.2. Meerwein–Eschenmoser Rearrangement. When an allylic alcohol is heated with an amide acetal (**9a**) or an *N,O*-ketene acetal (**9b**) at 110–140°C the product is the rearranged amide (**10**).[12a–d] This procedure is analogous to the Johnson-orthoester variant (see Section 15.1.2.B.6) except that the products are γ, δ-unsaturated amides instead of the esters. Two main explanations for the facility of this process have been given:[12a] (1) The presence of a nitrogen atom on the vinyl ether fragment increases the donor character of this moiety while lowering the activation energy of the process and (2) the *N,O*-ketene acetal has a lower ionization potential than the corresponding *O,O*-protocol and therefore is more reactive.

9a 9b 10

15.1.1.B.3. Marbet–Saucy Rearrangement. Allylic alcohols (**11**) may be heated (100–180°C; 1–48 h) with vinyl ethers (**12**)[13] in the presence of *p*-toluenesulfonic acid to give, after loss of carbonyl compounds and rearrangement, γ, δ-unsaturated aliphatic and alicyclic aldehydes and ketones (**2**) in good yields. A noteworthy example of the application of this procedure is the synthesis of squalene.

15.1.1.B.4. Dauben–Dietsche Rearrangement. Transetherification of an allyl alcohol (**11**) with a vinyl ether (**12**) using $Hg(OAc)_2$ as a catalyst[14] also leads to formation of the reactive allyl enol ether. This variant was employed for the

11 12

100 - 180 °C

2

stereospecific introduction of various functionalized angular methyl groups in octalin ring systems.

$Hg(OAc)_2$

11 12 2

15.1.1.B.5. Selenoxide Based Rearrangement. Allylic alcohols (**13**), on reaction[15,16] with the addition product (**14**) of phenylselenyl bromide and ethyl vinyl ether, give the selenide (**15**), which undergoes facile loss of phenylselenic acid on oxidation with $NaIO_4$ to yield the ketene acetal (**17**). Rearrangement to the γ,δ-unsaturated ester (**18**) occurs smoothly.

PhSeBr +

SePh OH Br OEt

13 14 15

PhSe OEt

Ph O Se OEt

16 17 18

15.1.1.B.6. Johnson Orthoester Rearrangement. In analogy to the amide acetal Claisen rearrangement this procedure involves the transformation of allylic alcohols (**11**) to γ,δ-unsaturated carboxylates (**22**)[17] via the orthoesters (**19**) (orthoacetate or orthopropionate). Loss of alcohol from the orthoester (**20**) catalyzed by propionic acid forms the reactive ketene acetal (**21**). The presence of the ethoxy group in (**21**) increases the donor character of the vinyl ether fragment as the ionization potentials of ketene acetals are lower than those of the nonsubstituted vinyl ethers.

OH + H⁺ 11 19 20

22 21 + CH_3CH_2OH

15.1.1.B.7. Ficini–Barbara or Ynamine–Claisen Rearrangement. In this method, an allylic *N*,*O*-ketene acetal (**24**) is generated by condensing an allyl alcohol with 1-(diethylamino)propyne (**23**).[18a] Bartlett and Hahne[18b] found that the stereochemistry of the ynamine–Claisen rearrangement can be controlled by selection of the reaction conditions. Lewis acid catalysis of the reaction at room temperature resulted in equilibration of the *N*,*O*-ketene acetals to form the thermodynamically favored (*Z*)-*N*,*O*-ketene acetal. On the other hand kinetic control of the reaction could be achieved by slow addition of the alcohol to a refluxing solution of the ynamine in xylene, which favors the rearrangement via the (*E*)-*N*,*O*-ketene acetal.

OH + N 11 23 24 25

15.1.1.B.8. Denmark Rearrangement. Based on Carpenter's qualitative[19b–d] theoretical analysis, Denmark and Harmata[20a–c] developed a new variant of the Claisen rearrangement utilizing inductively stabilized α-sulfonyl carbanions from (**26**) and (**27**) as π donors to accelerate the rearrangement. The process occurs under mild conditions (50°C or less) with excellent regioselectivity to form β-ketosulfones (**28**) and (**29**) in high yields (63–91%).

KH, 50 °C, 4 h

78%

26 **28**

KH, 20 °C, 0.25 h

91%

27 **29**

When 1,3,2-oxazaphosphorinanes are used as phosphorus-stabilized carbanions (**30**), the rearrangement is rapid and highly selective. The degree of relative asymmetric induction was as high as 90:10 for various substituent patterns and was dependent on the presence of lithium cations.[12d] The mild conditions required for these carbanion-accelerated rearrangements makes them synthetically very attractive.

30 **31**

15.1.1.B.9. Ireland Ester–Enolate Rearrangement. Allyl esters (**32**) were employed in the Claisen rearrangement by generation of the enolate by a strong base [e.g., lithium diisopropylamide (LDA)].[21a,b] Stereochemical control of enolate formation was possible by choice of the solvent systems. Quenching of the enolate at −78°C with trimethylchlorosilane (TMSCl)[21c,d] forms the trimethylsilyl ketene acetal (**33**), which rearranges cleanly at temperatures below 100°C to form (**34**).

O OTMS OTMS
1 LDA
2 TMSCl
32 **33** **34**

15.1.2. Transition State of the Claisen Rearrangement

As mentioned previously the Claisen rearrangement is a [3,3]-sigmatropic shift. The fragment **a′b′c′** in (**35**), which includes the oxygen atom, can be considered as a vinyl ether and the fragment **abc** as the alkene. If Ψ_1, Ψ_2, Ψ_3 are considered to be the MOs of the **a′b′c′** fragment (highest occupied molecular orbital, HOMO) and $\Psi'_1, \Psi'_2, \Psi'_3$ as the MOs of the alkene fragment (lowest unoccupied molecular orbital, LUMO), then the rearrangement can be explained by the interaction between the HOMO and the LUMO system. Out of the two possible transition states (chair or boat) the former does not involve the destabilizing nonbonding interaction between **b** and **b′** unlike the latter configuration (see Fig. 15.1).

The frontier molecular orbital (FMO) interactions that occur in the transition state of the Claisen rearrangement can also be looked upon as the interactions between two allylic radicals. Both $\Psi_1 - \Psi'_2$ and $\Psi_2 - \Psi'_3$ are symmetry forbidden. No energy is gained from the interactions of $\Psi_1 - \Psi'_1$ and $\Psi_3 - \Psi'_3$. Since the wave functions have nodes at **b** and **b′**, the interaction between Ψ_2 and Ψ'_2 provides no energy gain. Significant energy is gained from the overlap between Ψ_1 and Ψ'_3 favoring a suprafacial migration through a chairlike transition state (see Fig. 15.2).

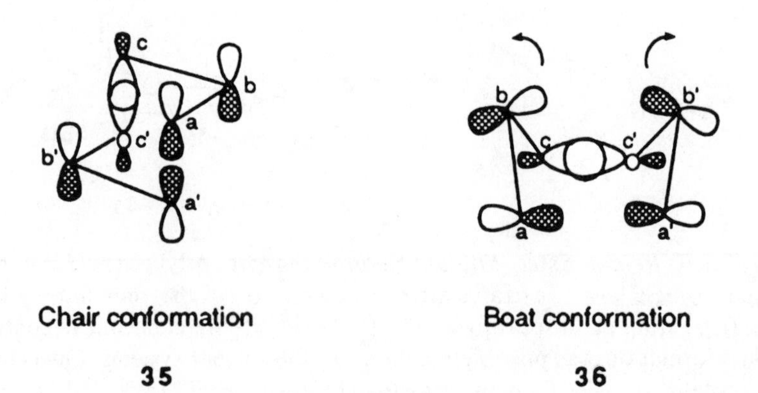

Figure 15.1.

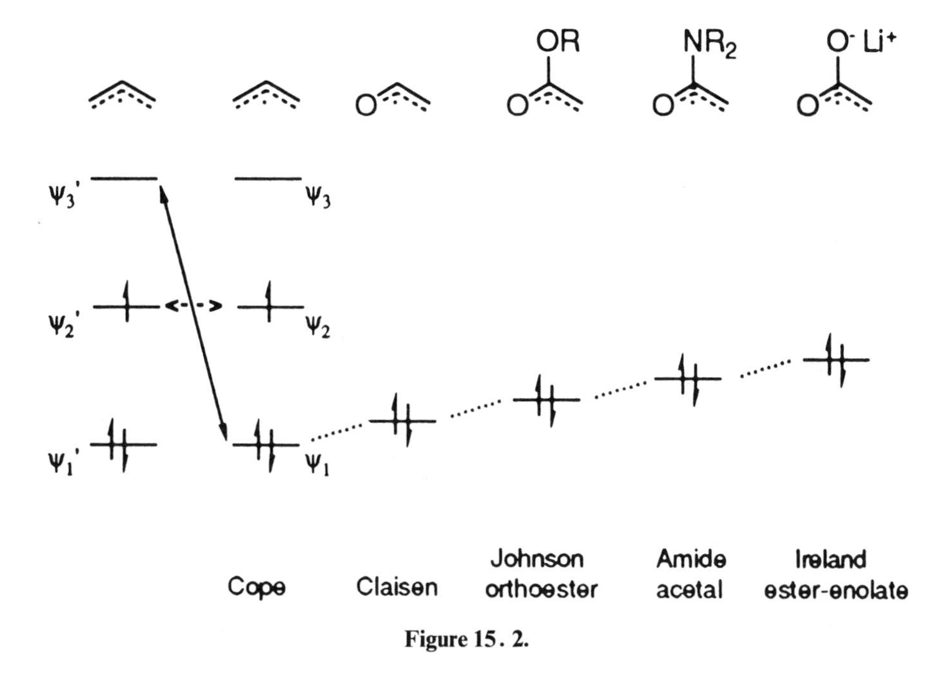

Figure 15. 2.

Experimental observation[19a] in the thermal rearrangement of (**37**) reveals that a chair transition state is preferred over the boat transition state by 5.7 kcal mol^{-1}, which is calculated from the product distribution depicted below.

225 °C

meso 37

(*E,Z*) 99.7 (*E,E*) 0.3 (*Z,Z*) 0

The Claisen process is facilitated by three factors: (1) Relief of strain energy in the substrate, (2) substitutents that can lower the energy gap between the HOMO and LUMO, and (3) favorable overlap of the interacting orbitals.

Sucrow and Richter[22] were the first to demonstrate that the controlling factor for the stereochemistry of the amide acetal Claisen rearrangement is the enol ether double-bond geometry. Reaction of (*E*)-allylic alcohols with 1-dimethylamino-1-methoxy-1-propene gave the *l*-amides (**38**) and the (*Z*)-allylic alcohols afforded the *u*-isomer (**39**) in ratios of 9 : 1 or greater. These results are interpreted as a preferential formation of (*Z*)-*N*,*O*-ketene acetal configuration (**40**) and implies that the rearrangement proceeded via a chairlike transition state.[22d–h]

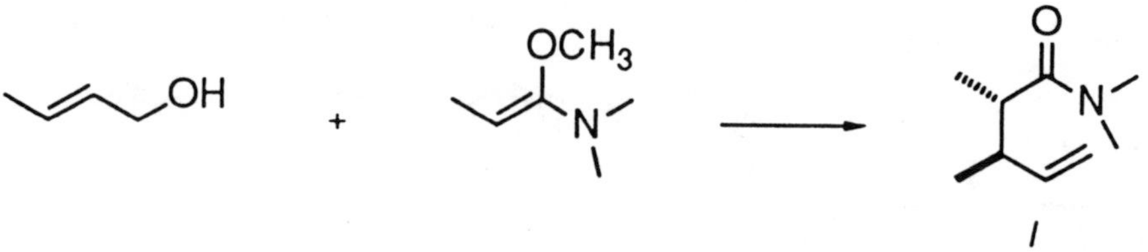

OH + OCH_3 N → O N

l

38

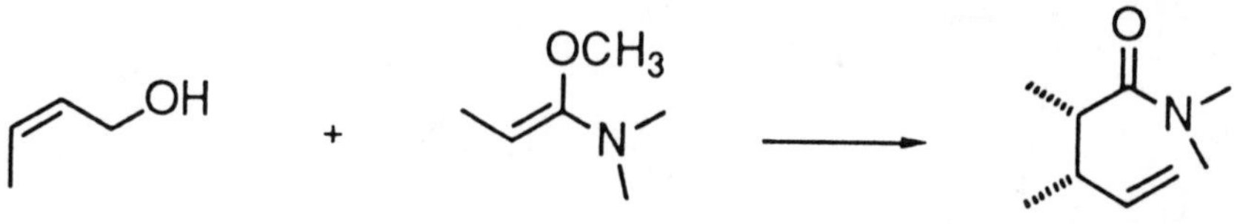

OH + OCH_3 N → O N

u

39

CH_3 O CH_3 —N ⇌ O CH_3 —N CH_3

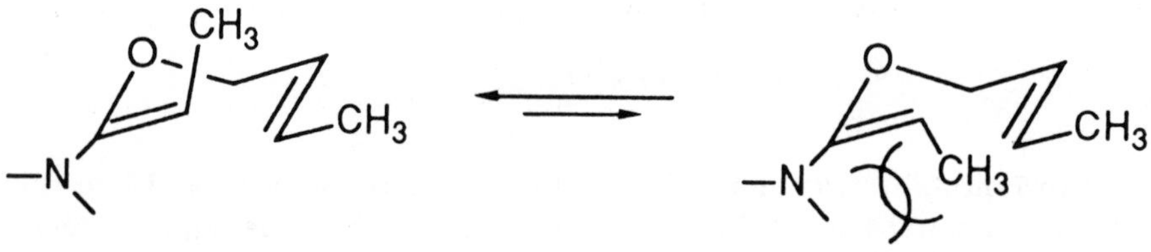

40 **41**

15.1.3. Product Stereochemistry

Sucrow's work clearly demonstrated that specific stereochemical relationships in the starting material are faithfully transformed to specific relationships in the product.[22,23] Unsymmetrically substituted termini of an allylic vinyl ether create two vicinal chiral centers upon Claisen rearrangement that can exist as a pair of *u* and *l* diastereomers, whose controlling factor is the geometry of ketene acetals as discussed above. By examining the rearrangement of the four stereoisomers of crotyl propenyl ethers, Schmid and his co-workers[24,25] confirmed that the (E, E) and (Z, Z) isomers gave predominantly *l* product while the (E, Z) and (Z, E) isomers afforded *u* isomer upon rearrangement (see Fig. 15.3).

A $\Delta\Delta G^{\ddagger}$ of 2.5–3.0 kcal revealed that the rearrangement showed a high propensity for the chairlike transition state, which minimizes 1,3-diaxial interactions. A $\Delta S^{\ddagger}$ of -10 to -15 eu indicated a high degree of order in the transition state. Kinetic studies showed that the (E, E) isomer rearranged nine times faster than the (Z, Z) isomer and hence an isomer with two methyl groups in an equatorial disposition was stereochemically favored. Dewar and Jie[26]

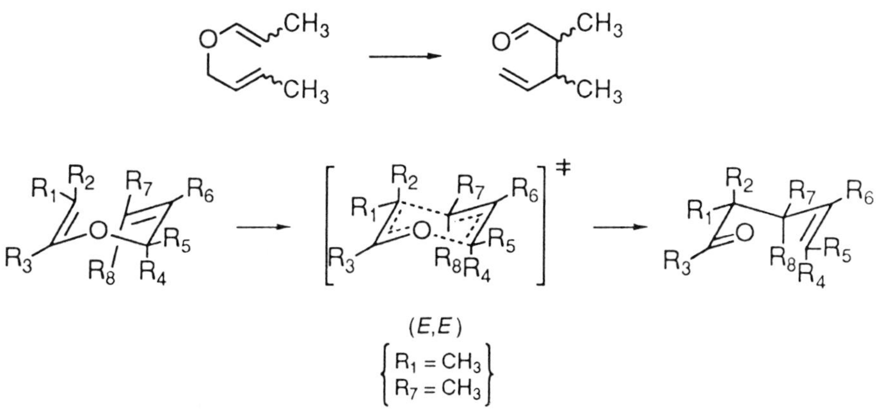

(E,E)
$R_1 = CH_3$
$R_7 = CH_3$

Figure 15.3.

recently reported AM1 calculations for the Claisen rearrangements of allyl vinyl ether and its derivatives, which are in agreement with the prediction that these reactions take place preferentially via chair-type transition states and tend to form (*E*) isomers.

15.2. ASYMMETRIC INDUCTION AND TRANSFER OF CHIRALITY

15.2.1. Background

From the above observations it is clear that two unsymmetrical double bonds and a chiral center in an allylic vinyl ether could be transformed into a product with two chiral centers and a stereoselectively formed double bond. Such a "transfer of chirality" in a sigmatropic process was first demonstrated by Hill and Gilman[27] in the Cope rearrangement. The (*R*)–(*E*) isomer (**43**) of the product could be obtained by invoking a chairlike transition state (**42**) with the bigger phenyl group in an equatorial position. Hill[28] showed that similar Claisen rearrangements proceeded suprafacially by 1,3-chirality transfer along the allylic array. Through the judicious control of the preferred transition states of the Claisen rearrangement,[29] high diastereoselection was observed in the formation of the butyrolactones by the use of optically active alcohols.

An α-hydroxyl[30a–e] group has also been employed to efficiently transfer chirality in chelation-controlled ester enolate Claisen rearrangements. The chelating lithium cation regulates the geometry of the intermediate enolate during the course of the rearrangement of an allylic glycolate.[30c]

With an optically active allylic ester, bearing an α-hydroxyl group a 1,3-transfer of chirality can occur along the allylic moiety. High diastereoselectivity (98 : 2) is coupled with efficient transmission of chirality [97–99% enantiomeric excess (ee)] when optically active allylic glycolate (**44**) undergoes ester enolate Claisen rearrangement to form 2-hydroxy-3-alkyl-4-hexenoic acids (**45**).[30b,d]

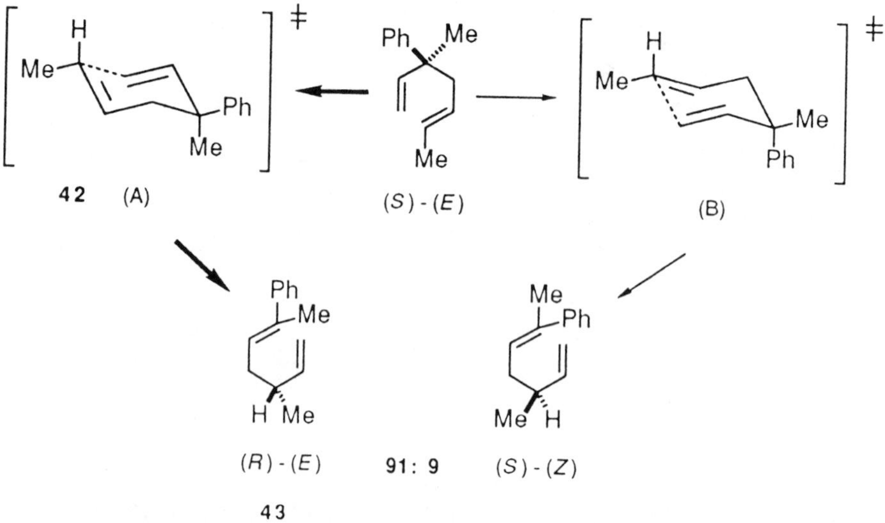

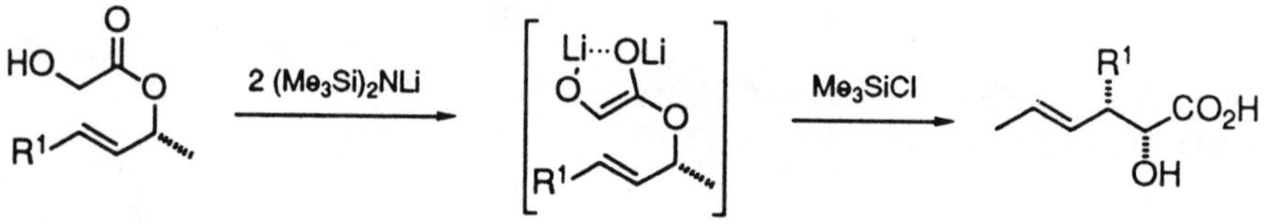

This type of modification is also effective for the Johnson orthoester and the amide acetal Claisen rearrangement. The requirement of optically pure starting materials is fundamental for the above asymmetric rearrangements. In each of the above examples, while the optical activity is conserved, the original chiral center in the reactant is destroyed during the creation of a new center in the product, a process Mislow[31] referred to as self-immolative asymmetric synthesis.

Transfer of chirality in Claisen rearrangements has also been achieved by using a remote chiral auxiliary,[32,33d] whose advantage is that it is unchanged after the sigmatropic rearrangement and recoverable. Effective induction of asymmetry by chiral centers in rigid cyclic systems like terpenes,[34] steroids,[35] butyrolactones,[29] and oxazolines[33c,d] has been observed. Placement of a chiral auxiliary at C-1 of the carbon framework has resulted in the induction of asymmetry in [2,3]-[36] as well as [3,3]-sigmatropic rearrangements.[37] Chiral amide enolates derived from (*S*)-prolinol and (*S*)-valinol were used by Nakai[36] in asymmetric [2,3]-Wittig rearrangements. High diastereoselectivity (95 : 5) was observed, but the relative asymmetric induction was moderate. Effectiveness of a chiral substituent at the hydroxyl group of the glycolate,[37e,f] was also proved in the Claisen rearrangement. Attachment of the chiral auxiliary at C-6 of the carbon framework has been effective in inducing asymmetry.[37a,g]

15.2.2. Proline Derivatives as Chiral Auxiliaries

A wide range of optically active natural compounds, for example, sugars, hydroxycarboxylic acids, and amino acids can be used as substrates in asymmetric synthesis. In recent years, the L-amino acids[38] have been used very frequently because of their great structural variety and ready availability, while the unnatural D-amino acids are employed less frequently. In particular various proline derivatives have been applied in enantioselective alkylations,[39] aldol reactions,[40] and asymmetric conjugate addition reactions[41] and strikingly high selectivities have been achieved, probably due to the rigidity of the five-membered ring, which provides a strong topological bias for diastereofacial selection in the bond construction.

15.2.3. The Amide Acetal Claisen Rearrangement

As mentioned earlier, the amide acetal Claisen rearrangement normally involves the heating of an allylic alcohol (120–180°C) with an amide acetal or a *N,O*-ketene acetal to form stereoselectively a γ,δ-unsaturated amide after rearrangement.[12b,c] The amide acetal Claisen rearrangement has been employed for the synthesis of terpene alcohols,[22b,c] sterols,[22f] alkaloids,[42] steroids,[22d,e] and other natural products.[43,3b].

The high temperatures required for the formation of the reactive *N,O*-ketene acetal are a potential drawback to the use of the amide acetal Claisen rearrangement in asymmetric syntheses. Therefore either highly restrictive chiral auxiliaries such as those developed by Kurth or the temperatures required for *N,O*-ketene acetal formation must be lowered. Recently, methods were developed[44] where propionamides and fluoroacetamides undergo diastereoselective amide acetal Claisen rearrangement at room temperature. The stereoselective synthesis of fluorinated molecules[45] was possible utilizing the rearrangement of fluorinated *N,O*-ketene acetals prepared at lower temperatures.

15.3.1. Fluoroacetamide Acetal Claisen Rearrangement

In our development of new methods for the stereoselective synthesis of fluorinated molecules,[45] rearrangement of a fluorinated *N,O*-ketene acetal was an attractive approach. However, the fluorinated ynamine necessary to take advantage of the Ficini–Barbara approach was not synthetically accessible. The higher temperatures required for the normal amide acetal rearrangement appeared to prohibit the stereoselective formation of the fluoro *N,O*-ketene acetal from available fluoroacetamides by that route.

The stereocontrol in the amide acetal Claisen rearrangement is thought to be derived from unfavorable steric interactions between the alkyl substituent on the nitrogen and the β substituent on the *N,O*-ketene acetal (**41**).[22] If the substituent is a methyl group, when equatorially placed, it will have a significant steric interaction with the planar delocalized C-1–C-3 residue (see Fig. 15.4) in the transition state. Hence, (*Z*)-*N,O*-ketene acetal (**40**) formation is favored over (*E*)-*N,O*-ketene acetal (**41**) formation. The (*Z*)-*N,O*-ketene acetal is known to be the thermodynamically favored conformation in the boron trifluoride–etherate promoted ynamine Claisen rearrangement.[18b] With the less sterically demanding fluorine, selectivity would most likely only be possible at lower temperatures.

15.3.2. Results and Discussion

It is possible to utilize fluoroacetamides in diastereoselective amide acetal Claisen rearrangements at room temperature.[44a] Three symmetric fluoroacetamides, *N,N*-dimethylfluoroacetamide, *N*-(fluoroacetyl)pyrrolidine, and *N,N*-diisopropylfluoroacetamide were prepared in excellent yields.[44b] The *O*-alkylation of the fluoroacetamides with dimethyl sulfate, following literature methods[46] for nonfluorinated amides, failed to yield the desired product. Presumably, fluorination diminishes the nucleophilicity of the carbonyl oxygen. With more reactive methyl triflate, the alkylation proceeded smoothly. The trifluoromethanesulfonate salts, often solids, were obtained in quantitative yields. With *N,N*-dimethylfluoroacetamide (**46a**), an undesired side product derived from methylation on nitrogen was obtained.

$CH_3SO_3CF_3$

46a → **47a** ($CF_3SO_3^-$) + **48a** ($CF_3SO_3^-$)

The kinetic alkylation product was apparently the imidate (**47a**), which equilibrates to (**48a**). When stoichiometric methyl triflate was used, after 3–4 h at 70 °C, (**47a**) was present in larger amounts than (**48a**). Using 20% excess of methyl triflate and heating the reaction mixture at 70 °C overnight suppressed the side reaction considerably. This *N*-alkylated product was not observed in the *N*-(fluoroacetyl)pyrrolidine (**46b**) and *N,N*-diisopropylfluoroacetamide (**46c**).

Our initial strategy was to heat the dimethyl acetals of the fluoroacetamides with the appropriate allylic alcohols to obtain the rearranged products. Treatment of the salt (**47a**) with sodium methoxide in methanol formed the desired dimethyl acetal (**49a**), whose separation from methanol was very tedious and difficult, and this intermediate did rearrange very cleanly by the *in situ* addition of allylic alcohols in refluxing benzene. In sharp contrast with (**46a**), on the other hand, dimethyl acetals could not be formed with (**46b**) and (**46c**).

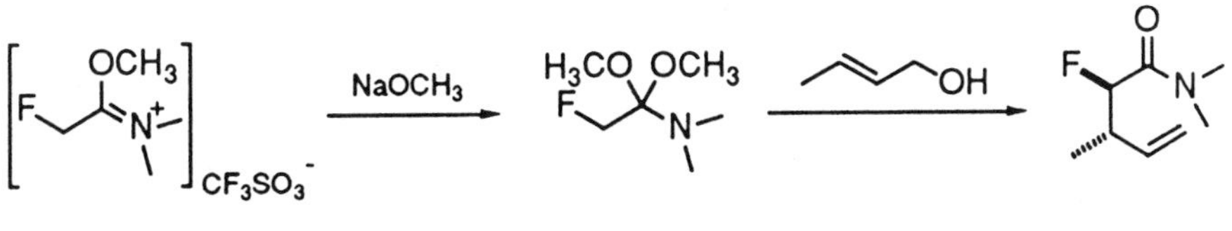

$NaOCH_3$

47a **49a**

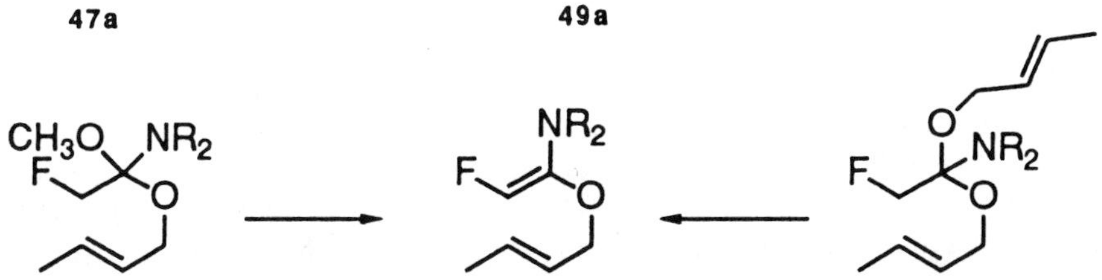

50 **51**

We assumed that formation of the *O*-crotyl-*N*,*O*-ketene acetal (**50**) was the slow step, with rearrangement occurring relatively easily; hence we attempted to prepare (**50**) directly without going through the dimethyl acetal intermediate like (**49a**). The triflate salt (**47a**) of *N*,*N*-dimethylfluoroacetamide, on treatment with the potassium alkoxide of (*E*)- or (*Z*)-2-buten-1-ol (crotyl alcohols), readily formed the bis(crotyl)acetal (**51**) in 35–40% yield even at $-80\,^{\circ}C$ in 30 min. The acetals could also be formed in 30% yields by reacting the salt (**47a**) with sodium alkoxides. Heating the acetals at $80\,^{\circ}C$ (refluxing benzene) for 7–24 h resulted in the formation of the rearranged amides (**52**) and (**53**) in 86–91% yields but the rearrangement did not occur in refluxing methylene chloride.

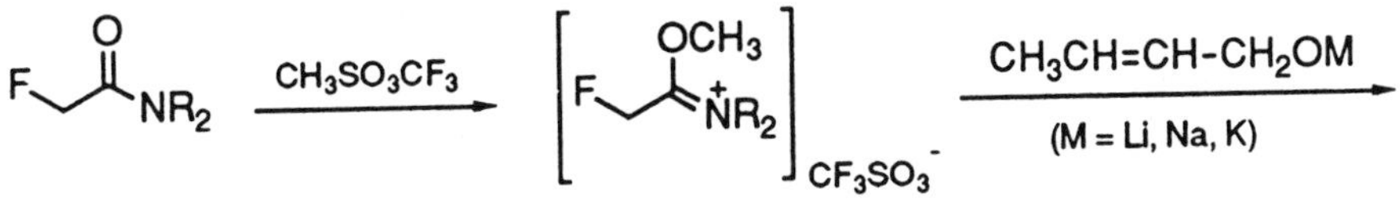

$CH_3SO_3CF_3$ $CH_3CH{=}CH{-}CH_2OM$ (M = Li, Na, K)

46 **47**

a. $R = CH_3$
b. $R = (CH_2)_4$
c. $R = CH(CH_3)_2$

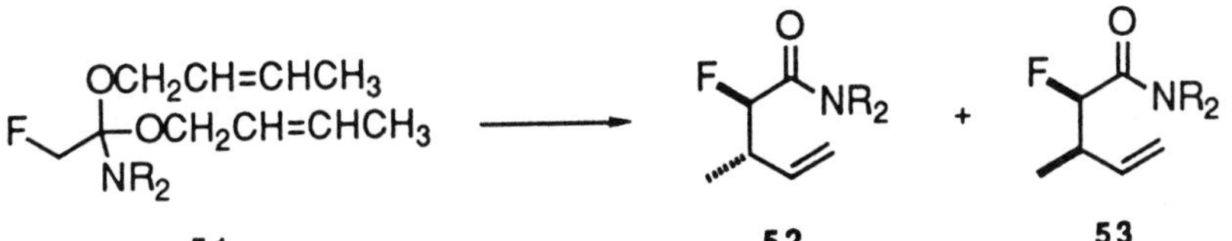

51 **52** **53**

Figure 15.4

The biscrotyl acetals (**51a**) could also be prepared in 74–78% yields when the salt (**47a**) was reacted with 3 equivalents of the allylic alcohol and 2.2–3 equivalents of methyllithium at room temperature overnight. However, treatment of 3 equivalents of the allylic alcohols with 4 equivalents of methyllithium, followed by the addition of the salts (**47a**) and stirring for 20 h at room temperature yielded the rearranged products (**52**) and (**53**) in 23–50% yields (Table 15.1). Thus, with apparent stoichiometric quantities of methyllithium, the acetals could be obtained but in the presence of greater than the theoretical amount of methyllithium only the rearranged products were observed.

The diastereoselectivity was determined by integration of the two sets of ^{19}F NMR resonances. The peaks assigned to the two diastereomers were separated by 1–2 ppm and these appeared as a doublet of doublets around δ-187 to δ-192, very distinct from the starting fluoroacetamides, which are triplets around δ-225 to δ-227. The ^{13}C NMR corroborated with the ^{19}F NMR. The diastereoselectivity of the rearrangement with *N*,*N*-dimethylfluoroacetamide (**46a**) was poor (1 : 1.6), although the rearrangement with (*Z*)-2-buten-1-ol exhibited a slight improvement in selectivity (2.9 : 1) over the (*E*) with the predominant formation of the opposite stereoisomer.

In order to increase the steric demand of the *N*-alkyl substituent and therefore to potentially increase the stereoselectivity of the *N*,*O*-ketene acetal formation, both pyrrolidinyl and disopropylfluoroacetamides (**46b** and **46c**) were *O*-alkylated with methyl triflate (Fig. 15.4). On treatment with 3 equivalents of the lithium alkoxides, prepared with greater than the theoretical amount of methyllithium, the rearranged products (**52**) and (**53**) were obtained in 26–

TABLE 15.1. Amide Acetal Claisen Rearrangement of Fluoroacetamides.

Entry	Amide	Alcohol	Time[a] (h)	Diastereoselectivity[b] (53 : 52)	Yield[c] (%)
1	46a	(*E*)	20	1 : 1.6	50
2	46a	(*Z*)	20	2.9 : 1	23
3	46b	(*E*)	14	1 : 2.7	87
4	46b	(*Z*)	18	5.1 : 1	77
5	46b	(*E*)	3[d]	1 : 1.4	96
6	46c	(*E*)	14	1 : 5.3	49
7	46c	(*Z*)	14	2.7 : 1	26
8	46c	(*E*)	70	1 : 2.6	78
9	46c	(*Z*)	70	3.1 : 1	27
10	46c	(*E*)	18[e]	1 : 1.3	89

[a]Temperature 25 °C.
[b]Determined by ^{19}F NMR at 7.04 T.
[c]Determined by gas chromatographic analysis.
[d]Temperature 67 °C.
[e]Temperature 42 °C.

87% yields after 12–70 h at room temperature (Table 15.1). The *N*-(fluoroacetyl)pyrrolidine (**46b**) seemed to be the optimum case with respect to yield (77–87%), stereoselectivity [5.1 : 1 with the (*Z*)-allylic alcohol], and time needed for rearrangement (14–18 h). Heating the salt (**47b**) with 2 equivalents of the lithium alkoxides at 40 °C for 2 h resulted in isolation of very little of the rearranged product. The reaction appeared to stop at formation of the intermediate acetal (**51b**). Heating at 67 °C [refluxing tetrahydrofuran (THF)] for 3 h resulted in 96% yield of the rearranged products but unfortunately the diastereoselectivity dropped to 1 : 1.4 with (*E*)-allylic alcohol (entry 5, Table 15.1). A similar drop in the diastereoselectivity and an increase in the yield was observed when the *N*,*N*-diisopropylfluoroacetamide salt (**47c**) was heated to 40 °C with the (*E*)-alkoxide (entry 10, Table 15.1). These observations with (**46b**) and (**46c**) were contradictory to that with (**46a**), which showed comparable selectivities at room temperature and in refluxing benzene. Clearly, the energy differences were small for the transition states leading to selective formation of the ketene acetals in this case. Thus when the isolated amides (**53c**) and (**52c**) were resubjected to the reaction conditions, there was no change in the ratio of the diastereomers formed.[47]

Although determination of diastereoselectivity by ^{13}C and ^{19}F NMR was straightforward, structure determination of the amide diastereomer formed preferentially was not possible. The product α-fluoro-γ,δ-unsaturated amides were converted to iodolactones and their spectra compared with those previously determined from the products of the ester enolate Claisen rearrangement.[45d] Treatment of the products with iodine in 50% aqueous THF at 0 °C (after Corey et al.)[48] or in 50% aqueous dimethoxyethane (DME) at room temperature for 2 days (after Yoshida and his co-workers)[49] failed to afford the iodolactones except with (**53a**) and (**52a**). However, when (**53b**) and (**52b**) obtained from the rearrangement with (*E*)-2-buten-1-ol in a ratio of 1 : 2.7 were heated with iodine in 1 : 1 DME–H_2O at 60 °C for 20 h, a mixture of four iodolactones, (**55**)–(**58**), were obtained in a ratio (by GC) of 1.9 : 1.7 : 1 : 1.

Similarly, the mixture of amides (**53b**) and (**52b**) obtained by the rearrangement of (**46b**) with (*Z*)-2-buten-1-ol in a ratio of 5.1 : 1 was lactonized. The four iodolactones were obtained in a ratio (by GC) of 1 : 4.0 : 10.8 : 8.5. Comparison of the ^{1}H and ^{13}C spectra of these lactones with those previously determined indicated that the (*E*)-*N*,*O*-ketene acetal (**54**) was being formed preferentially (see Fig. 15.4).

It was possible to infer from the chair transition states (Fig. 15.5) that when we reacted the (*E*)-*N*,*O*-ketene acetal with (*Z*)-(*E*)-allylic alcohol the *l* diastereomer (**52**) was formed and when we used the *z*-allylic alcohol, the *u* diastereomer (**53**) was formed preferentially.

The induction of asymmetry in the Claisen rearrangement of *N*,*O*-ketene acetals has been studied.[50,51] The diastereoselective chiral auxiliary-mediated aza-Claisen rearrangements of *N*-allyl-*N*,*O*-ketene acetals[51] was employed in the enantioselective preparation of chiral C(α)- and C(β)-substituted pent-4-enoic acids.[51] The stereochemical factors governing the stereoinduction at C(α)

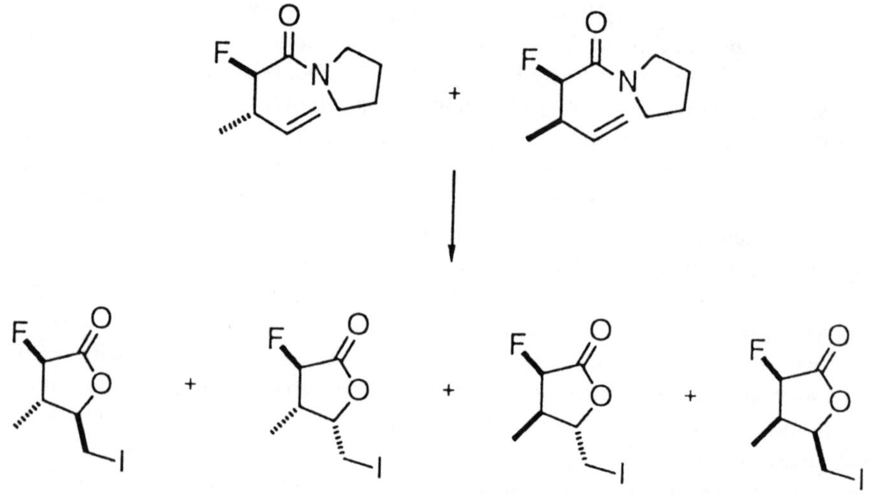

55 56 57 58

(92–94% ee) were the (*Z*)-*N*,*O*-ketene acetal alkene selectivity and the C(α)-*si*-face selectivity induced by the chiral auxiliary, oxazolidine[51b] (see Fig. 15.6).

Oxazolidine face selectivity along with chair–boat selectivity were the crucial transition state parameters responsible for C(β)-induction (52–94% ee).[51c] The above rearrangements yield two diastereomeric products at either C(α) or C(β). By this means, concomitant C(α),C(β)-asymmetric induction, with high enantioselectivity,[51d] was possible; the products being masked 2,3-substituted

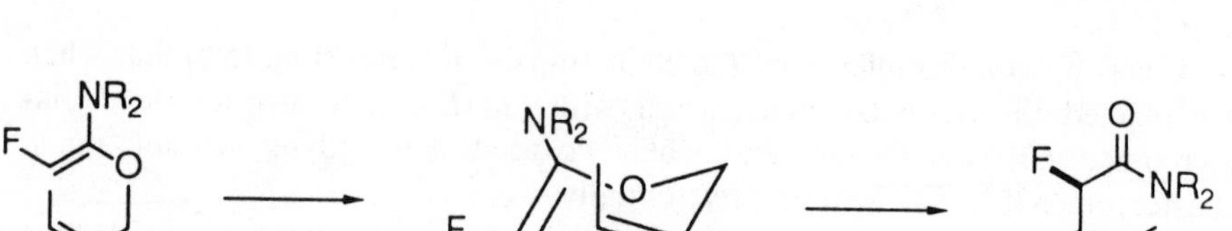

(*E*) - **54** → **52**

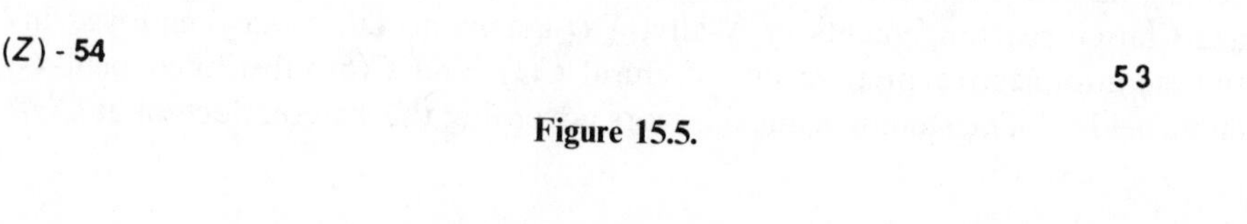

(*Z*) - **54** → **53**

Figure 15.5.

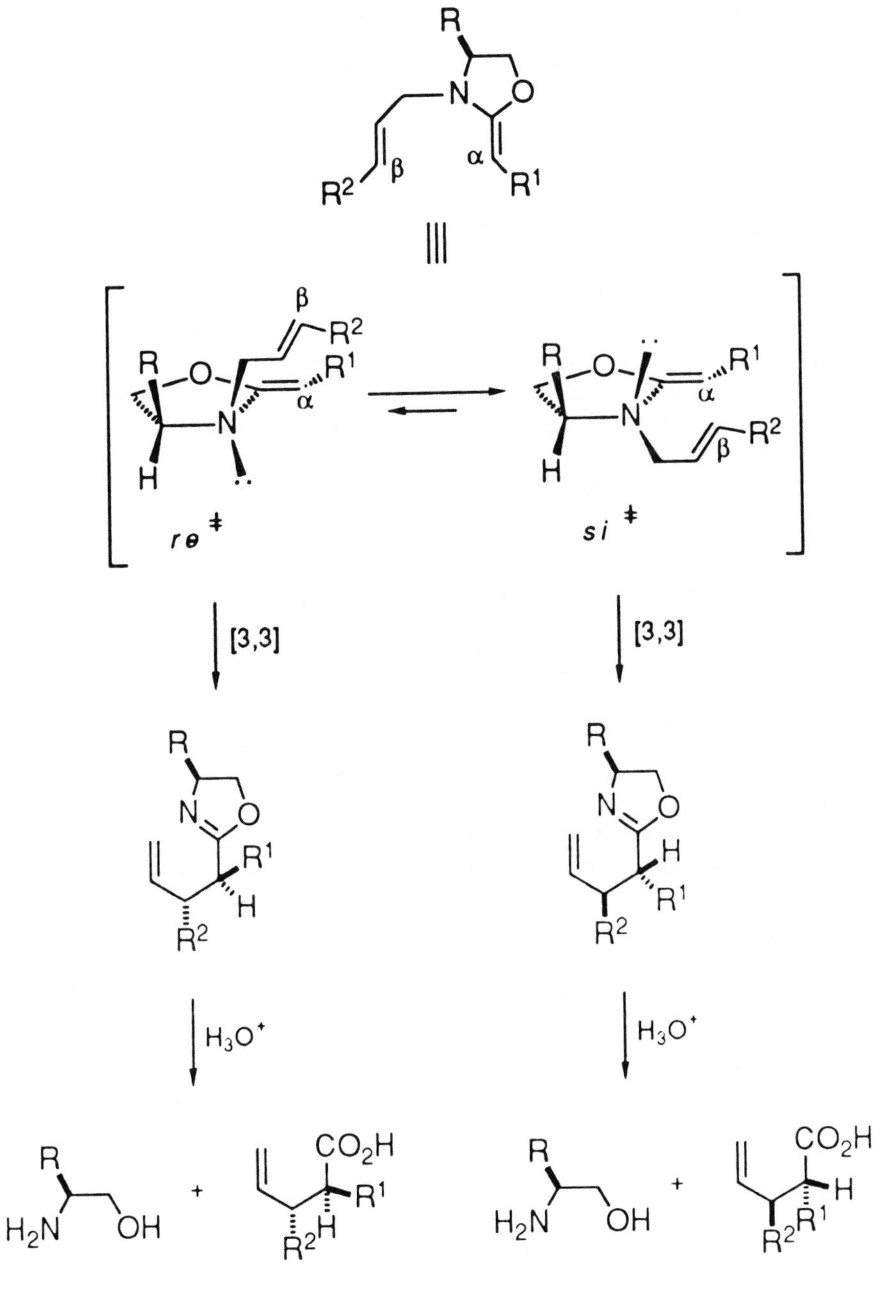

Figure 15.6.

pent-4-enoic acids. Each optically pure *N*-allyl-*N*,*O*-ketene acetal may yield any of the four possible diastereomeric products yet only one major product is produced in 97–98% ee in this highly stereoselective process. The overall yields for the process range from 73–87% in 79–92% diastereoselectivity. The *N*-allyl alkene geometry plays a pivotal role in determining the *l*–*u* stereoselectivity. Variation of the alkene geometry results in a reversal of the relative configuration at C(α),C(β) in the major product.[51] (See Fig. 15.6).

Because of the ready availability and structural variety, amino acids, particularly proline and its derivatives, have been successfully used as chiral auxiliaries.[52–54] The proline derivatives were also employed successfully in the amide acetal Claisen rearrangement[33a] with relative asymmetric induction ranging from 30–79%. Efficient diastereofacialselection was achieved due to the ability of the pyrrolidine ring to restrict the pericyclic framework to a single conformation. This was the first report of an ambient-temperature amide acetal Claisen rearrangement involving the induction of relative asymmetry by a remote asymmetric center outside the pericyclic framework.

15.4. THE ENANTIOSELECTIVE FLUOROACETAMIDE ACETAL CLAISEN REARRANGEMENTS OF *N*-FLUOROACETYL-(2*R*,5*R*)-2,5-DIMETHYLPYRROLIDINE

15.4.1. Introduction

We report below one effective solution to this problem by the enantioselective amide acetal Claisen rearrangement of *N*,*O*-ketene acetals prepared *in situ* from *N*-fluoroacetyl-(2*R*,5*R*)-2,5-dimethylpyrrolidine (**62**). The variety of approaches to the problem of the synthesis of optically active fluorinated compounds explored by other workers includes the development of asymmetric fluorinating reagents,[55] the use of enzymatic transformations,[56] the use of optically active sulfoxides[57] or imines,[58] and the resolution of fluorinated building blocks.[1]

The power of the Claisen rearrangement, which has been successfully applied to fluorinated substrates,[59,60] suggested that the judicious choice of a chiral auxiliary would allow the synthesis of optically active fluorinated materials from fluoroacetate precursors. Earlier studies showed the amide acetal Claisen rearrangement might be particularly well suited to this strategy. However, this type of rearrangement of (**59**), based on successful studies with the *N*-propionylprolinol derivative (**60**), yielded only a stereorandom mixture of products.[61]

It was postulated that the chiral auxiliary failed in the reaction of **59** because there was little control of the *N*,*O*-ketene acetal stereochemistry and only poor control of the relative configuration of the pyrrolidine ring relative to the *N*,*O*-ketene acetal as in (**61a**) and (**61b**).

While control of the *N*,*O*-ketene acetal geometry could be improved by careful selection of the reaction conditions, the latter problem, the necessity to

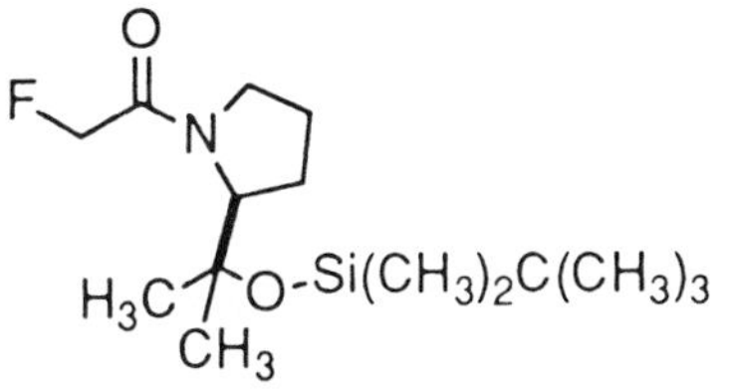

59

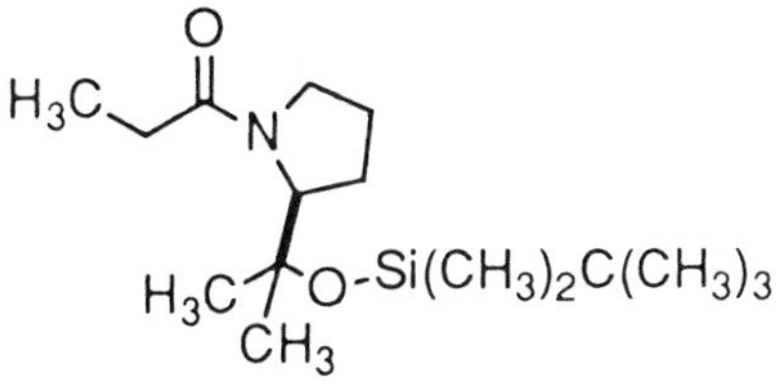

60

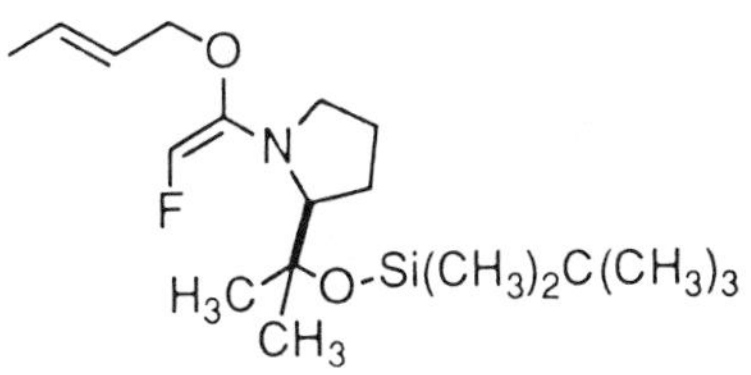

61a

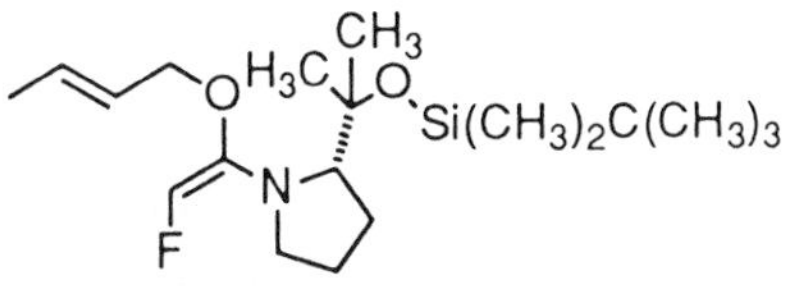

61b

control the carbonyl carbon–nitrogen bond geometry, could be avoided by employing a chiral auxiliary with a C_2-axis of symmetry.[53]

15.4.2. C_2 Symmetry and Asymmetric Induction

Enantiofacial differentiation, which provides asymmetric induction, can be achieved not only by an asymmetric compound but also by compounds that lack mirror symmetry, that is, are disymmetric.[53] The first C_2 chiral auxiliary to be used as a tool for asymmetric induction was 2,3-*O*-isopropylidene-2,3-dihydroxy-1,4-bis(diphenylphosphino)butane (DIOP).[54,62] It has been observed that chiral auxiliaries possessing C_2-symmetry elements often are superior stereochemical directors, permitting higher levels of asymmetric induction than those that are devoid of any symmetry. This occurs because the presence of a C_2 axis of symmetry within a chiral auxiliary serves the very crucial function of dramatically reducing the number of possible competing, diastereomeric transition states.[53] Among the group of C_2 symmetric molecules that have been efficiently employed as monodentate auxiliaries for asymmetric induction are the *trans*-2,5-disubstituted pyrrolidines. These include *trans*-2,5-dimethylpyrrolidine,[63,64] *trans*-2,5-bis-(methoxymethyl)pyrrolidine,[65] and *trans*-2,5-bis-(methoxymethoxymethyl)pyrrolidine.[66–69] *Trans*-2,5-dimethylpyrrolidine was the first monodentate C_2 auxiliary, initially introduced[63,64] as the amine component in enamine alkylation. In the past, the use of this auxiliary was limited due to the tedious resolution required for separation of the antipodes. However, Schlessinger largely obviated this problem by his synthesis of the compound in either optical series in 44% overall yield.[70–74] Recently, Masamune and his co-workers[75] reported the convenient synthesis of the (−)-(2*R*, 5*R*) enantiomer by enzymatic catalysis using baker's yeast. The power of *trans*-2,5-dimethylpyrrolidine[76] and its analogues as chiral auxiliaries for asymmetric induction has led Kurth and his co-workers[77] to look into the possibility of the higher ring homologue *trans*-2,6-dimethylpiperidine as a potential "C_2-symmetric" chiral reagent. (2*R*, 6*R*)-(−)-2,6-lupetidine and its 2,6-bis(benzyloxymethyl) analogue have been synthesized. Asymmetric induction can also be achieved by the discrimination of diastereotopic groups or faces. By diastereotopic alkene differentiation with concomitant face differentiation using a (2*R*, 5*R*)-bis(methoxymethyl)pyrrolidine moiety as a chiral auxiliary, enantioselective iodolactonization with high enantiomeric excess could be accomplished to afford chiral lactones.[65] *Trans*-2,5-bis(methoxymethoxy-methyl)-pyrrolidine has proved to be outstanding for the asymmetric monoalkylation of amide enolates,[66a,b] the asymmetric acylation of amide enolates,[67a,b] in aldol condensation,[66c] in the formation of a quarternary asymmetric center by dialkylation of cyanoacetamide Li enolate,[68] and in the [2,3]-Wittig rearrangement of alkenyloxyacetamides.[69]

15.4.3. Amide Acetal Claisen Rearrangements Employing (2*R*, 5*R*)-2,5-Dimethylpyrrolidine

The products of the amide acetal Claisen rearrangement of *N*,*O*-ketene acetals prepared *in situ* from *N*-fluoroacetyl-(2*R*, 5*R*)-2,5-dimethylpyrrolidine (**62**) were formed as a diastereomeric mixture (Table 15.2), as determined by ^{19}F NMR, in an overall yield of 61% in 75% conversion.[78] If a chairlike transition state was presumed for the rearrangement, then (**65**) and (**67**) would be formed from (*E*)-*N*,*O*-ketene acetal and in a like manner and (**64**) and (**66**) via the corresponding (*Z*) isomer.

$CH_3OSO_2CF_3$

62 **63** **64** + **65**

TABLE 15.2. ^{19}**F NMR Chemical Shifts for the Rearranged Products 64–67.**

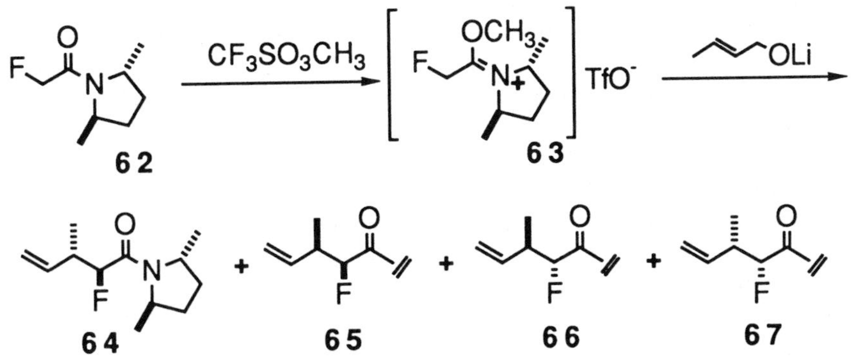

Compound	**64**	**65**	**66**	**67**
Ratio	3.0	13.4	1.1	1.0
^{19}F NMR	−190.80	−191.98	−188.62	−192.94
(δ ppm in $CDCl_3$)[a]	(−191.80)	(−195.76)	(−189.51)	(−197.96)

[a]Parentheses represent the ^{19}F NMR chemical shifts of each compound after hydrogenation.

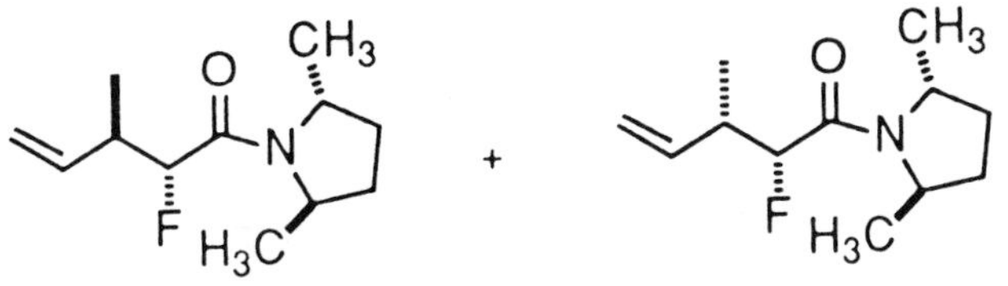

This relationship suggests that the (*E*)-*N*,*O*-ketene acetal must rearrange to form the products at least four times as often as the (*Z*)-*N*,*O*-ketene acetal, a selectivity that would be consistent with our earlier findings. In the rearrangement of the (*E*)-*N*,*O*-ketene acetal, the chiral auxiliary is effective in directing the allylic moiety to a single face of that *N*,*O*-ketene acetal with a selectivity of 13.4 : 1, which corresponds to the formation of (**65**) in an enantiomeric excess of 86%.

The product stereochemical assignments were made by synthesis of the correlation compounds in the following manner. The determination of relative configuration (that is, (**65**) and (**67**) had a *u* relative configuration and (**64**) and (**66**) had an *l* relationship) was based upon our previous studies of the fluoroacetate ester enolate Claisen.[63a,c] The racemic *u* and *l*-2-fluoro-3-methylpent-4-enoic acids were prepared in this way as a 9.0 : 1 mixture of *u*–*l* acids. The free acids were converted to the acid chlorides and condensed with the enantiomeric (2*S*, 5*S*)-dimethyl pyrrolidine. The absolute stereochemistry of the products was determined by the preparation of (**68**) and (**69**), as a 12.5 : 1 mixture, by the fluoro deamination and dediazoniation of (2*S*, 3*S*)-isoleucine (L-isoleucine).[79] This mixture was converted via the formation of active anhydrides[80] into (**70**) and (**71**), by condensation with (2*R*, 5*R*)-2,5-dimethylpyrrolidine, and into (**72**) and (**73**), by condensation with (2*S*, 5*S*)-2,5-dimethylpyrrolidine. In the absence of an additional stereogenic element, by NMR spectroscopy, (**72**) and (**73**) are indistinguishable from their mirror-image compounds (**74**) and (**75**). Close examination of the spectra of these materials left no doubt that the terminal double bond found in (**64**)–(**67**) had a pronounced effect on the fluorine chemical shift.

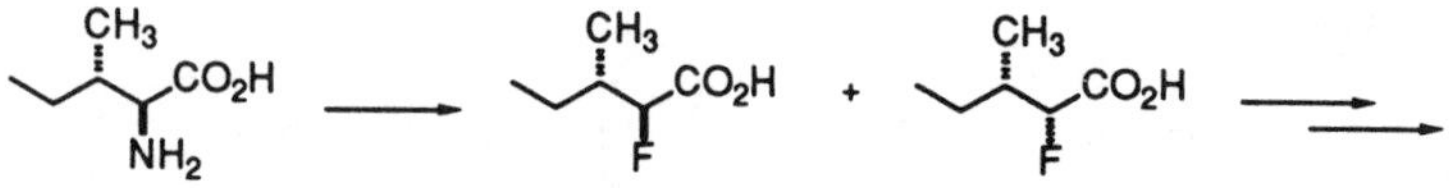

CH_3 CO_2H NH_2 → CH_3 CO_2H F **68** + CH_3 CO_2H F **69** ⇀

O CH_3 N F H_3C **70** + O CH_3 N F H_3C **71** O CH_3 N F H_3C **72** + O CH_3 N F H_3C **73**

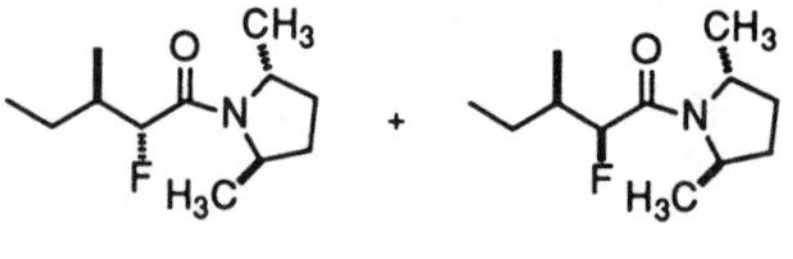

O CH_3 N F H_3C **74** + O CH_3 N F H_3C **75**

However, any uncertainty resulting from these chemical shift differences was easily resolved by the catalytic reduction of (**64**)–(**67**) over palladium on carbon to form (**70**), (**71**), (**74**), and (**75**) and the comparison of NMR data (Table 15.2). The reduction had little effect on the initially determined ratio of products and it was clearly concluded that the absolute stereochemistry of the products is as shown in each scheme and is also consistent with the chairlike transition state predicted by molecular mechanics. It was anticipated that the chiral auxiliary would act by restricting the motion of the crotyl fragment, allowing *ul* approach to only the *si* face of the *N*,*O*-ketene acetal. The *ul* trajectory, which would be required for bonding to the *re* face of the *N*,*O*-ketene acetal would be encumbered by the chiral auxiliary.

The effectiveness of *trans*-2,5-dimethylpyrrolidine in controling the diastereoselectivity of amide acetal Claisen rearrangements seems to be quite general. The rearrangement of the *N,O*-ketene acetal formed on reaction of *N*-propionyl-(2*S*, 5*S*)-2,5-dimethylpyrrolidine (**76**) with (*E*)-crotyl alcohol under the conditions described is extraordinarily selective forming in a ratio of $>10:1$, as determined by ^{1}H NMR, a single diastereomer (**77**) in 50% isolated yield in 60% conversion.

1 $CH_3OSO_2CF_3$

2. OLi

76 **77**

As our previous report established the consistent preference for the formation of the (*Z*)-*N,O*-ketene acetal in the reactions of propionamides, consideration of the diastereofacial influence of *trans*-2,5-dimethylpyrrolidine established above leads to the assignment of the stereochemistry of the major product to be *N*-(2*S*, 3*S*)-2,3-dimethylpent-4-enoyl-(2′*S*, 5′*S*)-2′,5′-dimethylpyrrolidine. Further studies to confirm this assignment and to explore the scope and limitations of this process are underway.

REFERENCES

1. Acs, M., V. D. Busche, C., and Seebach, D., *Chimia*, *44*, 90–92 (1990).
2. (a) Goldman, P., *Science*, *164*, 1123 (1969), (b) Smith, F. A., *Chem. Tech.*, 422 (1973). (c) Filler, R., *Chem. Tech.*, 752 (1974). (d) *Ciba Foundation Symposium, Carbon–Fluorine Compounds, Chemistry, Biochemistry and Biological Activities*; Elsevier: New York, 1972. (e) Filler, R., in *Organofluorine Chemicals and Their Industrial Applications*, (R. E. Banks Ed.), Horwood: Chichester, 1979. (f) *Biomedicinal Aspects of Fluorine Chemistry*, R. Filler and Y. Kobayashi (Eds.), Kodansha: Tokyo, 1982. (g) *Biochemistry Involving Carbon–Fluorine Bonds*, R. Filler (Ed.), American Chemical Society: Washington, DC, (1976). (h) Gerstenberger, M. R. C. and Haas, A., *Angew. Chem. Int. Ed. Engl.*, *20*, 647 (1981). (i) Welch, J. T., *Tetrahedron*, *43*, 3123–3197 (1987). (j) *Synthesis and Reactivity of Fluorocompounds*, N. Ishikawa, (Ed.), CMC: Tokyo, 1987, Vol. 3. (k) *Preparation, Properties and Industrial Applications of Organofluorine Compounds*, R. E. Banks (Ed.), Horwood: Chichester, 1982. (l) Kumadaki, I., *J. Synth. Org. Chem. Jpn.*, *42*, 786 (1984). (m) Bannai, K. and Kurozumi, S., *J. Synth. Org. Chem. Jpn.*, *42*, 794 (1984). (n) Mann, J., *Chem. Soc. Rev.*, *16*, 381 (1987). (o) Fujita, M. and Hiyama, T., *J. Synth. Org. Chem. Jpn.*, *45*, 664 (1987). (p) Kitazume, T. and Yamazaki, T., *J. Synth. Org. Chem. Jpn.*, *45*, 888–897 (1987). (q) Ojima, I., *L'Actualité Chimique, May*, 179 (1987). (r) Imperiali, B., *Biotechnol. Processes, 10 Synth. Pept. Biotechnol.* 97 (1988). (s) Bey, P., *Actual Chim. Ther. 16e Serie*, 111 (1989). (t) Shimizu, M. and Yoshioka, H., *J. Synth. Org. Chem. Jpn.*, *47*, 27–39 (1989).

3. (a) Smart, B., in *Chemistry of Functional Groups, Supplement D, The Chemistry of Halides, Pseudohalides and Azides*, S. Patai and Z. Rapoport (Eds.), Wiley: New York, 1983, pp. 603–655. (b) Hudlicky, M., *Chemistry of Organic Fluorine Compounds*, 2nd ed., Horwood: Chichester, 1976. (c) Chambers, R. D., *Fluorine in Organic Chemistry*, Wiley-Interscience: New York, 1973. (d) Sheppard, W. A. and Sharts, C. M., *Organic Fluorine Chemistry*, Benjamin: New York, 1968. (e) *Synthesis of Fluoroorganic Compounds* (I. L. Knunyants and G. G. Yacobson, Eds.) Springer-Verlag: New York, 1985. (f) J. F. Liebman, A. Greenberg and W. R. Dolbier, Jr., (Eds.), *Fluorine-Containing Molecules: Structure, Reactivity, Synthesis Application*, VCH: New York, 1988.

4. (a) Welch, J. T. and Eswarakrishnan, S., *Fluorine in Bioorganic Chemistry*, Wiley: New York, 1991. (b) Shimizu, M. and Yoshioka, H. J., *J. Synth. Org. Chem. Jpn., 47*, 27–39 (1989). (c) Takeuchi, Y. J., *J. Synth. Org. Chem. Jpn., 46*, 145–159 (1988). (d) J. F. Liebman, A. Greenberg, and W. R. Dolbier, Jr., (Eds.), *Fluorine-Containing Molecules: Structure, Reactivity, Synthesis, and Applications*, VCH Publishers: New York, 1988. (e) Welch, J. T., *Tetrahedron, 43*, 3123–3197 (1987). (f) R. Filler and Y. Kobayashi (Eds.), *Biomedicinal Aspects of Fluorine Chemistry*, R. Filler and Y. Kobayashi (Eds.) Kodansha: Tokyo, 1982.

5. (a) Walba, D. M., Razavi, H. A., Clark, N. A., and Parmar, D. S., *J. Am. Chem. Soc., 110*, 8686–8691 (1988). (b) Kitazume, T., Ohnogi, T., and Ito, K., *J. Am. Chem. Soc., 112*, in press (1990).

6. *L'Actualité-Chimique, May*, 135–188 (1987).

7. Topson, R. D., *Prog. Phys. Org. Chem., 12*, 1 (1976).

8. Levitt, L. S. and Widing, H. F., *Prog. Phys. Org. Chem., 12*, 119 (1976).

9. Fraga, A., Saxena, K. M., and Lo, B. W. N., *Atomic Data, 3*, 323 (1971).

10. (a) Desimoni, G., Tacconi, G., Barco, A., and Pollini, G. P., in *Natural Products Synthesis Through Pericyclic Reactions;* M. C. Caserio, (Ed.), ACS Monograph 180, American Chemical Society: Washington, DC, 1983; Chapt. 7. (b) Hill, R. K., in *Asymmetric Synthesis*, J. D. Morrison (Ed.), Academic: Orlando, FL, 1984, Vol. 3, Chapt. 8. (c) de Mayo, P., in *Molecular Rearrangements-2,* pp. 660. (d) Ziegler, F. E., *Acc. Chem. Res., 10*, 227–232 (1977). (e) Bennett, G. B., *Synthesis*, 589–606 (1977).

11. (a) Carroll, M. F., *J. Chem. Soc.*, 704 (1940). (b) *J. Chem. Soc.*, 1266. (c) *J. Chem. Soc.*, 1941, 507.

12. Meerwein, H., Florian, W., Schon, N., and Stopp, G., *Justus Liebigs Ann. Chem.*, 641, 1–39 (1961). (b) Felix, D., Gschwend-Steen, K., Wick, A. E., and Eschenmoser, A., *Helv. Chim. Acta, 52*, 1030–1042 (1969). (c) Wick, A. E., Felix, D., Steen, K., and Eschenmoser, A. *Helv., Chim. Acta, 47*, 2425–2429 (1964). (d) Jenkins, P. R., Gut, R., Wetter, H., and Eschenmoser, A., *Helv. Chim. Acta, 62*, 1922–1931 (1979).

13. (a) Marbet, R. and Saucy, G., *Helv. Chim. Acta, 50*, 2091–2095 (1967); *Helv. Chim. Acta.*, 2095–2100 (1967). (b) Faulkner, D. J. and Petersen, M. R., *J. Am. Chem. Soc., 95*, 553–563 (1973).

14. Dauben, W. G. and Dietsche, T. J., *J. Org. Chem., 37*, 1212–1216 (1972).

15. Petrzilka, M., *Helv. Chim. Acta, 61*, 2286–2289 (1978).

16. Petrzilka, M., *Helv. Chim. Acta, 61*, 3075–3078 (1978).

17. (a) Johnson, W. S., Werthmann, L., Bartlett, W. R., Brocksom, T. J., Li, T.-T., Faulkner, D. J., and Petersen, M. R., *J. Am. Chem. Soc., 92*, 741–743 (1970). (b)

Johnson, W. S., Gravenstock, M. B., Parry, R. J., Myers, R. F., Bryson, T. A., and Howard Miles, D., *J. Am. Chem. Soc.*, *93*, 4330–4332 (1971). (c) Johnson, W. S., Gravestock, M. B., and McCarry, B. E., *J. Am. Chem. Soc.*, *93*, 4332–4334 (1971).

18. (a) Ficini, J. and Barbara, C., *Tetrahedron Lett.*, 6425–6429 (1966). (b) Bartlett, P. A. and Hahne, W. F., *J. Org. Chem.*, *44*, 882–883 (1979).
19. (a) Doering, W. von E. and Roth, W. R., *Tetrahedron*, *18*, 67–74 (1962). (b) Carpenter, B. K., *Tetrahedron*, *34*, 1877–1884 (1978). (c) Burrows, C. J. and Carpenter, B. K., *J. Am. Chem. Soc.*, *103*, 6983–6984 (1981). (d) Burrows, C. J. and Carpenter, B. K., *J. Am. Chem. Soc.*, *103*, 6984–6985 (1981).
20. (a) Denmark, S. E. and Harmata, M. A., *J. Am. Chem. Soc.*, *104*, 4972–4974 (1982). (b) Denmark, S. E. and Harmata, M. A., *J. Org. Chem.*, *48*, 3369–3370 (1983). (c) Denmark, S. E. and Harmata, M. A., *Tetrahedron Lett.*, *25*, 1543–1546 (1984). (d) Denmark, S. E., Marlin, J. E., and Dorow, R. L., *Abstracts of Papers*, 191st National Meeting of the American Chemical Society, New York City, New York; American Chemical Society: Washington DC, 1986; ORGN 273.
21. (a) Ireland, R. E., Mueller, and R. H. Willard, A. K., *J. Am. Chem. Soc.*, *98*, 2868–2877 (1976). (b) Whitesell, J. K., Matthews, R. S., and Helbling, A. M., *J. Org. Chem.*, *43*, 784–786 (1978). (c) Ireland, R. E. and Mueller, R. H., *J. Am. Chem. Soc.*, *94*, 5897–5898 (1972). (d) Ireland, R. E. and Willard, A. K., *Tetrahedron Lett.*, 3975–3978 (1975).
22. Sucrow, W. and Richter, W., *Chem. Ber.*, *104*, 3679–3688 (1971). (b) Sucrow, W. and Richter, W., *Tetrahedron Lett.*, 3675–3676 (1970). (c) Sucrow, W., *Tetrahedron Lett.*, 4725–4726 (1970). (a) Sucrow, W., Schubert, B., Richter, W., and Slopianka, M., *Chem. Ber.*, *104*, 3689–3703 (1971). (e) Bryson, T. A., and Reichel, C. J., *Tetrahedron Lett.*, 2381–2384 (1980). (f) Sucrow, W. and Girgensohn, B., *Chem. Ber.*, *103*, 750–756 (1970). (g) Sucrow, W., Caldeira, P. P., and Slopianka, M., *Chem. Ber.*, *106*, 2236–2245 (1973). (h) Sucrow, W., Slopianka, M., and Caldeira, P. P., *Chem. Ber.*, *108*, 1101–1110 (1975).
23. For general reviews on the Claisen rearrangement see (a) Ziegler, F. E., *Chem. Rev.*, *88*, 1423–1452 (1988). (b) Blechert, S., *Synthesis*, 71–100 (1989). (c) Bartlett, P. A., *Tetrahedron*, *36*, 2–72 (1980). (d) Desimoni, G., Tacconi, G., Barco, A., and Pollini, G. P., in *Natural Products Synthesis Through Pericyclic Reactions*, M. C. Caserio (Ed.), ACS Monograph 180, American Chemical Society: Washington, DC, 1983; Chapt. 7.
24. Hansen, H. J. and Schmid, H., *Tetrahedron*, *30*, 1959–1969 (1974).
25. Vittorelli, P., Winkler, T., Hansen, H. J., and Schmid, H., *Helv. Chim. Acta.*, *51*, 1457–1461 (1968).
26. Dewar, J. J. S. and Jie, C., *J. Am. Chem. Soc.*, *111*, 511–519 (1989).
27. Hill, R. K. and Gilman, N. W., *J. Chem. Soc. Chem. Commun.*, 619 (1967).
28. Hill, R. K., Soman, R., and Sawada, S., *J. Org. Chem.*, *37*, 3737–3740 (1972).
29. Ziegler, F. E. and Thottathil, J. K., *Tetrahedron Lett.*, *23*, 3531–3534 (1982).
30. (a) Fujisawa, T., Tajima, K., Ito, M. and Sato, T., *Chem. Lett.*, 1169–1172 (1984). (b) Fujisawa, T., Kohama, H., Tsunekawa, H., and Sato, T., *Chem. Lett.*, 1553–1556 (1986). (c) Sato, T., Tajima, K., and Fujisawa, T., *Tetrahedron Lett.*, *24*, 729 (1983). (d) Fujisawa, T., Tajima, K., Ito, M. and Sato, T., *Chem. Lett.*, 1669–1672 (1984). (e) Kurth, M. J. and Yu, Ch.-M., *J. Org. Chem.*, *50*, 1840–1845 (1985).
31. Mislow, K., in *Introduction to Stereochemistry*, Benjamin: New York, 1965, p. 131.

32. Oda, J., Igarashi, T., and Inouye, Y., *Bull. Inst. Chem. Res., Kyoto Univ., 54*, 180 (1976); Chem. Abstr. 1977, *86*, 88836m.
33. (a) Welch, J. T. and Eswarakrishnan, S., *J. Am. Chem. Soc., 109*, 6716–6719 (1987). (b) Kurth, M. J. and Brown, E. G., *Synthesis*, 362–366 (1988). (c) Kurth, M. J., Decker, O. H. W., Hope, H., and Yanuck, M. D., *J. Am. Chem. Soc., 107*, 443–448 (1985). (d) Kurth, M. J. and Decker, O. H. W., *Tetrahedron Lett., 24*, 4535–4538 (1983).
34. (a) Church, R. F., Ireland, R. E., and Marshall, J. A., *J. Org. Chem., 27*, 1118–1125 (1962). (b) Church, R. F. and Ireland, R. E., *J. Org. Chem., 28*, 17–23 (1963).
35. (a) Morrow, D. F., Culvertson, T. P., and Hofer, R. M., *J. Org. Chem., 32*, 361–369 (1967). (b) Mikami, K., Kawamoto, K., and Nakai, T., *Chem. Lett.*, 115–118 (1985).
36. (a) Nakai, T., Kasuga, T., Fujimoto, K., and Mikami, K., *Tetrahedron Lett., 25*, 6011–6014 (1984). (b) Nakai, T., Kasuga, T., Takahashi, O., and Mikami, K., *Chem. Lett.*, 1729–1732 (1985).
37. (a) Cha, J. K. and Lewis, S. C., *Tetrahedron Lett., 25*, 5263–5266 (1984). (b) Kurth, M. J., Decker, O. H. W., Hope, H., and Yanuck, M. D., *J. Am. Chem. Soc., 107*, 443–448 (1984). (c) Kurth, M. J. and Decker, O. H. W., *J. Org. Chem., 50*, 5769–5775 (1985). (d) Kurth, M. J. and Soares, C. J., *Tetrahedron Lett., 28*, 1031–1034 (1987). (e) Kallmerten, J. and Wittman, M. D., *Tetrahedron Lett., 27*, 2443–2446 (1986). (f) Kallmerten, J. and Gould, T. J., *J. Org. Chem., 51*, 1152–1155 (1986). (g) Suzuki, T., Sato, E., and Unno, K., *J. Chem. Soc. Perkin Trans. 1*, 2263–2268 (1986).
38. Drauz, K., Kleeman, A., and Martens, J., *Angew. Chem. Int. Ed. Engl., 21*, 584–608 (1982).
39. (a) Sonnet, P. E. and Heath, R. R., *J. Org. Chem., 45*, 3137–3139 (1980). (b) Evans, D. A. and Takacs, J. M., *Tetrahedron Lett., 21*, 4233–4236 (1980). (c) Ahlbrecht, H., Bonnet, G., Enders, D., and Zimmermann, G., *Tetrahedron Lett., 21*, 3175–3178 (1980). (d) Enders, D., in *Asymmetric Synthesis*, J. D. Morrison, (Ed.), Academic: Orlando, FL, 1984; Vol. 3, Chap. 4. (e) Guoqiang, L., Hjalmarsson, M., Hogberg, H., Jernstedt, K., and Norin, T., *Acta Chem. Scand. Ser. B, 38*, 795–801 (1984). (f) Schultz, A. G. and Sundararaman, P., *Tetrahedron Lett., 25*, 4591–4594 (1984). (g) Enomoto, M., Ito, Y., Katsuki, T., and Yamaguchi, M., *Tetrahedron Lett., 26*, 1343–1344 (1985).
40. Iwasawa, N. and Mukaiyama, T., *Chem. Lett.*, 1441–1444 (1982).
41. Soai, K., Machida, H., and Ookawa, A., *J. Chem. Soc. Chem. Commun.*, 469–470 (1985).
42. Strauss, H. F. and Wiechers, A., *Tetrahedron, 34*, 127 (1978).
43. Muxfeldt, H., Schneider, R. S., and Mooberry, J. B., *J. Am. Chem. Soc., 88*, 3670–3671 (1966). (b) Sucrow, W., *Angew. Chem. Int. Ed. Engl., 7*, 629 (1968). (c) Ziegler, F. E., and Sweeny, J. G., *Tetrahedron Lett.*, 1097–1100 (1969).
44. (a) Welch, J. T. and Eswarakrishnan, S., *J. Org. Chem., 50*, 5909–5910 (1985). (b) Welch, J. T. and Eswarakrishnan, S., *J. Org. Chem., 50*, 5403–5405 (1985).
45. (a) Welch, J. T., Seper, K., Eswarakrishnan, S., and Samartino, J., *J. Org. Chem., 49*, 4720–4721 (1984). (b) Welch, J. T. and Eswarakrishnan, S., *J. Chem. Soc. Chem. Commun.*, 186–188 (1985). (c) Welch, J. T. and Seper, K. W., *Tetrahedron Lett., 25*, 5247–5250 (1984). (d) Welch, J. T. and Samartino, J. S., *J. Org. Chem., 50*, 3663–3665 (1985). (e) Welch, J. T. and Seper, K., *J. Org. Chem., 51*, 119–120 (1986).
46. (a) Bredereck, H., Effenberger, F., and Beyerlin, H. P., *Chem. Ber., 97*, 3081–3087 (1964). (b) Bredereck, H., Effenberger, F., and Simchen, G., *Chem. Ber., 96*, 1350–1355 (1963). (c) Salomon, R. G. and Raychaudhuri, S. R., *J. Org. Chem., 49*, 3659–3660 (1984).

47. Bordwell, F. G. and Fried, H. N., *J. Org. Chem.*, *46*, 4327–4331 (1981).
48. Corey, E. J., Shibasaki, M., and Knolle, J., *Tetrahedron Lett.*, 1625–1626 (1977).
49. (a) Takano, S., Yoshida, E., Hirama, M., and Ogasawara, K., *J. Chem. Soc. Chem. Commun.*, 776–777 (1976). (b) Takano, S., Hirama, M., Araki, T., and Ogawawara, K., *J. Am. Chem. Soc.*, *98*, 7084–7085 (1976).
50. Welch, J. T. and Eswarakrishnan, S., *J. Am. Chem. Soc.*, *109*, 6716–6719 (1987).
51. (a) Kurth, M. J. and Decker, O. H. W., *Tetrahedron Lett.*, *24*, 4535–4538 (1983). (b) Kurth, M. J., Decker, O. H. W.; Hope, H. and Yanuck, M. D., *J. Am. Chem. Soc.*, *107*, 443–448 (1985). (c) Kurth, M. J. and Decker, O. H. W., *J. Org. Chem.*, *50*, 5769–5775 (1985). (d) Kurth, M. J. and Decker, O. H. W., *J. Org. Chem.*, *51*, 1377–1382 (1986). (e) Kurth, M. J. and Soares, C. J., *Tetrahedron Lett.*, *28*, 1031–1034 (1987).
52. Sonnet, P. E. and Heath, R. R., *J. Org. Chem.*, *45*, 3137–3139 (1980). (b) Evans, D. A. and Takacs, J. M., *Tetrahedron Lett.*, *21*, 4233–4236 (1980). (c) Ahlbrecht, H., Bonnet, G., Enders, D. and Zimmerman, G., *Tetrahedron Lett.*, *21*, 3175–3178 (1980). (d) Enders, D., in *Asymmetric Synthesis*, J. D. Morrison (Ed.), Academic: Orlando, FL, 1984, Vol. 3, Chapt. 4. (e) Schultz, A. G. and Sundararaman, P., *Tetrahedron Lett.*, *25*, 4591–4594 (1984). (f) Enomoto, M., Ito, Y., Katsuki, T. and Yamaguchi, M., *Tetrahedron Lett.*, *26*, 1343–1344 (1985).
53. Whitesell, J. K., *Chem. Rev.*, *89*, 1581–1590 (1989).
54. Kagan, H. B. and Dang, T. P., *J. Am. Chem. Soc.*, *94*, 6429–6433 (1972).
55. Differding, E. and Lang, R. W., *Tetrahedron Lett.*, *29*, 6087–6090 (1988).
56. (a) Kitazume, T. and Yamazaki, T., *J. Synth. Org. Chem. Jpn.*, *46*, 888–897 (1987). (b) Kitazume, T., Okamura, N., Ikeya, T., and Yamazaki, T., *J. Fluorine Chem.*, *39*, 107–115 (1988). (c) Seebach, D., Renaud, P., Schweizer, W. B., Zuger, M. F. and Brienne, M.-J., *Helv. Chim. Acta*, *67*, 1843 (1984). (d) Seebach, D. and Renaud, P., *Helv. Chim. Acta*, *68*, 2342–2345 (1985). (e) Lin, J.-T., Yamazaki, T. and Kitazume, T., *J. Org. Chem.*, *52*, 3211–3217 (1987).
57. (a) Bravo, P., Piovosi, E., Resnati, G., and Fronza, G., *J. Org. Chem.*, *54*, 5171–5176 (1988). (b) Bravo, P. and Resnati, G., in *Perspectives in the Organic Chemistry of Sulfur*, B. Zwanenburg, and A. J. H. Klunder, (Eds.) Elsevier: Amsterdam, 1987. (c) Bravo, P., Piovosi, E., Resnati, G., and DeMunari, S., *Gazz. Chim. Ital.*, *118*, 115–122 (1988). (d) Bravo, P. Piovosi, E. and Resnati, G., *Synthesis*, 579–582 (1986). (e) Yamazaki, T., Ishikawa, N. Iwatsubo, H. and Kitazume, T., *J. Chem. Soc. Chem. Commun.*, 1340–1342 (1987).
58. Welch, J. T. and Seper, K. J., *J. Org. Chem.*, *53*, 2991–2999 (1988).
59. (a) Welch, J. T. and Samartino, J. S., *J. Org. Chem.*, *50*, 3663–3665 (1985). (b) Welch, J. T. and Eswarakrishnan, S., *J. Org. Chem.*, *50*, 5909–5910 (1985). (c) Welch, J. T., Plummer, J. S., and Chou, T.-S., *J. Org. Chem.*, *56* 353–359 (1991).
60. (a) Metcalf, B. W., Jarvi, E. T., and Burkhart, J. P., *Tetrahedron Lett.*, *26*, 2861–2864 (1985). (b) Camps, F., Messeguer, A., and Sanchez, F.-J., *Tetrahedron*, *44*, 5161–5167 (1988). (c) Yuan, W., Berman, R. J., and Gelb, M. H., *J. Am. Chem. Soc.*, *109*, 8071–8081 (1987). (d) Kolb, M., Gerhart, F., and Francois, J. P., *Synthesis*, 469–470 (1988). (e) Yokozawa, T., Nakai, T., and Ishikawa, N., *Tetrahedron Lett.*, *25*, 3991–3994 (1984). (f) Normant, J. F., Reboul, O., Sauvetre, R., Deshayes, H., Masure, D., and Villieras, J., *Bull. Chim. Soc. Fr.*, 2072–2078 (1974). (g) Yamazaki, T. and Ishikawa, N., *Bull. Chim. Soc. Fr.*, 937–943 (1986).
61. Unpublished results of Janet S. Plummer, The University at Albany.

62. Kagan, H. B., in *Asymmetric Synthesis*, J. D. Morrison (Ed.), Academic: Orlando, FL, 1984, Vol. 2, pp. 1–39.
63. Whitesell, J. K. and Felman, S. W., *J. Org. Chem.*, *42*, 1663–1664 (1977).
64. Whitesell, J. K., *Acc. Chem. Res.*, *18*, 280–284 (1985).
65. Fuji, K., Node, M., Naniwa, Y., and Kawabata, T., *Tetrahedron Lett.*, *31*, 3175–3178 (1990).
66. (a) Kawanami, Y., Ito, Y., Kitagawa, T., Taniguchi, Y., Katsuki, T., and Yamaguchi, M., *Tetrahedron Lett.*, *25*, 857–860 (1984). (b) Enomoto, M., Ito, Y., Katsuki, T., and Yamaguchi, M., *Tetrahedron Lett.*, *26*, 1343 (1985). (c) Katsuki, T. and Yamaguchi, M., *Tetrahedron Lett.*, *26*, 5807 (1985).
67. (a) Ito, Y., Katsuki, T. and Yamaguchi, M., *Tetrahedron Lett.*, *25*, 6105–6016 (1984). (b) Ito, Y., Katsuki, T., and Yamaguchi, M., *Tetrahedron Lett.*, *26*, 4643–4646 (1985).
68. Hanamoto, T., Katsuki, T., and Yamaguchi, M., *Tetrahedron Lett.*, *27*, 2463–2464 (1986).
69. Uchikawa, M., Hanamoto, T., Katsuki, T., and Yamaguchi, M., *Tetrahedron Lett.*, *27*, 4577–4580 (1986).
70. Harding, K. E. and Burks, S. R., *J. Org. Chem.*, *46*, 3920–3922 (1981).
71. Schlessinger, R. H. and Iwanowicz, E. J., *Tetrahedron Lett.*, *28*, 2083–2086 (1987).
72. Schlessinger, R. H., Iwanowicz, E. J., and Springer, J. P., *J. Org. Chem.*, *51*, 3070–3073 (1986).
73. Schlessinger, R. H., Iwanowicz, E. J., and Springer, J. P., *Tetrahedron Lett.*, *29*, 1489–1492 (1988).
74. Yamazaki, T., Gimi, R. H., and Welch, J. T., *Synlett*, 573–574 (1991).
75. Short, R. P., Kennedy, R. M., and Masamune, S., *J. Org. Chem.*, *54*, 1755 (1989).
76. Stafford, J. A., and Heathcock, C. H., *J. Org. Chem.*, *55*, 5433–5434 (1990).
77. Najdi, S. and Kurth, M. J., *Tetrahedron Lett.*, *31*, 3279–3282 (1990).
78. Yamazaki, T., Welch, J. T., Plummer, J. S., and Gimi, R. H., *Tetrahedron Lett.*, *32*, 4267–4270 (1991).
79. Olah, G. K., Prakash, G. K. S., and Chao, Y.-L., *Helv. Chim. Acta.*, *64*, 2528–2530 (1981).
80. Vaughan, Jr., J. R. and Osato, R. L., *J. Am. Chem. Soc.*, *74*, 676–678 (1952).

CHAPTER 16

Unusual Fluorinated Alkenes and Dienes, via Fluoride Ion Induced Processes

R. D. CHAMBERS

16.1. INTRODUCTION

Highly fluorinated alkenes are very electrophilic systems. The dominant theme in their chemistry is that of nucleophilic (NUC) attack, via intermediate carbanions (**1**),[1,2] which of course contrasts with the chemistry of alkenes, where ionic reactions are dominated by electrophilic attack, proceeding via car-

Synthetic Fluorine Chemistry,
Edited by George A. Olah, Richard D. Chambers, and G. K. Surya Prakash.
ISBN 0-471-54370-5

$$\text{e.g.}\quad Nuc^- + CF_2{=}C\langle \longrightarrow Nuc{-}CF_2{-}\bar{C}\langle \longrightarrow \text{Products}$$

(1)

$$\text{e.g.}\quad E^+ + CH_2{=}C\langle \longrightarrow E{-}CH_2{-}\overset{+}{C}\langle \longrightarrow \text{Products}$$

(2)

bocations (**2**). Indeed, it will become apparent that a mirror-image relationship exists between ionic reactions of unsaturated fluorocarbons and corresponding reactions of unsaturated hydrocarbons.[3]

16.2. FLUORINATED CARBANIONS

Since carbanions play such an important role in the chemistry of fluorinated alkenes, it is relevant to outline some of the factors that influence their stability. Andreades[4] studied base-catalyzed H–D exchange in NaOMe/MeOH solutions of systems **3**–**6** (Table 16.1), and it is clear that perfluoroalkyl groups, substituted at the developing carbanionic site, are strongly carbanion stabilizing (**7**). This may be attributed to the inductive effect, I_σ. However, the situation is more ambiguous for fluorine directly attached to the carbanionic site (**8**) because inductive electron withdrawal, I_σ, is offset by electron-pair repulsion, I_π, and the resultant of these two effects depends on the geometry of the system. Various data leads us to conclude that I_π is greater for a planar carbanion (**10**) than if the carbanion is tetrahedral (**9**), that is, because a fluorine atom attached to a developing planar site is significantly *less* acidifying than at a developing tetrahedral site. Table 16.1, dramatically illustrates that a perfluoroalkyl group is much more effective at carbanion stabilization than is a fluorine atom,

CF_3H (3) $\quad CF_3(CF_2)_5CF_2H$ (4) $\quad (CF_3)_2CFH$ (5) $\quad (CF_3)_3CH$ (6)

TABLE 16.1. Relative Acidities.[4]

Compound	**3**	**4**	**5**	**6**
Derived Anion	CF_3^-	$CF_3(CF_2)_5CF_2^-$	$(CF_3)_2CF^-$	$(CF_3)_3C^-$
Relative Reactivity	3	6	2×10^5	10^9
Approximate pK_a	31	30	20	11

I_σ $\bar{C}$-C ⟶ F (stabilizing) (**7**)

I_π $\bar{C}$⟶$\ddot{F}$ (destabilizing) (**8**)

109° sp^3c (**9**)

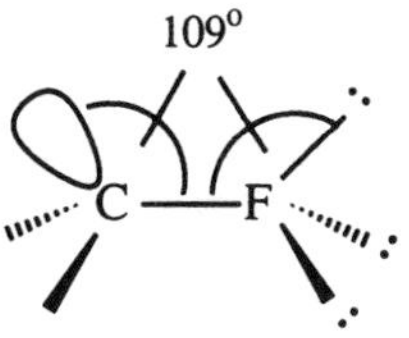

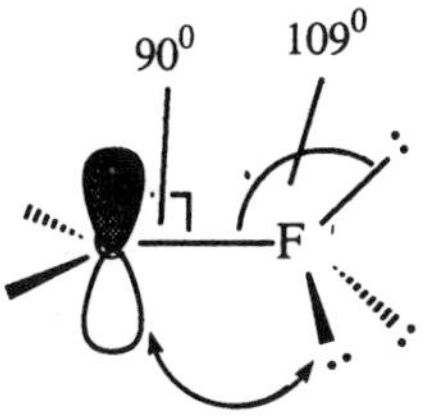

90^0 109^0 sp^2c (**10**)

although the attribution of this substantial stabilization to "C—F negative hyperconjugation," has been a matter for some debate.[3,5]

16.3. NUCLEOPHILIC ATTACK ON FLUORINATED ALKENES

Understanding relative reactivity orders and orientations of electrophilic attack on alkenes was an early outcome of mechanistic studies in organic chemistry and understanding of corresponding effects for nucleophilic attack on fluorinated alkenes is a significant challenge to our current theories of organic chemistry. The main facts to be accommodated are (a) attack occurs preferentially at a terminal $CF_2{=}$group; (b) introduction of perfluoroalkyl groups at the 2 position enhances reactivity, that is, $CF_2{=}CF_2 < CF_2{=}CFCF_3 \ll CF_2{=}C(CF_3)_2$; and (c) introduction of perfluoroalkyl at the *same* carbon (**11**) is more activating than at *opposite* sides of the double bond (**12**). We need to attribute a significant *polar contribution* to the

$$\text{i.e.} \quad \underset{(\mathbf{11})}{CF_2{=}C(CF_3)_2} \gg \underset{(\mathbf{12})}{CF_3CF{=}CFCF_3}$$

activating influence of fluorine at vinylic positions in order to account for the greater reactivity of alkenes bearing fluorine (**13a**) rather than chlorine (**13b**) at vinylic positions,

$$\text{i.e.} \quad \underset{(\mathbf{13a})}{=\overset{\delta+}{C} \longrightarrow \overset{\delta-}{F}} \gg \underset{(\mathbf{13b})}{=\overset{\delta+}{C} \longrightarrow \overset{\delta-}{Cl}}.$$

This implies a rate-limiting addition, in a two-step process, $k_2 \gg k_1$. Indeed, this simple approach accounts for many of the main features of nucleophilic attack on fluorinated alkenes.

$$Nuc^- + CF_2{=}C\langle \xrightarrow{k_1} Nuc\text{-}CF_2\text{-}\bar{C}\langle \xrightarrow{k_2} \text{Products}$$

Thus, the order of developing carbanion stabilities (see above) would account for both the orientation and reactivity orders shown.

$$Nuc^- + CF_2{=}CFCF_3 \longrightarrow Nuc\text{-}CF_2\text{-}\bar{C}\mathbf{F}CF_3$$

$$\not\longrightarrow Nuc\text{-}CF(CF_3)\text{-}CF_2^-$$

$$NucCF_2\text{-}\bar{C}F_2 < NucCF_2\bar{C}\mathbf{F}CF_3 \ll NucCF_2\bar{C}(CF_3)_2$$

(Strongly stabilizing fluorine atoms are **bold**)

Frontier orbital arguments may be used as an alternative approach.[6] This recognizes that HOMO–LUMO (highest occupied molecular orbital–lowest unoccupied molecular orbital) interaction between the nucleophile and *F*-alkene, respectively, will be important and that replacing fluorine in an alkene by a trifluoromethyl group reduces the LUMO energy, which correspondingly increases reactivity, that is, $CF_2{=}CF_2 < CF_2{=}CFCF_3 < CF_2{=}C(CF_3)_2$. It is in considering the effects on coefficients, however, that this approach is quite valuable and will account for the lower reactivity of **12** rather than **11**. Introduction of trifluoromethyl increases the coefficients in the LUMO at the carbon opposite **11a**; the effects of trifluoromethyl are enhanced in **11a** but for trifluoromethyl groups on opposite sides, that is, **12a**, their effect on coefficients and hence reactivity, is opposing. This would also account for the reactivity order $CF_2{=}CF_2 < CF_2{=}CFCF_3 > CF_3CF{=}CFCF_3$.

$$(CF_3)_2\text{Ⓒ}{=}\text{©} \quad \gg \quad CF_3\text{Ⓒ}{=}\text{Ⓒ}CF_3$$

(11a) (12a)

16.4. SOME OLIGOMERIZATIONS OF FLUORINATED ALKENES

A limitation in the development of organofluorine chemistry is, inevitably, the range of readily accessible building blocks for further synthesis. Therefore processes that form carbon–carbon bonds from industrially available fluoroalkenes, building up synthetically more sophisticated systems, are extremely important.

The oligomerization of *F*-ethene[7,8] Scheme I, illustrates how fluoride-induced processes may be used to build up trimer **19** to hexamer **23**, and we will refer to some of these systems later, especially the tetramer **20**. The latter may be formed from dimerization of **18**,[9] or via the trimer **19** by further reaction with anion **15**. The important mechanistic feature of this process is the fact that the extending anion (**16**) is not preserved for anionic polymerization but loses fluoride ion, to give an *F*-alkene (**17**), which rapidly reacts with fluoride ion, giving **18**, and so on. In this way, highly branched systems are obtained from fluorinated alkenes. It has been established[10] that the trimer forms a very stable tertiary anion (**21**), which could react with *F*-butene (**18**), to form a pentamer (**22**). The formation of hexamer **23** as the major isomer is particularly intriguing because it is a rare example of a terminal isomer being thermodynamically favored over an internal

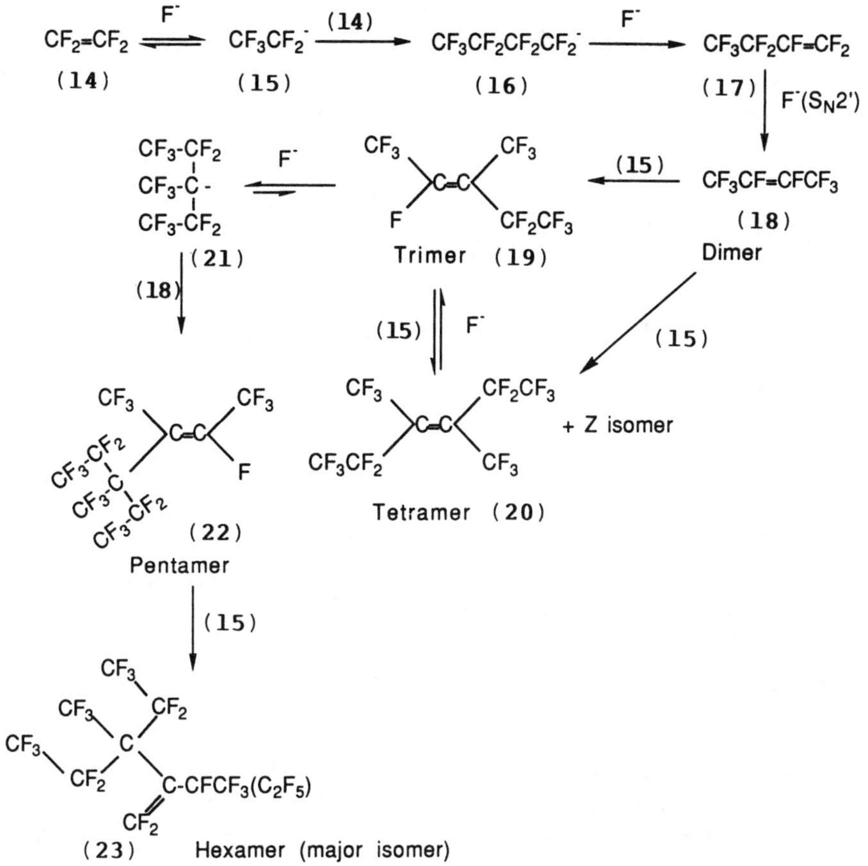

$CF_2{=}CF_2$ (14) $\xrightleftharpoons{F^-}$ $CF_3CF_2^-$ (15) $\xrightarrow{(14)}$ $CF_3CF_2CF_2CF_2^-$ (16) $\xrightarrow{F^-}$ $CF_3CF_2CF{=}CF_2$ (17)

(17) $\xrightarrow{F^-(S_N2')}$ $CF_3CF{=}CFCF_3$ (18) Dimer

(18) $\xrightarrow{(15)}$ Trimer (19): CF_3, CF_3, F, CF_2CF_3 on C=C

Trimer (19) $\xrightleftharpoons{F^-}$ $(CF_3CF_2)_2C^-CF_3$ (21)

(21) $\xrightarrow{(18)}$ Pentamer (22): CF_3, $(CF_3CF_2)_2C(CF_3)$–, CF_3, F on C=C

Trimer (19) $\xrightleftharpoons[(15)]{F^-}$ Tetramer (20): CF_3, CF_2CF_3, CF_3CF_2, CF_3 on C=C + Z isomer

Dimer (18) $\xrightarrow{(15)}$ Tetramer (20)

Pentamer (22) $\xrightarrow{(15)}$ (23) Hexamer (major isomer): $(CF_3CF_2)_2C(CF_3)$–$C(=CF_2)$–$CFCF_3(C_2F_5)$

Scheme I. Oligomerization of *F*-ethene.

isomer. Apparently, this arises from lesser steric interactions in this isomer but it is not obvious why this should be the case.

Perfluorocycloalkenes are also oligomerized to give some extremely interesting products; a dimer (**25**) is obtained from *F*-cyclopentene (**24**)[11] while a mixture of dimers (**27** and **28**), together with a trimer (**29**) is obtained from *F*-cyclobutene (**26**), although the dimers are best obtained by oligomerization using pyridine, which proceeds via an ylide-type process.[11,12]

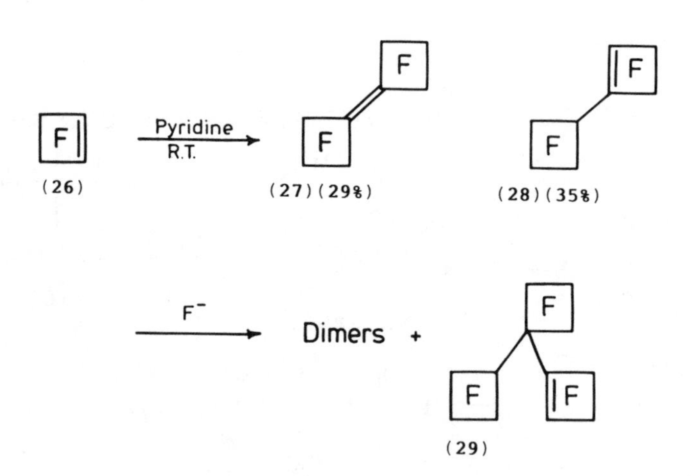

16.5. FORMATION OF DIENES

16.5.1. Thermal Reactions

The conversion of oligomers, described above, into corresponding dienes is a desirable objective, which has indeed been approached previously in our laboratories. Tatlow and his co-workers[13,14] successfully developed defluorination procedures for the synthesis of polyfluoroaromatic compounds by passing fluorinated cyclohexanes and cyclohexenes over hot iron and iron oxide. Applying similar procedures,[15] dienes **30** and **31** can be obtained from *F*-bicyclopentylidene (**25**) and *F*-bicyclobutylidene (**27**) and its isomer (**28**), respectively, although the yield of **31** is quite low, Scheme II. Furthermore, yields of products fall off rapidly as the defluorinating agent approaches the spent state.

Further complications arise in defluorination of the *F*-3,4-dimethyl-3-hexene

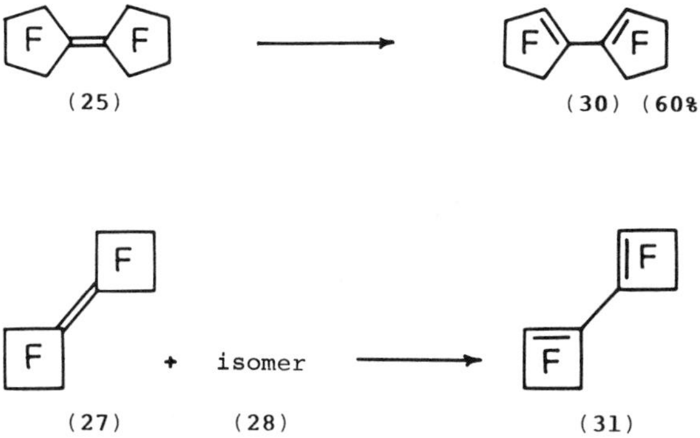

(25) → (30) (60%)

(27) + isomer (28) → (31)

Scheme II. Synthesis of *F*-dienes by defluorination (over Fe, 500 °C).[15]

(**20**) (the tetramer of *F*-ethene).[16] Here, the corresponding diene (**32**) is produced, but this is accompanied by a diene (**34**) that arises from fragmentation (loss of C_2F_6) of the starting material and, more serious, the desired diene (**32**) is partly converted to the more stable cyclobutene isomer (**33**). In separate experiments, we have shown that **32** is converted extensively to **33** by heating. In reactions of **20** over platinum (Scheme III) the principal product is **34**, that is, arising from fragmentation, together with the corresponding cyclobutene derivative **35**. Therefore we can rationalize the process as a competition with k_1 being enhanced by the presence of iron, while k_2 is the more favorable process over platinum, albeit at a higher temperature.

An interesting aspect of these experiments, however, is that the recovered starting material, after passing over iron, is a mixture of all three isomers **20**–**20b**.[17] The isomers **20a** and **b** are thermodynamically less stable than **20** (see earlier comment on the effect of CF_3 on orbital energies) and, in the presence of fluoride ion, are rapidly converted back to **20**. It is extremely interesting, however, that these have been isolated because it appears to be the only situation where all isomers in such a series have been obtained, that is, **20** $<$ **20a** $<$ **20b** (decreasing thermodynamic stability). There is ample evidence

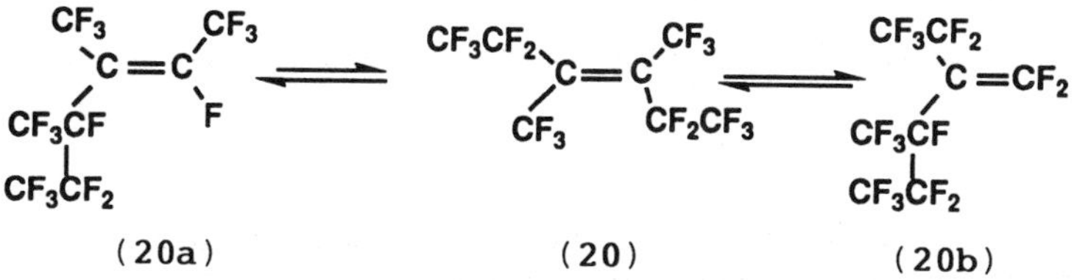

(20a) ⇌ (20) ⇌ (20b)

(20a) + **(20a)** + **(20b)** $\xrightarrow{\text{(i)}}$ products + **(20)** + **(20a)** $\xrightarrow{\text{(ii)}}$ products + **(20)**

(i), MeOH, room temp. ; (ii), MeOH, reflux.

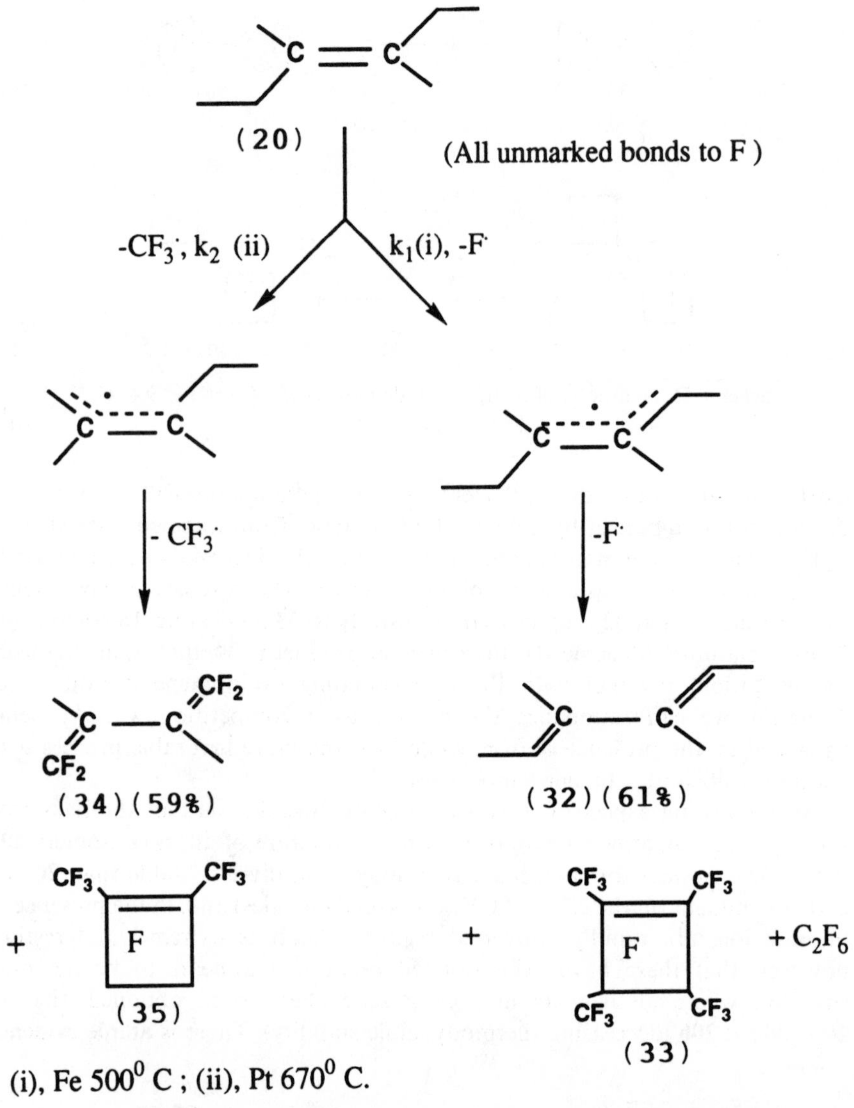

(All unmarked bonds to F)

(20)

$-CF_3^{\cdot}$, k_2 (ii) ; k_1(i), $-F^{\cdot}$

$-CF_3^{\cdot}$; $-F^{\cdot}$

(34) (59%) ; (32) (61%)

+ (35) + (33) + C_2F_6

(i), Fe 500° C ; (ii), Pt 670° C.

Scheme III. Pyrolysis reactions of *F*-ethene tetramer (20).

in the literature to illustrate that perfluorinated alkenes containing a terminal difluoromethylene group (**20b**) are more reactive toward nucleophiles than those with an internal double bond (**20a**). Also there is evidence to indicate that systems with a vinylic fluorine atom (**20a**) are more reactive than systems with four perfluoroalkyl groups attached to the double bond (**20**). The compounds

20–**20b** illustrate this reactivity order very clearly, suggesting that we may usually assume a reactivity order $(R_f)_2C{=}C(R_f)_2$ < $(R_f)_2C{=}CFR_f$ ≪ $(R_f)_2C{=}CF_2$ (where R_f = perfluoroalkyl). Thus, reaction of **20b** with neutral methanol occurred rapidly, even at room temperature, while reaction of **20a** occurred only at reflux and no reaction was detectable, with **20** under these conditions.

16.5.2. Fluoride-Ion Induced Reactions of Dienes

F-cyclobutene (**26**) is in equilibrium with *F*-butadiene (**36**) at high temperatures, although it is well known[18] that the equilibrium lies in the direction of the cyclic compound **26**. This equilibrium can, however, be displaced by the reaction with fluoride ion, leading to hexafluoro-2-butyne (**37**),[19] Scheme IV. In a related

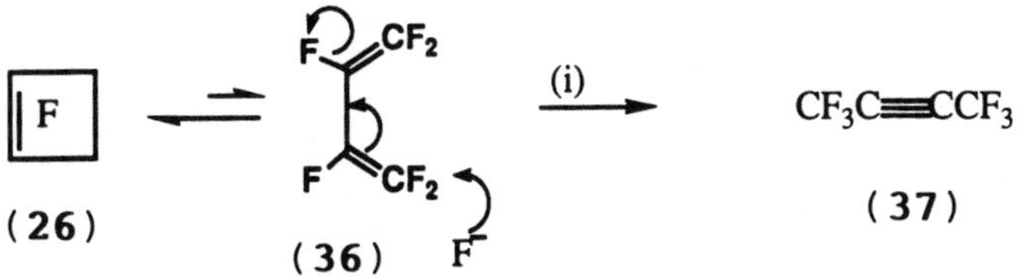

(26) ⇌ (36) —(i)→ $CF_3C{\equiv}CCF_3$ (37)

(i), CsF or KF, 510-590° C, flow system in N_2

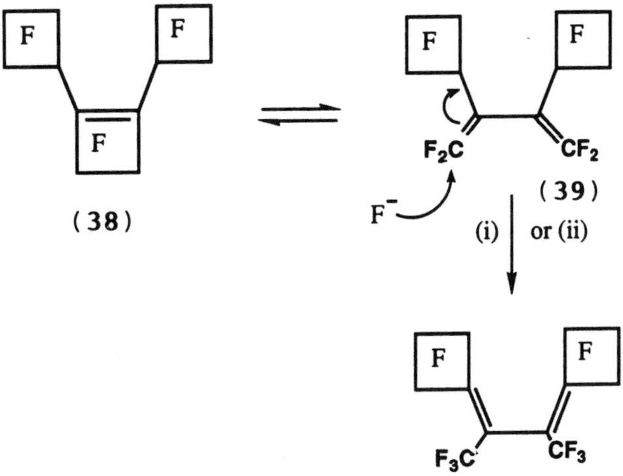

(38) ⇌ (39) —(i) or (ii)→ (40)

(i), CsF, 510° C, flow in N_2 (70%); (ii), CsF or KF, sealed tube, 300° C (quant.)

Scheme IV. Fluoride-ion induced rearrangements.

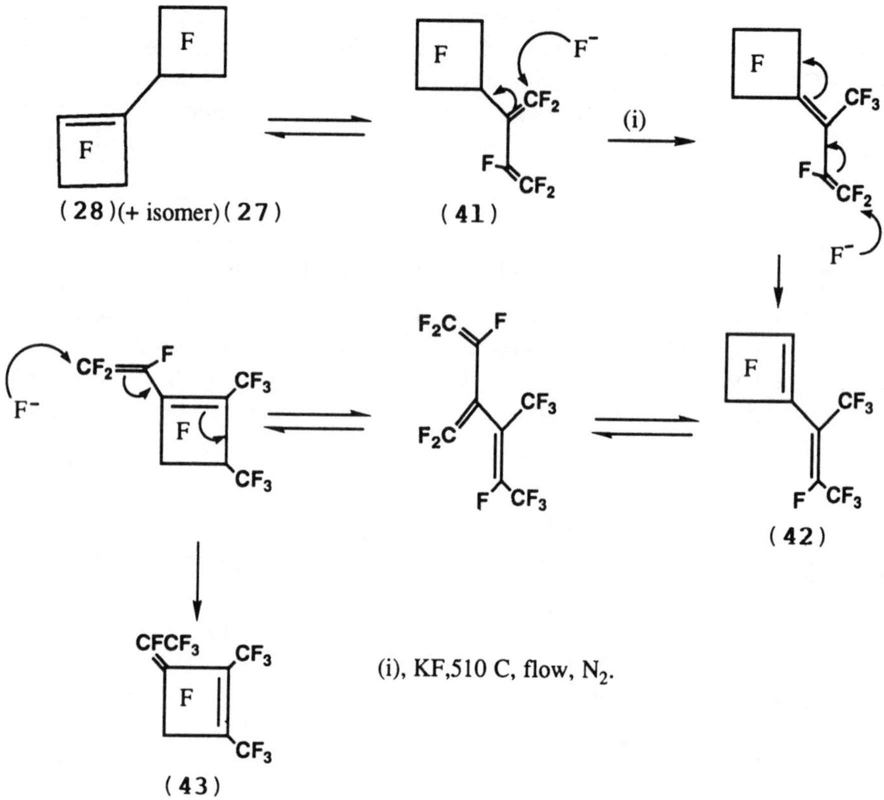

(28)(+ isomer)(27) (41) (42) (43)

(i), KF,510 C, flow, N_2.

Scheme V. A novel rearrangement!

manner, an unusual diene (**40**) has been obtained by passage of **38** over cesium fluoride, and the process almost certainly involves reaction of fluoride ion with an intermediate diene (**39**).

Probably the most remarkable diene-forming reaction is that involved in the formation of **43** from a mixture of **27** and **28**, Scheme V. In this process, all of the carbon atoms are originally forming four-membered rings in **28** while in the product (**43**), four of these carbon atoms are contained in trifluoromethyl groups! Nevertheless, a reasonable mechanism for this surprising conversion can be envisaged,[19] as shown in Scheme V, which involves a series of ring openings, for example, to **41**, rearrangements, for example, to **42**, and eventually leading to **43**.

Unusually, the diene (**34**) is converted by cesium fluoride to a mixture of dimers (**34** and **34b**), [19a] which can be rationalized as shown in Scheme VI.

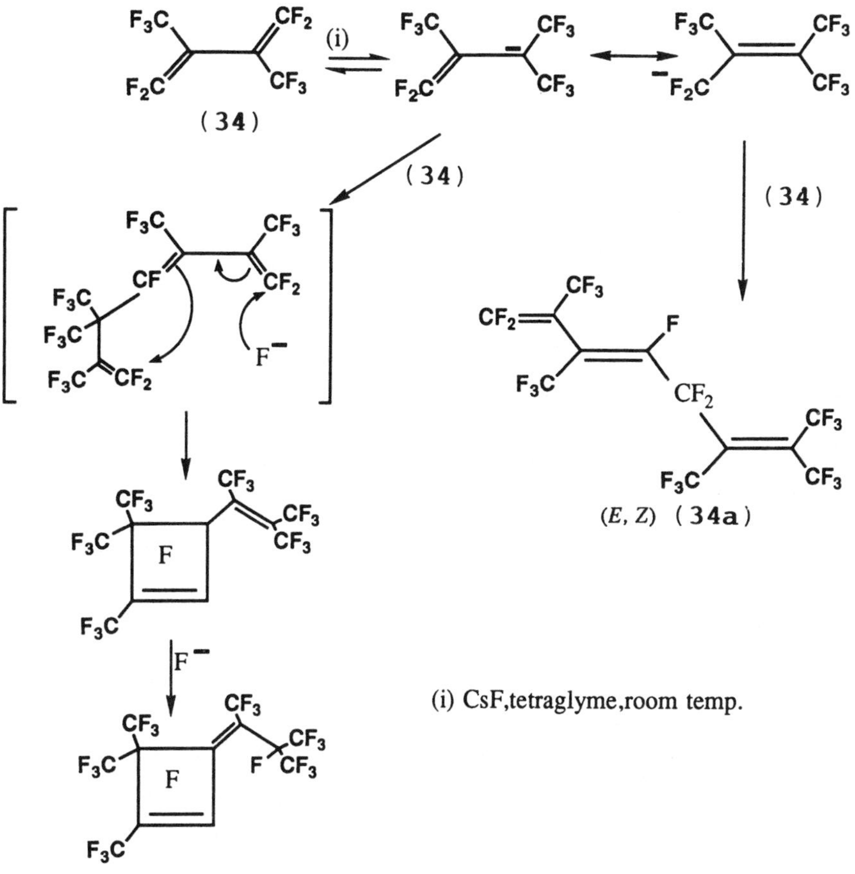

(i) CsF,tetraglyme,room temp.

Scheme VI. Fluoride-ion induced dimerization of *F*-2,3-dimethylbutadiene (**34**).

16.5.3. Electrochemical and Sodium Amalgam-Induced Defluorination

Electrochemical defluorination of *F*-cyclohexadiene (**44a** and **44b**) has been described by Tatlow and his co-workers,[26] which results in hexafluorobenzene (**45**). *F*-cyclohexene also gives hexafluorobenzene in an analogous process.[21]

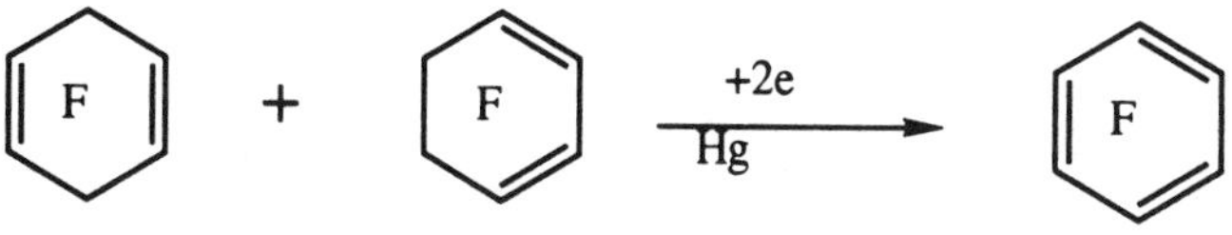

(**44a**) + (**44b**) $\xrightarrow[\text{Hg}]{+2e}$ (**45**)

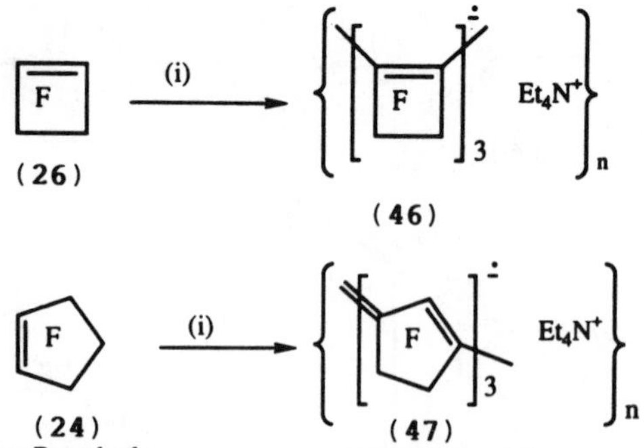

(26) → (i) → (46)

(24) → (i) → (47)

(i),Hg or Pt cathode. Blue-black solids.

Remarkably, electrochemical reductions of *F*-cyclobutene (**26**) and *F*-cyclopentene (**24**) give conjugated polymers that have been assigned structures **46** and **47**, respectively.[22] No simple dienes could be identified from these reactions but we have also explored electrochemistry for reductive defluorination of **20** and some success was achieved, giving diene **32**. However, preparative-scale electrochemistry

$$\mathbf{20} \xrightarrow{(i)(ii)} \mathbf{32}$$

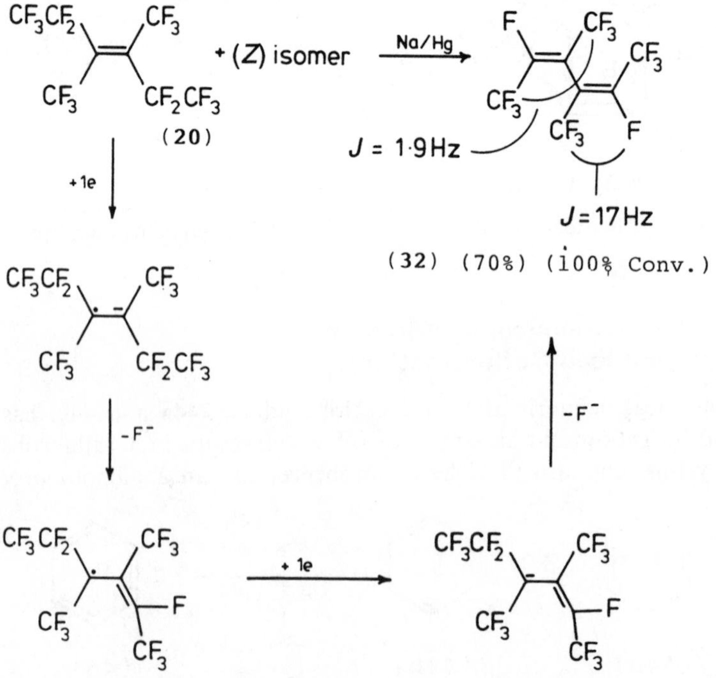

(20) + (*Z*) isomer —Na/Hg→ (32) (70%) (100% Conv.)

J = 1·9Hz; J = 17Hz

+1e; $-F^-$; +1e; $-F^-$

Scheme VII. Electromimetic reduction (defluorination).

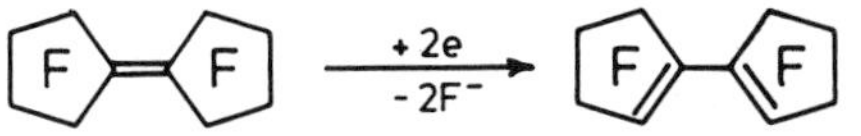

Scheme VIII. Defluorinations with sodium amalgam.[23]

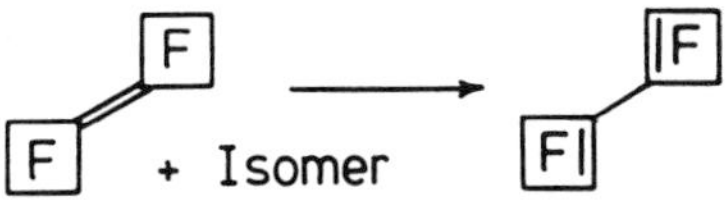

(i) divided Hg cell, CH_3CN, Et_4NBF_4 electrolyte, SCE.
(ii) About 1.5 V.

has its problems and we considered alternative electromimetic procedures. In particular, the use of mercury amalgam was explored and found to be surprisingly effective,[23] Scheme VII. Diene **32** is isolated as the *only* component of the product, while **30** and **31** are best obtained, Scheme VIII, from partial conversions and then the dienes are simply crystallized from excess of the starting *F*-bicycloalkylidines **25** and **27**, by cooling.

These dienes (**30**–**32**) will react further with sodium amalgam, if excess of the latter is used and it is somewhat puzzling that the dienes are isolated at all, from this powerfully reducing system, let alone in such high yield. However, we currently attribute this selectivity to a significant difference in reduction potentials (~ 1 V) between **20**, **25**, and **27** and their corresponding dienes.

It is quite reasonable to view both the electrochemical and the sodium–amalgam reductions as one-electron transfer processes, as outlined in Scheme VII, giving radical anions that subsequently lose fluoride ion and then further reduction of the subsequent radical allows a net defluorination to occur. Under the conditions used, tetramer (of *F*-ethene) **20** is converted to the all-trans isomer of the diene **32**, as indicated by ^{19}F NMR.

These studies have therefore presented us with a set of new electron deficient dienes to explore and this type of situation is one of the delights of the field of organofluorine chemistry. The following sections present some preliminary studies of these highly interesting systems.

16.6. REACTIONS OF NEW DIENES

A possibility that dienes (for example, **31**) could be intermediates in the process of electropolymerization of *F*-cyclobutene (**26**) or *F*-cyclopentene (**24**) (see above) can be discounted since no polymer was obtained on electrochemical reduction of either dienes **30** or **31**.[21]

16.6.1. Reactions with Nucleophiles

The dienes **30**–**32** are essentially electronically equivalent but differ massively in reactivity towards methanol.[23] The bicyclobutenyl derivative **31** is astoundingly reactive, Scheme IX, in that a spontaneous exothermic reaction occurs with

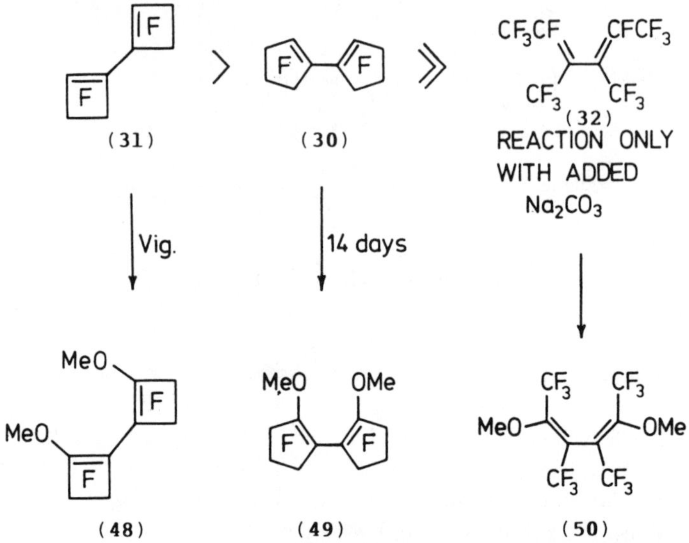

(31) > (30) ≫ $CF_3CF{=}C(CF_3)C(CF_3){=}CFCF_3$ (32)

REACTION ONLY WITH ADDED Na_2CO_3

Vig. — 14 days

MeO, MeO (48) — MeO, OMe (49) — MeO, OMe, CF_3, CF_3, CF_3, CF_3 (50)

Scheme IX. Reactions with neutral methanol (room temperature).

neutral methanol, on mixing at room temperature; the bicyclopentenyl **30** requires a long period at room temperature, while **32** did not react in the absence of a base.

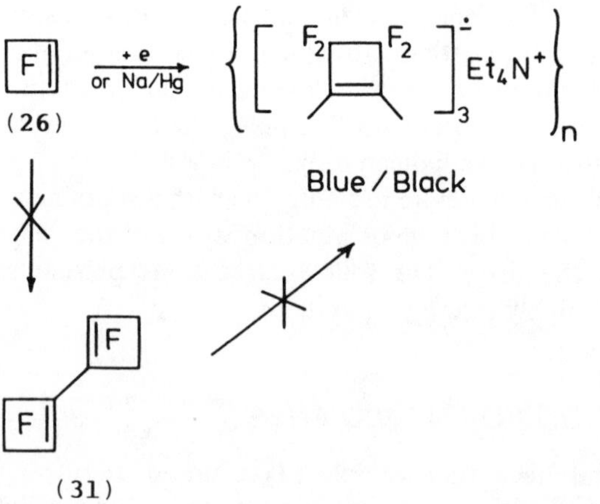

(26) $\xrightarrow[\text{or Na/Hg}]{+e}$ $\{[F_2\ F_2]^{\dot{-}}_3\ Et_4N^+\}_n$

Blue / Black

(31)

The dienes **30**–**32** are electronically very similar in that they can all be regarded, approximately, as *F*-tetra-alkylbutadienes (that is, $[R_fCF{=}C(R_f)]_2$, where R_f = perfluoroalkyl) and so, the extreme reactivity of **31** can be attributed to relief of angle strain during the formation of an intermediate carbanion, in addition of a nucleophile. This is a simple and yet spectacular illustration of the effect of angle strain on reactivity. It is to be noted that disubstitution products **48**–**50** are obtained in each case, implying that there is little interaction between the double bonds during the nucleophilic substitution process. Therefore, this presents quite a contrast with reaction of **34**, which gives a mixture containing only monosubstitution **51** and monoaddition **52** products.[24]

CF_3 CF_2 / CF_2 CF_3 (34) —MeOH→ CF_3 $CFOMe$ / CF_2 CF_3 (51) + CF_3 CF_2OMe H / CF_2 CF_3 (52)

16.6.2. Formation of Heterocycles

16.6.2.A. Furan Derivatives. Diene **32** is an excellent reactant for forming heterocycles. Hydrolysis of **32** occurs quantitatively to give the known furan derivative **53**. Indeed, it is difficult to avoid forming some of the furan unless solvents are kept scrupulously dry. Obviously, for the cyclization to proceed, there must be interconversion of isomers and we can think of alternate processes for forming **53**, as shown in Scheme X.

$[CF(CF_3){=}C(CF_3)]_2$ (32) —(i)→ CF_3 CF_3 CF_3 S CF_3 (54) + Furan

(30) —(i), R.T.→ (55) δ_F 102, 106, 124 ppm

(31) —(ii)→ INTERMOLECULAR REACTIONS ?

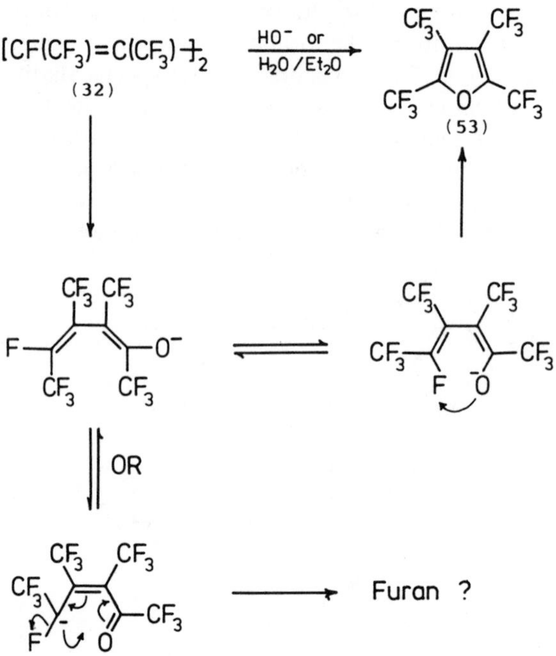

$[CF(CF_3){=}C(CF_3)]_2$ (32) $\xrightarrow[H_2O/Et_2O]{HO^- \text{ or}}$ (53)

Furan ?

Scheme X. Cyclization to form a furan derivative.

16.6.2.B. Thiophene Derivatives. An analogous process occurs with potassium sulfide to form the corresponding thiophene derivative (**54**), which had previously been obtained from the less accessible (to us) hexafluoro-2-butyne.[25,26] A thiophene derivative (**55**) has been obtained from the bicyclopentene derivative **30**, albeit in low yield, but the bicyclobutene derivative **31** gave only polymeric material where, apparently, a rapid intermolecular process competes with the sterically more strained intramolecular process, required for formation of a thiophene derivative.

16.6.2.C. Pyrrole Derivatives. We can also form a pyrrole derivative (**56**);[23] cesium fluoride is added to the system to inhibit loss of fluorine from an adjacent trifluoromethyl in the intermediate **58**. In the absence of cesium fluoride, the remarkable pyrrolo–quinoline product **57** is obtained along with the pyrrole derivative **56** and the formation of **57** is rationalized, as shown in Scheme XI. The pyrrole derivative (**56**) has previously been obtained[27] but in an approach that involved a valence isomer of *F*-tetramethylthiophene (**54**).

16.6.3. Formation of Cyclopentadienyl Derivatives

In procedures related to formation of the heterocycles described above, we have developed an analogous procedure for synthesis of a variety of otherwise

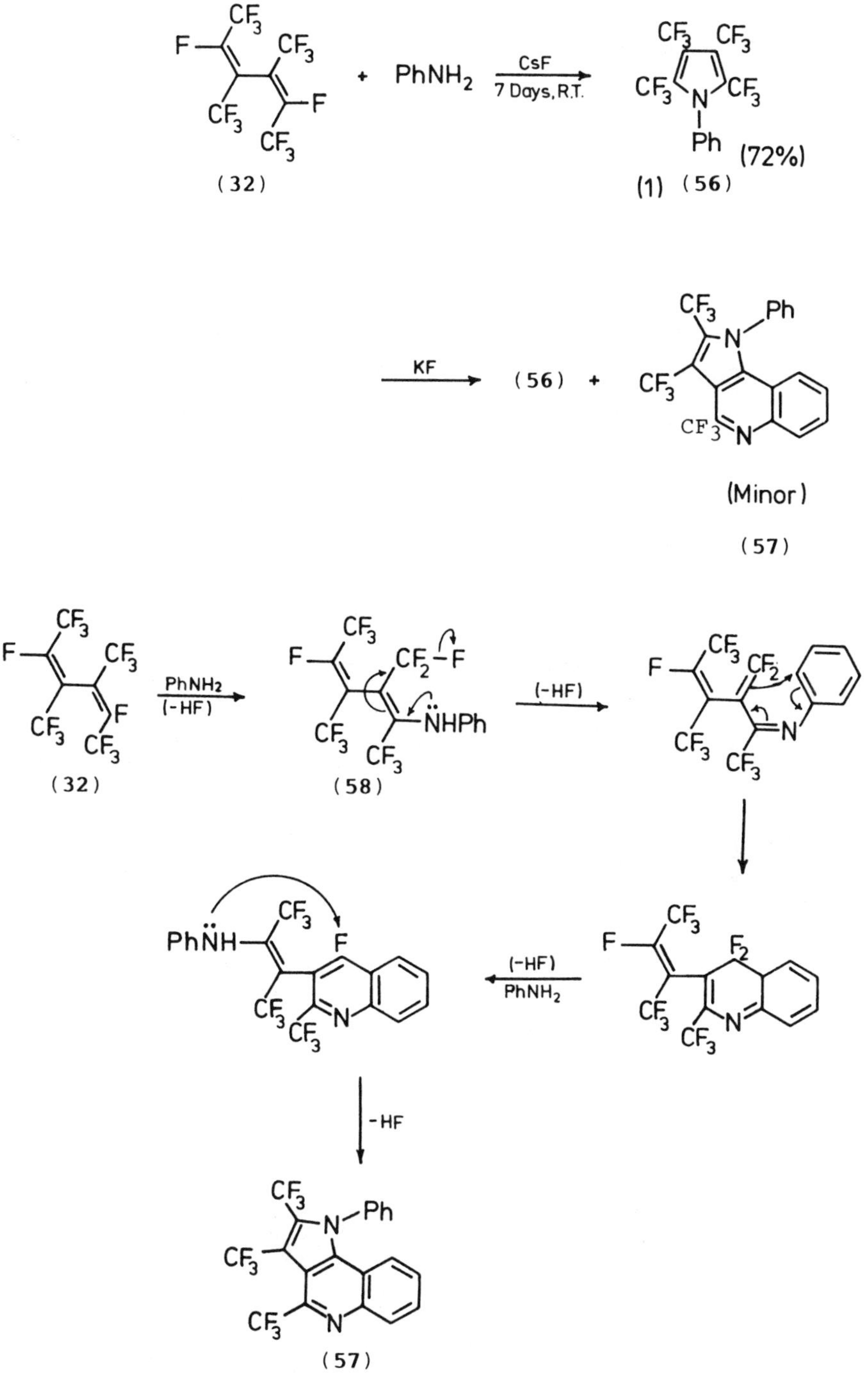

F CF3 CF3 CF3 F CF3 (32) + PhNH2 —CsF, 7 Days, R.T.→ CF3 CF3 CF3 N CF3 Ph (72%) (1) (56)

—KF→ (56) + CF3 N Ph CF3 CF3 N (Minor) (57)

F CF3 CF3 CF3 F CF3 (32) —PhNH2 (−HF)→ F CF3 CF2–F NHPh CF3 CF3 (58) —(−HF)→ F CF3 CF2 N CF3 CF3

F CF3 F2 N CF3 CF3 —(−HF) PhNH2→ PhNH CF3 F CF3 N CF3

—−HF→ CF3 N Ph CF3 CF3 N (57)

Scheme XI. Formation of pyrrolo–quinoline derivatives.

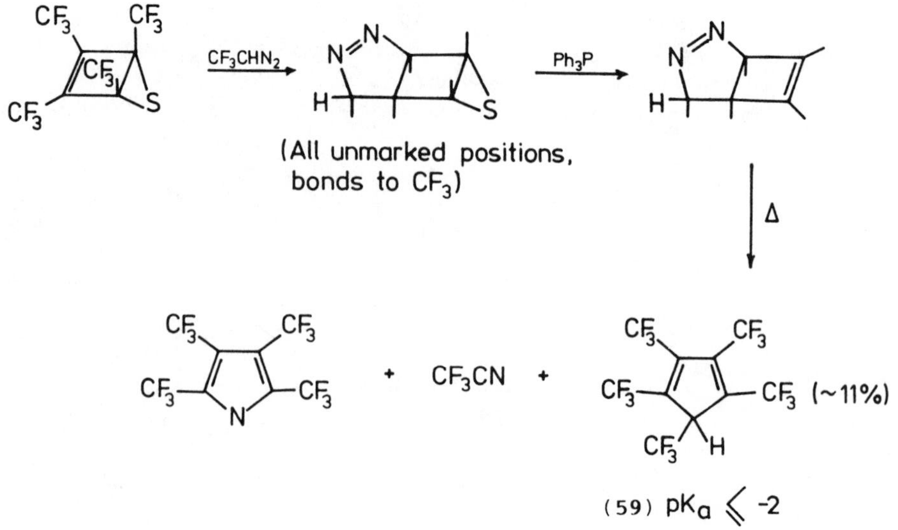

CF_3 CF_3 CF_3 CF_3 S

CF_3CHN_2

N=N H S

Ph_3P

N=N H

(All unmarked positions, bonds to CF_3)

Δ

CF_3 CF_3 CF_3 CF_3 N

+ CF_3CN +

CF_3 CF_3 CF_3 CF_3 (~11%) CF_3 H

(59) pK_a ≤ -2

Scheme XII. Synthesis of poly(trifluoromethyl) cyclopentadienes.[28]

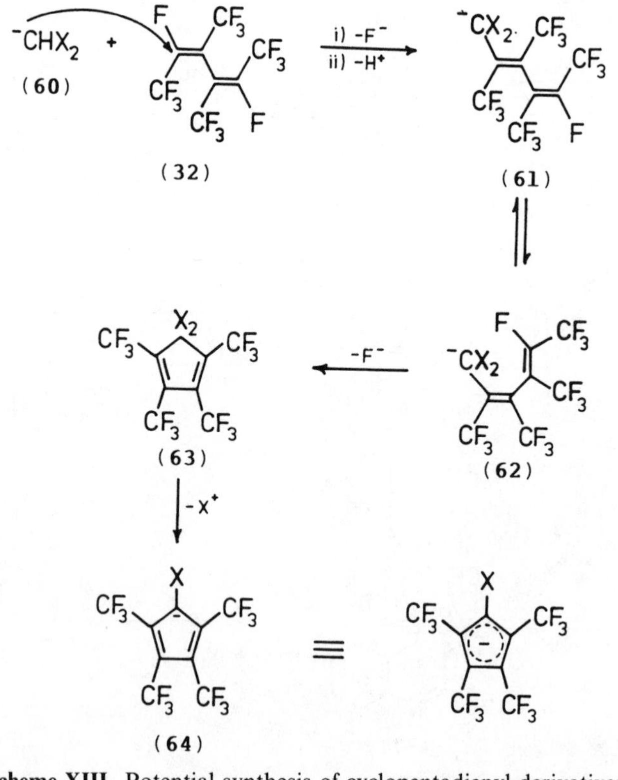

$^-CHX_2$ + F CF_3 CF_3 CF_3 CF_3 F

(60)

(32)

i) $-F^-$
ii) $-H^+$

$^-CX_2$ CF_3 CF_3 CF_3 CF_3 F

(61)

F CF_3 $^-CX_2$ CF_3 CF_3 CF_3

(62)

$-F^-$

X_2 CF_3 CF_3 CF_3 CF_3

(63)

$-X^+$

X CF_3 CF_3 CF_3 CF_3 ≡ X CF_3 CF_3 CF_3 CF_3

(64)

Scheme XIII. Potential synthesis of cyclopentadienyl derivatives.

relatively inaccessible cyclopentadienyl anions, derived from **32**. Laganis and Lemal[28] previously reported the extremely acidic diene **59**, obtained by an imaginative but low yield route, Scheme XII, and observed the derived cyclopentadienyl anion in solution. Also, Janulis and Arduengo[29] obtained derivatives using related methodology. However, a general approach to synthesis of various cyclopentadienyl systems could be envisaged, as shown in Scheme XIII. The approach involves reaction with a potentially difunctional carbon acid (**60**) with the diene (**32**) first to give the anion (**61**). It is known that barriers to rotation on allyl, *F*-allyl, and pentadienyl anions are low[30] and so interconversion between isomers **61** and **62** is to be anticipated. Then, if intramolecular internal vinylic displacement could occur, we would have formed cyclopentadienyl systems (**63**).[31] Furthermore, displacement of a group X from the system, would give the corresponding cyclopentadienyl salts (**64**).

A potential problem in this process would be the competitive reaction of base, used to produce anions (**60**), with the diene (**32**). However, it is now established that carbanions may be generated from suitably acidic systems using fluoride ion,[32] and the latter has been effectively used to generate the anions (**48**). Indeed, a variety of cyclopentadienyl systems have now been obtained.

16.6.3.A. From Malononitrile.[31] With malononitrile (Scheme XIV) we first see two discernible dienyl anions **64** and **65** formed after 30 min in the ratio

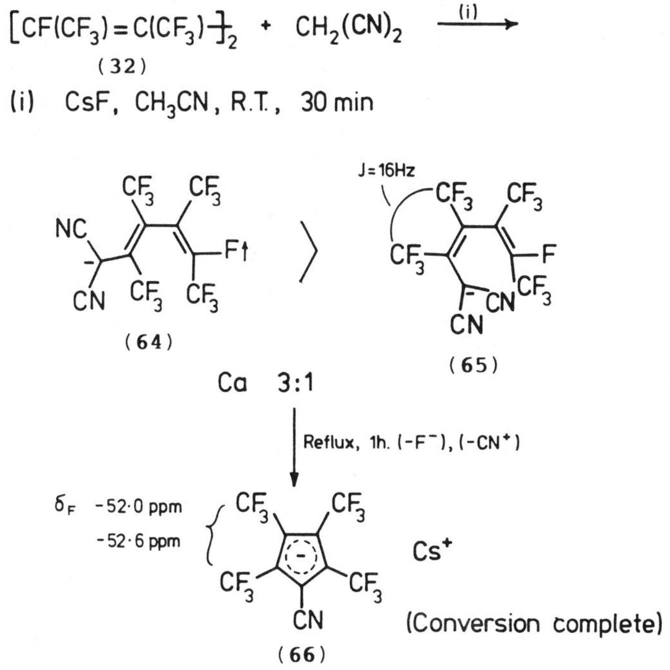

$[CF(CF_3)=C(CF_3)]_2 + CH_2(CN)_2 \xrightarrow{(i)}$

(32)

(i) CsF, CH_3CN, R.T., 30 min

(64) Ca 3:1 (65) J = 16Hz

Reflux, 1h. ($-F^-$), ($-CN^+$)

δ_F −52·0 ppm, −52·6 ppm; Cs^+ (66) (Conversion complete)

Scheme XIV. Reactions of **32** with $CH_2(CN)_2$.

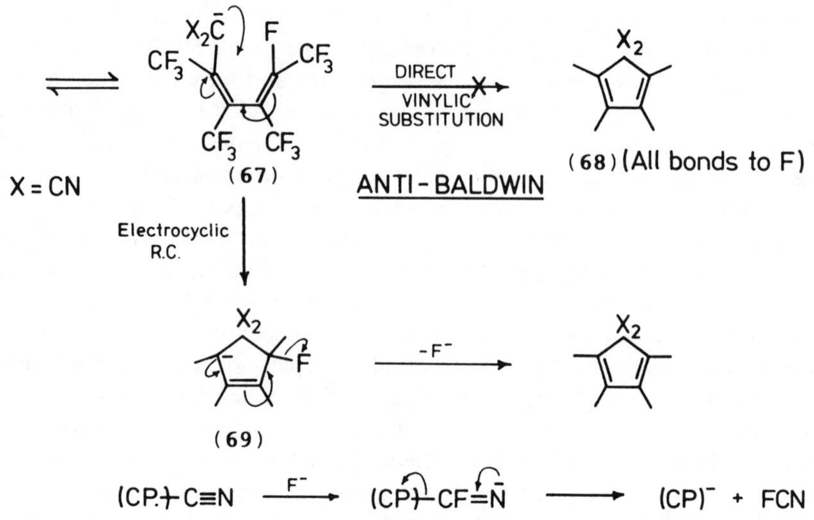

X = CN

(67)

DIRECT VINYLIC SUBSTITUTION

ANTI-BALDWIN

(68) (All bonds to F)

Electrocyclic R.C.

(69)

$-F^-$

(CP)–C≡N $\xrightarrow{F^-}$ (CP)–CF=N$^-$ ⟶ $(CP)^-$ + FCN

Scheme XV. Mechanism of ring closure.

about 3:1, favoring the trans isomer. However, ring closure on heating occurs but, surprisingly, this is accompanied by loss of the cyanide ion, to give directly the cesium cyclopentadienyl salt (**66**). Note that the chemical shift values are downfield from the 60-ppm region where we anticipate $CF_3C{=}C$ to occur.[33] This is consistent with the fact previously established that signals arising from fluorine attached to carbon atoms that are adjacent to a carbanionic center (that is, $F{-}C{-}\bar{C}$) are moved downfield, while those from fluorine atoms attached to the carbanion ($F{-}\bar{C}$) are moved upfield.[10] It is unlikely that direct intramolecular vinylic substitution (**67**–**68**) occurs, Scheme XV, but rather, that a process of electrocyclic ring closure (**67** to **69**) followed by loss of fluoride ion occurs. Analogous electrocyclic processes are very well known, leading to heterocycles,[34] but we are aware of only one probable example, leading to a carbocyclic system.[34,35] Loss of cyanide ion is easily appreciated, as an addition–elimination process and especially when the leaving group is such a stable anion, that is, a highly fluorinated cyclopentadienyl anion.

16.6.3.B. From Malonic Ester. Reaction with malonic ester occurs in a similar manner, Scheme XVI, but in this situation acyclic intermediates were not observed. Instead, the diene (**70**) was observed. However, on reflux, complete conversion to the cyclopentadienyl salt (**71**) was achieved and in this case, the process involves a fluoride induced decarboxylation step, Scheme XVII, promoted by such an effective leaving group.

In conclusion, it is hoped that this discussion has illustrated a number of features: (a) that the theories of organic chemistry can cope with the additional complexities of fluorine substitution, thus integrating fluorine compounds into

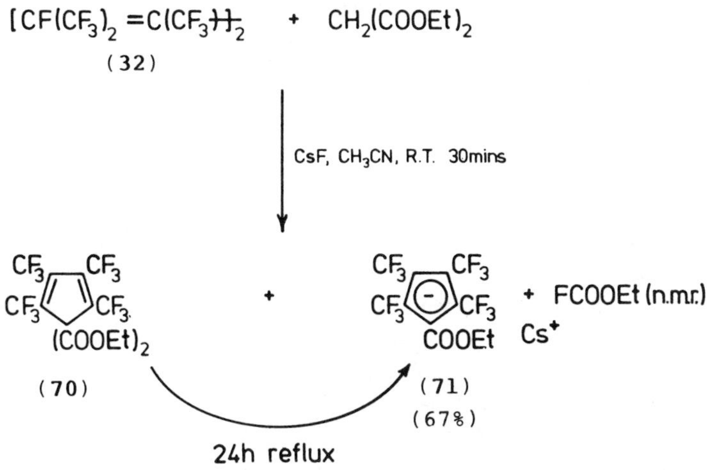

Scheme XVI. Reaction with $CH_2(COOEt)_2$.

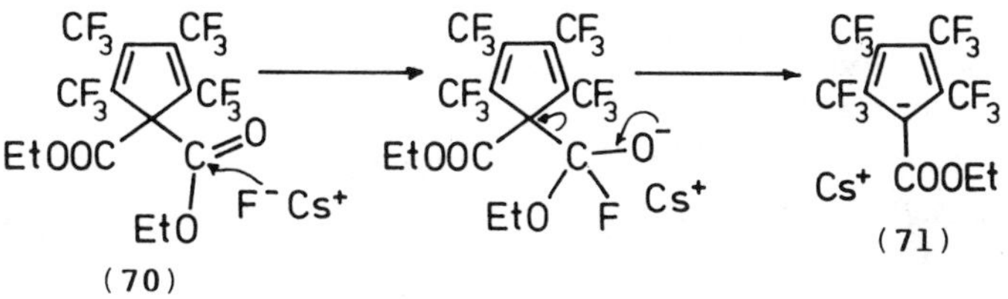

Scheme XVII. F^--induced decarboxylation.

the main stream of organic chemistry; (b) that fluorinated compounds have a fascinating chemistry; (c) that the latter provide excellent examples for demonstrating many of the principles of organic chemistry and should be used more often in basic textbooks to do so.

REFERENCES

1. Chambers, R. D., in *Fluorine in Organic Chemistry*, Wiley-Interscience, New York, 1973.
2. Chambers R. D. and Mobbs, R. H., *Adv. Fluorine Chem.*, *4*, 50 (1965).
3. Chambers, R. D. and Bryce, M. R., in *Comprehensive Carbanion Chemistry, Part C* (E. Buncel and T. Durst, Eds.), Elsevier, Amsterdam, 1987, 271.
4. Andreades, S., *J. Am. Chem. Soc.*, *86*, 2003 (1964).
5. Holtz D., *Prog. Phys. Org. Chem.*, *8*, 1 (1971).
6. Bryce, M. R., Chambers, R. D., and Taylor, G., *J. Chem. Soc. Perkin Trans.*, *1*, 509 (1984).

7. Graham, D. P., *J. Org. Chem.*, *31*, 955 (1966).
8. Fielding, H. C. and Rudge, A. J., *Br. Pat.* 1,082,127 1967.
9. Chambers, R. D., Jackson, J. A., Partington, S., Philpot, P. D., and Young, A. C., *J. Fluorine Chem.*, *6*, 18 (1975).
10. Bayliff, A. E., Bryce, M. R., Chambers, R. D., and Matthews, R. S., *J. Chem. Soc. Perkin Trans.*, *1*, 201 (1988).
11. Chambers, R. D., Taylor, G., Powell, R. L., *J. Chem. Soc. Perkin Trans.*, *1*, 429 (1980).
12. Pruett, R. L., Bahner, C. T., and Smith, H. A., *J. Am. Chem. Soc.*, *74*, 1638 (1952).
13. Coe, P. L., Patrick, C. R., and Tatlow, J. C., *Tetrahedron*, *9*, 240 (1960).
14. Harrison, D., Stacey, M., Stephens, R., and Tatlow, J. C., *Tetrahedron*, *19*, 1893 (1963).
15. Chambers, R. D. and Marper, E., unpublished observations.
16. Chambers, R. D., Lindley, A. A., Fielding, H. C., Moilliet, J. S., and Whittaker, G., *J. Chem. Soc. Perkin Trans.*, *1*, 1064 (1981).
17. Chambers, R. D., Lindley, A. A., and Fielding, H. C., *J. Fluorine Chem.*, *12*, 85 (1978).
18. Cheswick, J. P., *J. Am. Chem. Soc.*, *88*, 4800 (1966).
19. Chambers, R. D., Jones, C. G. P., Taylor, G., *J. Chem. Soc. Chem. Commun.*, 964 (1979).
19. (a) Chambers, R. D., Lindley, A. A., and Fielding, H. C., *J. Chem. Soc. Perkin Trans. 1*, 939 (1981).
20. Doyle, A. M., Pedler, A. E., and Tatlow, J. C., *J. Chem. Soc. (C)*, 2740 (1968). Doyle, A. M. and Pedler, A. E., *J. Chem. Soc. (C)*, 282 (1971).
21. Chambers, R. D. and Briscoe, M., unpublished observations.
22. Briscoe, M. W., Chambers, R. D., Silvester, M. J., and Drakesmith, F. G., *Tetrahedron Lett.*, *29*, 1295 (1988).
23. Briscoe, M. W., Chambers, R. D., Mullins, S. J., Nakamura, T., and Drakesmith, F. G., *J. Chem. Soc. Commun.*, 1127 (1990).
24. Bryce, M. R., Chambers, R. D., Lindley, A. A., and Fielding, H. C., *J. Chem. Soc. Perkin Trans. 1*, 2541 (1983).
25. Krespan, C. G., *J. Am. Chem. Soc.*, *83*, 3434 (1961).
26. Chambers, R. D., Jones, C. G. P., Silvester, M. J., and Speight, D. B., *J. Fluorine Chem.*, *25*, 47 (1984).
27. Kobayashi, Y., Ando, A., Kawada, K., Ohsawa, A., and Kumadaki, I., *J. Org. Chem.*, *45*, 2962 (1980).
28. Laganis, E. D. and Lemal, D. M., *J. Am. Chem. Soc.*, *102*, 6633 (1980).
29. Janulis Jr., E. P. and Ardvengo III, A. J., *J. Am. Chem. Soc.*, *105*, 3563 (1983).
30. Thompson, T. B. and Ford, W. T., *J. Am. Chem. Soc.*, *101*, 5459 (1979) and references contained. Farnham, W. B., Middleton, W. J., Fultz, W. C., and Smart, B. E., *J. Am. Chem. Soc.*, *108*, 3125 (1986).
31. Chambers, R. D. and Greenhall, M. P., *J. Chem. Soc. Chem. Commun.*, 1128 (1990).
32. Clark, J. H., *Chem. Rev.*, *80*, 429 (1980).
33. Chambers, R. D., Lindley, A. A., Philpot, P. D., Fielding, H. C., and Hutchinson, J., *Israel J. Chem.*, *17*, 150 (1978).
34. Huisgen, R., *Angew Chem. Int. Ed. Engl.*, *19*, 947 (1980).
35. Shoppee, C. W. and Henderson, G. N., *J. Chem. Soc. Perkin Trans. 1*, 765 (1975).

CHAPTER 17

Fluorinated Condensation Monomers

KURT BAUM

17.1. INTRODUCTION

Research on functionalized fluorocarbons was initiated at Fluorochem in 1978 with the objective of studying the synthesis and polymerization of perfluoroalkylene diacetylenes, $HC{\equiv}C(CF_2)_nC{\equiv}CH$. Acetylenes were known to polymerize to give thermally stable structures, which would be expected to provide useful linking groups for the fluorocarbon segments. Diacetylenes of this type were unknown, and several synthetic schemes for these materials were developed.[1] A practical method consisted of the thermal addition of perfluoroalkylene diiodides to bis(trimethylsilyl)acetylene to give the trimethylsilyl acetylenes, followed by desilylation with fluoride:

$$I(CF_2)_nI + Me_3SiC{\equiv}CSiMe_3 \xrightarrow[-SiMe_3I]{\Delta}$$

$$Me_3SiC{\equiv}C(CF_2)_nC{\equiv}CSiMe_3 \xrightarrow{KF} HC{\equiv}C(CF_2)_nC{\equiv}CH$$

Synthetic Fluorine Chemistry,
Edited by George A. Olah, Richard D. Chambers, and G. K. Surya Prakash.
ISBN 0-471-54370-5

Alternatively, it was found that addition of the perfluoroalkylene diiodides to trimethylsilylacetylene, followed by the elimination of HI from the resulting dialkene provided another route to the desired diacetylenes.

$$I(CF_2)_nI + HC{\equiv}CSiMe_3 \longrightarrow$$

$$(Me_3Si)IC{=}CH(CF_2)_nHC{=}CI(SiMe_3) \xrightarrow{DBU} HC{\equiv}C(CF_2)_nC{\equiv}CH$$

The diiodides needed for the synthesis of these diacetylenes were not commercially available, although Haszeldine[2] reported in a communication in 1951 that the materials can be prepared by the telomerization of tetrafluoroethylene with iodine. We developed this preparation to give perfluoroalkylene diiodides in kilogram quantities.[3] Tetrafluoroethylene was available sporadically in commercial cylinders at a high price and in an inconvenient stabilized form. Conditions for cracking polytetrafluoroethylene (PTFE) to the monomer without the formation of toxic by-products have been reported.[4] We have extended this reaction for the generation of kilogram quantities of tetrafluoroethylene by the vacuum pyrolysis of scrap PTFE.[5]

$$(CF_2CF_2)_n \xrightarrow{\Delta} CF_2{=}CF_2$$

$$CF_2{=}CF_2 + I_2 \longrightarrow I(CF_2CF_2)_nI \qquad n = 1\text{–}5$$

Fluorinated difunctional materials have applications in structural polymers, and polymers containing fluorine and nitro groups have potential applications as energetic binders. Access to these diiodides in large laboratory quantities facilitated research toward these applications.

17.2. FLUORINATED DIISOCYANATES

Fluorinated diisocyanates of the general structure

$$OCN(CH_2)_2(CF_2)_n(CH_2)_2NCO$$

were prepared from α,ω-diiodoperfluoroalkanes by insertion of ethylene into the CF_2—I bond, displacement of the iodo groups with azide, and reduction of the resulting azide groups to amines. Phosgenation of the diamine gave the diisocyanates.[6]

Perfluorinated diisocyanates, $OCN(CF_2)_nNCO$, have been prepared by the Curtius rearrangement of perfluorinated diacyl azides[7] or by a modified Lossen rearrangement involving pyrolysis of the silyl esters of perfluorohydroxamic acids.[8] These diisocyanates, however, are extremely sensitive to moisture and form polyurethanes that are hydrolytically and thermally unstable.

Fluorinated diisocyanates in which the isocyanate functionality is separated from the fluorinated segment by one intervening methylene group, $OCNCH_2\text{-}(CF_2)_n\text{-}CH_2NCO$ were prepared by either phosgenation of $\alpha,\alpha,\omega,\omega$-tetrahydroperfluoroalkylene diamines[7] or by the Curtius rearrangement of $\alpha,\alpha,\omega,\omega$-tetrahydroperfluorodiacyl azides.[9] Hollander and his co-workers[7] observed that $\alpha,\alpha,\omega,\omega$-tetrahydroperfluoroalkylene diisocyanates, in particular hexafluoropentamethylene diisocyanate, existed as a dimer and attempts to prepare high molecular weight polymers from this diisocyanate were unsuccessful; low melting solids or oils were obtained.

Fluorinated diisocyanates with two intervening methylene groups, $OCNCH_2CH_2\text{-}(CF_2)_n\text{-}CH_2CH_2NCO$, would be expected to be comparable in reactivity with aliphatic isocyanates and should be useful for preparation of fluorinated condensation polymers, such as poly(urethanes) and poly(ureas).[10]

α,ω-Diiodoperfluoroalkanes (**1a–c**), obtained from the telomerization of tetrafluoroethylene with iodine, were converted to ethylene adducts, **2a–c**, in high yields.[1,11] Displacement of the iodo groups with azide, reduction of azide groups to amines, and phosgenation of the diamine gave $OCNCH_2CH_2(CF_2)_nCH_2CH_2NCO$.

$$\underset{\textbf{1a-c}}{I\text{-}(CF_2)_n\text{-}I} + CH_2{=}CH_2 \xrightarrow{180°C}$$

$$\underset{\textbf{2a-c}}{ICH_2CH_2\text{-}(CF_2)_n\text{-}CH_2CH_2I} \xrightarrow[DMSO/25°C]{NaN_3}$$

$$\underset{\textbf{3a-c}}{N_3CH_2CH_2\text{-}(CF_2)_n\text{-}CH_2CH_2N_3} \xrightarrow[Pd(OH)_2,\ CH_3OH/\Delta]{NH_2NH_2}$$

$$\underset{\textbf{4a-c}}{H_2NCH_2CH_2\text{-}(CF_2)_n\text{-}CH_2CH_2NH_2} \xrightarrow[2.\ COCl_2]{1.\ HCl}$$

$$\underset{\textbf{5a-c}}{OCNCH_2CH_2\text{-}(CF_2)_n\text{-}CH_2CH_2NCO}$$

a: $n = 4$; **b**: $n = 6$; **c**: $n = 8$

Monofunctional fluorinated amines, $R_fCH_2CH_2NH_2$, were previously prepared by the reaction of $R_fCH_2CH_2I$ with ammonia[12] or by a two-step procedure[13] involving reaction of $R_fCH_2CH_2I$ with sodium azide followed by reduction of the azide. Our attempts to prepare 1,8-diamino-1,1,2,2,7,7,8,8-oxtahydroperfluorooctane (**4a**) by the Gabriel synthesis[14] or by the reaction of the iodo compound with ammonia gave mainly HI elimination products. Consequently, fluorinated diamines were prepared by the two-step approach involving diazides.

Displacement reactions of **2a–c** with sodium azide in 98% dimethyl sulfoxide (DMSO)–H_2O gave the diazides **3a–c** in 92–98% yields. High yields of **3a** were obtained in 98% DMSO–H_2O or 95% acetonitrile–water. In DMSO–H_2O the reaction proceeded at room temperature and reached completion in 4 h, whereas in acetonitrile–water, 16 h at reflux were needed. In moist alcoholic solvents, such as diethylene glycol, *tert*-butanol, *sec*-butanol, and *iso*-propanol, high temperatures (82–120 °C) and long reaction periods (48–60 h) were needed to reach completion. In *N,N*-dimethylformamide (DMF), elimination competed with displacement and the product was contaminated with about 10% of $CH_2{=}CH{-}(CF_2)_4{-}CH{=}CH_2$ and $N_3CH_2CH_2{-}(CF_2)_4{-}CH{=}CH_2$.

Diazides **3a**–**c** were reduced to diamines **4a**–**c** by conventional hydrogenation[15,16] or by the hydrazine-mediated catalytic-transfer hydrogenation.[17] Either procedure afforded fluorinated diamines **4a**–**c** in about 60% yield. Catalytic transfer hydrogenation, however, can be conducted in normal glassware at atmospheric pressure.[17] Reduction of **3a** with lithium aluminum hydride, sodium borohydride, titanium(III)chloride, or triphenylphosphine, by standard procedures[15,18–21] gave complex mixtures.

Diamines **4a**–**c** were converted to diisocyanates **5a**–**c** in 55–85% yield by reacting a slurry of the dihydrochloride salts in *o*-dichlorobenzene with phosgene at 135 °C. Direct displacement of **2a** with potassium cyanate[22] afforded mainly elimination products, whereas, reaction of **3a** with triphenylphosphine followed by carbon dioxide[23] gave **5a** in <5% yield.

$$\underset{\textbf{3a}}{N_3CH_2CH_2{-}(CF_2)_4{-}CH_2CH_2N_3} \xrightarrow[CH_2Cl_2]{Ph_3P}$$

$$[Ph_3P{=}NCH_2CH_2{-}(CF_2)_4{-}CH_2CH_2N{=}PPh_3] \xrightarrow[\text{benzene}]{CO_2}$$

$$\underset{\textbf{5a}}{OCNCH_2CH_2{-}(CF_2)_4{-}CH_2CH_2NCO}$$

17.3. FLUORONITRO COMPOUNDS BY NITRITE DISPLACEMENT

Previous syntheses of compounds containing the $-CF_2CH_2CH_2NO_2$ group have been limited to 1-nitro-1*H*,1*H*,2*H*,2*H*-perfluoropentane[24], prepared by the aldol condensation of nitromethane with perfluorobutyraldehyde, followed by acetylation, elimination, and reduction of the aldol adduct. With the availability of the iodo compounds, $ICH_2CH_2{-}(CF_2)_n{-}CH_2CH_2I$, it was of interest to examine the displacement of this type of iodo group with nitrite to provide a versatile source of difunctional fluorinated nitro compounds.[25] Both monofunctional (**6a**, **6b**) and difunctional substrates (**6c**, **6d**) were used.

The displacement of **6b** with sodium nitrite in DMF, in the presence of urea and phloroglucinol, gave 1-nitro-1*H*,1*H*,2*H*,2*H*-perfluorododecane (**7b**) in 52%

yield. The side product, 1*H*,1*H*,2*H*,2*H*-perfluorododecane-1-ol (**8b**), resulting from the hydrolysis of the corresponding nitrite ester, was isolated in 40% yield. Similarly, reactions of **6a**, **6c**, and **6d** with sodium nitrite in DMF gave 1-nitro-1*H*,1*H*,2*H*,2*H*-perfluorohexane (**7a**), 1,8-dinitro-1*H*,1*H*,2*H*,2*H*,-7*H*,7*H*,8*H*,8*H*-perfluorooctane (**7c**), and 1,10-dinitro-1*H*,1*H*,2*H*,2*H*,9*H*,9*H*,10*H*,10*H*-perfluorodecane (**7d**) in 24, 27, and 35% yields, respectively.

$$\underset{\textbf{6a,b}}{R_fCH_2CH_2{-}I} + NaNO_2 \xrightarrow[25^\circ C]{DMF} \underset{\textbf{7a,b}}{R_fCH_2CH_2{-}NO_2} + \underset{\textbf{8b}}{R_fCH_2CH_2{-}OH}$$

a: $R_f = C_6F_{13}$, **b**: $R_f = C_{10}F_{21}$

$$\underset{\textbf{6c,d}}{I{-}CH_2CH_2{-}(CF_2)_n{-}CH_2CH_2{-}I} \xrightarrow[DMF/25^\circ C]{NaNO_2} \underset{\textbf{7c,d}}{O_2NCH_2CH_2{-}(CF_2)_n{-}CH_2CH_2NO_2}$$

c: $n = 4$, **d**: $n = 6$

Attempts to convert the terminal nitroalkanes (**7a–d**) to terminal *gem*-dinitroalkanes by the Kaplan–Schechter oxidative nitration[26] or the potassium ferricyanide–sodium persulfate procedure[27] were unsuccessful. Alternatively, a modified oxidative nitration procedure, employing tetranitromethane (TNM),[28] was adopted. Nitro compounds **7a–d** were reacted with excess TNM in methanolic potassium carbonate solution to give the dinitro- and tetranitro potassium salts **9a–d**. Acidification of these salts with hydrochloric acid gave the *gem*-dinitro compounds **10a–d** in 36–68% yields. Formylation of the potassium salts with aqueous formaldehyde gave fluoronitro alcohols **11a–d** in 51–75% yields.

$$\underset{\textbf{7a,b}}{R_fCH_2CH_2NO_2} \xrightarrow[K_2CO_3]{TNM} \underset{\textbf{9a,b}}{R_fCH_2{-}C^-(NO_2)_2\,K^+} \xrightarrow{H_3O^+} \underset{\textbf{10a,b}}{R_fCH_2{-}C(NO_2)_2{-}H}$$

$$\textbf{9a,b} \xrightarrow[H_3O^+]{CH_2O} \underset{\textbf{11a,b}}{R_fCH_2{-}C(NO_2)_2CH_2OH}$$

a: $R_f = C_6F_{13}$, **b**: $R_f = C_{10}F_{21}$

$$O_2N{-}CH_2CH_2{-}(CF_2)_n{-}CH_2CH_2{-}NO_2 \xrightarrow[K_2CO_3]{TNM}$$

7c,d

$$K^{+\,-}C(NO_2)_2{-}CH_2{-}(CF_2)_n{-}CH_2{-}C(NO_2)_2^{-}\,K^+ \xrightarrow{H_3O^+} H{-}C(NO_2)_2{-}CH_2{-}(CF_2)_n{-}CH_2{-}C(NO_2)_2{-}H$$

9c,d **10c,d**

$$\downarrow CH_2O \mid H_3O^+$$

$$HOCH_2C(NO_2)_2{-}CH_2{-}(CF_2)_n{-}CH_2{-}C(NO_2)_2{-}CH_2OH$$

11c,d

c: $n = 4$; **d**: $n = 6$

The reaction of 1,1,6,6-tetranitrohexane with methyl acrylate using sodium hydroxide in aqueous methanol was reported by Nielsen and his co-workers[29] to give the corresponding diester in 83% yield. Extension of these reaction conditions to the fluorinated derivative **10c** gave the bis-adduct, 6,6,7,7,8,8,9,9-octafluoro-4,4,11,11-tetranitrotetradecanedioate (**12c**), in 17% yield. A 43% yield of **12c** was obtained using aqueous dioxane as the solvent at 60 °C in the presence of Triton-B. Similarly, **10d** was reacted with methyl acrylate to give **12d** in 20% yield.

$$H{-}C(NO_2)_2{-}CH_2{-}(CF_2)_n{-}CH_2{-}C(NO_2)_2{-}H + CH_2{=}CHCO_2CH_3 \xrightarrow[60\,^{\circ}C]{Triton\text{-}B}$$

10c,d

$$CH_3O{-}C(=O){-}CH_2CH_2{-}C(NO_2)_2{-}CH_2{-}(CF_2)_n{-}CH_2{-}C(NO_2)_2{-}CH_2CH_2{-}C(=O){-}OCH_3$$

12c,d

c: $n = 4$, **d**: $n = 6$

17.4. FLUORONITRO ALCOHOLS BY RADICAL–ANION COUPLING

Although perfluoroalkyl iodides do not readily undergo S_N1 or S_N2 displacement, the iodides have been replaced in a radical-chain nucleophilic substitution ($S_{RN}1$) reaction with salts of 2-nitropropane.[30] We have used this coupling reaction to produce fluoronitro alcohols[31] by starting with a functionalized secondary nitro compound.

Tris(hydroxymethyl)nitromethane acetonide[32] was deformylated with lithium hydroxide, using azeotropic removal of water by benzene, to give the anhydrous lithium salt of 2-nitropropane-1,3-diol acetonide. This salt was reacted under nitrogen in DMF with 1-iodoperfluoroheptane, 1-iodoperfluorooctane, or 1-iodoperfluorodecane to give the corresponding perfluoroalkyl-substituted nitro compounds in 77, 52, and 74% yields. Similarly, 1,4-diiodoperfluorobutane and 1,6-diiodoperfluorohexane were coupled to give the corresponding bis-adducts in 87 and 90% yields, respectively. The reactions were complete in 2–24 h as indicated by the disappearance of —CF_2—I absorbance in the ^{19}F NMR spectrum.

The perfluoroalkyl-substituted nitro acetonides were converted to the corresponding nitro diols and dinitro tetrols by transketalization with ethylene glycol and boron trifluoride etherate. Because of low solubility, the decyl derivative was unreactive under these conditions. Recrystallization of the alcohols from acetone regenerated the corresponding acetonides.

$$Li^+\ [\text{lithium 2-nitronate of 2,2-dimethyl-1,3-dioxane: } O_2N^-{=}C(CH_2O)_2C(CH_3)_2] + CF_3(CF_2)_xI \longrightarrow CF_3(CF_2)_x(NO_2)C(CH_2O)_2C(CH_3)_2 \qquad x = 6\text{–}8$$

$$\xrightarrow[(HOCH_2)_2]{BF_3} CF_3(CF_2)_x C(CH_2OH)_2 NO_2 \qquad x = 6, 7$$

$$Li^+\ [O_2N^-{=}C(CH_2O)_2C(CH_3)_2] + I(CF_2)_xI \longrightarrow (CH_3)_2C(OCH_2)_2C(NO_2){-}(CF_2)_x{-}C(NO_2)(CH_2O)_2C(CH_3)_2 \qquad x = 4,6$$

$$\xrightarrow[(HOCH_2)_2]{BF_3} O_2N{-}C(CH_2OH)_2{-}(CF_2)_x{-}C(CH_2OH)_2{-}NO_2 \qquad x = 4,6$$

17.5. MICHAEL ADDITION REACTIONS TO ACTIVATED β,β-DIFLUOROALKENES

Another potential route to *gem*-dinitro compounds with adjacent fluorine that was studied on this program is the Michael reaction of nitro compounds with activated β,β-difluoroalkenes.[33] The study of the Michael reactions of β,β-difluoro vinyl ketones or acrylic acid derivatatives has been limited because of the inaccessibility of these alkenes.[34] Reported synthetic routes to 3,3-difluoroacrylic acid are the zinc mediated reduction of 2,3-dichloro-3,3-difluoropropionic acid[35] and the reaction of carbon dioxide with 2,2-difluorovinyllithium.[36] Although, ethyl 3,3-difluoroacrylate was prepared by cautious dehydrohalogenation of ethyl 3-bromo-3,3-difluoropropionate, dehydrochlorination of the corresponding chloro derivative failed because of disproportionation leading to ethyl 3,3,3-trifluoropropionate.[37] The presence of the β-fluoro groups increased the reactivity of the alkene toward addition of fluoride ion.[38]

The Friedel–Crafts acylation of fluoro-substituted ethylenes with acid chlorides has been used to prepare 2-chloro-2-fluoroethyl alkyl ketones.[39] We extended this method to the ferric chloride catalyzed reaction of isobutyryl chloride with 1,1-difluoroethylene, which gave 1-chloro-1,1-difluoro-4-methyl-3-pentanone (**13**) in 41% yield.

Oxidation of **13** with *meta*-chloroperbenzoic acid gave isopropyl 3-chloro-3,3-difluoropropionate (**14**) in 33% yield. Although two products, 2-chloro-2,2-difluoroethyl isobutyrate and **14**, are possible in this oxidation, only **14** was observed.

$$CH_2{=}CF_2 + RCOCl \longrightarrow \underset{\mathbf{13}}{CF_2ClCH_2COR}$$

$$\mathbf{13} + \text{mCPBA} \longrightarrow \underset{\mathbf{14}}{CF_2ClCH_2COOR}$$

$$\mathbf{13} + Et_3N \longrightarrow \underset{\mathbf{15}}{CF_2{=}CHCOR}$$

$$\mathbf{14} + Et_3N \longrightarrow \underset{\mathbf{16}}{CF_2{=}CHCOOR} \qquad R = \text{—}CHMe_2$$

Dehydrochlorination of ketone **13** with triethylamine in diphenyl ether gave 1,1-difluoro-4-methyl-1-penten-3-one (**15**) in 55% yield. Simiarly, ester **14** was dehydrochlorinated with triethylamine in methylene chloride to give isopropyl 3,3-difluoroacrylate (**16**) in 49% yield. The dehydrohalogenation reaction was complete in less than 20 min at room temperature. The difluoroalkenes **15** and **16** were hydrolytically unstable; when solid sodium carbonate monohydrate was used as a base, chlorodifluoroketone (**13**) was converted to methyl isopropyl ketone. Dehydrochlorination of cyclohexy 3-chloro-3,3-difluoropropionate was reported to give an unresolvable mixture of cyclohexyl 3,3,3-trifluoropropionate and cyclohexyl 3,3-difluoroacrylate.[34] The formation of the trifluoromethyl group was not observed in the synthesis of difluoroalkenes **15** and **16**.

The reaction of the sodium salt of 1,1-dinitroethane (**17**) with unfluorinated methyl vinyl ketone or ethyl acrylate has been reported to give the C-Michael alkylation derivatives, 3,3-dinitrobutyl methyl ketone and ethyl 4,4-dinitrovalerate, in high yield.[40] The reactions of difluoroalkenes **15** and **16** were studied with 1,1-dinitroethane as a sodium salt or with KF or amine bases in THF, DMSO, DMF, or methylene chloride. Unlike the unfluorinated alkenes, **15** and **16** did not give the expected adducts under the conditions studied. Similarly, no adducts were formed from anions of nitroethane or trinitromethane. When the reaction of the dinitroethane salt **17** with **16** in methylene chloride using collidine as the base was followed by ^{19}F NMR, the appearance of a new absorption at δ -20 (possibly acyl fluoride) was observed. After water was added, the organic layer showed no organic ^{19}F NMR signals, and the dinitro salt (**17**) was converted to 2,5,5-trinitro-3-aza-4-oxa-2-hexene (**18**) in 85% isolated yield.

Formation of **18** from **17** was reported to occur in the presence of amine catalysts, but the reaction was slow and the yield poor.[41] In a control reaction, without the difluoroalkene **16**, the conversion of **17** to **18** in the presence of collidine in methylene chloride was 15% after 24 h. The reaction of **17** with acetyl chloride was reported to result in acylation of the oxygen of a nitro group to give the mixed nitronic–acetic anhydride, which subsequently gave **18** in high yield.[42]

The failure of nitronate ions to undergo Michael additions with difluoroalkenes **15** or **16** may be a result of irreversible addition at the oxygen of the nitro group. In nonfluorinated alkenes, such *O*-alkylation occurs rapidly but reversibly to regenerate the nitronate and the alkene. Michael adducts are formed only when the slower and irreversible *C*-alkylation occurs.[40] In the case of these fluoroalkenes with decreased electron density at the β carbon, irreversible *O*-alkylation of the nitro group would lead to formation of **18**.

$$\underset{\mathbf{17}}{CH_3C(NO_2){=}NO_2^-} + \underset{\mathbf{13}}{CF_2{=}CHCOR} \longrightarrow CH_3C(NO_2){=}N(\rightarrow O){-}O{-}CF_2CH_2COR \xrightarrow{\mathbf{17}}$$

$$\underset{\mathbf{18}}{CH_3C(NO_2){=}N{-}O{-}C(NO_2)_2CH_3} + FC(O)CH_2COR$$

In nonfluorine-containing systems, base-catalyzed Michael additions of β-nitro alcohols to activated alkenes generally give adducts of the nitronate ions resulting from deformylation of the alcohol.[43] In contrast, nitro alcohols underwent Michael addition in the presence of amine bases with difluoroalkenes **15** and **16** to give stable fluoroethers. Reaction of 2,2-dinitropropanol (**19**) with

15 and **16** gave 1,1-difluoro-1-(2,2-dinitropropoxy)-4-methyl-3-pentanone (**20**) and isopropyl 3,3-difluoro-3-(2,2-dinitropropoxy)propionate (**21**) in 72 and 33% yields, respectively. Reaction of 2-fluoro-2,2-dinitroethanol (**22**) similarly gave 1,1-difluoro-1-(2,2-dinitro-2-fluoroethoxy)-4-methyl-3-pentanone (**23**) and isopropyl 3,3-difluoro-3-(2,2-dinitro-2-fluoroethoxy)propionate (**24**) in 82 and 52% yields. Reaction of 2-fluoro-2-nitro-1,3-propanediol (**25**) with **15** gave the diadduct, 2,14-dimethyl-8-nitro-5,5,8,11,11-pentafluoro-6,10-dioxapentadecan-3,13-dione (**26**), in 60% yield. No adduct was formed in the reaction of 2,2-dinitro-1,3-propanediol (**27**) with **15** or **16**. Diol **27** has been shown to undergo deformylation under milder conditions than fluoronitrodiol (**25**).[44]

Another electronegatively substituted alcohol that was treated with an activated difluoroalkene was trifluoroethanol. Reaction of this alcohol with ketone **15** gave 1,1-difluoro-1-(2,2,2-trifluoroethoxy)-4-methyl-3-pentanone (**28**) in 71% yield.

$$\mathbf{15} + \underset{\mathbf{19}}{MeC(NO_2)_2CH_2OH} \longrightarrow \underset{\mathbf{20}}{MeC(NO_2)_2CH_2{-}O{-}CF_2CH_2COR}$$

$$\mathbf{16} + \mathbf{19} \longrightarrow \underset{\mathbf{21}}{MeC(NO_2)_2CH_2{-}O{-}CF_2CH_2COOR}$$

$$\mathbf{15} + \underset{\mathbf{22}}{FC(NO_2)_2CH_2OH} \longrightarrow \underset{\mathbf{23}}{FC(NO_2)_2CH_2{-}O{-}CF_2CH_2COR}$$

$$\mathbf{16} + \mathbf{22} \longrightarrow \underset{\mathbf{24}}{FC(NO_2)_2CH_2{-}O{-}CF_2CH_2COOR}$$

$$\mathbf{15} + \underset{\mathbf{25}}{HOCH_2CFNO_2CH_2OH} \longrightarrow$$

$$\underset{\mathbf{26}}{RCOCH_2CF_2{-}O{-}CH_2C(NO_2)FCH_2{-}O{-}CF_2CH_2COR}$$

$$\mathbf{15} + CF_3CH_2OH \longrightarrow \underset{\mathbf{28}}{CF_3CH_2{-}O{-}CF_2CH_2COR}$$

$$R = {-}CHMe_2$$

Because of the instability of **15** and **16**, these difluoroalkenes were typically prepared without isolation by reaction of chlorodifluoro ketone **13** or ester **14** with triethylamine or collidine in methylene chloride in the presence of an alcohol. When the fluoroalkenes were isolated prior to reaction with the nitro alcohols, the yields of the adduct were reduced.

The formation of nitro ethers by Michael addition of fluorodinitroethanol (**22**) had been observed previously only when the Michael acceptor was highly polarized as in methyl acetylenecarboxylate.[45] Otherwise, **22** deformylated in reactions with acrylates to give the corresponding *C*-addition products, 3-fluoro-3,3-dinitrobutyrates.[46]

$$R'C(NO_2)_2CH_2OH + \text{base} \longrightarrow R'C(NO_2)_2CH_2O^-$$

19, $R' = CH_3$
22, $R' = F$
$R = {-}CHMe_2$ or ${-}OCHMe_2$

$$R'C(NO_2)_2{-}CH_2O^- + CX_2{=}CHCOR \xrightarrow{X=F} R'C(NO_2)_2{-}CH_2{-}O{-}CF_2CH_2COOR$$

$$+CH_2O \downarrow -CH_2O$$

$$R'C(NO_2)_2^- + CX_2{=}CH{-}COR \xrightarrow{X=H} R'C(NO_2)_2{-}CH_2CH_2COR$$

ACKNOWLEDGMENTS

We thank the Office of Naval Research for financial support of this study. The work reported here is either being published in journals or has been published recently. Leading references in each section specify the technical contributors. Particular credit is due Dr. Thomas Archibald and Dr. Aslam Malik, who provided enthusiastic impetus for the entire project.

REFERENCES

1. Baum, K., Bedford, C. D., and Hunadi, R. J., *J. Org. Chem. 47*, 2251–2257 (1982).
2. Haszeldine, R. A., *Nature* (*London*), *167*, 139 (1951).
3. Bedford, C. D. and Baum, K., *J. Org. Chem.*, *45*, 347 (1980).
4. Lewis, E. E. and Naylor, M. A., *J. Am. Chem. Soc.*, *69*, 1968 (1947).
5. Hunadi, R. J. and Baum, K., *Synthesis*, 454 (1982).
6. Malik, A. A., Tzeng, D., Cheng, P., and Baum, K., *J. Org. Chem.*, *56*, 3043–3044 (1991).
7. (a) Hollander, J., Trischler, F. D., and Gosnell, R. B., *J. Polym. Sci. Part A-1, 5*, 2757 (1967). (b) Hollander, J., Trischler, F. D., and Harrison, E. S., *Polym. Prepr. Am. Chem. Soc.*, *8*, 1149 (1967). (c) Hollander, J., in *High Polymers*, *XXV*, Wall, L. (Ed.), 1972, Chap. 6, p. 195.
8. Middleton, W. J., *J. Org. Chem.*, *49*, 4541 (1984).
9. Takakura, T., Yamabe, M., and Kato, M., *J. Fluorine Chem.*, *41*, 173 (1988).
10. Ulrich, H., *J. Polymer Sci.*, *Macromol. Rev.*, *11*, 93 (1976).
11. Kunyants, I. L., Khrlakyan, S. P., Zeifman, Yu. V., and Shokina, V. V., *Izv. Akad. Nauk SSSR Ser. Khim.*, *2*, 384 (1964).
12. (a) Takao, H. and Hitoshi, H., *Japan Kokai 77,118,406* (1977) (Chem. Abstr. 88:61965f). (b) Foulletier, L. and Jean, P. L., *U.S. 4,059,629* (1977), (Chem. Abstr. 88:169572r).
13. Rondestvedt, C. S. and Thayer, G. L., *J. Org. Chem. 42*, 2680 (1977).
14. (a) Gibson, M. S. and Bradshaw, R. W., *Angew. Chem. Int. Ed. Engl.*, *7*, 919 (1968). (b) Sheehan, J. C. and Bolhofer, W. A., *J. Am. Chem. Soc. 72*, 2786 (1950). (c) Landini, D. and Rolla, F., *Synthesis*, 389 (1976). (d) Soai, K., Ookawa, A., and Kato, K., *Bull. Chem. Soc. Jpn. 55*, 1671 (1982).

15. (a) Scriver, E. F. V. and Turnbull, K., *Chem. Rev.*, *88*, 321 (1988). (b) Boyer, J. H. and Canter, F. C., *Chem. Rev.*, *54*, 1 (1954). (c) Hudlicky, M., in *Reduction in Organic Chemistry*, Halted Press, New York, 1984, 69.
16. (a) Corey, E. J., Nicolaou, K. C., Balanson, R. D., and Machida, Y., *Synthesis* 590 (1975). (b) Mungall, W. S., Greene, G. L. Heavner, G. A., and Lestinger, R. L., *J. Org. Chem.*, *40*, 1659 (1975).
17. Malik, A. A., Preston, S. B., Archibald, T. G., Cohen, M. P., and Baum, K., *Synthesis*, 450 (1989).
18. (a) Boyer, J. H., *J. Am. Chem. Soc.*, *73*, 5865 (1951). (b) Bose, A. K., Kistner, J. F., and Farber, L., *J. Org. Chem.*, *27*, 2925 (1962).
19. (a) Soai, K., Yokoyama, S., and Ookawa, A., *Synthesis*, 48 (1987). (b) Rolla, F., *J. Org. Chem.*, *47*, 4327 (1982). (c) Boyer, J. H. and Ellzey, S. E., *J. Org. Chem.*, *23*, 127 (1958).
20. Stanovnik, B., Tisler, M., Polanc, S., and Gracner, M., *Synthesis*, 65 (1978).
21. (a) Staudinger, H. and Hauser, E., *Helv. Chim. Acta*, *4*, 861 (1921). (b) Zimmer, H., Vaultier, M., Knouzi, N., and Carrie, R., *Tetrahedron Lett.*, *24*, 763 (1983). (c) Horner, L. and Gross, A., *Liebigs Ann. Chem.*, *591*, 117 (1955). (d) Kozirara, A., Osowska-Pacewicka, K., Zawadzki, S., and Zwierzak, A., *Synthesis*, *202* (1985).
22. (a) Vodop'yanov, V. G., Golov, V. G., Dvoeglazova, N. P., Ivanoc, M. G., Mysin, N. I., and Gerega, V. F., *Zh. Obshch. Khim.*, *51(8)*, 1885 (1981). (b) Schaeffer, W. D., *U.S. 3,017,420* (1962). (c) McMaster, A. L. and Park, L., *U.S. 3,440,269* (1969). (d) Argabright, P. A., Rider, H. D., and Sieck, R., *J. Org. Chem.*, *30*, 3317 (1965).
23. (a) Molina, P. and Arques, A. A., *Synthesis*, 596 (1982). (b) Tsuge, O., Kanemasa, S., and Matsuda, K., *J. Org. Chem.*, *49*, 2688 (1984).
24. Cook, D. J., Pierce, O. R., and McBee, E. T., *J. Am. Chem. Soc.*, *76*, 83 (1954).
25. Malik, A. A., Archibald, T. G., Tzeng, D., Garver, L. C., and Baum, K., *J. Fluorine Chem.*, *43*, 291–300 (1988).
26. Kaplan, R. B. and Shechter, H., *J. Am. Chem. Soc.*, *83*, 3535 (1961).
27. Garver, L. C., Grakauskas, V., and Baum, K., *J. Org. Chem.*, *50*,f 1699 (1985).
28. Bedford, C. D. and Nielsen, A. T., *J. Org. Chem.*, *43*, 2460 (1978); Altukhov, K. V. and Perekalin, V. V., *Russ. Chem. Rev.*, *45*, 1052 (1976).
29. Feuer, H., Leston, G., Miller, R., and Nielsen, A. T., *J. Org. Chem.*, *28*, 339 (1963).
30. Feiring, A. E., *J. Org. Chem.*, *48*, 347 (1983). Umemoto, T. and Kuriu, Y., *Tetrahedron Lett.*, *22*, 5197 (1981).
31. Archibald, T. G., Taran, C., and Baum, K., *J. Fluorine Chem.*, 243–248 (1989).
32. Linden, G. B. and Gold, M. H., *J. Org. Chem.*, *21*, 1175 (1956).
33. Archibald, T. G. and Baum, K., *J. Org. Chem.*, *55*, 3562–3565 (1990).
34. Leroy, J., Molines, H., and Wakselman, C., *J. Org. Chem.*, *52*, 290 (1987).
35. (a) Knunyants, I. L., Sterlin, R. N., and Bogachev, V. E., *Izv. Akad. Nauk. SSSR Ser. Khim.*, 425 (1958) (Engl. Transl. p. 407.) (b) Dickey, J. B. and McNally, G. U. S. Patent 2,571,678 (1951).
36. Gillet, J. P., Sauvetre, R., and Normant, J. F., *Synthesis*, 297 (1982).
37. Molines, H., and Wakselman, C., *J. Fluorine Chem.*, *25*, 447 (1984).
38. Molines, H. and Wakselman, C., *J. Fluorine Chem.*, *37*, 183 (1987).

39. (a) Knunyants, I. L., Sterlin, R. N., Pinkina, L. N., and Dyatkin, B. L., *Izv. Akad. Nauk. SSSR, Ser. Khim.*, 296 (1958) (Eng. Transl. p. 282.) (b) Ishikawa, N., Iwakiri, H., Edamura, K., and Kubota, S., *Bull. Chem. Soc. Jpn.*, *54*, 832 (1981). (c) Spawn, T. D. and Burton, D. J., *Bull. Soc. Chim. Fr.*, 876 (1986).

40. Shechter, H. and Zeldin, L., *J. Am. Chem. Soc.*, *73*, 1276 (1951).

41. (a) Belew, J. S., Grabiel, C. E., and Clapp, L. B., *J. Org. Chem.*, *77*, 1110 (1955). (b) Selvanov, V. F., Shchedrova, V. K., and Gidaspov, B. V., *Zh. Org. Khim.*, *8*, 1543 (1972) (Eng. Transl. p. 1574).

42. Shevelev, S. A., Erashko, V. I., and Fainzil'berg, A. A., *Izv. Akad. Nauk. SSSR, Ser. Khim.*, 1856 (1970) (Eng. Transl. p. 1747).

43. Noble, P., Borgardt, F. G., and Reed, W. L., *Chem. Rev.*, *64*, 19 (1964).

44. Berkowitz, P. T. and Baum, K., *J. Org. Chem.*, *45*, 4853 (1980).

45. Grakauskas, V. and Baum, K., *J. Org. Chem.*, *34*, 3927 (1969).

46. Grakauskas, V., *J. Org. Chem.*, **38**, 2999 (1973).

INDEX